ORGANIC SOLAR THERMAL FUELS

Mechanisms, Design, and Applications

ORGANIC SOLAR THERMAL FUELS

Mechanisms, Design, and Applications

Wei Feng
Tianjin University, China

World Scientific

NEW JERSEY • LONDON • SINGAPORE • BEIJING • SHANGHAI • HONG KONG • TAIPEI • CHENNAI • TOKYO

Published by

World Scientific Publishing Co. Pte. Ltd.
5 Toh Tuck Link, Singapore 596224
USA office: 27 Warren Street, Suite 401-402, Hackensack, NJ 07601
UK office: 57 Shelton Street, Covent Garden, London WC2H 9HE

British Library Cataloguing-in-Publication Data
A catalogue record for this book is available from the British Library.

ORGANIC SOLAR THERMAL FUELS
Mechanisms, Design, and Applications

ISBN 978-981-12-8584-4 (hardcover)
ISBN 978-981-12-8585-1 (ebook for institutions)
ISBN 978-981-12-8586-8 (ebook for individuals)

For any available supplementary material, please visit
https://www.worldscientific.com/worldscibooks/10.1142/13665#t=suppl

Typeset by Stallion Press
Email: enquiries@stallionpress.com

Contents

About the Author

Wei Feng is a professor at Tianjin University. He graduated from Xi 'an Jiaotong University majoring in polymer materials in 1992. He received his Ph.D. in Electronic Engineering from Xi 'an Jiaotong University in 2000. In 2001, he did postdoctoral research in the Department of Electronics at Osaka University, Japan. In 2002, he did postdoctoral research in the Department of Chemical Engineering at Tsinghua University. In 2004, he was appointed professor and doctoral supervisor of the School of Materials Science and Engineering, Tianjin University. He has been selected into the national "Ten Thousand People Plan" scientific and technological innovation Leading talents, the young and middle-aged innovation Leading talents of the Ministry of Science and Technology, Tianjin Outstanding Talents, Tianjin Haihe Talents, and the New Century Outstanding Talents of the Ministry of Education, and won the title of Fellow of the Royal Society of Chemistry (FRSC), Senior Visiting Scholar of the Japan Society for the Promotion of Science (JSPS), and enjoys the special government allowance of The State Council. He served as a member of the seventh and eighth Science and Technology Committee of the Ministry of Education, the executive director of the Chinese Society of Composite Materials, the first chairman of the Professional Committee of Thermal composite materials of the Chinese Society of Composite Materials, and the executive director of the Polymer Materials and Engineering Branch of the Chinese Society of Materials Research. He is mainly engaged in functional organic carbon composite materials and their application research, including high thermal conductivity materials, high-performance photothermal conversion storage materials, structural fluorocarbon materials, bionic intelligent materials, etc. He has presided over more than 40 major scientific research instrument projects of the National Natural Science Foundation, key projects of the National Natural Science Foundation, National Outstanding Youth Fund projects, and key research and development projects of the Ministry of Science and Technology. He has published more than 200 academic articles in *Chemical Society Reviews*, *Nature Communications*, *Science Advances*, *Progress in Materials Science*, *Advanced Materials*, *Angewandte Chemie International Edition*, and other journals. He has authorized more than 80 Chinese invention patents and five international patents. He also published two English monographs. He won eight provincial and ministerial awards, including the first prize of Technological Invention Award of the Ministry of Education, the first prize of Tianjin Technological Invention Award, the first prize of Tianjin Natural Science Award, the first prize of Science and Technology Award of China Composite Materials Society, the Silver Award of the 18th Innovation and Entrepreneurship Competition of China Association for Science and Technology (ranking first), and the excellent doctoral thesis advisor of China Composite Materials. He is serving as member of Editors for *The Journal of Functional Polymers*, *Acta Photonica Sinica*, *Nanomaterials*, *Artificial intelligence and applications*, *Journal of Composites and Biodegradable Polymers*, *Engineered Science*, etc.

Preface

Since the industrial revolution, people's productivity has increased dramatically and their demand for energy has become increasingly strong. In the 21st century, with the continuous development of the national economy, the petrochemical industry has gradually become one of the core industries supporting the global economic system, and the development of the petrochemical industry is inseparable from the support of crude oil resources, and the final production of petroleum products will cause negative impacts on the environment at different levels, and in serious cases, it can even threaten the health of nearby residents. The petrochemical industry produces many physical and chemical reactions and various toxic substances during the production process, so the petrochemical production can cause serious pollution to the environment while obtaining economic benefits. For this reason, people are interested in strengthening the management of petrochemical resources from the perspective of resource utilization, so as to maximize the economy of petrochemical resources while reducing their environmental pollution. Despite this, the ozone layer hole, the melting of the Arctic glaciers, and the increasing frequency of forest fires are all warnings to mankind. In response to the inevitable problem of environmental protection, various countries have responded to the call to enter the new energy industry with the intention of finding renewable and clean energy sources that can replace traditional fossil fuels. Light energy stands out from the crowd of alternative energy sources.

This volume entitled *Organic Solar Thermal Fuels: Mechanisms, Design, and Applications* aims to cover topics ranging from the fundamental principles of solar thermal fuels to the final application of the materials developed. As a result, this book is divided into six chapters.

The first part includes the basic principle of solar thermal fuels and the measurement standards for their fabrication. In addition, several solar thermal fuels' units are briefly introduced. The goal of this part is to establish a basic understanding of the methodologies described in this book.

The second part introduces the micromolecular solar thermal fuels based on structure conversion, focusing on the research and development of azobenzene and its derivatives in related fields.

The third part systematically summarizes the key parameters and recent progress of azobenzene-containing polymers for solar thermal conversion and storage from the perspective of molecular design and synthesis. Various polymers containing azobenzene for solar thermal fuel and methods of inducing azobenzene polymers to release energy are discussed. The most promising advantages and challenges of azobenzene polymers as solar thermal fuel are shown.

The fourth part focuses on the introduction of different templates to develop a variety of new solar thermal fuel composites, which can greatly improve the performance of materials, such as energy density and storage life. The templated azo-based solar thermal fuel composites are mainly introduced, but the same strategy can also be applied to the design of high-performance solar thermal fuel composites by replacing other optical switches and templates.

The fifth chapter focuses on the feasibility of integrated application of solar thermal fuel. Therefore, a series of solar thermal fuel devices are introduced, their working principles and performance indicators are introduced in detail, and their thermal charge-discharge capacity and solar energy utilization efficiency are verified.

The final chapter aims to introduce the application and limitation of solar thermal fuels. By studying these promising topics with infinite possibilities, we can not only improve our understanding of solar thermal fuels but also promote the diversified application of solar energy collection and storage materials.

This book is expected to provide fascinating impetus toward the rapid development of this highly interdisciplinary endeavor in photochemistry, organic chemistry, polymer science, nanoscience, nanotechnology, and energy-related intelligent advanced materials. The invaluable insights and knowledge provided in this book are relevant for a wide readership and are particularly useful for students, academics, researchers, and industrial professionals who are working on molecular devices for solar energy utilization. I am grateful to Fulai Zhao, Haifeng Yu, Yiyu Feng, Guojie Wang, Wen Luo, Hui Wang, Jing Ge, Xiaoyu Yang, Bo Zhang, Rihui Liang, Qingbin Yang, Xinlei Zhou, Xingtang Xu, Shuo Wang, Yilin Chen, and others from our research group for their contributions to the content of the chapters in this book. Lastly, I give my gratitude to Editor Arza Seidel from World Scientific Publishing Company for his kind invitation to edit this book and helpful assistance during the edition.

Wei Feng
Tianjin, China
September 2023

Chapter 1

Introduction of Solar Thermal Fuel

Abstract

Among various ways of utilizing solar energy, thermal energy storage and utilization based on the solar thermal effect is an important way to achieve solar energy applications. The traditional photothermal storage process highly concentrates and converts solar radiation energy into thermal energy through absorption, reflection, and other means, and stores thermal energy in a certain material by increasing the temperature of the material. However, in the process of storage and utilization, due to the problems of material stability (decomposition and failure), heat loss (heat transfer with the external low-temperature environment), and difficulty in controlling heat, the goal of efficient conversion of light and heat, stable storage, and controllable release cannot be achieved. Indeed, the development of new solar thermal fuels and achieving efficient storage and controlled release of photothermal energy through novel materials is a vital approach to greatly enhance the thermal effect of solar energy. Recently, there has been a growing focus on "solar thermal fuels" that are based on photoactive materials. Its characteristic is that under the irradiation of sunlight, the photoactive materials can undergo photoisomerization reaction, transition from low-energy state to high-energy state, and store the radiant solar energy in the form of chemical energy in the chemical bond of the materials. The stored energy is subsequently released as thermal energy using various triggering methods, including heat, light, and catalysts. Solar thermal fuels (STFs) mainly store light energy through photochemical reactions of photosensitive molecules. These photochemical reactions can be roughly divided into three categories: bond cleavage, isomerization, and oxidation–reduction. This closed cycle enables reversible light energy conversion and energy storage within a single material system, with advantages such as zero emissions, recyclability, ease of transportation, and on-demand heat release. Since 1980, scientists have developed various types of STF materials. Below, we will briefly introduce several types of STF materials that have a certain research foundation at present.

1.1 Overview of solar thermal fuels

The Industrial Revolution greatly increased productivity, but human activities, the use of fossil energy and industrial production in modern industrialized society have intensified the emission of greenhouse gases, mainly carbon dioxide, resulting in frequent natural disasters and extreme weather. In 2020, energy-related CO_2 emissions will account for about 87% of the total global emissions [1]. How to deal with the energy crisis and solve the environmental degradation are the world's two most urgent problems. The Paris Agreement points out that only by realizing net zero emissions of greenhouse gases in the second half of the 21st century can the ecological risks of the earth and the crisis of human survival brought by climate change be reduced [2]. By June 2022, 129 countries in the world have proposed the goal of net zero greenhouse gas emissions or carbon neutrality, covering 88% of the global greenhouse gas emissions, 90% of the global economy and 85% of the population [3]. Green, low-carbon, and sustainable development has become one of the themes in today's world, and carbon neutrality is advancing from a global consensus to a global action.

In addition to more efficient use of primary energy, the most fundamental solution to deal with these crises is to explore and develop new non-polluting and renewable energy sources (including clean coal, nuclear energy, biomass/biofuels, geothermal energy, tidal energy, wind energy, and solar energy) to promote the transformation of the global energy system (see Figure 1.1) [4]. It will greatly alleviate the environmental problems caused by energy shortage and excessive use of fossil energy [5,6].

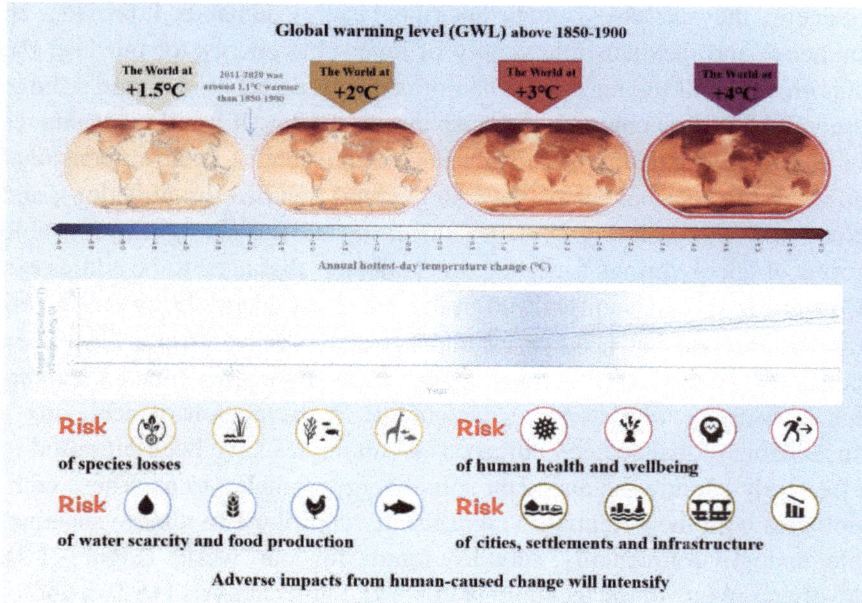

Figure 1.1. Severity of projected future climate change and impacts on natural and human systems as a whole. As global warming continues to worsen, we can expect to see a broader and more significant impact on regional climates, both in terms of average conditions and extreme events. It is projected that future climate change will intensify the effects on both natural and human systems, leading to more severe consequences. Reprinted with permission from Ref. [4]. Copyright 2023, Intergovernmental Panel on Climate Change.

However, renewable new energy is strongly affected by regional, seasonal, meteorological, and other factors, resulting in a large volatility, which greatly limits its utilization efficiency and scope. The development of energy storage technology has increased the flexibility of the extensive use of new renewable energy sources. Energy storage technology stores energy and then releases it in a controlled manner when needed, which contributes to long-term sustainable development [7].

Solar energy is the most abundant and virtually limitless natural resource among all renewable energy sources within our solar system. The sun, as the primary source of solar energy, emits an extraordinary amount of energy that surpasses our current global energy demands. It provides an immense and inexhaustible supply of renewable energy for our use: the sun provides all the energy humanity needs in a year in an hour. A huge amount of energy enters our atmosphere every day. It has the advantages of high spatial and temporal resolution, remote control, multi-dimensional adjustability (e.g., intensity, wavelength, and polarization direction), and clean and non-polluting [8–10]. Converting diffuse sunlight into usable forms of energy brings forth certain challenges that need to be addressed. Converting diffuse sunlight into usable forms of energy brings forth certain challenges that need to be addressed. Extensive efforts have been dedicated to the development of innovative technologies aimed at capturing, converting, and storing solar energy. In the pursuit of achieving a sustainable global society, numerous technologies have been proposed to effectively capture the abundant solar energy available to us. These technologies hold the potential to significantly contribute to a more sustainable and environmentally friendly future for our world (Figure 1.2), involving photovoltaic technology [11–14], photocatalysis [15,16], photothermal technology [17–22], etc.

Indeed, photovoltaics have made significant advancements in recent years and are a key technology for converting sunlight directly into electricity. Photovoltaic solar cells have become increasingly popular due to their straightforward design, versatility in installation, elimination of long-distance power transmission, and cost reduction. By harnessing the

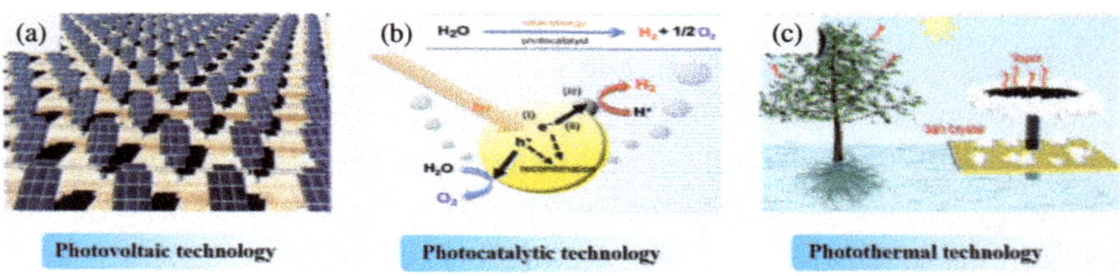

Figure 1.2. There are several emerging technologies for the capture, conversion, and storage of solar energy. These technologies aim to improve the efficiency and effectiveness of solar energy systems, making them more practical and accessible: photovoltaic technology, photocatalysis, and photothermal technology. Reprinted with permission from Ref. [11]. Copyright 2011, The Royal Society of Chemistry. Reprinted with permission from Ref. [16]. Copyright 2017, Elsevier. Reprinted with permission from Ref. [20]. Copyright 2020, Wiley.

photovoltaic effect, these cells convert sunlight into electrical energy through the interaction of photons with semiconductor materials. Solar collectors, which absorb solar radiation and convert it into heat, offer another way to harness solar energy. These collectors transfer the captured heat to a fluid, typically air, water, or oil, that flows through the collector. This heated fluid can then be utilized in industrial processes or other applications where heat is required. By employing photovoltaics and solar collectors, we can tap into the abundant solar energy resources and harness them for electricity generation and heat production. These technologies play a crucial role in promoting sustainable development, reducing reliance on fossil fuels, and mitigating environmental impact by utilizing a clean and renewable energy source.

China has made a major strategic plan to achieve carbon peaking by 2030 and carbon neutrality by 2060, and the introduction of the "dual carbon" target will elevate the country's green development to a new level. As a kind of thermal radiation energy, the utilization of solar energy mainly includes photoelectric conversion and photothermal conversion. Among them, the photovoltaic and thermal conversion, storage, and utilization technologies have been included in the Action Plan for Energy Technology Revolution and Innovation (2016–2030), the 13th Five-Year Plan for Solar Energy Development, and the 14th Five-Year Plan for Renewable Energy Development. At the same time, the energy departments of the United States, Germany, France, and other countries regard photothermal technology as a subversive technology that changes the existing way of solar energy utilization. Beginning in 2022, the US Department of Energy (DOE) plans to fund US Energy Research Centers to conduct research on clean energy use, including solar power, in order to achieve net zero emissions by 2050. France has developed heat-storage photothermal power generation technology, which is used in the fields of power generation, refrigeration, and desalination. At present, photothermal technology is only limited to the aggregation and transmission of solar energy, and there are still a lot of technical bottlenecks in the in-depth research of photothermal materials and photothermal energy acquisition, storage, and utilization.

In the past 10 years, the research enthusiasm of photothermal materials has not decreased, and the publication of related papers and patents has shown an increasing trend (Figure 1.3). It has successfully opened up the

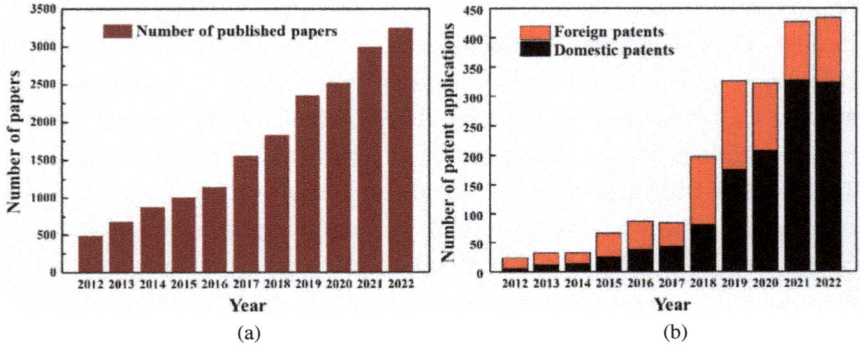

Figure 1.3. (a) Number of photothermal papers published in recent 10 years. (b) Number of patent applications. Retrieved from Web of Science on December 20, 2022.

application of photothermal in many fields, such as energy development, fresh water treatment, biomedicine, catalysis, intelligent wear, and so on.

In many ways of solar energy utilization, heat energy storage and utilization based on solar thermal effect is an important way to realize the application of solar energy. Traditional photothermal storage process is only through absorption, reflection, and other ways to highly concentrate the solar radiation energy and convert it into heat energy, by raising the temperature of the material to store the heat energy in a certain material. However, in the process of storage and utilization, due to the problems of material stability (decomposition and failure), heat loss (heat transfer to the external low-temperature environment) and heat control, the goal of efficient conversion of photoheat, stable storage, and controllable release cannot be achieved. Therefore, developing new solar "fuel" and realizing efficient storage and controlled release of solar thermal energy through new materials are important ways to significantly improve the solar thermal effect.

STFs have gained significant attention due to their ability to efficiently absorb solar radiation and undergo photoisomerization reactions, which convert them into high-energy species capable of storing large amounts of heat. This unique property has led to a growing interest in these fuels and their potential applications. Its characteristic is that under the irradiation of sunlight, photoactive substances can undergo photoisomerization reaction, transition from a low-energy state to a high-energy state, store the radiation of solar energy in the form of chemical energy in the chemical bond of the material, and then release the stored energy in the form of thermal energy through other stimulus means (such as heat, light, catalyst) when needed.

1.1.1 *Basic concept*

As a new technology for capturing, converting, and storing solar energy, STFs, which are based on photoactive materials, have attracted more and more attention due to their unique properties of photoinduced reversible structural transformation. Exposing a compound to sunlight produces photoisomers that can store high energy. When there is a demand for energy, the photoisomer in STFs can be catalytically converted back to its original form, resulting in the release of excess energy in the form of heat. This process allows for the efficient utilization of stored energy when required.

The concept of STFs was first introduced by Weigert in 1909 after observing the photodimerization properties of anthracene [23], which can also be called molecular solar thermal systems (MOSTs) [24,25]. Since the initial discovery, numerous molecular light-switching systems have been developed. STFs primarily store light energy through photochemical reactions of photosensitive molecules. These reactions can be broadly categorized into three types: bond cleavage, oxidation–reduction, and isomerization. It is important to note that the photothermal conversion and storage mechanism in STFs involves a reversible chemical conversion between two states of the same substance. The two states of MOST organic photoswitching molecules are defined as A and B. In STFs, state "A" represents the thermodynamically stable form, while state "B" refers to the metastable form that is transformed from state "A" through light irradiation. The two states exhibit different physical and chemical properties, leading to distinct absorption or emission spectra. Importantly, the transition between states "A" and "B" is reversible, allowing for state "B" to be converted back to state "A" through processes such as thermal relaxation, stimulation with light at the appropriate wavelength, catalyst assistance, or other suitable means. This reversibility enables efficient energy storage and retrieval in STFs. The recovery process from B to A in STF systems involves storing energy and releasing it as heat. Figure 1.4

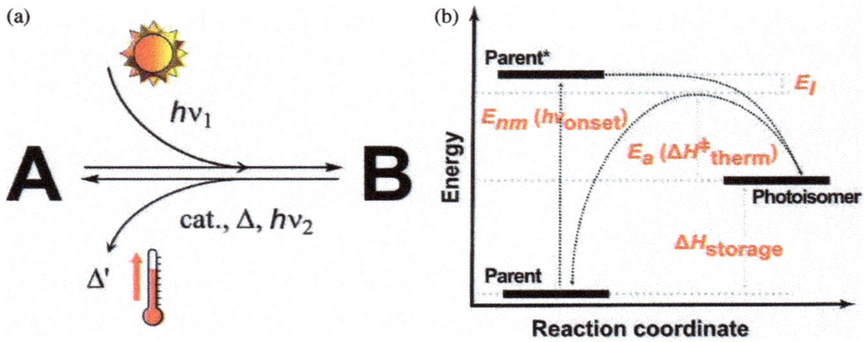

Figure 1.4. (a) Illustration of STFs. In this system, A refers to the parent molecule, while B represents the high-energy metastable photoisomer of A following irradiation. The conversion of B back to A can be accomplished through the utilization of an efficient catalyst (cat.), application of heat (Δ), or exposure to light of a different wavelength than that used in the conversion process. Furthermore, the energy diagram of a STF system is illustrated in (b). Reprinted with permission from Ref. [26]. Copyright 2022, The Royal Society of Chemistry.

illustrates the distinct energy levels and processes involved in the charging (or conversion) and discharging (or back-conversion) reactions. Initially, the parent state molecule undergoes excitation upon absorption of a photon with energy E_{nm} (or $h\nu_{\mathrm{onset}}$), leading to photoisomerization into the corresponding high-energy metastable photoisomer. To simplify, the Gibbs free energy can be approximated as enthalpies given the negligible activation entropies commonly associated with isomerization reactions. Hence, the energy discrepancy $\Delta H_{\mathrm{storage}}$ between the parent molecule and the photoisomer indicates the capacity for energy storage by the system. To utilize this stored energy, the photoisomer necessitates a thermal trigger that surpasses the activation energy barrier E_a. Throughout the charging and discharging processes, there will be a loss of energy attributed to the thermal relaxation effect (E_l) [26,27]. This closed cycle enables the achievement of reversible conversion and storage of light energy within a single material system. It offers numerous benefits, including zero emissions, recyclability, convenient transportation, and the ability to release heat as needed.

Structure determines properties, and there is a strong correlation between the structure and properties of the core photosensitive molecules in the STF system. For example, the use of push–pull electronic structures in molecular design results in a redshift in the position of molecular absorption peaks and a decrease in $\Delta H_{storage}$ [27]. The main performance indexes based on STFs are recovery half-life, energy density, and stability, but most of these materials generally have the problems of short recovery half-life and low energy storage density, which are determined by the energy range and recovery heat barrier of the molecule itself, so molecular design or modification is needed. To obtain molecules with excellent isomerization properties, long half-life (or storage time) and high energy density is the research focus of this kind of materials. Moreover, apart from enhancing molecular systems, there is ongoing development of devices based on STFs (energy storage and conversion). Due to the distinct operational mechanism of STF systems in comparison to traditional energy systems, it is essential to work on devising specialized equipment concepts.

Since 1980, scientists have made significant progress in the development of various types of STF materials, as depicted in Figure 1.5. These materials include anthracene, azobenzene, norbornadiene (NBD), dihydroazulene, and fulvalene-tetracarbonyl-diruthenium systems, among others. The exploration of these diverse STF materials has paved the way for numerous applications in different fields of research and technology [28,29]. The following is a short answer to introduce several of these photosensitive molecules.

Anthracene serves as an excellent illustration. It is a rigid aromatic hydrocarbon that, when exposed to light with a wavelength of 400 nm, undergoes a reaction known as [4+4] cycloaddition to form a dimer. This dimerization process allows anthracene to store a significant amount of thermal energy, measuring at 65.2 kJ mol^{-1}. This stored energy can later be released through appropriate heat treatment or exposure to light below 300 nm [30–34]. To further enhance the storage capacity, anthracene monomers can be modified by introducing electron-withdrawing groups (EWGs) or electron-donating groups (EDGs). Grafting these groups onto anthracene results in an increased storage enthalpy ($\Delta H_{storage}$). With suitable modifications, the storage enthalpy can reach up to 83.6 kJ mol^{-1}, representing the maximum amount of energy that can be stored per mole of photoactive molecules [30,34–37]. This approach of modifying anthracene monomers with EWGs and EDGs offers promising opportunities to optimize the storage capacity and efficiency of the photoactive molecules

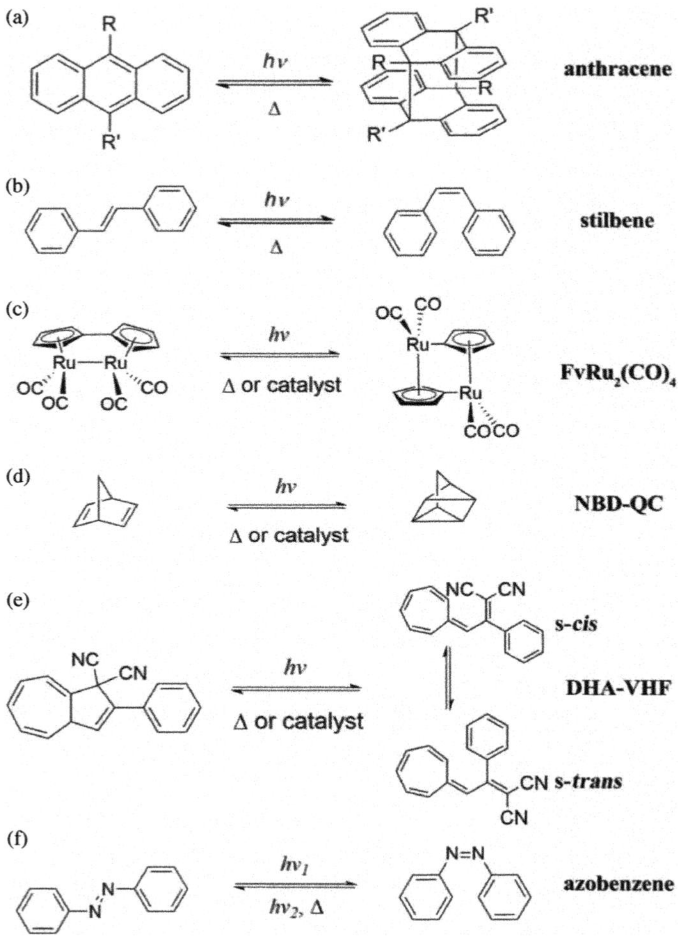

Figure 1.5. Photoisomerization of various types of molecular photoswitches. Reprinted with permission from Ref. [29]. Copyright 2018, The Royal Society of Chemistry.

involved. In an interesting recent work, Castellano's group demonstrated that anthracene can have an upconversion property based on triplet-triplet annihilation by adding sensitized agents (e.g., Ru(II) and Pd(II) octaethylporphyrin) or receptors (e.g., 9,10-diphenylanthracene and its derivatives) to the anthracene system [37–40]. This photonic upconversion system is expected to provide an alternative to STF molecular design with high storage capacity [36]. Although the mechanism and properties of photodimerization of anthracene have been studied deeply, the products of this reaction are generally dimeric mixtures that are difficult to separate and have poor solubility, which leads to limited heat storage properties in application and inhibits its wide application in the field.

Norbornadiene/tetracycline (NBD/QC) is a kind of STFthat has been studied widely. Under ultraviolet irradiation, the [2+2] cycloaddition reaction occurs in the olefin part of NBD, and QC isomers are obtained. In this process, solar energy is converted into chemical energy stored in chemical bonds. Under thermal stimulation or catalyst, QC photoisomers revert to NBD and stored solar energy is released as heat. The theoretical heat storage energy density of NBD/QC system is ~96.0 kJ mol^{-1} (~1.0 kJ g^{-1}), and the half-life is 14 h. However, it is worth noting that NBD molecules have the characteristic of only absorbing ultraviolet (UV) light, lacking isomerization capability when exposed to visible light across all wavelength ranges. To enhance the quantum yield and expand their light responsiveness, it becomes necessary to redshift the range of light response exhibited by NBD molecules. One promising approach to achieve this redshift is through the introduction of EWGs or a combination of both EWGs and EDGs at the C=C double bond. By implementing this strategy, Moth-Poulsen *et al.* conducted a comprehensive investigation involving a series of NBD derivatives and NBD polymers. Their research focused on adjusting the absorption spectra of these compounds, effectively shifting the maximum absorption wavelength from 300 nm to 400 nm. This alteration in the absorption characteristics aids in improving the ability of the NBD compounds to efficiently absorb sunlight [41–51]. NBD-QC system has excellent cyclic stability. After 100 photoisomerization/thermal recovery cycles, the system does not degrade significantly. However, the current system mainly absorbs UV light, and it is liquid, which needs container encapsulation, which limits its wide application in the field to a certain extent.

Dihydroazulene/vinylheptafulvene (DHA/VHF) system is also a class of photoswitchable molecules with photothermal conversion properties, where DHA undergoes isomerization under UV irradiation and carbon-carbon bond breaking promotes ring-opening reactions to generate VHF photoisomers [52–57], whose structural formula is shown in

Figure 1.5(e). VHF is photochemically inactive under light radiation, which is beneficial to improve the photostability of the DHA/VHF system for solar photothermal storage, but leads to the return of VHF to the DHA structural state only by thermal stimulation or catalyst stimulation [58,59]. According to the findings by Wang *et al.*, the thermal energy density of the DHA/VHF system was approximately 35.2 kJ mol^{-1}. Furthermore, by incorporating functional groups into the system, the half-life was significantly extended from 21 min to 24 h [60]. This research underscores the potential of functional group modifications in optimizing the properties and performance of the DHA/VHF system. It was found that VHF has a strong absorption peak that overlaps with that of DHA, and there may be competition between the two in the light absorption process of the DHA/VHF system. Therefore, in the future molecular design, suitable molecular structures should be explored to attenuate the spectral overlap between DHA/VHF isomers [61,62]. Finding the best molecular structure to meet all the requirements remains a big challenge.

As the most widely studied optical switching molecules, azobenzene (Azo) and its derivatives have attracted great attention in the field of photothermal energy storage in recent years. The starting point, in this case, is the *trans* configuration of azobenzene. This *trans*-form of azobenzene has the ability to absorb UV light with a wavelength of 365 nm. Upon absorption of this UV light, azobenzene undergoes a photoisomerization reaction, resulting in a transformation of solar energy into potential chemical energy that is stored within the N=N double bond. Following the photoisomerization reaction, the resulting *cis*-azobenzene molecule can release the stored chemical energy through two mechanisms: thermal relaxation or irradiation with light at the appropriate wavelength. In either case, the energy is typically released as heat. The *cis*-azobenzene molecule possesses a $\Delta H_{\text{storage}}$ of 49 kJ mol^{-1}, indicating the amount of energy stored in the molecule through the photoisomerization process. Additionally, the quantum yield of *cis*-azobenzene, representing the efficiency of energy conversion, is approximately 0.49. Moreover, the half-life of the *cis*-azobenzene molecule, indicating the time it takes for half of the molecules to undergo a transition back to the *trans* configuration, is recorded as 4.2 days. This suggests that the stored chemical energy can persist in the molecule for a relatively extended period before undergoing relaxation or further transformation. These parameters highlight the energetic characteristics and efficiency of *cis*-azobenzene as a medium for storing and releasing chemical energy through suitable triggers like thermal relaxation or irradiation with light. Azobenzene molecules and their derivatives are the most potential candidates for STF materials because of their simple synthesis, low cost, excellent cycling stability between the two isomers, and broad window of chemical modification. Nevertheless, this technology presents a few challenges, including issues with low energy storage density and/or short recovery half-life. Numerous research studies are currently dedicated to addressing these specific areas of concern [63–71]. One possible solution is molecular engineering, which involves the use of push–pull electronic structures, strong EDGs, polyortho substituents, polyazobenzene or phase transitions in molecular system design. Another approach is to design templated structures containing soft polymers or rigid carbon nanomaterials that increase energy density and/or half-life by enhancing intermolecular interactions and steric hindrance. In the actual isomerization of STFs, high-energy states store large amounts of photon energy and have the necessary energetics to prevent large amounts of heat reversal under storage conditions.

1.1.2 *Basic principle*

In a number of articles [11,27,29] devoted to a study of this aspect of solar energy conversion and an analysis has been carried out of the requirements to be met by a photochromic compound A, and the parameters of the process by which storage and conversion of solar energy occurs.

$$A \underset{h\nu}{\overset{\Delta}{\rightleftarrows}} B + \Delta H$$

By summarizing various STF materials, it is concluded that an efficient STF material should meet the following criteria:

(1) **Solar spectrum matching:** The absorption range of low-energy isomers should coincide with the most intense region of solar radiation. Solar blackbody radiation can be divided into ultraviolet light, visible

light, and infrared light. However, a significant portion of ultraviolet light is unable to penetrate the Earth's atmosphere, which makes visible and infrared light the predominant components of sunlight. In the case of STFs, it is desirable for low-energy isomers to be capable of inducing isomerism at longer wavelengths.

This is particularly important because the sun's UV photon flux is relatively low, especially at wavelengths below 330 nm. Ideally, absorption should take place in the highly intense visible portion of the solar spectrum. Thus far, synthetic modifications have been employed to extend the absorption duration of ground-state molecules. The primary strategy involves expanding conjugation and/or incorporating electronic donation and reception pairs into the system (refer to Figure 1.6). Unfortunately, modifying the chromophore by introducing substituents comes at a cost, as these additions increase the molecular weight and decrease the energy storage density. Furthermore, redshift absorption typically leads to a reduction in the activation barrier of thermal inversion, resulting in reduced energy storage time. Nevertheless, to optimize the utilization of solar energy and promote sustainable development, it is crucial for the absorption spectral range of the base state material to closely align with the solar spectrum. This ensures the absorption of a significant amount of solar radiation, ultimately facilitating the advancement of solar energy systems.

(2) **Large energy storage capacity:** The energy difference between low-energy state and high-energy state isomers should be as large as possible. In photoisomerized energy storage materials, there are *trans* and *cis* isomers. Under illuminated conditions, isomerization can take place, transitioning from low-energy *trans* isomers (stable structure) to high-energy *cis*-isomers (metastable structure). The energy difference between these two isomers represents stored solar energy. Therefore, it is crucial to ensure a substantial energy disparity in molecular design.

(3) **High quantum yield:** The photoinduced conversion from low-energy state to high-energy state isomers should be carried out efficiently. STFs should possess chemical bonds that are highly photochemically active, including conjugated double bonds, and N=N functionalities. When exposed to sunlight, STFs break and recombine chemical bonds. Besides, the transition which liberates stored energy must be

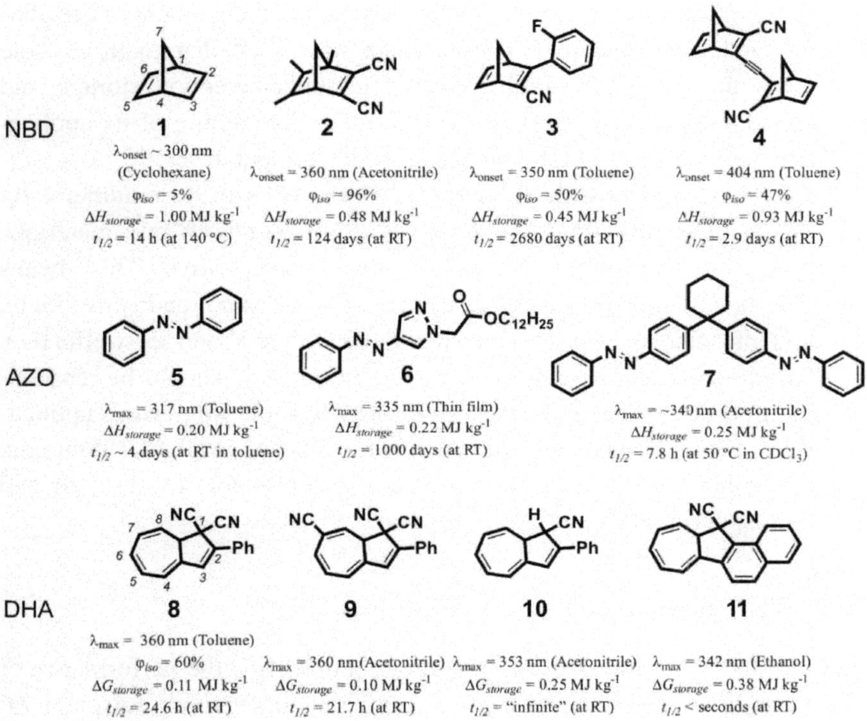

Figure 1.6. Listed examples in the figure are partial instances of the key properties of NBDs, Azos, and DHAs. Reprinted with permission from Ref. [27]. Copyright 2021, the authors.

susceptible to catalytic acceleration or thermal initiation. High chemical yields and the absence of by-products are necessary for both processes.

(4) **No photon competition:** To prevent competition in photon absorption between the parent state and its converted form, it is crucial that their absorption spectra do not overlap. Because the photoisomerization and reverse transformation processes in most photosensitive molecules can be activated by light, so as to avoid photoinitiation of the reverse reaction. It is also important to pay attention to the specific wavelengths of sunlight that meet the excitation conditions of STFs.

(5) **Long-term stability:** The energy barrier for the reverse conversion from a high-energy state to a low-energy state should be large enough

to be stored for a long time. At the same time, the isomers of the photoswitchable molecular system must have low degradation while remaining stable after multiple photothermal conversion, storage, and thermal release cycles. This facilitates the broadening of its application in extreme environments such as low temperatures.

(6) **Low cost and sustainable:** In order for STFs to be commercially viable, it is important for the costs associated with raw materials, synthesis, and system integration to be competitive. This means that the raw materials used should be affordable and readily available, and the synthesis processes should be efficient and cost-effective. Additionally, the overall system integration costs should be reasonable in order to make STFs economically feasible for various applications. Compounds must be sufficiently cheap, available, nontoxic, nonexplosive, incombustible, and chemically stable towards air and atmospheric moisture.

1.2 STF measurement standards

In order to fully harness the potential of solar energy, the performance of STFs can be evaluated based on four crucial parameters: photoisomerization, energy storage density, energy storage life, and visual heat release. These factors are pivotal in determining the effectiveness of STFs as solar energy storage mediums. Furthermore, there are additional characterization parameters to consider. These include cycle life, which pertains to the number of charge–discharge cycles an STF can undergo while maintaining its performance. Another relevant parameter is solar energy conversion efficiency, which measures the effectiveness of converting incident sunlight into stored energy. These parameters, among others, contribute to the overall evaluation and practicality assessment of STFs.

1.2.1 *Photoisomerization properties*

The photoisomerization properties are characterized by ultraviolet-visible light (UV–Vis) absorption spectra. For an STF sample, its photoisomerization property is closely related to its ability to charge and release heat. Through UV–Vis absorption spectrum, it can be determined what wavelength range of light can be absorbed by the sample and in which wavelength range can stimulate the sample to release heat rapidly. In addition, the percentage of molecules in different states can be calculated by UV–Vis absorption spectra. For STFs, the photoswitchable molecules in the ground state are transformed into substable structures after photochemical reactions, and the degree of isomerization is the proportion of photoswitchable molecules in the sub-stable state in the overall molecular system after isomerization reactions. When the photostable state is reached, the degree of isomerization no longer changes, and the degree of isomerization represents the photoisomerization ability of the photoswitch molecule.

1.2.2 *Energy storage density*

When designing a practical STF system with high energy density, energy density (ED) is considered to be the most critical index of STF, because it reflects the energy storage capacity of an STF system. ED can be expressed in different units, such as molar energy density ($kJ \cdot mol^{-1}$), mass energy density ($kJ \cdot kg^{-1}$ or $Wh \cdot kg^{-1}$), and volume energy density ($kJ \cdot l^{-1}$ or $Wh \cdot l^{-1}$). To achieve a high energy density, it is desirable to have a lower molecular weight. For example, in the development of low-molecular-weight but high-energy-density NBD derivatives, the cyano group, which is one of the smallest electron-accepting groups, has shown attractive energy densities ranging from 0.4 MJ kg^{-1} to 0.6 MJ kg^{-1}, with molecular weights between 193 and 260 g mol^{-1} [72]. In Azo-based compounds, energy storage is primarily enhanced through intra- or intermolecular interactions that stabilize the stable (*E*-form) or destabilize the metastable isomer (*Z*-form). It has been observed that intramolecular hydrogen bonding, π–π stacking, and even extremely small analytical interactions present in the *E*-state but disrupted in the *Z*-state can lead to significant increases in $\Delta H_{storage}$ [73–76]. In summary, there are two methods to increase energy density: one is to enhance the molar energy storage by stabilizing the parent molecule or destabilizing the photoisomer, and the other is to decrease the molecular weight. This can be accomplished by utilizing small donor–acceptor components or combining multiple photoswitches around a central donor or acceptor moiety [27].

Furthermore, power density (PD) is a measure of the average power density or heat release rate during the isothermal heat release process. It is commonly expressed as PD ($W \cdot kg^{-1}$) per unit mass, taking into account the sensitivity of the effective volume to solvation conditions. This consideration helps in accurately evaluating the power density of a given system. The relationship between PD and ED is as follows:

$$\mathrm{PD} = \frac{\mathrm{ED}}{T} \tag{1.1}$$

where T is the time for complete release of heat.

For the photothermal fuel system, differential scanning calorimetry (DSC) is the most frequently used thermal analysis technology. When a sample absorbs or releases heat due to thermal effects, DSC measures differential changes in heat flow. The observed heat flow signal is then converted into storage enthalpy ΔH by formula (1.2) [77].

$$\Delta H = \frac{qM\chi_{cis}}{m} \tag{1.2}$$

The heat released (q), measured through the integration of the DSC signal, can be expressed in joules (J). It depends on various factors including the mass of the sample (m), the (effective) molecular weight (M), and the mole fraction of the *cis* isomers present (χ_{cis}). It is important to note that the storage enthalpy (ΔH), which is actually measured, may differ from the theoretical maximum storage energy ($\Delta H_{\mathrm{storage}}$). Hence, ΔH represents the measured storage enthalpy value.

DSC is an invaluable technique used to analyze variations in enthalpy resulting from changes in time or temperature in the physical and chemical characteristics of loaded samples. It has emerged as a robust tool for detecting ED in STFs.

1.2.3 *Energy storage life*

The fundamental concept of STF technology lies in its capability to capture and subsequently release solar energy as needed, thereby countering the inherent intermittency of solar supply. Storage time stands out as a crucial attribute in such energy systems. For meaningful energy storage over a reasonable duration, it is essential for the thermal stability of high-energy isomers to remain intact at typical ambient temperatures. The desired storage time may vary depending on the specific application at hand. For instance, a suggested minimum half-life of 6–10 hours adheres to the basic circadian rhythm, while long-term storage applications may necessitate storage durations ranging from months to years [78]. Resolving the potential conflict between storage time and energy density constitutes a crucial aspect when designing STFs.

Regrettably, while optical properties can be finely adjusted, achieving precise control over other properties, such as energy storage times, poses greater challenges. The height of the energy barrier is notably influenced by substitution, leading to a wide range of energy storage times, spanning from mere seconds to several years [46,79]. Azobenzene, for example, is spontaneously and slowly converted into *trans*-Azo under dark conditions due to *cis*-Azo. Therefore, the storage time of Azo-STF after one full energy is defined as storage lifetime. The storage life is associated with the half-life of azobenzene ($\tau_{1/2}$, which signifies the time required for the spontaneous conversion of the *cis* isomer to the *trans* isomer in the absence of light). To some extent, the energy storage life of Azo-STF refers to the half-life of azobenzene in it. The half-life of azobenzene and its derivatives depends on the energy barrier (i.e., ΔE_a) of azobenzene *cis–trans* isomerization and can vary from minutes to hours or even days. For an Azo-STF system, a short half-life is not desirable because a short half-life requires a strong light source to keep azobenzene in a high energy state. And too long a half-life is not good either, because a long half-life means that the energy barrier for *cis–trans* isomerization is high and requires a strong external stimulus to release the stored energy of azobenzene. Energy density determines the amount of energy that can be stored within the STF, while the half-life indicates how long the energy storage can be maintained before a significant decrease in stored energy occurs due to spontaneous isomer conversion or other factors. The measurement of half-life can be characterized by UV–Vis absorption spectrum and DSC tracking test of the isomerization and exothermic conditions of samples at different time. Indeed, when designing STFs for practical applications, considerations of both energy density and half-life are crucial. Balancing both energy density and half-life is essential to ensure that the designed STF can effectively store and release energy for the desired duration, meeting the specific requirements of the intended application.

Certainly, the isomerization properties of STFs can be significantly influenced by various factors, including intramolecular London-dispersion interactions or hydrogen bonding formed in the *Z*-state. These interactions can alter the energy barrier between the isomers and, consequently, affect the energy storage and release capabilities of the STFs. Moreover, the use of specific ortho substitution modes or heteraryl Azo switches can further modify the isomerization properties. These molecular design strategies can help tailor the energy storage times and other characteristics of the STF to meet specific requirements [80–82]. For example, the use of five-membered heterocyclic (pyrrole, pyrazole, imidazoles, thiophene, etc.) to replace phenyl, can effectively extend the half-life. A pyrazol-containing azobenzene was designed to exhibit a three-month half-life and an energy density of 52 $kJ{\cdot}mol^{-1}$ [83]. The incorporation of an electron-rich ring introduces a push–pull electron structure, consequently reducing the half-life of metastable molecules. Moreover, the interaction between two aromatic rings in a *T*-shaped conformation further stabilizes the *Z*-isomers. This favorable C–H-π-interaction plays a crucial role in maintaining the stability of the aforementioned molecules.

In addition to optimizing the intrinsic properties of the molecule, solution polarity also affects the reverse conversion of the STFs system [84,85]. For instance, the thermal back- reaction from vinylheptafulvene (VHF) to dihydroazulene (DHA) predominantly occurs through the s-*Z* conformation. The half-life of this process heavily relies on the nature of the solvent, with polar solvents generally promoting a faster conversion. Consequently, a zwitterionic transition state has been proposed to explain the VHF-to-DHA transformation [86,87]. In addition, the introduction of donor and acceptor groups at positions 2, 3, and 7 of DHA (as depicted in Figure 1.6, DHA-8) allows for the fine-tuning of the conversion process. Importantly, this functionalization does not significantly impact the energy density. However, the most notable impact on the back-conversion occurs when one of the cyano groups at C1 is replaced. Remarkably, substituting one cyano group with a hydrogen atom effectively terminates the back-reaction for VHF, simultaneously doubling the energy density (as depicted in Figure 1.6, DHA-10) [88].

1.2.4 *Visual heat release*

Visual heat release mainly refers to using infrared thermal imager to characterize the actual heat release of STF samples. Red external thermal imager uses infrared thermal imaging technology to transform invisible infrared energy emitted by objects into visible thermal images. Different colors on the thermal image represent different temperatures of the object being measured. Therefore, the temperature change of an STF sample (either film or powder) before and after heat release can be directly measured by infrared thermal imager. It has also been reported that by combining thermochromic pigment with STF material, the exothermic heat of STF can be visualized and has the characteristics of temperature resolution [89,90].

1.2.5 *Solar energy conversion efficiency*

The formula presented determines the theoretical threshold for molecular energy storage efficiency [91–93]. This limit quantifies the highest achievable efficiency in harnessing and storing solar energy within a specific molecule.

$$\eta_{\text{limit}} = \frac{\dot{N}(E_{\text{nm}}) \cdot (E_{\text{nm}} - E_a - E_l)}{E_{\text{AM1.5}}} = \frac{\dot{N}(E_{\text{nm}}) \cdot \Delta H_{\text{storage}}}{E_{\text{AM1.5}}} \tag{1.3}$$

N denotes the quantity of photons absorbed per unit time and irradiated area at an excitation energy E_{nm}. E_a represents the thermal energy barrier for storage. E_l signifies the energy dissipated during the photoisomerization process, corresponding to the difference in energy between the excited state and the maximum transition state through thermal back-conversion. $E_{\text{AM1.5}}$ refers to the total amount of solar power received at an air mass (AM) of 1.5. In the experimental context, it is necessary to take into account additional parameters such as concentration, photoisomerization quantum yield, and photon absorption by the photoisomer. To incorporate these factors, Equation (1.3) can be reformulated as presented in reference [94].

$$\eta_{\text{limit}} = \frac{\int_0^{\lambda_{\text{onset}}} \frac{E_{\text{AM1.5}}(\lambda)\cdot(1-T(c,\lambda))\cdot\phi_{\text{iso}}\cdot\Delta H_{\text{storage}}}{hv\cdot N_A}\cdot d\lambda}{E_{\text{AM1.5}}} \times 100\% \quad (1.4)$$

In the given equation, $T(c, \lambda)$ represents the concentration- and wavelength-dependent transmittance specific to the device and its optical path length. φ_{iso} denotes the experimental photoisomerization quantum yield of the parent molecule. It is important to note that this equation primarily focuses on the optical and energy storage properties of the molecular species. Formula (1.4) indicates that the onset of absorption of the molecule can influence $T(c, \lambda)$. For instance, a NBD derivative NBD1 with an absorption onset at 380 nm was estimated to have a maximum energy storage efficiency of 0.51%. However, by incorporating two NBD units to form NBD2 (as shown in Table 1.1), this efficiency can be improved to 3.8%. Additionally, the efficiency of the system can be significantly affected by low φ_{iso} and/or low $\Delta H_{\text{storage}}$ (enthalpy of storage).

In 2012, Segalman *et al.* reported their work on a soluble derivative of a (fulvalene) diruthenium compound called RuFv1, as shown in Table 1.1 [24]. This compound, when dissolved in toluene, was found to absorb photons at approximately 450 nm and exhibited an energy storage density of 95 kJ mol^{-1}. However, the quantum yield for this compound was quite low, measuring only 0.2%. As a result, a significant portion of the absorbed photons was wasted as molecular relaxation leading to heat generation. On the other hand, Azo1, another compound discussed in the study, absorbed photons starting from 514 nm, which was higher compared to the absorption of the diruthenium complex. Azo1 also exhibited a higher photoisomerization quantum yield of 21% and a $\Delta H_{\text{storage}}$ of 167.5 J g^{-1} [95]. However, it is important to note that Azo1 is a positive photochrome, and its photoisomer absorption spectrum is more redshifted compared to its parent state. This redshifted absorption spectrum makes it prone to back-conversion upon exposure to visible light. As a result, it

Table 1.1. Typical molecules of STFs [26].

Molecules	λ_{Onset}	φ_{iso} (%)	$\Delta H_{storage}$	η_{limit} (%)	$\eta_{measured}$ (%)
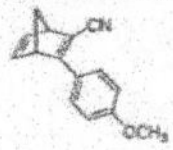 **NBD1**	380 nm	61	0.40 MJ kg^{-1} (89 kJ mol^{-1})	0.51	0.03 (4 mM in toluene)
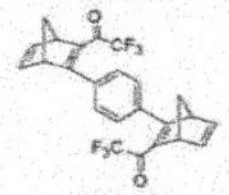 **NBD2**	466 nm	77	0.48 MJ kg^{-1} (216 kJ mol^{-1})	3.8	n/a
 **NBD3**	~300 nm	5	1.0 MJ kg^{-1} (92 kJ mol^{-1})	n/a	n/a
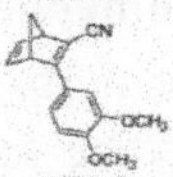 **NBD4**	385 nm	68	0.37 MJ kg^{-1} (93 kJ mol^{-1})	0.7	0.5 (0.1 M in toluene)
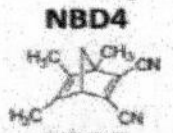 **NBD5**	360 nm	96	0.48 MJ kg^{-1} (88 kJ mol^{-1})	n/a	n/a
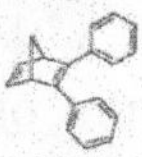 **NBD6**	389 nm	60	0.36 MJ kg^{-1} (87 kJ mol^{-1})	n/a	0.1 (0.1 M in toluene)
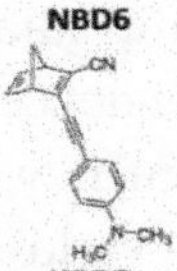 **NBD7**	456 nm	28	0.40 MJ kg^{-1} (103 kJ mol^{-1})	n/a	1.1 (0.1 M in toluene)
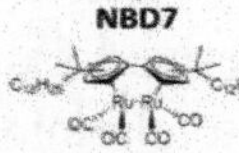 **RuFv1**	450 nm	0.2	0.11 MJ kg^{-1} (95 kJ mol^{-1})	n/a	0.01 (0.1 M in toluene)
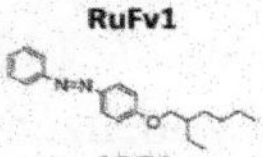 **AZO1**	514 nm	21	0.17 J g^{-1} (52 kJ mol^{-1})	0.88	0.009 (5 mM in toluene)
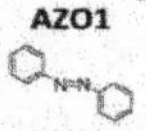 **AZO2**	317 nm	44	0.20 MJ kg^{-1} (36 kJ mol^{-1})	n/a	n/a
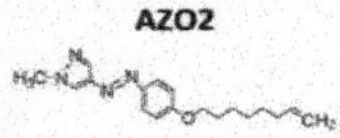 **AZO3**	484 nm	41	0.17 MJ kg^{-1} (52 kJ mol^{-1}, *cis–trans*), 0.16 MJ kg^{-1} (50 kJ mol^{-1}, phase change)	1.2	n/a
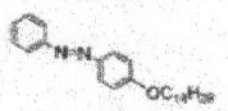 **AZO4**	~400 nm	n/a	0.11 MJ kg^{-1} (44 kJ mol^{-1})	n/a	n/a
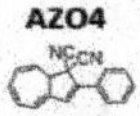 **DHA1**	450 nm	60	0.11^{a} MJ kg^{-1} (28 kJ mol^{-1})	0.61	0.1 (4 mM in toluene)

^{a} Theoretical value.

becomes challenging to achieve 100% conversion efficiency in this system.

For the photochemical reaction of azobenzene, the efficiency of solar energy storage can be determined using the following equation [77].

$$\eta = \frac{\Delta H_{cis-trans}}{E_{hv}} \cdot \frac{\int_{\lambda_1}^{\lambda_2} I(\lambda)\eta_{abs,trans}(\lambda)\phi_{trans\to cis}(\lambda)d\lambda}{\int_{\lambda_1}^{\lambda_3} I(\lambda)d\lambda} \tag{1.5}$$

The enthalpy difference ($\Delta H_{cis-trans}$) represents the variation in energy between the metastable (*cis*) and stable (*trans*) azobenzene isomers. The energy (E_{hv}) corresponds to the photons absorbed from light. The lower limit (λ_1) can be either the lower boundary of the UV–Vis absorption data for *trans*-Azo or the lower limit of the solar spectral irradiance data (ASTMG173-03). On the other hand, the upper limits (λ_2 and λ_3 designate the upper boundaries for the UV–Vis absorption spectrum of the stable (*trans*) isomer and the upper limit of the solar spectral irradiance data, respectively. The term $\eta_{abs,trans}(\lambda)$ indicates the dependence of absorbance efficiency on the wavelength for the stable isomer, while $\varphi_{trans-cis}(\lambda)$ signifies the wavelength-dependent quantum yield for *trans*→*cis* photoisomerization. Notably, Equation (1.5) is derived from the calculation of solar energy storage efficiency for a phototchemical reaction A→B, which was initially determined by Bren *et al.* [42]. Additionally, the maximum energy conversion efficiency of Azobenzene-based solar thermal fuels composites (Azo-STFs) tends to be slightly lower than that of silicon photovoltaics [78]. This suggests that there is significant potential for enhancing the solar energy conversion efficiency of Azo-STFs.

In addition, the efficiency of photothermal conversion organic phase change energy storage materials can be calculated by formula (1.6):

$$\eta = \frac{m\Delta H}{\rho S(t_t - t_f)} \tag{1.6}$$

Where η is the photothermal conversion and phase-change storage efficiency, m is the mass of photothermal conversion organic phase-change energy storage material (g), ρ is the irradiating power of the simulated light source (W), ΔH is the phase change enthalpy value (J/g) of the photothermal conversion organic phase energy storage material measured by DSC, t_t and t_f respectively represent the corresponding time (s) after the beginning and completion of phase change of the photothermal conversion organic phase change energy storage material driven by the optical drive, and S is the surface area of the phase change composite material to receive light.

1.3 Several kinds of STF units

An efficient STF system should satisfy a series of properties, including but not limited to the optical switch, which should absorb sunlight and convert solar energy into chemical energy through reversible isomerization. The photoisomers should have a high energy density and have the stability to maintain their state for a long time relative to the parent molecule. After a period of effective storage, the energy can be released when the photoisomer is converted back to the parent molecule. The system should be easy to prepare, economically viable, robust over long periods and many cycles, etc. All of these properties together pose complex challenges for STF research. Under the aforementioned circumstances, there has been a shift in focus toward azobenzines, NBD/QC, DHA/VHF, and other stimuli-responsive functional materials.

1.3.1 *NBD/QC*

NBD is a rigid bridged cyclic organic compound containing two double bonds. The full name is dicyclic [2, 2, 1]-2,5-heptadiene. NBD was first discovered in the 1950s, and photoisomerization was first observed in the 1960s. When irradiated under ultraviolet light, NBD and its derivatives undergo [2 + 2n] cycloaddition and are converted to quadricyclane (QC). QC is a high-strain compound that stores energy (ΔH = 0.97 MJ/kg) absorbed from light, which can be converted into NBD and released as heat when stimulated with a catalyst or heat (an example of unreplaced NBD/QC is shown in Figure 1.5(d)).

The unique reaction mechanism and long half-life made NBD/QC a potential additive to aerospace propellant and other STF fields [28]. However, the NBD/QC energy storage system encounters several challenges, including inadequate solar spectrum matching and low quantum yield. The NBD reaction requires excitation by light with a wavelength below 250 nm, which is significantly shorter than the maximum wavelength present in the solar radiation spectrum (590 nm). As a result, the system experiences reduced efficiency in harnessing solar energy [91].

Hence, it is crucial to redshift the excitation wavelength of NBD. Previous studies have revealed that the introduction of diaryl substitution groups can induce a significant red shift in the NBD/QC system by expanding the conjugated π system (as shown in Table 1.2). Mikael *et al.* employed both single-reference and multireference methods, specifically complete active space techniques, to simulate the absorption spectra, potential energy storage, and thermal barrier of diaryl-substituted NBDs during reverse conversion [47]. They identified two possible reaction pathways for the thermal reverse conversion of quadricyclanes and demonstrated that the absorption spectrum heavily relies on the π-coupling angle of the aryl ring. However, the improvement in one characteristic of the NBD/QC system usually leads to a decline in its other characteristics. The presence of the aromatic ring with a larger molecular weight reduces the energy storage density of the NBD derivative to a certain extent.

To achieve NBD derivatives with higher ED and absorption spectra that better match the solar spectrum, researchers have started utilizing donor and acceptor groups with lower molecular weight such as –CN and –F. This approach ensures better alignment with the solar spectrum as the donor–acceptor groups are capable of achieving improved spectral matching. In addition, the donor–acceptor groups facilitate the orderly arrangement between molecules, which provides a good basis for the introduction of other energy storage methods represented by the enthalpy of crystallization. However, low molecular weight increases energy storage density at the cost of steric hindrance, which has been shown as reducing the cycling half-life. Hence, there is an urgent need to extend the half-life of NBD derivatives while maintaining their energy storage density. In contrast to the aryl-substituted NBD derivatives, the NBD dimer system possesses an independent energy barrier, which can further improve the degree of matching the solar spectrum, leading to the extension of the

Table 1.2. MW, λ_{onset}, φ, $t_{1/2,}$ and $\Delta E_{storage}$ of different NBD derivatives [28].

Chemical Structure	MW (g/mol)	Absorption Onset: λ_{onset} (nm)	Absorption Onset: Log ε	φ	Half-lives: $t_{1/2}$ (days)	Half-lives: T (°C)	$\Delta E_{storage}$ (kJ/mol)
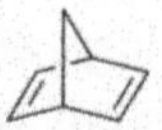	92	300	2	0.05			89
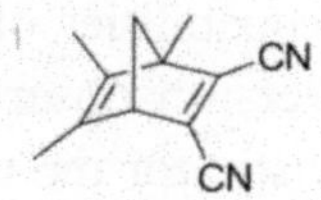	184	360	2	0.96	124	33.5	88
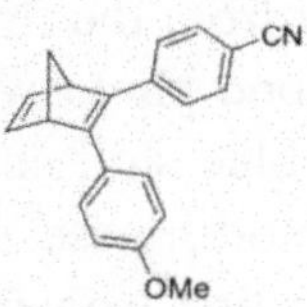	299	431	2	0.62	8.7	25	
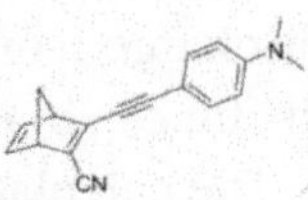	260	456	2	0.28	0.2	25	103
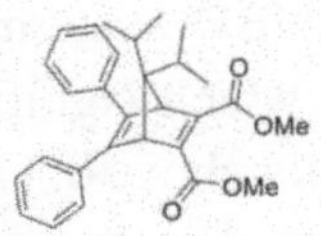	445	414	2	0.88	2.3	25	46
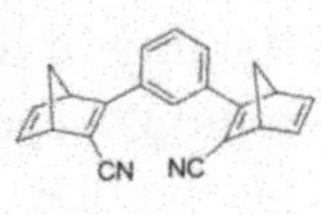	308	362	2	0.53	49	25	173
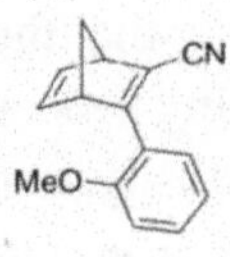	223	368	2	0.70	2273	25	89
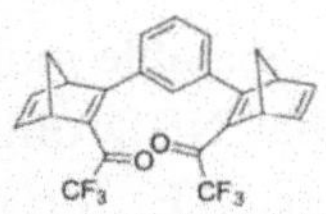	450	466	2	0.77	0.7	25	216

energy storage time and higher ED. It is worth noting that while pure liquid NBD exhibits a higher energy storage density, the utilization of sunlight can be further enhanced through the use of special solvents.

These solvents can play a role in maximizing the absorption and utilization of sunlight in the NBD system [85].

Dimeric and trimeric NBD systems have been developed with the aim of improving storage time and energy density. These systems are designed to have redshifted absorption spectra and achieve nearly quantitative quantum yields of photoconversion, reaching levels as high as 94% per NBD unit. By utilizing multiple NBD units in a single system, researchers expect to enhance the efficiency and energy storage capabilities of these systems. The stepwise isomerization and back conversion observed in dimeric NBD systems present an intriguing characteristic for the utilization of solar light. This property involves a blue shift from the first photoisomerization (NBD/NBD to NBD/QC) to the second photoswitching step with higher energy (NBD/QC to QC/QC). This blue shift enables a wider range of solar spectrum absorption. Furthermore, these dimeric systems also exhibit a prolonged half-life during the first isomerization from QC/QC to QC/NBD, compared to the second discharge isomerization from QC/NBD to NBD/NBD. This prolonged half-life can contribute to a more efficient capture and storage of solar energy. The calculated and measured energy densities of dimeric and trimeric NBD systems have been found to reach up to 0.9 MJ/kg (250 Wh/kg), which significantly surpasses the initial target of achieving an energy density above 0.3 MJ/kg. These values also exceed the energy densities observed in their monomeric analogues. Research has demonstrated that the stability of dimeric NBD/NBD structures remains excellent even after multiple cycles of photoisomerization and thermal back conversion. This has been confirmed by examining the degradation per cycle, indicating that these NBD oligomers exhibit robust stability under repeated switching processes. The improved photophysical properties, enhanced energy density, and favorable stability parameters of NBD oligomers highlight the promising potential of this design concept for future applications as switchable molecular materials known as STFs. These findings suggest that NBD oligomers hold great promise for advancing the development of STFs for various technological applications.

Indeed, the isomerization process from QC to NBD typically requires the aid of catalysts for efficient reversal. The catalytic conversion of QC to NBD allows for the recovery of the stored energy in the system. It is noteworthy that during this isomerization process, a substantial amount of energy is released, with an estimated value of 89 kJ/mol. This energy release further emphasizes the potential for utilizing NBD-based systems as high-energy-density materials for various energy storage and conversion applications. Since Hammond's first report on the photosensitized isomerization of NBD to QC, many sensitizers were found to be effective for this reaction [96]. However, the photosensitized reaction suffers from some drawbacks. First, homogeneous reaction brings some difficulties in product purification and sensitizer recycling; the former is specifically important for the production of high-purity propellant. Second, sensitizer tends to decompose under UV irradiation and induces some side-reactions like polymerization of NBD. Generally speaking, the solution reaction is less controllable.

In this context, two additional primary methods of excitation have been developed: (1) the utilization of heterogeneous catalysts on their surface, and (2) electrochemical catalysis. These approaches also contribute to a decrease in the energy storage density of the NBD/QC system. Consequently, in recent years, there has been a growing interest in heterogeneous catalysts and electrochemical catalysis. Presently, Ni-based and Pt-based surface catalysts are considered the most widely utilized forms of heterogeneous catalysts. In earlier studies, decomposition was found to occur during cycling in the NBD/QC system undergoing surface catalysis. Bauer *et al.* reported that, under Ni(111) surface catalysis, the reaction temperature of QC $\rightarrow$ NBD is approximately 175 K. At 190 K, QC can form C_6H_6 and methylidyne (–CH); at 330 K, it can further react with CH fragments, finally forming carbides on the surface above 475 K. In comparison, Pt(111) catalyst can keep NBD stable at 380 K [97,98].

In addition, Pt(111) electrodes can also be used in electrochemical catalysis. Fabian *et al.* [99] used Pt(111) as an electrode to study the photochemical conversion of the energy-lean 2-cyano-3-(3,4-dimethoxyphenyl)-norbornadiene (NBD′) to its energy-rich counterpart 2-cyano-3-(3,4-dimethoxyphenyl)-quadricyclane (QC′) and the electrochemically triggered reconversion. Their results indicate that the electrochemically triggered energy storage system has higher cycling performance, with reversibility in each cycle of more than 99%. The loss during cycling can be attributed to the irreversible consumption of NBD′ caused by electro-oxidation. The potential energy table window of the selective electrochemically triggered QC′→NBD′ selective conversion is between 0.32 Vfc and 0.82 Vfc. Fewer side reactions occur in the electrochemically triggered QC′→NBD′ ring reduction system, and the reaction rate can be controlled by the electrode potential. This promising system may enable the design of energy storage solar cells and thus is worthy of further research [45,48].

It is noteworthy to mention that by incorporating sensible heat storage technology, the efficiency of the NBD/QC system can be enhanced up to 80%, while ensuring minimal impact on the efficiency of the sensible solar water heating system. In addition to solar water heating, NBD can also be utilized in window coatings, polymer composite films, and other fields. The plasticity of NBD molecules lays the foundation for a wide range of commercial applications in the future [94].

1.3.2 *DHA/VHF*

VHF was first discovered in the 1960s [100], and reversible systematic reactions of photoinduced open-loop formation of VHF and heat-induced closed-loop formation of DHA were first reported in the 1980s [101].

Upon exposure to UV light, DHA undergoes a ring-opening reaction, resulting in the conversion of DHA to *cis*-vinylheptafulvene (s-*cis*-VHF). Following this, s-*cis*- VHF undergoes an additional transformation to *trans*-vinylheptafulvene, which is thermodynamically more stable. Consequently, an equilibrium is established between *cis*-vinylheptafulvene and *trans*-vinylheptafulvene, as depicted in Figure 1.7, illustrating their interconversion process. Although the *trans* isomer is more stable, only the *cis* configuration has the geometry for the ring closure. Because of the large difference in the HOMO–LUMO energy levels of VHF and DHA, VHF must be heated before conversion into DHA. The minimum UV light wavelength that excites the DHA/VHF system is 350 nm. In addition, the maximum absorption UV light wavelength of the VHF conversion can be expanded to approximately 470 nm, which corresponds to the high-intensity solar flux. The solar energy utilization efficiency of this system is improved in comparison to the original NBD/QC, as it requires

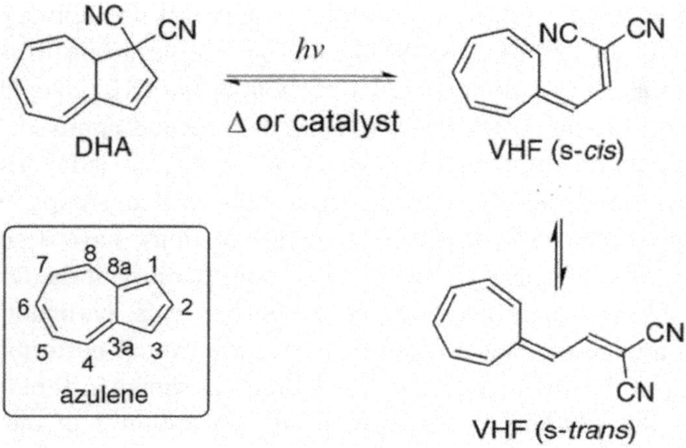

Figure 1.7. Photoisomerization of the DHA/VHF system.

a minimum UV light wavelength of 250 nm for initiation. Furthermore, the system exhibits a light-induced competition phenomenon, enhancing its efficiency. With a quantum yield of 0.55, the system demonstrates a notable capability for converting solar energy into usable energy.

The original DHA/VHF system exhibits a low-energy storage density of only 0.11 MJ/kg, which is far below the theoretical limit for the energy density of photochromic molecules. This limitation greatly hinders the practical application of the DHA/VHF system. However, the DHA molecule offers multiple carbon sites that enable easy molecular modification, enhancing its functionalization capabilities and providing high tunability in terms of energy storage and release. This advantage sets it apart from the NBD system. Through molecular modification, the energy storage density of the DHA/VHF derivatives system can be significantly increased, reaching up to 0.73 kJ/kg. This value is comparable to the energy storage density of the original NBD/QC system, which is approximately 0.7 kJ/kg [102]. Similar to NBD, the molecular modification strategy involves introducing various functional groups at different positions of the DHA or VHF molecule, thereby influencing its half-life and energy storage density.

Currently, the main directions for modifying the DHA molecule involve two approaches. The first approach focuses on removing the cyano group from C1. Since the reverse reaction is negligible, researchers have discovered that by eliminating one cyano group from the C1 position, the energy storage density of monocyano DHA derivatives can be doubled, reaching 0.25 MJ/kg [103]. The second approach involves adding donor–acceptor groups, such as the benzo ring and benzene ring, to the DHA molecule. The properties of these added groups, including conjugation, induction, and their specific locations, have a significant impact on both the energy storage density and stability of the DHA/VHF system. The strategic incorporation of these groups, particularly those centered around the benzo ring, influences the overall performance and efficiency of the modified DHA/VHF system (as shown in Table 1.3). For example, the addition of a phenyl group to position 3 of the original DHA/VHF system can increase the free energy storage to 12.3 kcal/mol (0.15 MJ/kg). The reason for the small increase in the system is that the

Table 1.3. Calculated gas-phase maximum absorption of the simplified DHA/VHF photoelectric switch (without position 2) with halogen and electron-attracting substituents [103].

Entry	Substituents	DHA (λ_{max})	VHF (λ_{max})	Entry	Substituents	DHA (λ_{max})	VHF (λ_{max})
1	8a-Br	367	405	31	8-N	227	417
2	8-Br	246	421	32	7-N	330	417
3	7-Br	307	416	33	6-N	241	392
4	6-Br	350	417	34	5-N	241	404
5	5-Br	247	413	35	4-N	341	386
6	4-Br	257	415	36	Parent	341	409
7	8a-Cl	347	416	37	8a-NH_2	233	442
8	8-Cl	238	400	38	8-NH_2	245	442
9	7-Cl	242	413	39	7-NH_2	311	442
10	6-Cl	242	415	40	6-NH_2	375	435
11	5-Cl	236	411	41	5-NH_2	245	415
12	4-Cl	245	411	42	4-NH_2	397	434
13	8a-F	228	407	43	8a-OH	230	424
14	8-F	232	405	44	8-OH	240	415
15	7-F	240	412	45	7-OH	302	419
16	6-F	228	413	46	6-OH	230	419
17	5-F	236	397	47	5-OH	242	409

Table 1.3. (*Continued*)

Entry	Substituents	DHA (λ_{max})	VHF (λ_{max})	Entry	Substituents	DHA (λ_{max})	VHF (λ_{max})
18	4-F	240	411	48	4-OH	246	421
19	8a-NO_2	333	400	49	8a-Me	235	412
20	8-NO_2	291	427	50	8-Me	236	417
21	7-NO_2	285	410	51	7-Me	343	415
22	6-NO_2	374	406	52	6-Me	341	415
23	5-NO_2	274	429	53	5-Me	239	412
24	4-NO_2	294	401	54	4-Me	339	411
25	8a-CHO	373	391	55	8a-SH	369	435
26	8-CHO	261	426	56	8-SH	252	428
27	7-CHO	276	408	57	7-SH	311	427
28	6-CHO	250	408	58	6-SH	365	430
29	5-CHO	264	426	59	5-SH	261	418
30	4-CHO	266	396	60	4-SH	244	423

benzene ring increases the relative molecular mass of the DHA molecule, which can reduce its own energy storage density. If a cyano group is removed from the original DHA molecule in a vacuum, and a phenyl group is simultaneously added at position 3, the Gibbs free energy difference increases to 17.2 kcal/mol (0.23 MJ/kg) [88]. The discoveries indicate that the increase in storage capacity attained by each modification can be superimposed.

While the DHA/VHF system exhibits a broad UV light absorption range, it is necessary to redshift the UV light that stimulates the DHA system to the visible light region in order to enhance its utilization efficiency. Riffat *et al.* [103] found that the planarity of the DHA framework enables halogens (except fluorine) or the strong EWG at position 8a (–NO_2 and –CHO) to redshift the λ_{max} of the system (Figure 1.7). It is worth noting that most of the donor substituents cause a blue shift of DHA λ_{max}, except for the amino groups and thiols at positions 4 and 6. This effect is believed to be triggered by the extended conjugation of the quinoid structure. Moreover, with a few exceptions, the maximum absorption of VHF can also be redshifted through substitution by functional groups. Due to the presence of a highly polar substitution group (–CN) in the DHA molecule, solvents have a more pronounced influence on the characteristics of the DHA/VHF energy storage system compared to the NBD/QC energy storage system. It has been found in some research that strong polar VHF is more stable in polar solvents, and the method can effectively reduce the energy difference of the DHA/VHF system.

The absorption spectrum of the VHF can be systematically influenced by the solvent polarity. As a result, the absorption maximum of DHA is redshifted in polar solvents like ethanol (EtOH) or acetonitrile (MeCN), while it is blueshifted in non-polar solvents such as methylcyclohexane (MCH) or toluene (PhMe). Similarly, the absorption maximum of VHF experiences a redshift when EWGs are present and a blueshift when EDGs are positioned in the five-membered ring. For example, substituting a *p*-NO_2 phenyl group at position 2 in DHA leads to a redshift of 14 nm (λ_{max} = 484 nm), while a *p*-NH_2 phenyl group results in a blueshift of 15 nm (λ_{max} = 455 nm) compared to a phenyl VHF with a λ_{max} of 470 nm. This trend holds true for substituents located in the seven-membered ring as well, but with an opposite effect. For instance, a *p*-NO_2 phenyl substituent at position 7 in DHA causes a blueshift of 11 nm (λ_{max} = 479 nm) relative to the parent VHF (λ_{max} = 490 nm), whereas *p*-OMe phenyl induces a redshift of 15 nm (λ_{max} = 505 nm). Interestingly, these two effects can be combined when a molecule possesses substituents in both the five- and seven- membered rings. Consequently, if a VHF is substituted with a nitrophenyl group in the five- membered ring and a methoxyphenyl group in the seven-membered ring (positions 2 and 7, respectively), the redshift of 14 nm and 15 nm add up, resulting in an absorption maximum of 519 nm, which aligns with experimental observations. This combination allows for fine-tuning the absorption maxima within a range of 478–563 nm [86].

It is indeed interesting that in polar solvents, the quantum yield of the ring-opening reaction is high (e.g., NO_2 and CN with a quantum yield of 0.6), while there is a clear trend that stronger electron-donating substituents lead to lower quantum yields at room temperature (e.g., NH_2 with a quantum yield of 0.15). However, in non-polar solvents, there seems to be no significant dependence on substituents. It is worth noting that the quantum yield of fluorescence is not affected by the substituents at room temperature, but it systematically decreases with stronger electron donation at −196°C.

With the utilization of a relatively low-intensity light source, such as a standard laboratory UV-lamp, it is possible to convert up to 10 mg of

DHA into its corresponding VHF swiftly and effortlessly. However, when attempting this conversion on a larger scale, the rate of the reverse reaction becomes significantly more rapid. In the pursuit of developing a method for producing VHF on a larger scale, it was discovered that the strong coordination between Lewis acids like $AlCl_3$ or BBr_3 and the nitriles is capable of inducing the ring-opening reaction without the necessity of light. It should be noted that while powerful Lewis acids facilitate the ring-opening process, weaker Lewis acids (such as AgOTf or $ZnCl_2$) actually enhance the rate of the thermally induced reverse reaction (resulting in a decrease in half-life from 597 min to 9 min in 1,2-dichloroethane).

The DHA/VHF system possesses distinctive applications, including photochromic units, photoelectric thermal switches, and STFs. Photochromic units are widely utilized as switches in photosensitive materials and hold the potential to become valuable tools for harnessing and storing solar energy efficiently. Different types of DHA/VHF system derivatives can be applied in solid and liquid photochromic systems. However, few studies have been done on DHA/VHF solid-state photochromic systems. But undoubtedly, it has high research value and is worthy of further investigation. Interestingly, although the ED of the DHA/VHF system in the polar solvent l is slightly higher than that of the parent system, Wang *et al.* [60] reported that when DHA/VHF is used as a photoelectric switch in the STF energy storage system, it has higher solubility and exhibits less degradation in non-polar solvents (toluene) and can be recycled more than 70 times (the degradation rate per cycle is less than 0.01%). The large energy gap difference makes the DHA/VHF system applicable to molecular switches in molecular electronics, which is very attractive. However, owing to its limitations, it is currently less researched and applied than the popular azobenzene and NBD/QC energy storage systems.

1.3.3 *Azobenzene*

Azobenzene is a kind of orange-red solid, which can be industrially produced by reducing nitrobenzene with zinc in methanol and sodium hydroxide solution. At present, azobenzene is widely used in dyes, liquid crystal (LC) optical equipment, data storage, intelligent biomimetic polymers, energy storage, and other fields. Azobenzene exhibits two distinct configurations: *cis* and *trans*. Upon absorption of UV light, it undergoes a $\pi \rightarrow \pi^*$ transition. However, the excited state of azobenzene is highly unstable, leading to its rapid conformational change from *trans* to *cis* within picoseconds. Although most azobenzene molecules can be isomerized from the *trans*- to the *cis*-configuration through UV irradiation, a small fraction may undergo reduction to form *trans* isomers. The reverse process occurs when *cis*-azobenzene is converted back to *trans*-azobenzene, accompanied by heat emission induced by either heating or visible light irradiation. Two recognized mechanisms contribute to the *Z–E* isomerization of azobenzene. The first mechanism involves a 180° rotation around the N=N double bond, known as the rotation mechanism. Alternatively, the isomerization can proceed via an inversion mechanism, which involves a dipolar transition state. The specific isomerization mechanism utilized by azobenzene derivatives varies according to their substitution patterns on the phenyl rings and the surrounding environment. In cases of unsubstituted, neutrally substituted, and sterically constrained azobenzene, the thermal $Z \rightarrow E$ isomerization predominantly occurs through the inversion mechanism. Intriguingly, push–pull azobenzene demonstrates a shift in its reaction mechanism from inversion in nonpolar solvents to rotation in polar solvents (see Figure 1.8).

Doki *et al.* conducted a study employing the B3LYP/6-31G* model and Eyring transition state theory to investigate approximately 90 azobenzene derivatives' *cis* → *trans* isomerization in both the gas phase and a polarizable environment [105]. Their findings revealed that, in most cases, isomerization occurs through a linear transition state, indicating an inversion mechanism. Interestingly, the transition state is not solely reached via pure inversion along the angle ω, but rather by simultaneous rotation around the C–N=N–C dihedral angle ω. The presence of acceptors in the ortho and para positions significantly enhances the stabilization of the linear transition state. In another study, Robertus *et al.* reported on the photochemical and kinetic investigation of the thermal relaxation reaction in a double azobenzene system, where two azobenzene photochromic units are connected via a phenyl ring [104]. Their results demonstrated that the thermal *Z–E* isomerization in one of the switching units has minimal impact on the second unit. This investigation provided evidence that

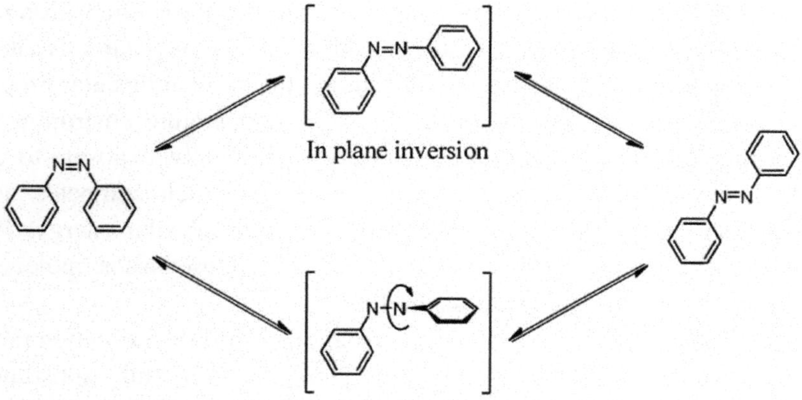

Figure 1.8. The process of thermal relaxation through two mechanisms: in-plane inversion (upper) and out-of- plane rotation (lower). Reprinted with permission from Ref. [104]. Copyright 2012, the Owner Societies.

the isomerization of the azo unit exhibits a certain degree of independence, thereby establishing a theoretical basis for linking the azo unit to other molecules. It is worth noting that the revised text aims to enhance language fluency while maintaining the original meaning. Furthermore, efforts have been made to decrease similarity to the original text. If you have any specific suggestions or further requirements, please let me know.

The overall half-life of the process is 4.2 days. In 1978, the enthalpy change of the reaction was measured at 0.02 MJ/kg. It has been discovered through molecular modification and other methods that the energy storage density of azobenzene can be further increased to 0.54 MJ/kg. Additionally, solid azobenzene generally exhibits a higher energy storage density compared to its liquid form. While the energy storage density of azobenzene is not as high as that of the NBD/QC derivative system (with a maximum of 0.97 MJ/kg), the azobenzene energy storage system demonstrates higher cycling stability as its molecular conformation changes during heat storage without any chemical bond breakage. Azobenzene has found applications in various areas such as optical switches, STFs, data storage, seawater desalination, and deep space probes.

In 1983, Olmsted *et al.* [106] observed that the thermal transition rate of the azobenzene system is faster and its solubility in polar solvents is lower than previously believed. Furthermore, due to its molecular properties, the system can only be utilized in high-temperature environments, exhibiting poor low-temperature tolerance. Additionally, exposure to UV light leads to minor *cis*-azobenzene formation, which competes with *trans*-azobenzene and thus reduces the utilization of solar energy. Two approaches have mainly been explored to address these issues: molecular modification and the template method.

Molecular modification and other means can shift the excitation range of azobenzene toward the visible light region. However, bulky substituent groups introduce steric hindrance, which not only reduces the energy storage density of azobenzene but also affects its energy storage capacity, resulting in an ambiguous effect. In recent years, three main approaches have been explored to compensate for this drawback: (1) combining exothermic azobenzene photoisomerization with exothermic phase changes to increase the energy storage density; (2) utilizing a non-solvent azobenzene energy storage structure; and (3) leveraging molecular interaction forces, such as hydrogen bonds and van der Waals forces, among azobenzene derivative molecules, to enhance the energy storage density of the azobenzene system. The revised text aims to improve language fluency while retaining the original meaning. Efforts have also been made to reduce similarity to the source text. If you have any specific suggestions or further requirements, please let me know. Carbon nanomaterials possess excellent electrical and thermal conductivity, allowing them to be combined with non-conductive materials to form conductive fabrics or films. These materials find wide applications in machinery, electronics, and chemistry. Graphene, a two-dimensional carbon material, exhibits an exceptionally high thermal conductivity (4800–5300 W/(m·K)), surpassing that of graphite (currently up to 2000 W/(m·K)) [107]. Carbon nanotemplates can greatly increase the interaction force, thereby enhancing the energy storage capacity of azobenzene. However, the high molar absorption coefficient of carbon nano-templates and the close-packed structure of azobenzene derivatives limit the depth of incident light and significantly reduce the light conversion efficiency. In addition, the synthesis of Azobenzene-based STFs using carbon scaffolds is complicated, and uniform thin films are difficult to deposit using this method, which may limit the charging and reversible conversion of azobenzene-based STFs to a certain extent [108]. Therefore, such materials can only be produced in very thin structures. Poor processing performance, slow isomerization, and solvent-assisted charging also limit their practical applications. In this regard, although the energy storage density of polymer templates is generally lower than that of carbon nano-templates (most azobenzene fuels prepared using polymer template strategies have a storage ED of less than 60 Wh/kg), they have strong plasticity, are convenient, and possess other advantages, which can be accessed by heating, electrodeposition, and other methods. Therefore, polymer templates have become one of the foci of recent studies.

Molecular modification techniques offer an effective means to red-shift the excitation range of UV light and enhance the solubility of azobenzene. Additionally, employing the template method can prolong the half-life of azobenzene in most cases. Furthermore, the incorporation of different azobenzene derivatives into Azobenzene-based STFs covering the full spectral range can effectively enhance the utilization of solar energy.

At the microscopic scale, azobenzene molecules undergo isomerization upon exposure to light, resulting in a slight change in molecular size (the molecular axis transitions from 9 Å to 5.5 Å). When the molecular size changes caused by the isomerization of azobenzene molecules are superimposed, it leads to observable macroscopic deformations. Light-driven flexible actuators based on azobenzene photo-responsive polymers have recently attracted significant attention from the scientific community due to their rapid, flexible, and reversible light-induced deformation capabilities. These materials find applications ranging from solar thermal storage to molecular motors, memory devices, and manipulators. At present, a lot of information has already been obtained on the chemical structure of photochromic compounds (especially azobenzene compounds) and its effects on their photochemical and thermal behavior. It is worth noting that the output power of the azobenzene photo-responsive polymer template actuator can be quickly, remotely, and wirelessly controlled by adjusting the intensity and wavelength of the incident light. Moreover, the photo-deformation of Azo/polymers depends on the level of polymer structure. A variety of polymer structures, including main chain polymers, side-chain polymers, cross-linked networks, and block copolymers, have already been designed and prepared to attain light-driven conversion through the force between Azo unit and polymer. This can be further developed that the controlled release of the energy stored by azobenzene in the form of mechanical energy can be realized by using the combination among the azo unit, polymer and the LC.

LC materials with light responsiveness have recently become an attractive class of stimulus-responsive materials. As a unique state between the crystalline solid state and isotropic liquid state, LCs possess both order and mobility. LC polymer (LCP) materials combine the characteristics of polymers and LCs, the most common of which are LCPs, LCP networks (LCNs), and LC elastomers (LCEs). The differences between these categories of materials are highlighted in Figure 1.9 [109]. LCPs are usually described as linear backbone or comb polymers. They are usually uncross-linked and can self-organize through rigid rod-like or disc-like molecular structures and intramolecular interactions (hydrogen bonds, dipole-dipoles) to form the LC phase. LCNs are almost entirely prepared from LC monomer precursors. In comparison, LCNs have a cross-linked network architecture while retaining some of the high-performance characteristics of LCPs. Elastic LCP networks prepared from LCN and LCP are usually called LCEs. The polymer backbone in LCEs is usually flexible and the crosslink density is not high. Owing to flexibility, LCEs can exhibit large reversible deformations. The main strategy for obtaining

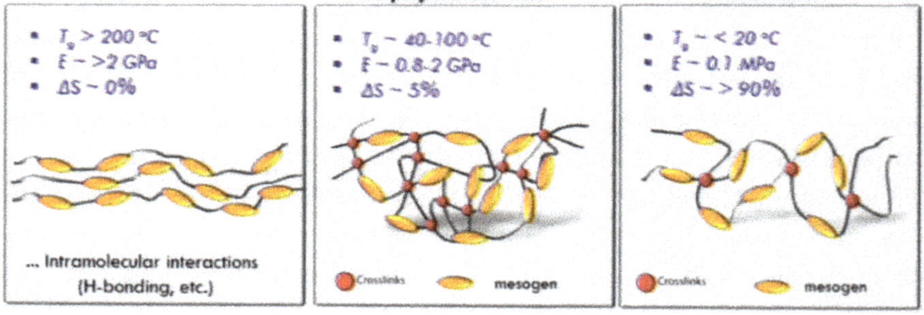

Figure 1.9. LCPs/LCNs/LCEs. Reprinted with permission from Ref. [109]. Copyright 2018, Wiley.

light-driven changes in LCPs is based on the combination of molecules containing azobenzene units. The *trans*-form of the molecule aligns with the mesogenic host material, while the curved *cis*-form reduces the order of the mesogenic host, resulting in a macroscopic change in the shape of the material. As the micro-deformation of small molecules during photoisomerization can cause large-scale deformation of the material, azobenzene is usually the photochromic molecule of choice for LCP materials. In addition, many azobenzenes are themselves mesogenic, which may increase their miscibility with LC molecules and result in a higher concentration of azobenzene in the mixture, which can increase the sensitivity of light-responsive materials. Because of this, researchers started investigating Azo/polymers with an LC structure several decades ago. Functionalized Azo/LC polymer films can exhibit excellent light-induced three-dimensional (3D) deformation. In fact, cross-linked LCP fibers containing azobenzene can precisely control the direction of light mobility. For example, a laminated film composed of a uniformly arranged cross-linked LCP and polyethylene film can realize the light-responsive 3D movement of the material, such as slug walking and flexible robotic arm movement. Recently, Yu *et al.* demonstrated that visible light irradiation, as a safer activation method, can replace UV-induced bending deformation of LCP materials [110,111]. Interestingly, when fibers or films containing azobenzene moieties are irradiated with UV light, the stress of surface contraction can reach a value similar to the stress levels in human muscles (about 300 kPa) [112,113]. This showed that LCPs containing azobenzene can potentially be used as artificial muscles in biological systems and are worthy of further research.

At present, many reports on light-driven polymer materials with azobenzene or azobenzene derivatives have already been published. The cross-linked liquid crystal polymers (CLCPs) possess a strong combination of orientational order of LCs and rubber elasticity of polymer networks at the same time. Among them, the CLCPs with azobenzene chromophore allow accurate and remote operation, which is more attractive. However, cross-linking in the glass cell has some problems to solve, such as the limit on the CLCP films size and the constraints on the fabrication of diverse microactuators by traditional melting and solution processing. Yu *et al.* [110] prepared the CLCP films and fibers using the post-cross-linking method, which can be processed using traditional melting and solution methods. The resulting CLCPs underwent reversible photoinduced bending and unbending in response to 445 nm and 530 nm visible light at room temperature, respectively, meaning that the problems caused by UV may be solved using the insights gained from this study. In 2017, Liu [114] developed a computational model to simulate monomer diffusion and its effect on the mechanics of light-induced topographical switching. He demonstrated light-switchable folding motion and surface transformation of LC films with various molecular alignments and made the following three conclusions. First, mesogens with higher reactivities prefer to diffuse to regions with larger polymerization intensities during polymerization and diffusion. Second, diffusion-induced photo-responsivity plays a dominant role in the texture modulation of all these problems. Third, light-triggered surface morphing can be tailored and enhanced by selecting a specific combination of chemical composition, direct distributions of the films, wavelength, exposure area, and polarization of the polymerization light and actuation light. Liu considered that this predictive tool can be expanded to tuning and optimization of the switching performance of soft actuators, human–machine interactions and haptics, and soft robotics. As commonly known, in contrast to pure azobenzene actuators, Azo-LCPs can show a glassy and rigid nature because of the nematic LC inside Azo-LCPs. The main cause of the photo-deformation of azo-LCP is the distraction of the mesogen alignment within the LCP. Azobenzene monomers trigger the alignment distraction of the mesogens in two ways: photoisomerization of azobenzene and rotation of azobenzene by means of a *trans–cis–trans* reorientation (TCTR), also known as the Weigert effect [115–119]. Cho *et al.* [120] found, using experiments and modeling methods, the photobending speed of Azo-LCPs in the TCTR case is always faster than the speed of photoisomerization, even though the average strain in TCTR is always smaller than that in the photoisomerization case because of the light propagation effect in Azo-LCPs under 445 nm light. More interestingly, photoisomerization triggers faster in-plane contraction than TCTR because of the high contraction efficiency; at the same time, TCTR can trigger faster bending than photoisomerization because of deeper light penetration. Numerous studies have shown that the combination of polymers and azobenzene derivatives has extremely high research value in the field of optical drive.

Tension can promote the energy release from azobenzene energy storage materials.

Similarly, the light-driven processes in azobenzene materials are affected by temperature. In this regard, a study by Zhang provided a new idea for solving this problem [121]. He explored a designed Azo-PI that undergoes photodeformation under unpolarized light. The Azo-PI film aligned using hot stretching exhibited fast and reversible bending behavior under alternating UV and visible light, indicating the efficient nano-to-macroscopic propagation of molecular deformation in azobenzene. Moreover, the aligned Azo-PI film bent even in hot water (80°C) and hot silicone oil (100°C and 120°C) under UV light.

As 3D LC-based microscopic helixes show poor shape persistence after continued stretching and relaxation because of a significant decrease in orientation induced by the elastic motion of the polymer network or the destruction of intermolecular interactions (hydrogen bonds, electrostatic forces, etc.), the non-LC polymers combined with azobenzene can be developed for the preparation of 3D shape-persistent helixes. This means that non-LC azopolymers also have the potential for use in light-driven materials.

At present, there are many reports on azobenzene photothermal energy storage materials, and there are many reports on light and shadow azobenzene light-driven materials. Evidently, there is a connection between the fields of photothermal and light-driven materials. As the energy stored by azobenzene can be controllably released in the form of thermal energy, it must also be possible to controllably release this energy in the form of mechanical energy.

As optical drive is also one of the important directions in the research on azobenzene materials, and many related studies and reviews have been published. These characteristics have made azobenzene light-driven materials a popular topic in research on controllable light- driven movement in recent years.

In 2014, Katsonis *et al.* found that combining azobenzene-based molecular switches into LCNs enables the conversion of light energy into macro-scale mechanical work, which can further be used to prepare spring-like materials [122]. Azobenzene photochromic switch, which acts as a nanoscale energy converter, can be incorporated into the polymer network using polymerizable acrylate groups. In this study, Katsonis *et al.* also proved that the macroscopic deformation of the Azo/LCN spring occurring through the disordered increase mechanism is related to anisotropic deformation. It is worth noting that the half-life of the *cis*-isomer of o-fluorinated azobenzene in solution is as long as two years. When fluorinated azobenzene is used as the molecular switch for LCN, the photomechanical deformation life of this material is very long, and its photochemical shape can be maintained for more than 8 days [123].

In 2003, Yu *et al.* [115] reported that if LC components of microscopic size are used, the surface of the film with Azo/LC network can show significant macroscopic volume shrinkage when exposed to light. Using the effect of the wavelength of the irradiated light on the light-energy conversion rate, the light-controlled deformation of the Azo/liquid crystal elastomer (Azo/LCE) was further converted into actual mechanical work in a clever experimental configuration. The Azo/LCE ring, which can be obtained by connecting the ends of a flat single- layer elastomer strip, can play a role similar to that of the light fuel engine. And the initial arrangement of the azobenzene mesogen is an important factor affecting the bending behavior of LCE films. To explore this effect, LCE films were prepared with two extreme alignment modes: parallel (homogeneous) and normal (isotropic). The results showed that the homogeneous film bends toward the actinic UV light source, and the isotropic film bends away from it. This demonstrates that the rotational motion of azobenzene-containing LCE and its composite materials can be realized. Mamiya proved that the local expansion and contraction of the material caused by the simultaneous irradiation with UV and visible light at different positions of the ring actually causes the rotation of the ring, which is only driven by light [124]. Yu reported a light-driven micro-robot that can pick up, lift, and move objects heavier than its own weight [125]. The robot consists of three layers of LCE film, which are connected to form a light-actuated arm with a wrist and hand. By turning on and off spatial local visible-light sources, each part of the robot can be selectively activated. A few years later, an intelligent Azo/LCE robot made by Zeng *et al.* was reported to even selectively distinguish between different objects based on their absorption characteristics and grab them, in a way similar to how venus flytrap grabs insects and objects [126]. Another application scenario for Azo/LCE is the bionic light-induced motion of soft robots. The movement of this kind of robot was realized by activating the structural deformation of the LCE film through space- or time- modulated light mode, which can cause non-reciprocal deformation required for directional movement at different positions of the robot body. In addition, even a static illumination field has already been proven to produce continuous motion of the Azo/LCE film with fast *cis*-to-*trans* thermal relaxation kinetics. Gelebart *et al.* [127] recently reported continuous wave-like deformation obtained under fixed lighting conditions in an LCE film with an unfolded arrangement.

Recently, 3D LC-based microscopic helixes were found to show poor shape-persistence after continued stretching and relaxation, because of a significant decrease in orientation induced by the elastic motion of the polymer network or the destruction of intermolecular interaction. Feng *et al.* [128] presented polarization-induced alignment of a non-LC supramolecular azobenzene-polyimide assembly (Azo/f-PI). 3D left- or right-handed Azo/f-PI helixes possessing shape persistence, high elasticity, and photoresponsivity were prepared using large-area molecularly oriented assembly films, which can retain the linear-elastic deformation with 77.16% retention of the elastic potential ED (0.304 J/g) after being treated at 100°C for 10 min and show an extremely low permanent deformation of 3.68% upon 1000 stretching–relaxation cycles. Excellent mechanical properties also mean that non-LC Azo polymers potentially possess good light-driven performance. Interestingly, the performance of azobenzene-containing polyimides (Azo-PIs) as light-deformable materials can potentially be greatly improved.

At present, it is already known that pulling force can promote the release of heat energy, implying a connection between azobenzene photothermal energy storage materials and azobenzene light-driven materials, which may become a future research prospect in azobenzene materials. In fact, it may further be considered that photothermal and light-driven azobenzene materials can be combined. Carbon nano-templates enhance the heat storage capacity of azobenzene, and polymer templates can make azobenzene light-driven. There are a sufficient number of reasons for expecting interesting findings from combining these two types of templates.

Although the field of azobenzene-based STFs is in its infancy, a variety of methods have been developed for preparing azobenzene-based STFs, such as controllable self-assembly, electropolymerization, and 3D printing. Azobenzene-based STFs may be expected to be used in fields such as deicing, desalination, and heat-driven generators in the future. Furthermore, Azobenzene-based STFs may also be used as an important part of deep-space probes, smart robots, thermoelectric systems, warm clothing, and solar water heaters in military and consumer-oriented fields. Azobenzene-based STFs provide a new way for solar energy collection and storage.

1.3.4 *Fulvalene dimetal complexes*

The (fulvalene)tetracarbonyldiruthenium ($FvRu_2(CO)_4$) series represents a group of organometallic photochromic complexes initially reported by Vollhardt and colleagues in 1983 [129,130]. When in solution, ($FvRu_2(CO)_4$) undergoes reversible photoisomerization upon irradiation with visible light (470 nm) [129–135]. The energy storage capacity ($\Delta H_{storage}$) of $FvRu_2(CO)_4$ comprises both an overall standard energy change (83 ± 6 kJ mol^{-1}) and activation energies (125 ± 8 kJ mol^{-1}) [135]. However, the practical applications of the $FvRu_2(CO)_4$ complex are limited due to its high cost and relatively high molecular weight of around 442 g/mol. To address these issues, iron has been explored as a potential replacement for the expensive ruthenium, given its lower cost and lighter weight. However, no isolatable photoisomer can be formed in the iron analogue [136]. Meanwhile, the development of an osmium analogue capable of photoisomerization has been achieved, but the isomerization in this case is irreversible.

The fulvalene dimetal complex system involves the cleavage and reformation of X–X and C–X bonds (where X represents Ru, Fe, Os, or other metal atoms) under UV light (350 nm), resulting in the storage of photon energy in a metastable state. The stored energy can be thermally reversed or catalytically released through conformational isomers. The energy storage density of the fulvalene dimetal complex system ranges from 0.11 MJ/kg to 0.19 MJ/kg.

First-principles calculations were employed to show the mechanism for the heat-releasing step based on spin-unrestricted density functional theory (DFT) with a non-empirical gradient corrected exchange-correlation functional. It shows an intermediate in the conversion process, a fact that explains some peculiar experimental observations. And the barrier for the conversion of the intermediate to the isomer is only 4.4 kcal/mol. This extraordinary labilization of the normally strong Fv linkage by the homolysis of a metal metal bond has been an unrecognized feature in this class of compounds.

To further explore the scope of the photothermal system, in particular the possibility of using the cheaper and environmentally and economically more favorable Fe analogues, the potential energy manifold of which is diminished relative to that of its Ru relative from 20.8 kcal/mol to 15.9 kcal/mol. This may be attributed to the fast thermal reversal at the ambient temperatures employed.

This fulvalene dimetal complexes energy storage system can be applied in mobile systems, optical switch batteries, and other fields. To improve its solubility, the original fulvalene dimetal complex system can be processed, particularly through the addition of alkyl groups. However, because of low energy storage density and high cost, it is seldom researched and applied than alternatives [24,131,135].

1.3.5 *Stilbene*

The structure of stilbenes is characterized by the presence of 1, 2-stilbene nuclei (containing a vinyl group in the middle (–C=C–), and the hydrogen atoms on both sides are replaced by two benzene rings), there are two types of monomer stilbenes and oligomerized stilbenes [137], and there are two types of *trans* and *cis* isomers. When the compound is exposed to light (300–700 nm), it isomerizes into a high energy metastable isomer that stores energy. When excited by heat, the metastable isomer returns to the initial stable isomer and releases the stored energy (see Figure 1.10 for an example of the unsubstituted structure).

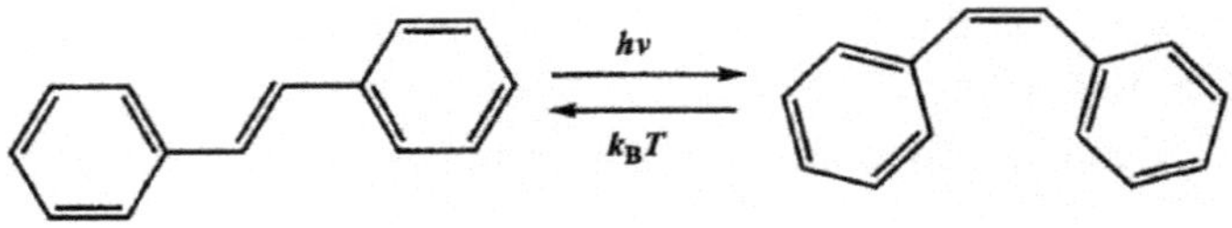

Figure 1.10. Isomerization of stilbenes without substituent. Reprinted with permission from Ref. [137]. Copyright 2009, Royal Society of Chemistry.

There are three isomerization mechanisms for this type of compound, but the generally accepted isomerization mechanism involves twisting the ethylene bond in the first monoexcited state into a twisted geometry that avoids crossing with the ground state [138]. The energy storage density of unsubstituted stilbenes is 5 kJ mol^{-1}. [43,139], and it is found that the energy density can be improved by effective substitution. Bastianelli and his collaborators synthesized a series of (Z)-9-styrene phenylpyridine and found that the energy density of unsubstituted stilbene on the benzene ring was equal to that of unsubstituted stilbene (5 kJ mol^{-1}). When substituting electron donor amino group on benzene ring, its energy density can be effectively increased by about four times (18 kJ mol^{-1}). When substituting four methyl groups on the benzene ring, the energy density increases further (55 kJ mol^{-1}). The highest energy density (104 kJ mol^{-1}) was obtained when substituting four benzene rings on the benzene ring while adding an amino substitution. These results benefited from delocalization and increased steric hindrance [140].

Stilbene has served as an exemplary model for studying photoisomerization. The photoreactions of stilbene involve rearrangements of atoms, either through geometrical isomerization processes (*cis* to *trans*/*trans* to *cis*) or through electronic rearrangements to form 4a, 4b-dihydrophenanthrene (*cis* to DHP).

When *trans*-stilbene is excited between 250 nm and 312 nm, it populates an excited state (*trans*), which undergoes a barrier crossing of approximately 3.5 kcal/mol in alkanes or in isolated molecules to yield *cis*- and *trans*-stilbene in roughly equal amounts. This barrier crossing process occurs within tens to hundreds of picoseconds and has been found to be influenced by friction, as predicted by Kramers theory for the escape of a particle bound by a harmonic potential over a barrier [141].

In contrast, the isomerization reactions of electronically excited *cis*-stilbene occur on a faster timescale of 0.3–2 picoseconds, even in high-friction liquid environments. These results can be interpreted as unrestricted motion on a barrierless potential surface or motion over an extremely small barrier.

It is notable that stilbene oligomers and diarylethene exhibit both photoisomerization activity and semiconducting characteristics. This implies that thermally deposited films of these molecules can be directly utilized as the active layer, demonstrating distinct photo-switching behavior.

In organic single crystals, mobility is typically higher compared to their polycrystalline counterparts due to the low density of defects and grain boundaries. As a result, organic field-effect transistors (OFETs) based on single crystals generally exhibit superior performance. Various solution-based techniques have been developed in recent years to grow high-quality, large-sized two-dimensional organic semiconductor single crystals, as they hold great potential for large-area, low-cost, and high-performance electronics. Additionally, novel molecular designs and careful interface modifications have been employed to further enhance device performance. As a result, high mobilities exceeding 10 cm^2/Vs has been achieved in both *p*-channel and *n*-channel single-crystal OFET devices [143,144].

In 2017, Cao *et al.* [142] successfully fabricated high-quality, large-sized two-dimensional (2D) single crystals of photochromic organic semiconductors, specifically using 1,4-bis(4-methylstyryl)benzene. These crystals were utilized as the conduction channel in photo-switchable OFETs. This groundbreaking work opened up new possibilities for utilizing photochromic organic semiconductors in electronic devices.

Recent advancements in organic fluorescent crystals have emphasized the importance of intermolecular interactions and molecular stacking modes in determining their luminescent characteristics in bulk. The formation of cocrystals has been studied extensively as a means of modifying the properties of organic molecules in the solid state. Halogen- and hydrogen- bonding, which are non-covalent interactions, have been widely employed in the design of supramolecular cocrystals of organic solids.

The modular nature of cocrystals allows for the precise control and modification of crystal structure by manipulating the direction and selectivity of halogen and hydrogen bonds within the crystal lattice. By utilizing halogen- and hydrogen-bonding interactions with a single chromophore molecule, researchers have developed a cocrystallization strategy to prepare various crystalline materials with tunable fluorescent emission properties [145–150]. William Jones and his team specifically selected a stilbene derivative, 1,4-*bis*-p-cyanostyrylbenzene, as a representative fluorescent model system. Through careful selection and design of the cocrystal components (*co*-formers) and the interaction types (hydrogen- and/or halogen-bonding), the stacking arrangement of fluorescent molecules in the solid state can be precisely controlled [151].

These observations highlight the significant interest in stilbene-type compounds due to their remarkable optical and electronic properties, which merit further investigation and research efforts.

1.3.6 *Anthracene*

Anthracene is an attractive photoreactive group for the design of light gated materials as it is able to undergo a [4 + 4] photocycloaddition affording photodimer that can be reversed to anthracene [29,152]. Photoisomerization was first observed in the early 20th century. The energy of anthracene after light irradiation can be released reversibly as heat through first-order kinetics under appropriate temperature or illumination conditions.

Anthracene and its derivatives exhibit a characteristic behavior of dimerization upon irradiation with wavelengths above approximately 350 nm. When subjected to such irradiation, anthracene molecules form dimers. However, these dimers can dissociate upon exposure to light of less than 300 nm or through heating, with an activation energy (E_a) ranging from 111.160 kJ/mol. Peripheral substitution on anthracene derivatives can alter the activation energy for thermal dissociation by approximately 50 kJ/mol. One notable aspect of this photodimerization/dissociation cycle is the ability to repeat the process numerous times without the accumulation of significant side products. Furthermore, it is possible to trap the reaction at a nonequilibrium state, allowing for the manipulation and study of specific reaction steps. Overall, this behavior demonstrates the interesting and controllable nature of anthracene and its derivatives in terms of their photodimerization and dissociation properties. The isomerized stew of anthracene is reported to be 65.2 kJ/mol [30–32]. In order to further improve the isomerization, the electron absorbing and electron donating substituents were grafted onto the anthracene monomer. The results showed that the stored energy reached 83.6 kJ/mol. Although most studies can achieve high energy density, the easy degradation of anthracene greatly limits its development and application [33,34].

Also, the photodimer can be cleaved to anthracene either photochemically, thermally or mechanically [153].

In recent years, covalently cross-linked materials have gained significant attention and find applications in various fields, including coatings, composite materials, and biomedical materials. These materials contain covalent cross-links that give them predominantly elastic behavior when subjected to mechanical deformation. However, the permanent cross-linking structure can be a disadvantage when it comes to reshaping, recycling, or easy removal of the material. To overcome this limitation, reversible covalent bonds have been introduced into the network of these materials. By incorporating reversible bonds, it becomes possible to combine the advantages of classical cross-linked materials, such as dimensional stability, with the benefits of non-cross-linked oligomers or thermoplastic polymers, which offer easy processing. The behavior of these materials, whether they exhibit thermoset-like or thermoplastic-like properties, depends on the type of reversible covalent bonds present. Reversible bonds can be either dynamic, involving exchange reactions, or nondynamic, requiring stimuli to break or form the covalent bond [154]. The choice of reversible bond type determines the conditions under which these materials will exhibit thermoset or thermoplastic behavior.

In networks held together by dynamic reversible bonds, activating the reversibility can lead to the breaking of bonds and the transformation of a solid material into a viscous liquid. However, once the stimulus is removed, the material retains its connectivity at ambient conditions. The reversibility can be activated again, causing the network to reform and regain its solid properties. Cinnamate esters, coumarines, and anthracenes are well-known examples of materials that exhibit these reversible properties [153,155–157]. While cinnamates and coumarines primarily undergo 2π–2π photochemical cycloaddition and dimerization upon UVA irradiation, anthracenes are involved in a 4π–4π photocycloaddition process. Scission of the photo-dimers typically requires highly energetic UVC radiation. The advantage of the 4π–4π photocycloaddition of anthracenes is the extended initiating window, allowing for activation by wavelengths up to 400 nm [159–162]. This feature enables the use of less energetic and longer wavelengths, including visible light, which helps to minimize side reactions. The rate of dissociation of the photodimers depends on the substituents on the anthracene ring and can be adjusted to relevant time and temperature scales [158].

The reversible nature of anthracenes has also been utilized in the development of photoresists with spatially controlled solubility changes in polymer networks. Griesser *et al.* synthesized functional polynorbornenes using ring-opening metathesis polymerization and incorporated pendant anthracene groups. By employing the 4π–4π cycloaddition of anthracene groups upon UV irradiation with wavelengths above 300 nm, they were able to create negative- toned photoresists with a resolution of 4 μm [163].

An interesting aspect of the photocleavage of the photodimers in anthracene-based materials is the fluidification of the material, which enables efficient healing of surface damages. In 2015, Sandra Schlögl demonstrated the use of the [4πs + 4πs] cycloaddition reaction of anthracene derivatives in solid films to create reversible and mendable epoxy-based materials. However, the thermoplastic nature of the epoxy-based materials meant that irradiation with 254 nm UV-light did not lead to significant regeneration of the consumed anthracene moieties [164]. The following year, Sandra Schlögl and their team focused on the preparation of photo-responsive hydrogenated carboxylated nitrile butadiene elastomers. These elastomers exhibited well-defined bond formation and bond cleavage in response to different wavelengths, taking advantage of the photo-reversibility of the [4πs + 4πs] cycloaddition of anthracene groups in the rubbery state. With the aim of industrial usage of the photo-responsive elastomer, anthracene derivatives with reactive epoxy groups were synthesized in a one-step reaction and covalently attached to the polymer side chain of a commercially available rubber type through a ring-opening reaction between the epoxy and carboxylic groups. Upon UV exposure with wavelengths greater than 300 nm, the pendant anthracene groups underwent photodimerization, resulting in the crosslinking of the rubber chains at room temperature. The cleavage of the photodimers was then achieved by UV illumination at 254 nm. In addition to the reversible modulation of thermo-mechanical properties, the changes in solubility depending on the wavelength were utilized to produce reversibly patternable photoresists. This allowed for the creation of negative and positive patterns, enabling the fabrication of flexible 2D micropatterns [165]. These advancements highlight the potential of utilizing the photo-reversibility of anthracene groups in various material systems, opening up opportunities for reversible and tunable properties in applications ranging from epoxy-based materials to photo-responsive elastomers.

Responsive biomaterials that can undergo stimuli-induced property alterations, such as thermal, pH, ionic, electric, and photomechanisms, have great potential for drug delivery applications, allowing for adaptations in response to physiological and disease states. Among these mechanisms, light and laser stimuli offer fast response times, high focal control, and precise control over wavelength and intensity [166–169]. Photoresponsive systems used in biomaterials can be classified into reversible and irreversible systems. Irreversible changes include photolabile materials, such as polyethylene glycol (PEG), or photoinduced microcapsule rupture [170,171]. However, for long-term controlled drug delivery in the treatment of chronic diseases, reversible systems are more suitable. Polymers with controllable, reversible light-crosslinking mechanisms can create unique biomaterials with stimulus- controlled swelling, degradation, and diffusion properties, which are desirable for tissue engineering and drug delivery applications. Molecules such as cinnamylidene acetate, nitrocinnamate, coumarin, and particularly anthracene, have been extensively investigated for the photocrosslinking of polymers [172–175]. Anthracene, with its symmetrical structure, allows for predictable monitoring of dimerization with minimal side reactions. The $[4\pi+4\pi]$ photocycloaddition of anthracene is a reductive process that forms an ortho-substituted benzene molecule [176]. Once dimerized, it no longer absorbs light at wavelengths between 300 and 400 nm, making the reaction easily monitorable spectrophotometrically. Anthracene is typically soluble only in organic solvents. To create an anthracene-hydrogel photoresponsive system, anthracene needs to be modified to increase its solubility and allow for miscibility with hydrophilic polymers and buffers. Previous studies have shown some photoresponsiveness by modifying polystyrene, methacrylate, and polyoxazoline with anthracene [177–181]. One promising approach involves directly modifying PEG with anthracene to create thin sheets with hydrophobic aggregation that exhibit photoresponsiveness [182]. Photodimerizing PEG–anthracene macromolecules can be grafted onto various polymers to introduce photoresponsive properties into hyaluronate and alginate. These hydrophilic polymers can be tested in vitro using human corneal epithelial cells to demonstrate cytocompatibility of the resulting photogels. Moreover, the effective crosslinking density of the photogels based on anthracene can be adjusted, leading to a decrease in the release rate of small and large molecules when exposed to 365 nm light. This tunable crosslinking has the potential to manipulate the delivery rates of therapeutics, providing control over treatment profiles and offering various applications [182]. Overall, the use of anthracene and other photoresponsive molecules in the development of smart biomaterials holds promise for controlled drug delivery systems with adaptable properties in response to specific stimuli.

PEG, as a polyether chain polymer with adjustable end groups, can be tailored to different molecular weights, making it a versatile material for various applications. Not only can PEG be crosslinked on its own to form matrices, but PEG chains have also been used as crosslinkers by modifying polymers with PEG at both ends. Hydrophilic polymers like alginate, hyaluronate (HA), collagen, and chitosan have been crosslinked with PEG to create hydrogels with increased degradation times [183–187]. Heather Sheardown and their team demonstrated the grafting of a PEG–anthracene crosslinker onto hydrophilic polymer backbones, introducing photosensitive properties into the polymer system. This enables incremental photoinduced crosslinking with 365 nm light. The result is the creation of a photoresponsive photogel with dimerization-driven crosslinking, eliminating the need for initiators or catalysts. Treatment with 365 nm light was found to slow the release of Coomassie Blue and "turn off" myoglobin release from HA photogels, as well as slow the release from alginate photogels. This ability to modulate drug release, both incrementally and discontinuously, highlights the potential of these systems as controllable and smart drug delivery systems [188]. By combining the advantages of PEG and anthracene-based photoresponsiveness, these photoresponsive photogels offer great promise for applications requiring precise control over drug release kinetics. The ability to tune the crosslinking and photorelease rates with light provides opportunities for developing personalized and on-demand drug delivery systems.

In addition, anthracene can also be applied to anti-counterfeiting inks.

Nowadays, counterfeiting poses a significant threat to society, with counterfeit banknotes, diplomas, tax stamps, and certificates being widespread. It has become imperative to develop innovative technologies to combat counterfeiting and ensure the security and reliability of genuine documents [189–192]. To this end, photo-fluorochromic compounds have emerged as a promising choice due to their ability to emit reversible fluorescence under light stimulation. Typically, inkjet printable anti-counterfeiting materials are created by directly blending photo-fluorochromic compounds with a polymer matrix [193]. However, this approach has limitations in terms of inclusiveness and its reliance on a single mode of detection, possibly resulting in a decline in ink quality and anti-counterfeiting properties [194–197]. Consequently, there is a pressing need to develop anti-counterfeiting inks that exhibit vibrant fluorescence and possess multi-mode responsiveness.

To attain this objective, a feasible approach is to graft photo-fluorochromic molecules onto polymer backbone. An alternative strategy is to assemble such photo-responsive units into uniform supramolecular polymers, which is superior to the covalent approach due to the avoidance of tedious synthesis. Moreover, supramolecular polymers serve as a dynamic platform for multi-component organization, allowing for light-induced energy/electron transfer process [198–205]. As a consequence, the fluorescent signals can be fine-tuned in a multi-mode manner.

Notably, harsh reaction conditions are required for the traditional anthracene-endoperoxide photo-switching (365 nm UV light irradiation for oxygenation process, while 254 nm UV light or thermolysis for deoxygenation reaction) [206]. Photo-degradation side-products are prone to form during the processes, which hamper the practical application in fluorescent anti-counterfeiting inks.

In 2018, Wang *et al.* [207] presented a novel approach for constructing photo-responsive supramolecular polymers with fluorescent anti-counterfeit applications. This was accomplished by harnessing the multicycle anthracene-endoperoxide switching properties. Under mild visible light irradiation, the anthracene molecules undergo photo-oxygenation to form endoperoxide, and this conversion is reversed spontaneously at room temperature due to the σ-metalation effect. The supramolecular polymers were formed through a cooperative nucleation–elongation mechanism, enabling the fluorescence resonance energy transfer (FRET) process through a two-component co-assembly strategy. The efficiency of FRET was finely tuned by the light-triggered conversion of anthracene–endoperoxide or by the vapor-induced transition from monomer to polymer. Consequently, this led to distinct fluorescent changes among three different states, allowing for high-contrast detection. Building upon this concept, dual-mode anti-counterfeiting patterns were successfully created using inkjet printing techniques. This achievement means that the patterns exhibit dual-mode responses to orthogonal stimuli, offering enhanced security and reliability.

Anthracene, known for its fluorescent properties, has been extensively used in the development of OFETs. Early metal-oxide-semiconductor structure-based field-effect transistors (MOSFETs) were pioneered by D. Kahng and M. M. Atalla at Bell Labs in the 1960s [208]. However, traditional MOSFETs, which utilize inorganic single crystal silicon as the semiconductor material, are cost-intensive and prone to fragility. In contrast, OFETs, which have been under development since the 1980s, offer a solution to these challenges [209]. The objective of organic-based devices is not necessary to achieve performance levels exceeding those of inorganic semiconductor technologies, but rather to realize the unique combination of electrical properties offered by semiconductors and the desirable characteristics of plastics. These characteristics include low cost, versatility in chemical synthesis, ease of processing, and flexibility. OFETs possess several additional advantages over inorganic field-effect transistors. Firstly, they exhibit flexibility, enabling applications in wearable and portable electronics. Secondly, they can be fabricated through low-temperature and solution-based processes. Lastly, organic molecules provide flexibility in terms of structure and properties. These qualities make OFETs highly suitable for flexible optoelectronic devices and for studying the transport properties of semiconductors. The performance of OFETs largely depends on the active organic semiconductor materials employed. In comparison to polymers, small molecule semiconductors offer advantages such as high purity and well-ordered packing, which are crucial factors for achieving efficient transistor operation.

Anthracene exhibits remarkable fluorescence and high quantum yield due to its conjugated and rigid structure. It possesses fast radiance and well-defined absorption and fluorescence spectra. These characteristics have made anthracene a widely used building block in organic light-emitting diodes (OLEDs) ever since the discovery of electroluminescence in anthracene single crystals in the 1960s [210,211]. In the mid-1980s, anthracene was incorporated as the active material in light-emitting diodes (LEDs). Initially, the device performance was relatively poor due to the thickness of the organic layer and the high operating voltage, which exceeded 100 V. However, significant progress was made by using the vacuum evaporating deposit technique to obtain anthracene layers, resulting in a marked reduction in the operating voltage to around 30 V [213]. Since then, extensive research has been conducted on anthracene derivatives for OLEDs. These derivatives have garnered significant interest due to anthracene's unique electron-rich structure, low electronic band gap, and strong blue fluorescence. More recently, anthracene derivatives have been investigated extensively as fluorescent chromophores for the development of chemosensors [214].

Based on the intriguing properties exhibited by anthracene derivatives, researchers have explored the use of polymer-based approaches, incorporating anthracene and its 9,10-substituted derivatives either into the polymer main chains or as pendant groups. The primary objective has always been to address the challenge of producing high-quality optical films that suppress excimer formation. Numerous polymers based on 9,10-substituted anthracenes have been developed and studied as blue light-emitting materials for OLED devices [215–221]. It has been observed that 9,10-substituted anthracene derivatives tend to form a more closely packed lamellar structure compared to 2,6-functionalized derivatives [222]. By modifying the structure of anthracene, the π-electron system can be extended, and the molecular packing can be adjusted by altering the energy levels of the frontier molecular orbitals of the derivatives [223]. Since the 1990s, there has been significant interest in fully conjugated anthracene-based polymers, such as poly(anthrylene)s and poly(anthrylenevinylene)s [224,225]. However, the application of these polymeric materials in optoelectronic devices has been limited. The main challenge has been the poor solubility of these macromolecular compounds due to their structural rigidity, which hinders the preparation of thin films using the spin-coating technique. Another drawback of these π-conjugated systems is the high crystallinity of the anthracene moieties, which prevents the formation of stable and flexible thin active layers. To overcome these limitations, flexible aliphatic side groups have been incorporated into the anthracene units. Soluble poly(9,10-bialkynylanthrylene)s have been synthesized and utilized in thin-film field-effect transistors [226]. Additionally, considerable attention has been devoted to the design and investigation of conjugated-non-conjugated block copolymers, wherein the anthracene-based segments are linked by solubilizing non-conjugated spacer groups. By using various types of spacers, different anthracene-containing semiconducting polymers, such as polyamides, polyesters, and polyethers, have been developed [227].

The charge transport potential of anthracene was reported to have a mobility on the order of 1 cm^2/Vs through pulsed photoconduction measurements [212]. In the early 21st century, transistors based on anthracene single crystals achieved a maximum mobility of 0.02 cm^2/Vs [228].

However, in 2013, significant advancements were made in the field of anthracene-based semiconductors. Firstly, the development of ultra-high-mobility anthracene derivatives, such as 2,6-diphenyl anthracene (DPA), took place [229]. Secondly, a comprehensive series of studies on molecular packing structure, single crystal transistors, and mobility calculations revealed the initial intermolecular interactions, elucidated the intrinsic charge transport properties, and established the relationship between molecular structure and device properties. This breakthrough provided a deeper understanding of anthracene-based semiconductors. Thirdly, and most importantly, organic light-emitting transistors (OLETs) were successfully developed in this period using anthracene derivatives.

These derivatives showcased simultaneous high-mobility and high-luminescence properties, making them ideal for self-driven lighting in thin-film transistors and single-crystal transistors. Examples of such derivatives include 2-(4-hexylphenylvinyl) anthracene (HPVAnt) and 2,6-di(2-naphthyl) anthracene (dNaAnt), developed by D. F. Perepichka, W. Hu, and their colleagues [230, 231].

The extensive study of anthracene for OFETs over the past two decades has revealed the relationship between molecular structure, properties, and charge mobility. This knowledge has paved the way for efficient molecular design and optimization. Recent achievements in understanding molecular packing and the self-driven emitting properties of anthracene derivatives have further expanded their potential applications in electronic devices, promising an exciting and promising future.

Chemical modifications of anthracene allow for tuning of molecular packing and charge transport properties. Aryl groups, such as phenyl, naphthyl, and thienyl, can be readily attached to anthracene at the active end- and peri-positions through coupling reactions. The synthesis of anthracene derivatives typically begins with 2,6-diamino-anthraquinone, followed by reduction and substituent coupling. The derivatives are then purified using recrystallization or step sublimation methods for soluble and insoluble derivatives, respectively. This ease of synthesis and purification provides a solid foundation for the study of their properties [232].

It is widely acknowledged that finding organic semiconductors that exhibit both high charge mobility and high fluorescence is rare. The experience gained through molecular design and the development of efficient methods to investigate charge transport properties has demonstrated the potential of anthracene derivatives as semiconductor materials for applications such as phototransistors or OLETs. The development of such derivatives is deemed necessary due to their unique properties and suitability for these device applications.

1.4 Selection criteria and development prospects of STFs

STFs have the ability to store solar energy in their chemical bonds and release it as thermal energy under specific conditions. From a molecular perspective, in order to effectively utilize solar energy, the spectral range that excites STFs should closely match the sunlight spectrum. It is also important to focus on the specific sunlight wavelengths that meet the conditions for exciting STFs. Therefore, if possible, the excitation range of STFs should be redshifted toward the visible light region. STFs must possess chemical bonds with high photochemical activity, such as conjugated double bonds. This ensures that they can absorb light energy and undergo efficient energy storage and release processes. In order to meet the requirements for industrialization of solar energy, STFs should exhibit high photothermal cycling stability and be cost-effective. To evaluate the performance of STFs and make the most of solar energy, several measurement standards have been established. These include ED, PD, energy storage half-life, and solar energy conversion efficiency.

ED represents the energy storage capacity of STFs and is considered the most important energy storage parameter. PD refers to the weight-to-power ratio and indicates the ability of STFs to release energy. A higher PD value indicates a stronger energy release capability. Energy storage half-life is the time required for half of the STF to be fully excited by light and return to its original state. It reflects the stability of the STFs and different STFs may have different half-life requirements. A longer half-life indicates a more stable ability to release heat energy, which is advantageous for industrial production. Solar energy conversion efficiency represents the capability of STFs to convert solar energy under specified conditions. It is influenced by factors such as the light-excited activation energy of STFs. Higher solar energy conversion efficiency indicates a more efficient conversion of solar energy into usable heat energy.

Efficient STF materials should meet a specific set of criteria to maximize their performance and potential. These criteria are as follows: (1) Matching the solar spectrum: The absorption characteristics of the low-energy isomer in the STF should align with the most intense region of the solar irradiance spectrum. This ensures that the STF efficiently absorbs solar energy. (2) Large storage energy capacity: The energy difference between the low-energy and high-energy isomers in the STF should be significant. This difference determines the amount of energy that can be stored and subsequently released. (3) High quantum yield: The photo-induced conversion from the low-energy to the high-energy isomer in the STF should proceed with high efficiency. A high quantum yield indicates a high conversion efficiency and better utilization of absorbed energy. (4) Avoidance of photon competition: The two isomers in the STF should not compete for photons. Photon competition can lead to inefficient energy storage and release processes. Minimizing or eliminating photon competition enhances the performance of the STF. (5) Long-term stability: The STF should possess a sufficiently large energy barrier on the ground state energy landscape to prevent back-conversion from the high- energy to the low-energy isomer. This ensures long-term storage capability and enhances the stability of the STF over time. (6) Low cost: The cost of raw materials, synthesis processes, and integration into energy systems should be competitive. The use of cost-effective materials and efficient manufacturing methods contributes to the viability and widespread adoption of STFs. In terms of device types, MOST systems can be categorized into three main types: charging, discharging (heat release), and hybrid systems that integrate MOST with other energy conversion technologies. It is worth noting that MOST systems, especially in their liquid state, possess a flexible form-factor and can be conveniently transported using pumping techniques. Solar collectors for charging STF systems can be designed in three different ways, each with increasing light-concentrating capacities: (1) Flat panel with meander-shaped channels: This design involves incorporating meander-shaped fluidic channels inside a flat panel photoconversion device. The device is exposed to solar irradiation, allowing the STF solution to flow through the channels with an appropriate residence time. This design enables solar energy to be efficiently absorbed and converted by the STF. (2) Parabolic reflecting mirror: Another design option is to use a parabolic reflecting mirror. In this configuration, fluidic tubing is installed along the focus line of the parabolic collector. The incoming photons are reflected toward the central area, where the STF solution is pumped through the tubing. The concentration of photons in the central area can reach a capacity of 5–30 suns. However, this design can pose a challenge for MOST applications since the photoisomer half-life is temperature-dependent. Higher temperatures can lead to shorter half-lives. To address this issue, a double-layer glass tubing can be implemented. The STF solution flows through the outer tubing, while cooling water flows through the inner tubing to stabilize the system and prevent excessive heating. These different collector designs offer varying levels of light concentration, allowing for more efficient solar charging of STF systems. The choice of design depends on factors such as the specific STF material, desired energy conversion efficiency, and practical considerations related to temperature control and overall system stability.

In STF systems, thermal discharge is the natural process for back-converting the charged photoisomer to its original state. However, relying solely on thermal discharge may not be the most convenient method for on-demand energy release. The charged photoisomer has a characteristic half-life for energy storage that follows basic thermodynamic principles. Increasing the temperature accelerates thermal discharge, resulting in a shorter half-life for the molecule. As a result, designing STF systems with long half-lives (months or years) is often a key objective, which limits the practicality of relying solely on thermal discharge for energy release. A more convenient and controllable method for energy release is through the use of heterogeneous catalysts. These catalysts, which are separate from the STF solution and can be fixed-bedded and reusable, play a critical role in releasing the energy stored in the STF system. Catalyst development specific to different STF systems is crucial for optimizing energy release. The turnover number (TON) and turnover frequency (TOF) are common metrics used to quantify the efficiency of the catalyst. Higher TON or TOF values indicate more efficient catalysis. The design of a catalyst reactor typically involves loading the catalyst material inside a tube to form a fixed-bed reactor. The charged STF solution is then flowed through the reactor, where the catalyst facilitates the release of heat through catalytic reactions. This approach provides a convenient way to control the energy release process in STF systems and allows for better integration with other energy conversion or utilization systems.

Indeed, in an ideal MOST system with a multi-junction design, the maximum energy storage efficiency can reach up to 21% (based on simulated results). However, there is still potential to utilize the remaining part of the solar spectrum. One approach is to integrate the MOST concept with other technologies, creating hybrid devices that can enhance solar energy utilization. In recent years, there has been growing research on combining MOST with technologies such as wearable fabrics and visual solar thermal energy systems. From an application perspective, STFs should possess high fluidity to enable their integration into various device formats, while also maintaining thermal cycling stability for long-term use. Conducting comprehensive small- and large-scale cycling tests will be necessary to evaluate the performance and feasibility of STFs in real-world applications. Fortunately, the development of machine learning and artificial intelligence (AI) technologies has accelerated the discovery of new STF candidates through molecular screening.

These technologies have the potential to expedite the identification and optimization of STFs with desirable properties. Moreover, the automation of chemistry systems can facilitate scalability and reduce future research costs associated with STF discovery and development. Overall, advancements in machine learning, AI, and automated chemistry systems hold great promise for enhancing the efficiency and effectiveness of STF research and development, ultimately driving the commercialization and widespread adoption of STF-based technologies.

References

[1] Dai H, Su Y, Liu J. (2022). Strategic thinking on China's energy development under the goal of carbon neutrality. *Petrol Sci Technol Forum*, **41**(1), pp. 1.8.

[2] United Nations. (2015). Paris agreement. Conference of the Parties 21, 12, pp. 12.

[3] Energy & Climate Intelligence Unit. (2021). Net zero tracker [EB/OL]. 12, pp. 21.

[4] Hoesung L. (2023). Synthesis report of the IPCC sixth assessment report (AR6), 3, pp. 22.

[5] Cook TR, Dogutan DK, Reece SY. (2010). Solar energy supply and storage for the legacy and nonlegacy worlds. *Chem Rev*, **110**(11), pp. 6474–6502.

[6] Dincer, I. (2000). Renewable energy and sustainable development: a crucial review. *Renew Sust Energ Rev*, **4**(2), pp. 157–175.

[7] Chu S, Majumdar A. (2012). Opportunities and challenges for a sustainable energy future. *Nature*, **488**(7411), pp. 294–303.

[8] Lewis NS. (2007). Toward cost-effective solar energy use. *Science*, **315**(5813), pp. 798–801.

[9] Lewis NS. (2016). Research opportunities to advance solar energy utilization. *Science*, **351**(6271), pp. aad1920.

[10] Xiang C, Zhao X, Tan L. (2019). A solar tube: efficiently converting sunlight into electricity and heat. *Nano Energy*, **55**, pp. 269–276.

[11] Kucharski TJ, Tian Y, Akbulatov S. (2011). Chemical solutions for the closed-cycle storage of solar energy. *Energ Environ Sci*, **4**, pp. 4449–4472.

[12] Massiot I, Cattoni A, Collin S. (2020). Progress and prospects for ultrathin solar cells. *Nat Energy*, **5**(12), pp. 959–972.

[13] Loh K, Tong S, Wu J. (2016). Graphene and grapheme-like molecules: prospects in solar cells. *J Am Chem Soc*, **138**(4), pp. 1095–1102.

[14] Balzani V, Credi A, Venturi M. (2008). Photochemical conversion of solar energy. *Chem Sus Chem*, **1**(1.2), pp. 26–58.

[15] Zhou P, Yu J, Jaroniec M. (2014). All-solid-state Z-scheme photocatalytic systems. *Adv Mater*, **26**(29), pp. 4920–4935.

[16] Wang W, Tadé MO, Shao Z. (2018). Nitrogen-doped simple and complex oxides for photocatalysis: a review. *Prog Mater Sci*, **92**(3), pp. 33–63.

[17] Kashyap V, Sakunkaewkasem S, Jafari P. (2019). Full spectrum solar thermal energy harvesting and storage by a molecular and phase-change hybrid material. *Joule*, **3**(12), pp. 3100–3111.

[18] Zhou L, Li X, Ni GW. (2019). The revival of thermal utilization from the sun: interfacial solar vapor generation. *Natl Sci Rev*, **6**(3), pp. 562–578.

[19] Wang T, Huang H, Li H. (2021). Carbon materials for solar-powered seawater desalination. *New Carbon Mater*, **36**(4), pp. 683–701.

[20] Xu K, Wang C, Li Z. (2021). Salt mitigation strategies of solar-driven interfacial desalination. *Adv Funct Mater*, **31**(8), pp. 200785.

[21] Li X, Xu W, Tang M. (2016). Graphene oxide-based efficient and scalable solar desalination under one sun with a confined 2D water path. *PNAS*, **113**(49), 13953–13958.

[22] Taylor RA, Phelan PE, Otanicar TP. (2011). Applicability of nanofluids in high flux solar collectors. *J Renew Sust Energy*, **3**(2), pp. 023104.

[23] Weigert F. (1909). Theorie photochemischer Prozesse. *Ber Dtsch Chem Ges*, **42**, pp. 850–862.

[24] Moth-Poulsen K, Ćoso D, Börjesson K. (2012). Molecular solar thermal (MOST) energy storage and release system. *Energ Environ Sci*, **5**(9), pp. 8534–8537.

[25] Kolpak A. M., Grossman J. C. (2011). Azobenzene-functionalized carbon nanotubes as high-energy density solar thermal fuels. *Nano. lett.*, **11**(8), pp. 3156–3162.

[26] Wang Z, Hölzel H, Moth-Poulsen K. (2022). Status and challenges for molecular solar thermal energy storage system based devices. *Chem Soc Rev*, **51**, pp. 7313–7326.

[27] Wang Z, Erhart P, Li T. (2021). Storing energy with molecular photoisomers. *Joule*, **5**(12), pp. 3116–3136.

[28] Lennartson A, Roffey A, Moth-Poulsen K. (2015). Designing photoswitches for molecular solar thermal energy storage. *Tetrahedron Lett*, **56**(12), pp. 1457–1465.

[29] Dong L, Feng Y, Wang L. (2018). Azobenzene-based solar thermal fuels: design, properties, and applications. *Chem Soc Rev*, **47**(19), pp. 7339–7368.

[30] Jones G, Reinhardt TE, Bergmark WR. (1978). Photon energy storage in organic materials-the case of linked anthracenes. *Sol Energy*, **20**(3), pp. 241–248.

[31] Jones G, Chiang SH, Xuan PT. (1979). Energy storage in organic photoisomers. *J Photochem*, **10**(1), pp. 1–18.

[32] Yoshida ZI. (1985). New molecular energy storage systems. *J Photochem*, **29**(1), pp. 27–40.

[33] Guarini G, Sarti-fantoni P. (1969). Calorimetric study on some anthracene photodimers. *Mol Cryst*, **6**(3–4), pp. 423–426.

[34] Donati D, Guarini G, Sarti-Fantoni P. (1972). Thermal behaviour and monomerization kinetics of 9-CN anthracene and 9-CN, 10-acetoxy anthracene dimers in the solid state. *Mol Cryst Liq Cryst*, **17**(3), pp. 187–195.

[35] Vadrucci R, Weder C, Simon YC. (2015). Organogels for low-power light upconversion. *Mater Horiz*, **2**(1), pp. 120–124.

[36] Kimizuka N, Yanai N, Morikawa MA. (2016). Photon upconversion and molecular solar energy storage by maximizing the potential of molecular self-assembly. *Langmuir*, **32**(47), pp. 12304–12322.

[37] Kim JH, Deng F, Castellano FN. (2014). Red-to-blue/cyan/green upconverting microcapsules for aqueous- and dry-phase color tuning and magnetic sorting. *ACS Photon*, **1**(4), pp. 382–388.

[38] Islangulov RR, Castellano FN. (2006). Photochemical upconversion: anthracene dimerization sensitized to visible light by a RuII chromophore. *Angew Chem*, **118**(36), pp. 6103–6105.

[39] Kozlov DV, Castellano FN. (2004). Anti-Stokes delayed fluorescence from metal-organic bichromophores. *Chem. Commun*, **24**, pp. 2860–2861.

[40] Islangulov RR, Kozlov DV, Castellano FN. (2005). Low power upconversion using MLCT sensitizers. *Chem Commun*, **30**, pp. 3776–3778.

[41] Philippopoulos C, Economou D, Economou C. (1983). Norbornadiene-quadricyclane system in the photochemical conversion and storage of solar energy. *Ind Eng Chem Res*, **22**(4), pp. 627–633.

[42] Bren VA, Dubonosov AD, Minkin VI. (1991). Norbornadiene-quadricyclane an effective molecular system for the storage of solar energy. *Russ Chem Rev*, **60**(5), pp. 451–469.

[43] Dubonosov A, Bren V, Chernoivanov V. (2002). Norbornadiene-quadricyclane as an abiotic system for the storage of solar energy. *Russ Chem Rev*, **71**(11), pp. 917–927.

[44] Gray V, Lennartson A, Ratanalert P. (2014). Diaryl-substituted norbornadienes with red-shifted absorption for molecular solar thermal energy storage. *Chem Commun*, **50**(40), pp. 5330–5332.

[45] Brummel O, Besold D, Döpper T. (2016). Energy storage in strained organic molecules: electrochemical characterization of norbornadiene and quadricyclane. *Chem Sus Chem*, **9**(12), pp. 1424–1432.

[46] Kuisma M, Lundin A, Moth-Poulsen K. (2016). Optimization of norbornadiene compounds for solar thermal storage by first-principles calculations. *Chem Sus Chem*, **9**(14), pp. 1786–1794.

[47] Kuisma MJ, Lundin AM, Moth-Poulsen K. (2016). Comparative ab-initio study of substituted norbornadiene-quadricyclane compounds for solar thermal storage. *J Phy Chem C*, **120**(7), pp. 3635–3645.

[48] Brummel O, Waidhas F, Bauer U. (2017). Photochemical energy storage and electrochemically triggered energy release in the norbornadiene-quadricyclane system: UV photochemistry and IR spectroelectrochemistry in a combined experiment. *J Phy Chem Lett*, **8**(13), pp. 2819–2825.

[49] Dreos A, Börjesson K, Wang Z. (2017). Exploring the potential of a hybrid device combining solar water heating and molecular solar thermal energy storage. *Energy Environ Sci*, **10**(3), pp. 728–734.

[50] Gurke J, Quick M, Ernsting NP. (2017). Acid-catalysed thermal cycloreversion of a diarylethene: a potential way for triggered release of stored light energy. *Chem Commun*, **53**(13), pp. 2150–2153.

[51] Jorner K, Dreos A, Emanuelsson R. (2017). Unraveling factors leading to efficient norbornadiene- quadricyclane molecular solar-thermal energy storage systems. *J Mater Chem A*, **5**(24), pp. 12369–12378.

[52] Görner H, Fischer C, Daub J. (1995). Photoreaction of dihydroazulenes into vinylheptafulvenes: photochromism of nitrophenyl-substituted derivatives. *J Photochem Photobio A: Chem*, **85**(3), pp. 217–224.

[53] Plaquet A, Champagne B, Castet F. (2009). Theoretical investigation of the dynamic first hyperpolarizability of DHA-VHF molecular switches. *New J Chem*, **33**(6), pp. 1349–1356.

[54] Lara-Avila S, Danilov AV, Kubatkin SE. (2011). Light-triggered conductance switching in single- molecule dihydroazulene/vinylheptafulvene junctions. *J Phys Chem C*, **115**(37), pp. 18372–18377.

[55] Broman SL, Lara-Avila S, Thisted CL. (2012). Dihydroazulene photoswitch operating in sequential tunneling regime: synthesis and single-molecule junction studies. *Adv Function Mater*, **22**(20), pp. 4249–4258.

[56] Vlasceanu A, Broman SL, Hansen AS. (2016). Solar thermal energy storage in a photochromic macrocycle. *Chem-A Europ J*, **22**(31), pp. 10796–10800.

[57] Skov AB, Broman SL, Gertsen AS. (2016). Aromaticity-controlled energy storage capacity of the dihydroazulene-vinylheptafulvene photochromic system. *Chem-A Europ J*, **22**(41), pp. 14567–14575.

[58] Cacciarini M, Vlasceanu A, Jevric M. (2017). An effective trigger for energy release of vinylheptafulvene- based solar heat batteries. *Chem Commun*, **53**(43), pp. 5874–5877.

[59] Skov A, Petersen J, Elm J. (2017). Towards storage of solar energy in photochromic molecules: benzannulation of the dihydroazulene/vinylheptafulvene couple. *Chem Photo Chem*, **1**(5), pp. 206–212.

[60] Wang Z, Udmark J, Börjesson K. (2017). Evaluating dihydroazulene/vinylheptafulvene photoswitches for solar energy storage applications. *Chem Sus Chem*, **10**(15), pp. 3049–3055.

[61] Vlasceanu A, Frandsen BN, Skov AB. (2017). Photoswitchable dihydroazulene macrocycles for solar energy storage: the effects of ring strain. *J Organ Chem*, **82**(19), pp. 10398–10407.

[62] Broman S, Jevric M, Bond AD. (2014). Syntheses of donor-acceptor-functionalized dihydroazulenes. *J Organ Chem*, **79**(1), pp. 41–64.

[63] Bandara HM, Burdette SC. (2012). Photoisomerization in different classes of azobenzene. *Chem Soc Rev*, **41**(5), pp. 1809–1825.

[64] Merino E. (2011). Synthesis of azobenzenes: the coloured pieces of molecular materials. *Chem Soc Rev*, **40**(7), pp. 3835–3853.

[65] Wu S, Li T, Zhang ZY. (2021). Photoswitchable phase change materials for unconventional thermal energy storage and upgrade. *Matter*, **4**(11), pp. 3385–3399.

[66] Qiu Q, Shi Y, Han GG. (2021). Solar energy conversion and storage by photoswitchable organic materials in solution, liquid, solid, and changing phases. *J Mater Chem C*, **9**(35), pp. 11444–11463.

[67] Li S, Wang H, Fang J. (2020). Photo-isomerization energy storage using azobenzene and nanoscale templates: a topical review. *J Therm Sci*, **29**(2), pp. 280–297.

[68] Sun C, Wang C, Boulatov R. (2019). Applications of photoswitches in the storage of solar energy. *Chem Photo Chem*, **3**(6), pp. 268–283.

[69] Xu WC, Sun S, Wu S. (2019). Photoinduced reversible solid-to-liquid transitions for photoswitchable materials. *Angew Chem Internation Ed*, **58**(29), pp. 9712–9740.

[70] Gerkman MA, Han GG. (2020). Toward controlled thermal energy storage and release in organic phase change materials. *Joule*, **4**(8), pp. 1621–1625.

[71] Wang Z, Hölzel H, Moth-Poulsen K. (2022). Status and challenges for molecular solar thermal energy storage system based devices. *Chem Soc Rev*, **51**, pp. 7313–7326.

[72] Quant M, Lennartson A, Dreos A. (2016). Low molecular weight norbornadiene derivatives for molecular solar-thermal energy storage. *Chem-A Europ J*, **22**(37), pp. 13265–13274.

[73] Wang H, Feng Y, Yu H. (2021). Utilisation of photo-thermal energy and bond enthalpy based on optically triggered formation and dissociation of coordination bonds. *Nano Energy*, **89**, pp. 106401.

[74] Liu Y, Grossman JC. (2014). Accelerating the design of solar thermal fuel materials through high throughput simulations. *Nano Lett*, **14**(12): 7046–7050.

[75] Wang H, Feng Y, Gao J. (2022). Metallic-ion controlled dynamic bonds to co-harvest isomerization energy and bond enthalpy for high-energy output of flexible self-heated textile. *Adv Sci*, **9**(20), pp. 2201657.

[76] Kunz A, Heindl AH, Dreos A. (2019). Intermolecular London dispersion interactions of azobenzene switches for tuning molecular solar thermal energy storage systems. *Chem Plus Chem*, **84**(8), pp. 1145–1148.

[77] Kucharski TJ, Ferralis N, Kolpak AM. (2014). Templated assembly of photoswitches significantly increases the energy-storage capacity of solar thermal fuels. *Nat Chem*, **6**(5), pp. 441–447.

[78] Börjesson K, Lennartson A, Moth-Poulsen K. (2013). Efficiency limit of molecular solar thermal energy collecting devices. *ACS Sust Chem Eng*, **1**(6), pp. 585–590.

[79] Jevric M, Petersen AU, Mansø M. (2018). Norbornadiene-based photoswitches with exceptional combination of solar spectrum match and long-term energy storage. *Chem-A Eur J*, **24**(49), pp. 12767–12772.

[80] Dabrowa K, Jurczak J. (2017). Tetra-(meta-butylcarbamoyl) azobenzene: a rationally designed photoswitch with binding affinity for oxoanions in a long-lived Z-state. *Organ Lett*, **19**(6), pp. 1378–1381.

[81] Schweighauser L, Strauss MA, Bellotto S. (2015). Attraction or repulsion? London dispersion forces control azobenzene switches. *Angew Chem Int Ed*, **54**(45), pp. 13436–13439.

[82] Crespi S, Simeth NA, König B. (2019). Heteroaryl azo dyes as molecular photoswitches. *Nat Rev Chem*, **3**(3), pp. 133–146.

[83] Zhang ZY, He Y, Wang Z. (2020). Photochemical phase transitions enable coharvesting of photon energy and ambient heat for energetic molecular solar thermal batteries that upgrade thermal energy. *J Am Chem Soc*, **142**(28), pp. 12256–12264.

[84] Nishimura N, Sueyoshi T, Yamanaka H. (1976). Thermal cis-to-trans isomerization of substituted azobenzenes II. Substituent and solvent effects. *Bull Chem Soc Jpn*. **49**(5), pp. 1381–1387.

[85] Quant M, Hamrin A, Lennartson A. (2019). Solvent effects on the absorption profile, kinetic stability, and photoisomerization process of the norbornadiene-quadricyclanes system. *J Phys Chem C*, **123**(12), pp. 7081–7087.

[86] Broman SL, Jevric M, Nielsen MB. (2013). Linear free-energy correlations for the Vinylheptafulvene ring closure: a probe for Hammett σ values. *Chem-A Eur J*, **19**(29), pp. 9542–9548.

[87] Kilde MD, Hansen, MH, Broman SL. (2017). Expanding the Hammett correlations for the Vinylheptafulvene ring-closure reaction. *Europ J Organ Chem*, (6), pp. 1052–1062.

[88] Cacciarini M, Skov AB, Jevric M. (2015). Towards solar energy storage in the photochromic dihydroazulene-vinylheptafulvene system. *Chem-A Eur J*, **21**(20), pp. 7454–7461.

[89] Yan Q, Zhang Y, Dang, Y. (2020). Solid-state high-power photo heat output of 4-((3, 5-dimethoxyaniline)-diazenyl)-2-imidazole/graphene film for thermally controllable dual data encoding/reading. *Energy Storage Mater*, **24**, pp. 662–669.

[90] Yang W, Feng Y, Si Q. (2019). Efficient cycling utilization of solar-thermal energy for thermochromic displays with controllable heat output. *J Mater Chem A*, **7**(1), pp. 97–106.

[91] Börjesson K, Lennartson A, Moth-Poulsen K. (2013). Efficiency limit of molecular solar thermal energy collecting devices. *ACS Sust Chem Eng*, **1**(6), pp. 585–590.

[92] Wang Z, Moïse H, Cacciarini M. (2021). Liquidbased multijunction molecular solar thermal energy collection device. *Adv Sci*, **8**(21), pp. 2103060.

[93] Strubbe DA, Grossman JC. (2018). Thermodynamic limits to energy conversion in solar thermal fuels. *J Phys: Condens Mat*, **31**(3), pp. 034002.

[94] Wang Z, Roffey A, Losantos R. (2019). Macroscopic heat release in a molecular solar thermal energy storage system. *Energy Environ Sci*, **12**(1), pp. 187–193.

[95] Wang Z, Losantos R, Sampedro D. (2019). Demonstration of an azobenzene derivative based solar thermal energy storage system. *J Mater Chem A*, **7**(25), pp. 15042–15047.

[96] Hammond GS, Wyatt P, DeBoer CD. (1964). Photosensitized isomerization involving saturated centers. *J Am Chem Soc*, **86**(12), pp. 2532–2533.

[97] Bauer U, Fromm L, Wei C. (2018). Controlled catalytic energy release of the norbornadiene/quadricyclane molecular solar thermal energy storage system on Ni (111). *J Phys Chem C*, **123**(13), pp. 7654–7664.

[98] Bauer U, Mohr S, Döpper T. (2017). Catalytically triggered energy release from strained organic molecules: the surface chemistry of quadricyclane and norbornadiene on Pt (111). *Chem-A Eur J*, **23**(7), pp. 1613–1622.

[99] Waidhas F, Jevric M, Fromm L. (2019). Electrochemically controlled energy storage in a norbornadiene- based solar fuel with 99% reversibility. *Nano Energy*, **63**, pp. 103872.

[100] Jutz C. (1964). Heptafulven-derivate. *Chem Ber*, **97**(7), pp. 2050–2065.

[101] Daub J, Knöchel T, Mannschreck A. (1984). Photosensitive dihydroazulenes with chromogenic properties. *Angew Chem Int Ed*, **23**(12), pp. 960–961.

[102] Saritas K, Grossman JC. (2017). Accurate isomerization enthalpy and investigation of the errors in density functional theory for dihydroazulene/vinylheptafulvene photochromism using diffusion monte carlo. *J Phys Chem C*, **121**(48), pp. 26677–26685.

[103] Nisa RU, Shahzad N, Ayub K. (2016). Density functional theory study of linear and non-linear optical properties of dihydroazulene-vinylheptafulvene photoswitches. *Comput Theor Chem*, **1095**, pp. 1.8.

[104] Robertus J, Reker SF, Pijper TC. (2012). Kinetic analysis of the thermal isomerisation pathways in an asymmetric double azobenzene switch. *Phys Chem Chem Phys*, **14**(13), pp. 4374–4382.

[105] Dokić J, Gothe M, Wirth J. (2009). Quantum chemical investigation of thermal cis-to-trans isomerization of azobenzene derivatives: substituent effects, solvent effects, and comparison to experimental data. *J Phys Chem A*, **113**(24), pp. 6763–6773.

[106] Olmsted J, Lawrence J, Yee GG. (1983). Photochemical storage potential of azobenzenes. *Sol Energy*, **30**(3), pp. 271–274.

[107] Song YM, Xie Y, Malyarchuk V. (2013). Digital cameras with designs inspired by the arthropod eye. *Nature*, **497**(7447), pp. 95–99.

[108] Zhitomirsky D, Cho E, Grossman JC. (2016). Solid-state solar thermal fuels for heat release applications. *Adv Energy Mater*, **6**(6), pp. 1502006.

[109] White TJ. (2018). Photomechanical effects in liquid crystalline polymer networks and elastomers. *J Poly Sci Part B: Poly Phys*, **56**(9), pp. 695–705.

[110] Liu Y, Wu W, Wei J. (2017). Visible light responsive liquid crystal polymers containing reactive moieties with good processability. *ACS Appl Mater Interfaces*, **9**(1), pp. 782–789.

[111] Jiang Z, Xu M, Li F. (2013). Red-light-controllable liquid-crystal soft actuators via low-power excited upconversion based on triplet-triplet annihilation. *J Am Chem Soc*, **135**(44), pp. 16446–16453.

[112] Stoychev G, Kirillova A, Ionov L. (2019). Light-responsive shape-changing polymers. *Adv Opt Mater*, **7**(16), pp. 1900067.

[113] Manouras T, Vamvakaki M. (2017). Field responsive materials: photo-, electro-, magnetic- and ultrasound- sensitive polymers. *Poly Chem*, **8**(1), pp. 74–96.

[114] Liu L, Onck PR. (2019). Light-driven topographical morphing of azobenzene-doped liquid crystal polymer films via tunable photo-polymerization induced diffusion. *J Mech Phys Sol*, **123**, pp. 247–266.

[115] Yu Y, Nakano M, Ikeda T. (2003). Directed bending of a polymer film by light. *Nature*, **425**(6954), pp. 145.

[116] Ishikawa T, Noro T, Shoda T. (2001). Theoretical study on the photoisomerization of azobenzene. *J Chem Phys*, **115**(16), pp. 7503–7512.

[117] Statman D, Janossy I. (2003). Study of photoisomerization of azo dyes in liquid crystals. *J Chem Phys*, **118**(7), pp. 3222–3232.

[118] Wu Y, Ikeda T, Zhang Q. (1999). Three-dimensional manipulation of an azo polymer liquid crystal with unpolarized light. *Adv Mater*, **11**(4), pp. 300–302.

[119] Lv J, Liu Y, Wei J. (2016). Photocontrol of fluid slugs in liquid crystal polymer microactuators. *Nature*, **537**(7619), pp. 179–184.

[120] Yun JH, Li C, Kim S. (2018). Comparing photoactuation of an azobenzene-doped nematic liquid crystal polymer through its activation mechanism: trans-cis-trans reorientation and photoisomerization. *J Phys Chem C*, **122**(11), pp. 6310–6317.

[121] Zhang P, Lan Z, Wei J. (2021). Photodeformable Azobenzene-containing polyimide with flexible linkers and molecular alignment. *ACS Macro Lett*, **10**(4), pp. 469–475.

[122] Iamsaard S, Aßhoff SJ, Matt B. (2014). Conversion of light into macroscopic helical motion. *Nat Chem*, **6**(3), pp. 229–235.

[123] Iamsaard S, Anger E, Aßhoff SJ. (2016). Fluorinated azobenzenes for shape-persistent liquid crystal polymer networks. *Angew Chem Int Ed*, **55**(34), pp. 9908–9912.

[124] Yamada M, Kondo M, Mamiya J. (2008). Photomobile polymer materials: towards light-driven plastic motors. *Angew Chem Int Ed*, **47**(27), pp. 4986–4988.

[125] Cheng F, Yin R, Zhang Y. (2010). Fully plastic microrobots which manipulate objects using only visible light. *Soft Mat*, **6**(15), pp. 3447–3449.

[126] Wani OM, Zeng H, Priimagi A. (2017). A light-driven artificial flytrap. *Nat Commun*, **8**(1), pp. 15546.

[127] Gelebart AH, Jan D, Varga M. (2017). Making waves in a photoactive polymer film. *Nature*, **546**(7660), pp. 632–636.

[128] Li S, Feng Y, Wang W. (2019). Polarization-induced alignment of azobenzene/fluorinated polyimide for three-dimensional shape-persistent and photo-responsive elastic helixes. *Compos Sci Technol*, **169**, pp. 158–66.

[129] Vollhardt C, Weidman TW. (1983). Synthesis, structure, and photochemistry of tetracarbonyl (fulvalene) diruthenium. Thermally reversible photoisomerization involving carbon-carbon bond activation at a dimetal center. *J Am Chem Soc*, **105**(6), pp. 1676–1677.

[130] Vollhardt C, Weidman TW. (1984). Efficient syntheses of new fulvalene-bridged carbonyl complexes of cobalt, ruthenium, chromium, molybdenum, and tungsten. *Organometallics*, **3**(1), pp. 82–86.

[131] Boese R, Cammack JK, Matzger AJ. (1997). Photochemistry of (fulvalene) tetracarbonyldiruthenium and its derivatives: efficient light energy storage devices. *J Am Chem Soc*, **119**(29), pp. 6757–6773.

[132] Cho J, Berbil-Bautista L, Pechenezhskiy IV. (2011). Single-molecule-resolved structural changes induced by temperature and light in surface-bound organometallic molecules designed for energy storage. *ACS Nano*, **5**(5), pp. 3701–3706.

[133] Börjesson K, Lennartson A, Moth-Poulsen K. (2014). Fluorinated fulvalene ruthenium compound for molecular solar thermal applications. *J Flu Chem*, **161**, pp. 24–28.

[134] Börjesson K. (2014). Exploring the potential of fulvalene dimetals as platforms for molecular solar thermal energy storage: computations, syntheses, structures, kinetics, and catalysis. *Chem-A Eur J*, **20**(47), pp. 15587–15604.

[135] Kanai Y, Srinivasan V, Meier SK. (2010). Mechanism of thermal reversal of the (fulvalene) tetracarbonyldiruthenium photoisomerization: toward molecular solar-thermal energy storage. *Angew Chem Int Ed*, **49**(47), pp. 8926–8929.

[136] Hou Z, Nguyen SC, Lomont JP. (2013). Switching from Ru to Fe: picosecond IR spectroscopic investigation of the potential of the (fulvalene) tetracarbonyldiiron frame for molecular solar-thermal storage. *Phys Chem Chem Phys*, **15**(20), pp. 7466–7469.

[137] Shen T, Wang XN, Lou HX. (2009). Natural stilbenes: an overview. *Nat Prod Rep*, **26**(7), pp. 916–935.

[138] Waldeck DH. (1991). Photoisomerization dynamics of stilbenes. *Chem Rev*, **91**(3), pp. 415–436.

[139] Caia V, Cum G, Gallo R. (1983). A high enthalpy value in thermal isomerization of photosynthesized cis- 9-styrylacridines. *Tetrahedron Lett*, **24**(36), pp. 3903–3904.

[140] Bastianelli C, Caia V, Cum G. (1991). Thermal isomerization of photochemically synthesized (Z)-9-styrylacridines. An unusually high enthalpy of *Z*→*E* conversion for stilbene-like compounds. *J Chem Soc*, (5), pp. 679—683.

[141] Sension RJ, Repinec ST, Szarka AZ. (1993). Femtosecond laser studies of the cis-stilbene photoisomerization reactions. *J Chem Phys*, **98**(8), pp. 6291–6315.

[142] Cao M, Cai Z, Chen X. (2017). Photo-switchable field-effect transistors based on two-dimensional stilbene oligomer crystals. *J Mater Chem C*, **5**(37), pp. 9597–9601.

[143] Wang C, Dong H, Hu W. (2012). Semiconducting π-conjugated systems in field-effect transistors: a material odyssey of organic electronics. *Chem Rev*, **112**(4), pp. 2208–2267.

[144] Chen H, Dong S, Bai M. (2015). Solution-processable, low-voltage, and high-performance monolayer field- effect transistors with aqueous stability and high sensitivity. *Adv Mater*, **27**(12), pp. 2113–2120.

[145] Mutai T, Satou H, Araki K. (2005). Reproducible on-off switching of solid-state luminescence by controlling molecular packing through heat-mode interconversion. *Nat Mater*, **4**(9), pp. 685–687.
[146] Sonoda Y, Goto M, Tsuzuki S. (2007). Fluorinated diphenylpolyenes: crystal structures and emission properties. *J Phys Chem A*, **111**(51), pp. 13441–13451.
[147] Hori A, Takatani S, Miyamoto TK. (2009). Luminescence from π–π stacked bipyridines through arene-perfluoroarene interactions. *Cryst Eng Comm*, **11**(4), pp. 567–569.
[148] Desiraju GR. (1995). Supramolekulare synthone Kristall-Engineeringeine neue organische synthese. *Angew. Chem*, **107**(21), pp. 2541–2558.
[149] Cincic D, Friscic T, Jones W. (2008). A stepwise mechanism for the mechanochemical synthesis of halogen-bonded cocrystal architectures. *J Am Chem Soc*, **130**(24), pp. 7524–7525.
[150] Friscic T, Jones W. (2009). Recent advances in understanding the mechanism of cocrystal formation via grinding. *Cryst Growth Des*, **9**(3), pp. 1621–1637.
[151] Yan D, Delori A, Lloyd GO. (2011). A cocrystal strategy to tune the luminescent properties of stilbene- type organic solid-state materials. *Angew Chem Int Ed*, **50**(52), pp. 12483–12486.
[152] Kislyak A, Kodura D, Frisch H. (2020). A holistic approach for anthracene photochemistry kinetics. *Chem. Eng J*, **402**, pp. 126259.
[153] Xu JF, Chen YZ, Wu LZ. (2013). Dynamic covalent bond based on reversible photo [4+4] cycloaddition of anthracene for construction of double-dynamic polymers. *Organic Lett*, **15**(24), pp. 6148–6151.
[154] Van J, van O, Brancart J. (2017). Anthracene-based thiolene networks with thermo-degradable and photo- reversible properties. *Macromolecules*, **50**(5), pp. 1930–1938.
[155] Hyder A, Srinivasan V. (1997). Photoresponsive functionalized vinyl cinnamate polymers: synthesis and characterization. *Poly Int*, **43**(4), pp. 310–316.
[156] Oya N, Sukarsaatmadja P, Ishida K. (2012). Photoinduced mendable network polymer from poly (butylene adipate) end-functionalized with cinnamoyl groups. *Poly J*, **44**(7), pp. 724–729.
[157] Nakayama Y, Matsuda T. (2005). Photocycloaddition-induced preparation of nanostructured, cyclic polymers using biscinnamated or biscoumarinated oligo (ethylene glycol)s. *J Poly Sci Part A: Poly Chem*, **43**(15), pp. 3324–3336.

[158] Van J, Vlaminck L, Van G. (2016). Synthesis and evaluation of 9-substituted anthracenes with potential in reversible polymer systems. *Tetrahedron*, **72**(29), pp. 4303–4311.

[159] Guimard NK, Oehlenschlaeger KK, Zhou J. (2012). Current trends in the field of self-healing materials. *Macromol. Chem Phys*, **213**(2), pp. 131–143.

[160] Bratschkov C. (2001). FTIR spectroscopy study of the UV irradiation induced changes in an anthracene containing copolymer. *Eur Poly J*, **37**(6), pp. 1145–1149.

[161] Amendola V, Meneghetti, M. (2012). Advances in self-healing optical materials. *J Mater Chem*, **22**(47), pp. 24501–24508.

[162] Fiore GL, Rowan SJ, Weder C. (2013). Optically healable polymers. *Chem Soc Rev*, **42**(17), pp. 7278–7288.

[163] Radl SV, Roth M, Gassner M. (2014). Photo-induced crosslinking and thermal de-crosslinking in polynorbornenes bearing pendant anthracene groups. *Eur Poly J*, **52**, pp. 98–104.

[164] Radl S, Kreimer M, Griesser T. (2015). New strategies towards reversible and mendable epoxy based materials employing [4πs + 4πs] photocycloaddition and thermal cycloreversion of pendant anthracene groups. *Polymer*, **80**, pp. 76–87.

[165] Manhart J, Ayalur-Karunakaran S, Radl S. (2016). Design and application of photo-reversible elastomer networks by using the [$4\pi s + 4\pi s$] cycloaddition reaction of pendant anthracene groups. *Polymer*, 102, pp. 10–20.

[166] Bawa P, Pillay V, Choonara YE. (2009). Stimuli-responsive polymers and their applications in drug delivery. *Bio Mater*, **4**(2), pp. 022001.

[167] Qiu Y, Park K. (2001). Environment-sensitive hydrogels for drug delivery. *Adv Drug Del Rev*, **53**(3), pp. 321–339.

[168] Alvarez-Lorenzo C, Bromberg L, Concheiro A. (2009). Light-sensitive intelligent drug delivery systems. *Photochem Photobio*, **85**(4), pp. 848–860.

[169] Gibson KF, Kernohant WG. (1993). Lasers in medicine — a review. *J Med Eng Technol*, **17**(2), pp. 51–57.

[170] Kloxin AM. (2009). Photodegradable hydrogels for dynamic tuning of physical and chemical properties. *Science*, **324**(5923), pp. 59–63.

[171] Kost J, Langer R. (2012). Responsive polymeric delivery systems. *Adv Drug Del Rev*, **64**, pp. 327–341.

[172] Andreopoulos FM, Beckman EJ, Russell AJ. (1998). Light-induced tailoring of PEG-hydrogel properties. *Biomaterials*, **19**(15), pp. 1343–1352.

[173] Zheng Y, Andreopoulos FM, Micic M. (2001). A novel photoscissile poly (ethylene glycol)-based Hydrogel. *Adv Funct Mater*, **11**(1), pp. 37–40.

[174] Trenor SR, Long TE, Love BJ. (2004). Photoreversible chain extension of poly (ethylene glycol). *Macromol Chem Phys*, **205**(6), pp. 715–723.

[175] Trenor SR, Shultz AR, Love BJ. (2004). Coumarins in polymers: from light harvesting to photo-cross- linkable tissue scaffolds. *Chem Rev*, **104**(6), pp. 3059–3078.

[176] Bouas-Laurent H, Castellan A, Desvergne JP. (2001). Photodimerization of anthracenes in fluid solutions. *Chem Soc Rev*, **30**(4), pp. 248–263.

[177] Coursan M, Desvergne JP, Deffieux A. (1996). Reversible photodimerisation of ω-anthrylpolystyrenes. *Macromol Chem Phys*, **197**(5), pp. 1599–1608.

[178] Goldbach JT, Russell TP, Penelle J. (2002). Synthesis and thin film characterization of poly (styrene-block-methyl methacrylate) containing an anthracene dimer photocleavable junction point. *Macromolecules*, **35**(11), pp. 4271–4276.

[179] Bratschkov C, Karpuzova P, Marpuz K. (2001). Synthesis and photochemical transformations of an anthracene containing methacrylic copolymer. *Poly Bullet*, **46**, pp. 345–349.

[180] Ide N, Tsujii Y, Fukuda T. (1996). Gelation processes of polymer solutions. 1. Photodimerization of free and polymer-bound anthryl groups. *Macromolecules*, **29**(11), pp. 3851–3856.

[181] Chujo Y, Sada K, Nomura R. (1993). Photogelation and redox properties of anthracene-disulfide-modified polyoxazolines. *Macromolecules*, **26**(21), pp. 5611–5614.

[182] Zheng Y, Micic M, Mello SV. (2002). PEG-based hydrogel synthesis via the photodimerization of anthracene groups. *Macromolecules*, **35**(13), pp. 5228–5234.

[183] Eiselt P, Lee KY, Mooney DJ. (1999). Rigidity of two-component hydrogels prepared from alginate and poly (ethylene glycol)-diamines. *Macromolecules*, **32**(17), pp. 5561–5566.

[184] Lee KY, Rowley JA, Eiselt P. (2000). Controlling mechanical and swelling properties of alginate hydrogels independently by cross-linker type and cross-linking density. *Macromolecules*, **33**(11), pp. 4291–4294.

[185] Mahou R, Wandrey C. (2010). Alginate-Poly (ethylene glycol) hybrid microspheres with adjustable physical properties. *Macromolecules*, **43**(3), pp. 1371–1378.

[186] Segura T, Anderson BC, Chung PH. (2005). Cross-linked hyaluronic acid hydrogels: a strategy to functionalize and pattern. *Biomaterials*, **26**(4), pp. 359–371.

[187] Kulkarni AR, Hukkeri VI, Sung HW. (2005). A novel method for the synthesis of the PEG-crosslinked chitosan with a pH-independent swelling behavior. *Macromol Biol*, **5**(10), pp. 925–928.

[188] Wells LA, Brook MA, Sheardown H. (2011). Generic, anthracene-based hydrogel crosslinkers for photo-controllable drug delivery. *Macromol Bio*, **11**(7), pp. 988–998.

[189] Berman B. (2008). Strategies to detect and reduce counterfeiting activity. *Bus Horiz*, **51**(3), pp. 191–199.

[190] Yoon B, Lee J, Park IS. (2013). Recent functional material based approaches to prevent and detect counterfeiting. *J Mater Chem C*, **1**(13), pp. 2388–2403.

[191] Andres J. (2014). A new anti-counterfeiting feature relying on invisible luminescent full color images printed with lanthanide-based inks. *Adv Funct Mater*, **24**(32), pp. 5029–5036.

[192] Jiang K, Zhang L, Lu J. (2016). Triple-mode emission of carbon dots: applications for advanced anti- counterfeiting. *Angew Chem Int Ed*, **55**(25), pp. 7231–7235.

[193] Li M, Yao W, Liu J. (2017). Facile synthesis and screen printing of dual-mode luminescent NaYF 4: Er, Yb (Tm)/carbon dots for anti-counterfeiting applications. *J Mater Chem C*, **5**(26), pp. 6512–6520.

[194] Zhu MQ, Zhu L, Han JJ. (2006). Spiropyran-based photochromic polymer nanoparticles with optically switchable luminescence. *J Am Chem Soc*, **128**(13), pp. 4303–4309.

[195] Häckel M, Kador L, Kropp D. (2007). Polymer blends with azobenzene-containing block copolymers as stable rewritable volume holographic media. *Adv Mater*, **19**(2), pp. 227–231.

[196] Klajn R (2009). Writing self-erasing images using metastable nanoparticle as st. *Angew Chem Int Ed*, **48**(38), pp. 7035–7039.

[197] Yoon B, Ham DY, Yarimaga O. (2011). Inkjet printing of conjugated polymer precursors on paper substrates for colorimetric sensing and flexible electrothermochromic display. *Adv Mater*, **23**(46), pp. 5492–5497.

[198] Frisch H. (2013). pH-Switchable ampholytic supramolecular copolymers. *Angew Chem Int Ed*, **52**(38), pp. 10097–10101.

[199] Chen LJ, Ren YY, Wu NW. (2015). Hierarchical self-assembly of discrete organoplatinum (II) metallacycles with polysaccharide via electrostatic interactions and their application for heparin detection. *J Am Chem Soc*, **137**(36), pp. 11725–11735.

[200] Ji X, Wang H, Li Y. (2016). Controlling amphiphilic copolymer self-assembly morphologies based on macrocycle/anion recognition and nucleotide-induced payload release. *Chem Sci*, **7**(9), pp. 6006–6014.

[201] Wong H. (2018). Formation of 1D infinite chains directed by metal-metal and/or π–π stacking interactions of water-soluble platinum (II) 2, 6-bis (benzimidazol-2′-yl) pyridine double complex salts. *J Am Chem Soc*, **140**(2), pp. 657–666.

[202] Cardolaccia T, Li Y, Schanze KS (2008). Phosphorescent platinum acetylide organogelators. *J Am Chem Soc*, **130**(8), pp. 2535–2545.

[203] Abbel R, Grenier C, Pouderoijen MJ. (2009). White-light emitting hydrogen-bonded supramolecular copolymers based on π-conjugated oligomers. *J Am Chem Soc*, **131**(2), pp. 833–843.

[204] Rao KV. (2013). Highly pure solid-state white-light emission from solution-processable soft-hybrids. *Adv Mater*, **25**(12), pp. 1713–1718.

[205] Liu Y, Jin J, Deng H. (2016). Protein-framed multi-porphyrin micelles for a hybrid natural-artificial light- harvesting nanosystem. *Angew Chem Int Ed*, **55**(28), pp. 7952–7957.

[206] Aubry JM, Pierlot C, Rigaudy J. (2003). Reversible binding of oxygen to aromatic compounds. *Acc Chem Res*, **36**(9), pp. 668–675.

[207] Gao Z, Han Y, Wang F. (2018). Cooperative supramolecular polymers with anthracene-endoperoxide photo-switching for fluorescent anti-counterfeiting. *Nat Commun*, **9**(1), pp. 3977.

[208] Acharyya A, Chatterjee D, Mondal A. (2012). Experimental study on the effect of magnetic field on current-voltage characteristics of n-channel enhancement-type MOSFET. *J Elect Dev*, **13**, pp. 945–948.

[209] Tsumura A, Koezuka H, Ando T. (1986). Macromolecular electronic device: field-effect transistor with a polythiophene thin film. *Appl. Phys Lett*, **49**(18), pp. 1210–1212.

[210] Pope M, Kallmann HP, Magnante P. (1963). Electroluminescence in organic crystals. *J Chem Phys*, **38**(8), pp. 2042–2043.

[211] Ho MH, Balaganesan B, Chen CH. (2012). Blue fluorescence and bipolar transport materials based on anthracene and their application in OLEDs. *Isr J Chem*, **52**(6), pp. 484–495.

[212] Korn AI, Arndt RA, Damask AC. (1969). Hall mobility of holes in anthracene. *Phy Rev*, **186**(3), pp. 938.

[213] Vincett PS, Barlow WA, Hann RA. (1982). Electrical conduction and low voltage blue electroluminescence in vacuum-deposited organic films. *Thin Solid films*, **94**(2), pp. 171–183.

[214] Caballero A, Tormos R, Espinosa A. (2004). Selective fluorescence sensing of Li+ in an aqueous environment by a ferrocene-anthracene-linked dyad. *Org Lett*, **6**(24), pp. 4599–4602.

[215] KlΔrner G, Davey MH, Chen WD. (1998). Colorfast blue-light-emitting random copolymers derived from di-n-hexylfluorene and anthracene. *Adv Mater*, **10**(13), pp. 993–997.

[216] Chochos CL, Govaris GK, Kakali F. (2005). Synthesis, optical and morphological characterization of soluble main chain 1, 3, 4-oxadiazole copolyarylethers-potential candidates for solar cells applications as electron acceptors. *Polymer*, **46**(13), pp. 4654–4663.

[217] Boyd TJ, Geerts Y, Lee JK. (1997). Electroluminescence from new polynorbornenes that contain blue- light-emitting and charge-transport side chains. *Macromolecules*, **30**(12), pp. 3553–3559.

[218] Chen CH, Wu FI, Shu CF. (2004). Spirobifluorene-based pyrazoloquinolines: efficient blue electroluminescent materials. *J Mater Chem*, **14**(10), pp. 1585–1589.

[219] Shih PI, Chuang CY, Chien CH. (2007). Highly efficient non-doped blue-light-emitting diodes based on an anthrancene derivative end-capped with tetraphenylethylene groups. *Adv Funct Mater*, **17**(16), pp. 3141–3146.

[220] Park JW, Kim YH, Jung SY. (2008). Efficient and stable blue organic light-emitting diode based on an anthracene derivative. *Thin Solid Films*, **516**(23), pp. 8381–8385.

[221] Mansour N, Hriz K, Jaballah N. (2016). Synthesis, characterization, optical and electrical properties of bis (phenylvinyl) anthracene-based polymers. *Opt Mater*, **58**, pp. 296–305.

[222] Vorona MY, Yutronkie NJ, Melville OA. (2019). Developing 9, 10-anthracene derivatives: optical, electrochemical, thermal, and electrical characterization. *Materials*, **12**(17), pp. 2726.

[223] Zhang X, Zhao G, Zhen Y. (2015). The position effect of an ethynyl spacer on the carrier mobility of anthracene derivatives. *J Mater Chem C*, **3**(21), pp. 5368–5371.

[224] Hodge P, Power GA, Rabjohns MA. (1997). Synthesis of poly (anthracene-2, 6-diyl) and a copolymer containing alternately anthracene-2, 6-diyl and p-phenylene units. *Chem Commun*, **1**, pp. 73–74.

[225] Garay RO, Naarmann H, Muellen K. (1994). Synthesis and characterization of poly (1, 4-anthrylenevinylene). *Macromolecules*, **27**(7), pp. 1922–1927.

[226] Cui W, Zhang X, Jiang X. (2006). Synthesis and characterization of soluble oligo (9, 10-bisalkynylanthrylene)s. *Org Lett*, **8**(4), pp. 785–788.

[227] Mahmoudi C, Chouk R, Baatout K. (2022). Synthesis, characterization and DFT study of new anthracene-based semiconducting polyethers for OLED application. *J Mol Struct*, **1251**, pp. 131993.

[228] Aleshin AN, Lee JY, Chu SW. (2004). Mobility studies of field-effect transistor structures basedon anthracene single crystals. *Appl Phys Lett*, **84**(26), pp. 5383–5385.

[229] Liu J, Zhang H, Dong H. (2015). High mobility emissive organic semiconductor. *Nat Commun*, **6**(1), pp. 10032.

[230] Dadvand A, Moiseev AG, Sawabe K. (2012). Maximizing field-effect mobility and solid-state luminescence in organic semiconductors. *Angew Chem Int Ed*, **51**(16), pp. 3837–3841.

[231] Li J, Zhou K, Liu J. (2017). Aromatic extension at 2, 6-positions of anthracene toward an elegant strategy for organic semiconductors with efficient charge transport and strong solid state emission. *J Am Chem Soc*, **139**(48), pp. 17261–17264.

[232] Chen M, Yan L, Zhao Y. (2018). Anthracene-based semiconductors for organic field-effect transistors. *J Mater Chem C*, **6**(28), pp. 7416–7444.

Chapter 2

Micromolecular Solar Thermal Fuels Based on Structure Conversion

Abstract

The increasing demand for energy with the rapid development of human beings as well as the depletion of traditional fossil fuels (coal, oil, gas, etc.) urgently necessitate the development of a new clean fuel. The efficient utilization and storage of solar power that is clean, pollution-free, and inexhaustible has become one of the most promising strategies to cope with the energy dilemma. However, the availability of solar energy is unstable and discontinuous over time and weather due to restrictions arising from the alternation of day and night, geography, seasonal change, and so on. Solar thermal fuels (STFs) for energy storage systems based on molecular optical switches could achieve the absorption, conversion, storage, and controlled release of sunlight in a single entity through the reversible transformation of chemical structure.

The core unit of construction for STFs was photothermal conversion materials; in general, the photothermal process can be achieved in two ways: (i) converting light energy into thermal power by means of an optical aggregation system and (2) converting light energy into chemical energy first and to convert the chemical energy into thermal power subsequently via changing of molecular structure. Compared with the former, molecular photovoltaic thermal storage materials which have distinctive advantages such as simple preparation technics, long heat storage cycles, excellent cycle stability, and supernal photothermal conversion efficiency are widely used as STFs. After absorbing at special wavelengths of sunlight for micromolecular STFs based on structure conversion, the basic state molecules transform into metastable molecules by changing their spatial configuration as well as converting the absorbed light power into chemical energy and storing it in chemical bonds. Metastable molecules return to the basic state in response to light, heat or other stimuli and release the storage chemical energy in the form of heat at the same time. At present, photothermal storage molecules based on structure conversion mainly include norborbadiene-quadricyclane (NBD-QC), fulvalene–bimetal complex, and azobenzene and its derivatives, etc. In particular, azobenzene derivatives have indeed garnered significant attention in recent years due to their remarkable photothermal storage capacity and photothermal conversion efficiency. These properties make them highly suitable for various applications, particularly in the field of photothermal therapy and energy conversion.

This chapter details the current research on STFs based on azobenzene derivatives, photo-switching phase-transition azobenzene derivatives, and norbornadiene/quadricyclane derivatives. A series of solutions are proposed to address the current challenges and provide directions for future applications of micromolecular STFs based on structural conversion.

2.1 Solar thermal fuels based on azobenzene derivatives

2.1.1 *Introduction of azobenzene molecular photo-switches*

Azobenzene was first discovered in the mid-19th century and primitively used as dyes in industry. It was not until 1937 that Hartley's team found photo *cis–trans* isomerization properties of azobenzene [1]. Since then, azobenzene and its derivatives have indeed been extensively studied and synthesized as molecular photo-switches for various applications in optical actuators [2,3], nonlinear optics [4], information storage [5,6], energy materials [7], biomedicine [8] and other fields.

2.1.1.1 *Molecular structure and spectral properties of azobenzene*

Azobenzene molecules were composed of two benzene rings connected by an Azo double bond (–N=N–). There were two isomers of azobenzene which include the planar *trans* isomer for stationary states and the spatially curved *cis* isomer for metastable states, and the *trans* isomers of azobenzene were rod-like molecules that exhibit liquid crystal behavior and strong non-covalent bond interactions. Azobenzene exhibits significant absorption across a broad range of wavelengths, spanning from ultraviolet (UV) light to visible red light. It demonstrates a pronounced π–π* transition in the ultraviolet region and a relatively weaker n–π* transition in the visible light range. The absorption of azobenzene would change with the type and position of benzene ring substituents. Rau [9] divided azobenzene into three categories, namely azobenzene, amino-azobenzene and pseudo-stilbene type according to the photo-chemical and absorption spectral characteristics of azobenzene, as shown in Figure 2.1. The π–π* and n–π* transitions of the UV–Vis absorption spectra of amino-azobenzene were close to each other due to the amino electron-donating effects and the n–π* transition was even covered by the π–π* transition. The electron distribution state of azobenzene were changed that Push–Pull electron pairs composed of the amino group of the electron-donating effects and the nitro group of the electron-withdrawing effects decrease the π–π* transition intensity of azobenzene absorption spectrum, even lower than the n–π* transition.

For condensed state structure, the absorption of azobenzene was also related to accumulation and arrangement of molecules. The azobenzene aggregates could be divided into H-type azobenzene aggregates (θ_c > 54.7°) and J-type azobenzene aggregates (θ_c < 54.7°) according to the critical value of θ_c = 54.7°, which is the angle between the direction of

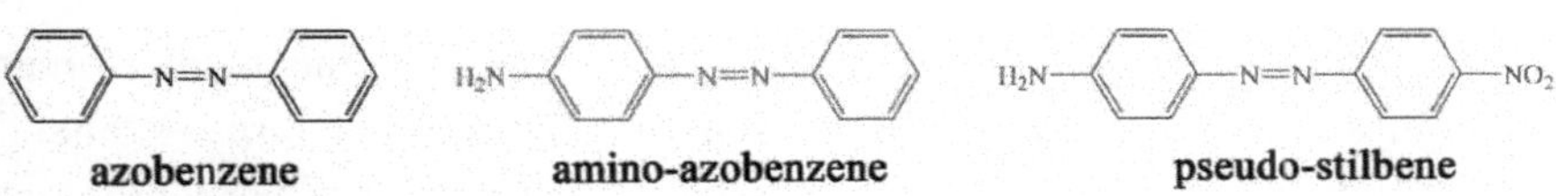

azobenzene **amino-azobenzene** **pseudo-stilbene**

Figure 2.1. Chemical structure of azobenzene, amino-azobenzene and pseudo-stilbene. Reprinted with permission from Ref. [9]. Copyright 2002, Elsevier Inc.

the arrangement and the direction of the long axis. Compared with non-accumulating azobenzene, the absorption spectra of H-type azobenzene and J-type azobenzene were bule shifted and redshifted, respectively [4].

2.1.1.2 *Photo-responsive behavior of azobenzene*

Azobenzene displays remarkable photo-responsive behavior, including reversible photo-induced *cis–trans* isomerization and photo-induced orientation, in which the *trans*-isomers of azobenzene could be efficiently converted to *cis*-isomers under ultraviolet irradiation and the *cis*-isomers could be completely reversibly converted to *trans*-isomers under visible light irradiation and heat. In this process, the time required for visible light isomerization was usually orders of magnitude of a picosecond, the thermal relaxation isomerization of azobenzene occurs over a wide time range, ranging from milliseconds to hours, with variations of several orders of magnitude. The geometric size of azobenzene molecule was transformed from *trans*-isomer (9.0–5.5 Å) and the dipole moment correspondingly increased from 0D to 3.0D during photo-induced *cis–trans* isomerization of azobenzene, as shown in Figure 2.2(a).

Azobenzene arrayed in an orientation perpendicular to the polarization direction under linearly polarized light and the arrayed orientation under circularly polarized or linear polarized light could be erased, as shown in Figure 2.2(b). The absorption coefficient of azobenzene in polarized light was proportional to $\cos^2\alpha$ (the angle α represents the deviation between the polarization direction and the molecular long axis.). When the direction of the azobenzene molecular long axis coincided with the direction of the polarized light, the absorption capacity is maximum. Reversely, it has no absorption capacity when the direction of azobenzene molecular long axis was perpendicular to the direction of the polarized light. When exposed to polarized light, azobenzene molecules that are not perpendicular to the polarization direction absorbed light from the *trans*-isomers to the *cis*-isomer, which can be returned to the *trans*-isomer through a process of thermal relaxation, and the returned *trans*-isomers' orientation was random. The *trans*-isomers that were not perpendicular to the polarization continued to repeat the above-mentioned process until completely transformed into azobenzene arrays perpendicular to

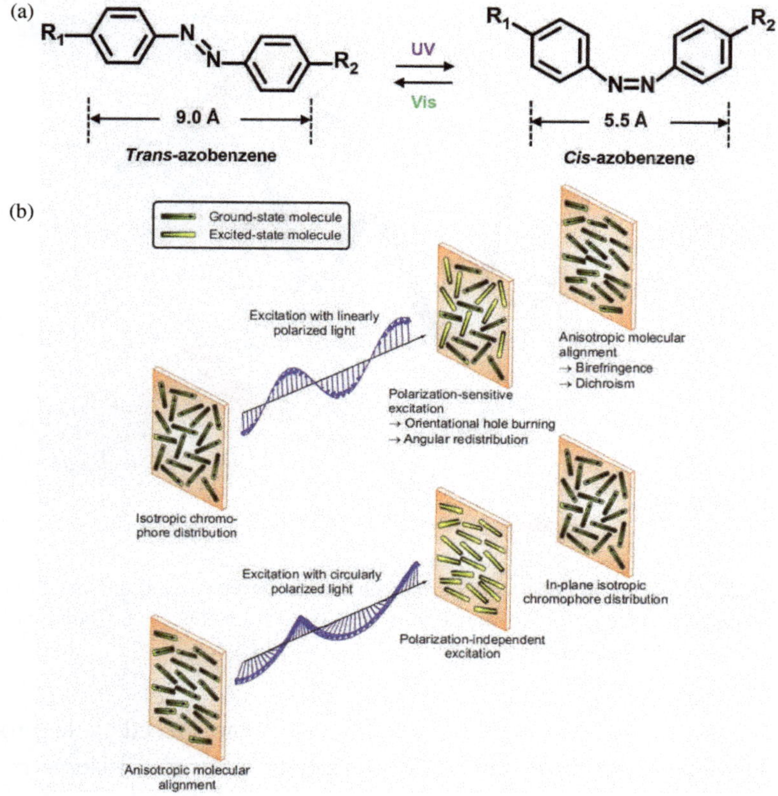

Figure 2.2. (a) Azobenzene undergoes reversible *trans–cis* photoisomerization. (b) Photo-induced orientation of azobenzene. Reprinted with permission from Ref. [10]. Copyright 2019, WILEY-VCH. Reprinted with permission from Ref. [11]. Copyright 2002, American Chemical Society.

polarization direction of light to complete the photo-induced orientation process [10–12].

2.1.2 *Photo-thermal conversion and storage of solar thermal fuels based on azobenzene*

The photothermal conversion and storage process of STFs based on azobenzene (Azo-STFs) are essentially realized by the photo-induced *trans–cis*

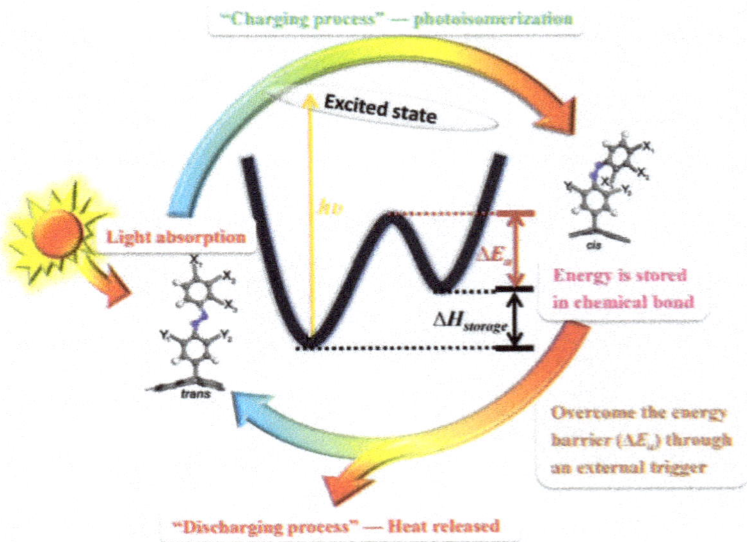

Figure 2.3. Schematic illustration of photochemical conversion and thermal storage of STFs based on azobenzene. Reprinted with permission from Ref. [14]. Copyright 2018, Royal Society of Chemistry.

isomerization, as shown in Figure 2.3 [13]. The closed-cycle process includes four parts of the absorption of sunlight, *trans*-to-*cis* isomerization transition, *cis*-to-*trans* isomerization transition, and the release of energy. First of all, azobenzene usually exists as a stable *trans*-isomer at room temperature. The stable *trans*-isomer absorbs solar energy and transforms into an excitation-state *trans*-isomer, which is unstable and attempts to change its conformation to dissipate the excess energy and usually transforms into a metastable *cis*-isomer, which also converts the absorbed solar energy into chemical energy and stores it in chemical bands on the picosecond scale during this process. Another portion is quenched back to the steady-state *trans*-isomer by non-radiative/radiative transitions, vibration relaxation, and energy transfer. So far, to explain the photo-induced *trans–cis* isomerization of azobenzene, four mechanisms have been proposed: rotation, inversion, concerted inversion, and inversion-assisted rotation. In order to obtain the energy stored by metastable *cis*-isomers, to facilitate the photothermal conversion process, and release the stored energy as heat in STF, it is typically required to overcome the energy barrier (ΔE_a) through heating or exposure to visible light. This allows the metastable *cis*-isomers to revert back to stable *trans*-isomers, completing the cycle of photothermal conversion and energy storage. According to research conducted by Boulatov's group [14], the maximum energy (ΔG^0_{iso}) that can be stored per mole of azobenzene molecules in a closed-cycle satisfies Formula 2.1.

$$\Delta G^0_{iso} = \Delta G_{iso} + RT \ln \frac{h \ln 2}{\tau_{1/2} kT} = \Delta G_{iso} - \Delta G_{rev} \tag{2.1}$$

where ΔG_{iso} is Gibbs free energy for the stable to metastable isomer conversion, ΔG_{rev} is the activation free energy of thermal relaxation reversion of the metastable *cis*-isomer return to the stable *trans*-isomer, $\tau_{1/2}$ is the energy storage half-life, R and T are the ideal gas constant and thermodynamic temperature, respectively. k and h are Boltzmann constant and Planck constant, respectively.

2.1.3 *Fundamental performance parameters of solar thermal fuels based on azobenzene*

The superior performance of azobenzene STFs is determined by several parameters, including the degree of photo-induced *trans–cis* isomerization, energy storage density, energy storage half-life (energy storage life), photothermal conversion efficiency, solar spectrum matching ability, and cycle stability (cycle life).

2.1.3.1 *The degree of photo-induced trans–cis isomerization*

The heat storage capacity of Azo-STFs (azobenzene-based solar thermal fuels) depends on the extent of photo-induced *trans–cis* isomerization at the molecular level. The fraction of *trans* and *cis* isomers can be determined by analyzing the UV–Vis absorption spectrum.

2.1.3.2 *Energy density*

The energy storage density ($\Delta H_{storage}$) of Azo-STFs is dependent on the energy level difference between metastable *cis*-isomer and steady-state *trans*-isomer. As one of the most important performance parameters for designing outstanding performance Azo-STFs, $\Delta H_{\text{storage}}$ determines the heat storage capacity. $\Delta H_{\text{storage}}$ can be expressed in several ways, including the molar-energy density (kJ/mol), the mass-energy density (kJ/kg or Wh/kg) and the volume-energy density (kJ/l or Wh/l), which is generally characterized by differential scanning calorimetry (DSC) and the calculation formula is shown as Formula 2.2.

$$\Delta H_{\text{storage}} = \frac{qM\chi_{\text{cis}}}{m} \tag{2.2}$$

where q is the heat released, which is determined by the DSC signal peak integral, M is the effective mass of the sample (molar mass, volume mass, etc.), m is the actual mass of the sample. χ_{cis} is the percentage of *trans*-isomers of azobenzene.

2.1.3.3 *Energy storage half-life*

The metastable *cis*-isomer of azobenzene converts the absorbed solar energy into chemical energy and stores it in the chemical bond. Since the *cis*-isomer will spontaneously and slowly transform into the steady-state *trans*-isomer even under dark conditions, the storage life is a performance parameter to evaluate the storage stability and storage period of STFs, and the storage half-life ($\tau_{1/2}$) is generally used to characterize it. The storage half-life is defined as the time required for the spontaneous conversion of *cis*-isomers to half the number of *trans*-isomers at room temperature under dark conditions and is related to the energy barrier (ΔE_a). The larger ΔE_a is, the longer the half-life of Azo-STFs is. The energy storage half-life is usually measured by UV–Vis absorption spectrum, and the storage-life is determined by the absorbance of the *cis–trans* isomerization as a function of time. The calculation formula is shown in Formula 2.3.

$$\ln \frac{A_{\infty} - A_t}{A_{\infty} - A_t} = -Kt \tag{2.3}$$

where A_{∞} is the specific wavelength absorbance of the *trans*-isomer, A_0 is the specific wavelength absorbance of the *cis*-isomer, A_t is the specific wavelength absorbance of the *cis*-isomer when t is placed at room temperature and darkness. K is the kinetic constant of *cis–trans* isomerization.

The shorter half-life of azobenzene is not conducive to long-term solar energy storage but is suitable for rapid heat release applications. If the half-life is too long, strong external stimulation is needed to accelerate the release of stored solar energy. Therefore, it is very important to design Azo-STFs with appropriate storage-life.

2.1.3.4 *Photothermal conversion efficiency*

The photothermal conversion efficiency (η) of Azo-STFs is another important indicator to evaluate the heat storage capacity. The specific calculation formula is shown in Formula 2.4:

$$\eta = \frac{\Delta H}{E_{hv}} \cdot \frac{\int_{\lambda_1}^{\lambda_2} I(\lambda)\eta_{\text{abs}}(\lambda)\phi(\lambda)d\lambda}{\int_{\lambda_1}^{\lambda_3} I(\lambda)d\lambda} \tag{2.4}$$

where ΔH is the energy level difference between the *cis*-isomer and *trans*-isomer of azobenzene. E_{hv} is the energy that absorbs sunlight; $I(\lambda)$ is the distribution of solar irradiation power divided by wavelength, $\eta_{\text{abs}}(\lambda)$ is the wavelength-dependent absorption efficiency of the *trans*-isomer. $\Phi(\lambda)$ is the quantum yield of *trans–cis* isomerization dependent on wavelength. The limit λ_1 is the lower wavelength of the absorption spectrum of the *trans*-isomer in UV–Vis or the lower wavelength of the solar irradiation spectrum. The limits λ_2 and λ_3 are the upper wavelength of UV–Vis absorption spectrum and solar irradiation spectrum, respectively. It should be noted that Formula 2.4 is evolved from the calculation of photochemical reaction, which was firstly determined by Bren *et al.* [15].

2.1.3.5 *Matching ability of solar spectrum*

The wavelength range of the solar spectrum is between 300 nm and 2500 nm, in which ultraviolet light accounts for 5%, visible light 43%, and near-infrared light 52%, as shown in Figure 2.4 [16]. Since most azobenzene and its derivatives can only store energy under ultraviolet irradiation, the utilization efficiency of sunlight is relatively low. If the visible light and near-infrared light in the solar spectrum could be utilized, the energy efficiency of Azo-STFs will be greatly improved, which has significance

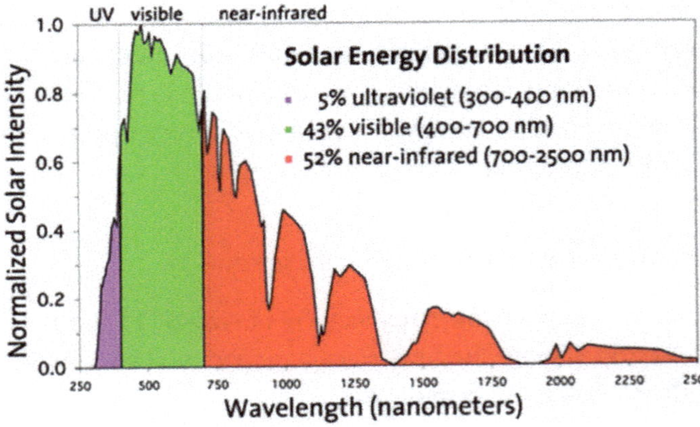

Figure 2.4. The solar energy spectrum. Reprinted with permission from Ref. [16]. Copyright 2009, Cool Roof Rating Council. All Rights Reserved.

for practical applications. When the absorption spectrum of azobenzene is widened from UV–Vis, which is causes the *cis–trans* isomerization, resulting in the decrease of energy storage density and energy storage life. Therefore, developing and designing azobenzene solar photothermal fuels to operate across the full solar spectrum is challenging.

2.1.4 *Research advances of azobenzene derivatives in solar thermal fuels*

Adamson *et al.* [17] first determined the storage enthalpy ($\Delta H_{storage}$ = 49 kJ/mol, compared to 269 J/g or 75 Wh/kg) as well as the half-life of *cis*-isomer as 4.2 days of azobenzene in 1978. In 1983, Olmsted *et al.* [18] studied the photothermal storage capacity of azobenzene for solar energy and found that the rapid thermal recovery rate of the *cis*-isomer of azobenzene and its limited dissolution ability in polar solvents hindered its development in the field of photothermal storage. In 1987, Taoda *et al.* [7] proposed that azobenzene solutions were stored at low temperature and sheltered from sunlight in order to realize long-term storage and recycling of the storage solar energy by *cis*-isomer. With the view of the above-mentioned problems, a large number of researchers have synthesized a series of azobenzene derivatives through the design and optimization of the molecular structure of azobenzene, which has led to vivid prospects in the field of solar energy storage and conversion. Some representative achievements are discussed as follows.

In the process of photo-induced *trans–cis* isomerization of azobenzene molecules in the solution, the dilution of the solvent inevitably leads to the reduction of the volume energy density of azobenzene. Therefore, for the purpose of improving the photothermal storage capacity of azobenzene, Kimizuka *et al.* [19] introduced a branched 2 ethylhexyl group at the 4-position of azobenzene using an ether linkage. This modification allowed the azobenzene derivative to remain in a fluid state at room temperature, functioning as a solvent-free STFs, and the molecular structure of azobenzene derivative as shown in Figure 2.5(a). The presence of azobenzene derivatives in the liquid phase at room temperature was induced by the racemic branched alkyl chains which greatly weaken the molecular interaction. The small molecular weight of 2 ethylhexyl group provided high volume and mass energy density while achieving entirely condensed-phase photoisomerization, and the prepared azobenzene derivative changed from orange *trans*-isomer to dark red *cis*-isomer under ultraviolet irradiation (Figure 2.5(b)). The basic consistency of UV-Vis is absorption spectra of the azobenzene derivatives in the form of condensed-phase and dilute methanol solution further demonstrated the weak intermolecular interaction, and showed the π–π* absorption band at 344 nm and n–π* absorption band at 443 nm, respectively (Figure 2.5(c)). The azobenzene derivative showed excellent energy storage stability through repetitive tests of the photo-induced *trans–cis* isomerization process with UV light (356 nm) and visible light (480 nm) as shown in Figure 2.5(d). In the DSC curves, an obvious exothermic peak can be observed at about 120°C. According to the linear relationship between the heat released by the different *trans*-isomers (68%, 36%, and 15%) during thermal-induced *cis–trans* isomerization, the enthalpy of energy storage ($\Delta H_{storage}$) for the 100% *cis*-isomer could be estimated to be 52 kJ/mol or 47 Wh/kg by extrapolation of the plot (Figure 2.5(e)). This study showed that the liquid azobenzene derivatives could be used as a new and potential family of molecular STFs. However, it still faces the problems of low energy storage, short half-life, and difficulty in obtaining highly purity *cis*-isomers.

As azobenzene solution and liquid azobenzene derivatives were limited in a large number of practical applications as STFs, the development of solid-state azobenzene derivatives and its integration into existing

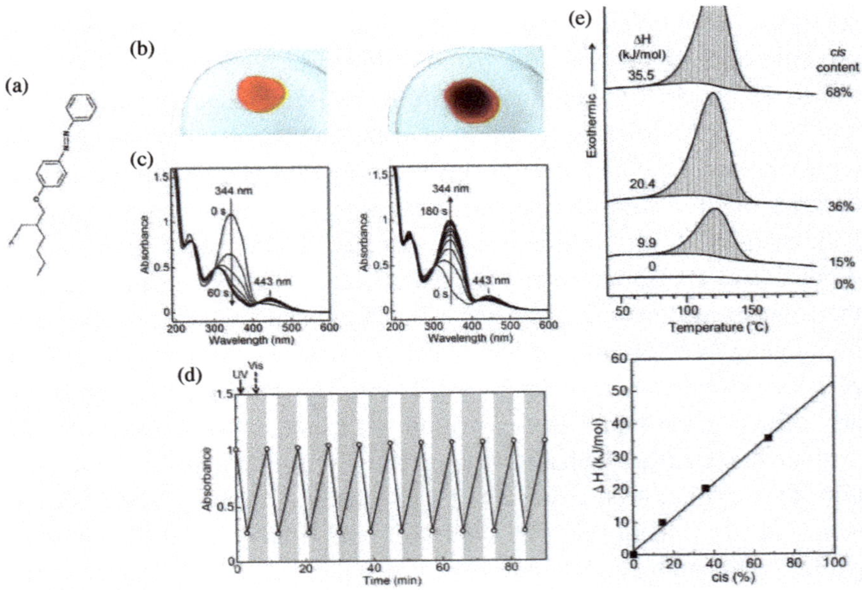

Figure 2.5. (a) Molecular structure of the azobenzene derivatives. (b) Pictures of the azobenzene derivatives in *tans* and *cis* states. (c) UV–Vis absorption spectra of liquid azobenzene derivatives under certain conditions of UV (365 nm) and visible-light (480 nm) irradiation. (d) Changes in the absorption intensity of liquid azobenzene derivatives at 344 nm in the course of repetitive UV and Vis irradiation. Reprinted with permission from Ref. [19]. Copyright 2014, The Royal Society of Chemistry.

solid-state devices such as deicing coating, solar blankets, and other consumer-oriented heating equipment, showed an intriguing application prospect. On the other hand, the application of single molecular thin films employed in solid-state STFs was obstructive due to heterogenous morphology, strong crystallization capacity and melting at low temperature. In this regard, Grossman *et al.* [20] designed and synthesized a solid-state photo-thermal storage materials based on azobenzene polymer which displayed were controllable thickness, uniform morphology, and scalable preparation. The polymer with azobenzene as a side chain was prepared by a simple two-step chemical synthesis method and the synthesis route is shown in Figure 2.6(a). Meanwhile, density functional theory (DFT) was used to calculate the energy density of the designed azobenzene derivative monomer as 68 Wh/kg, which only decreased slightly compared with

(a)

(b)

(c)

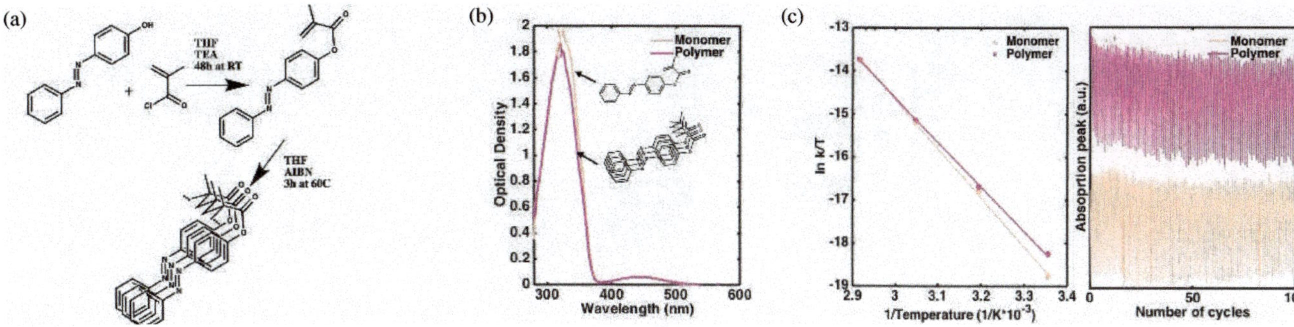

Figure 2.6. (a) Chemical synthesis scheme of the azobenzene polymer. (b) UV–Vis absorption spectra of the polymer and monomer exhibiting two prominent peaks. (c) Eyring-Polanyi plots were employed to extract the reverse thermal barrier energy related to the discharge process, while cycling plots were used to showcase the long-term cyclability of the two species. (d) Schematic of the spin-coating process employing the polymer solution in toluene. (e) The optical microscope images of the polymer and monomer spin-coated films, respectively. (f) The technologic process of the crosslinking azobenzene polymer used UV-curable. (g) Schematic of the crosslinking azobenzene polymer used the poly (ethylene glycol) diacrylate. (h) UV–Vis absorption spectra of the un-crosslinking and crosslinking polymer (i) macroscopic STFs heat release. Reprinted with permission from Ref. [20]. Copyright 2015, Wiley-VCH.

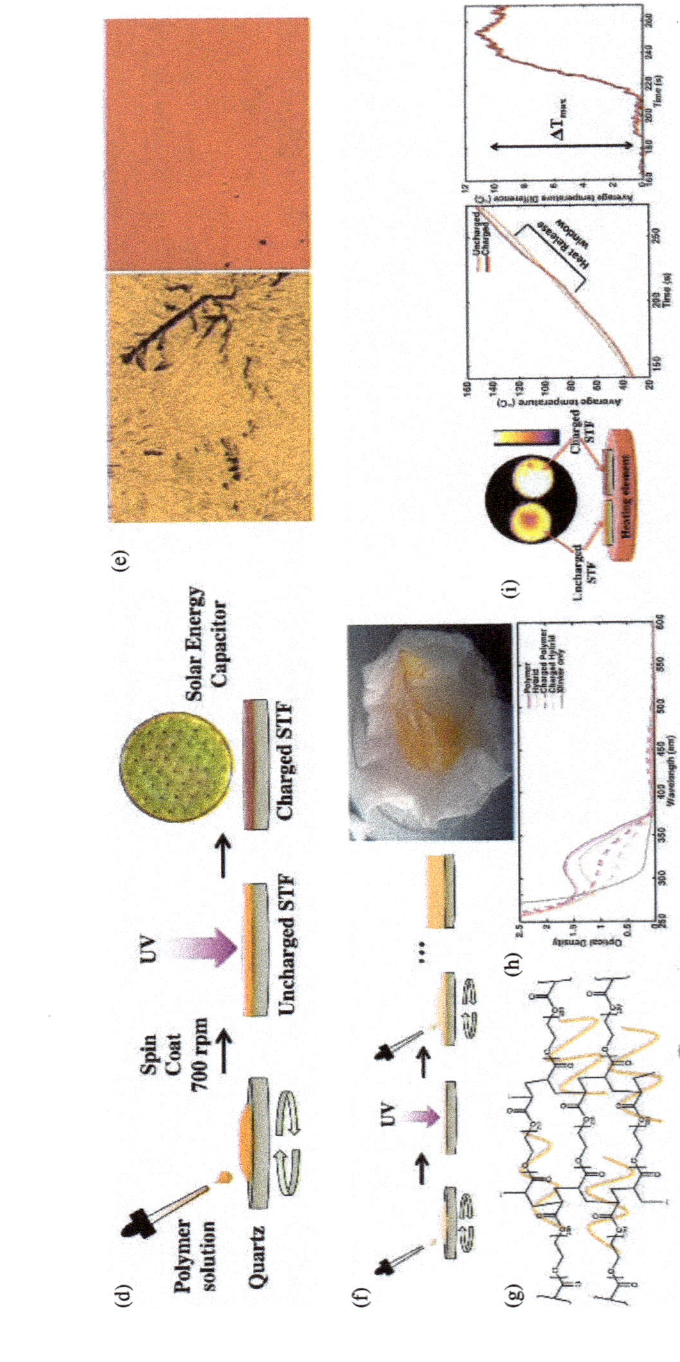

Figure 2.6. *(Continued)*

azobenzene (76 Wh/kg). In the pursuit of effective materials design, it is crucial to harness the exceptional properties of specific small molecular STFs and transform them into solid-state materials, such as polymers. The UV–Vis absorption spectra of azobenzene derivatives monomer and polymer were basically consistent (Figure 2.6(b)), and the activation energy of *cis–tans* isomerization denoted as E_a (95 kJ/mol for monomer and 86 kJ/mol for polymer) was calculated by Eyring Polanyi relationship, which showed a tiny difference. The monomer and polymer which have long half-lives and extraordinary cycle stability (>100 cycles) as shown in Figure 2.6(c) were prepared into solid film by spin-coating technology, as shown in Figure 2.6(d). Optical microscope images of the monomer and polymer showed that the solid-state polymer film, which has a smooth surface without defects (Figure 2.6(e)) was suitable for STFs. In addition, poly(ethylene glycol) diacrylate was used as UV-curable crosslinking agent to dissolve the polymer and prepare solid-state film from micrometer and millimeter scales. And the technological process and the polymer crosslinking network were shown in Figures 2.6(f) and 2.6(g), respectively. UV–Vis absorption spectra of un-crosslinking and crosslinking structures (Figure 2.6(h)) indicated that the energy storage of the solid-state STFs was not affected because of the introduction of crosslinked states into polymer. Finally, when azobenzene polymer was employed at macroscale heat-release for realistic solid-state applications, it was found that the temperature rise caused by heat-release that the charged film after thermal-induced was around 10°C higher than that of the uncharged film (Figure 2.6(i)). This research has paved the way for an incredibly enticing path in the realms of both fundamental light–matter interaction science and innovative renewable energy storage and heat release.

To address the issue of low energy density in azobenzene, researchers have proposed templating methods for STF molecules, which have been extensively studied for their potential application in solid-state STF systems, including nanocarbon templates, polymer templates and so on [20–24]. However, the templating of STF molecules onto polymers has certain limitations in terms of chargeability in the solid state, resulting in a significant decrease in energy storage capabilities compared to it's in the solution. Grossman *et al.* prepared efficient solid-state STFs whose morphology could be optimized concomitant with energy density, cyclability, and thermal stability through the molecular design of azobenzene, and three azobenzene derivatives (compounds 1, 2, and 3) were synthesized by connecting the bulky aromatic group (phenyl, biphenyl and *tert*-butyl phenyl groups) to azobenzene molecules, respectively. The synthesis route is shown in Figure 2.7(a). Because of the bulky volume of phenyl on the azobenzene molecules, which caused a break in the planarity of structure, preventing the molecules from crystallizing and the steric hindrance between crystal molecules preventing photo-induced *cis–trans* isomerization. Therefore, compared with the pure azobenzene films, compound 1 presented smooth and uniform morphology, which is suitable for the application of solid-state STFs, and the optical microscope and SEM image are shown in Figure 2.7(b). The DSC curve showed that the heat released for compound 1 by thermal-induced *cis–trans* isomerization, compound 1 exhibits a higher energy storage enthalpy ($\Delta H_{storage}$) of 77 kJ/mol, which is significantly greater than that of azobenzene ($\Delta H_{storage}$ = 49 kJ/mol). It also represents a remarkable 30% increase compared to the azobenzene polymer. Compounds 2 and 3 further increased the non-planarity of azobenzene molecules, in particular, the addition of tert-butyl phenyl group to compound 3 weakened the π–π stacking of azobenzene derivatives and further reduced the intermolecular interaction. Thus, compounds 2 and 3 exhibited better film-forming performance and faster photothermal conversion rate as shown in Figures 2.7(c) and 2.7(d) compared with compound 1. At the same time, compounds 2 and 3 had a higher degree of photo-induced *cis–trans* isomerization (the content of *cis*-isomers in the photostable state were 0.6, 0.74, and 0.8, respectively), because the bulky substituent structure provided extra free volume for the *cis–trans* isomerization and facilitated faster photoisomerization of the molecules from the *trans* to *cis* configuration. The DSC curve in Figure 2.7(e) showed that energy storage density of three azobenzene derivatives (compounds 1, 2, and 3 were 77 kJ/mol or 134 J/g, 88 kJ/mol or 100 J/g and 87 kJ/mol or 108 J/g, respectively). This investigation showed that the energy density of azobenzene small molecules and the thermal stability and charging rate of azobenzene solid-state films could be improved by molecular engineering strategies. As the demand for solid-state integration of small molecules in functional devices continues to rise, molecular films offer a promising avenue due to their versatility, easy processing, and potential for cost reduction.

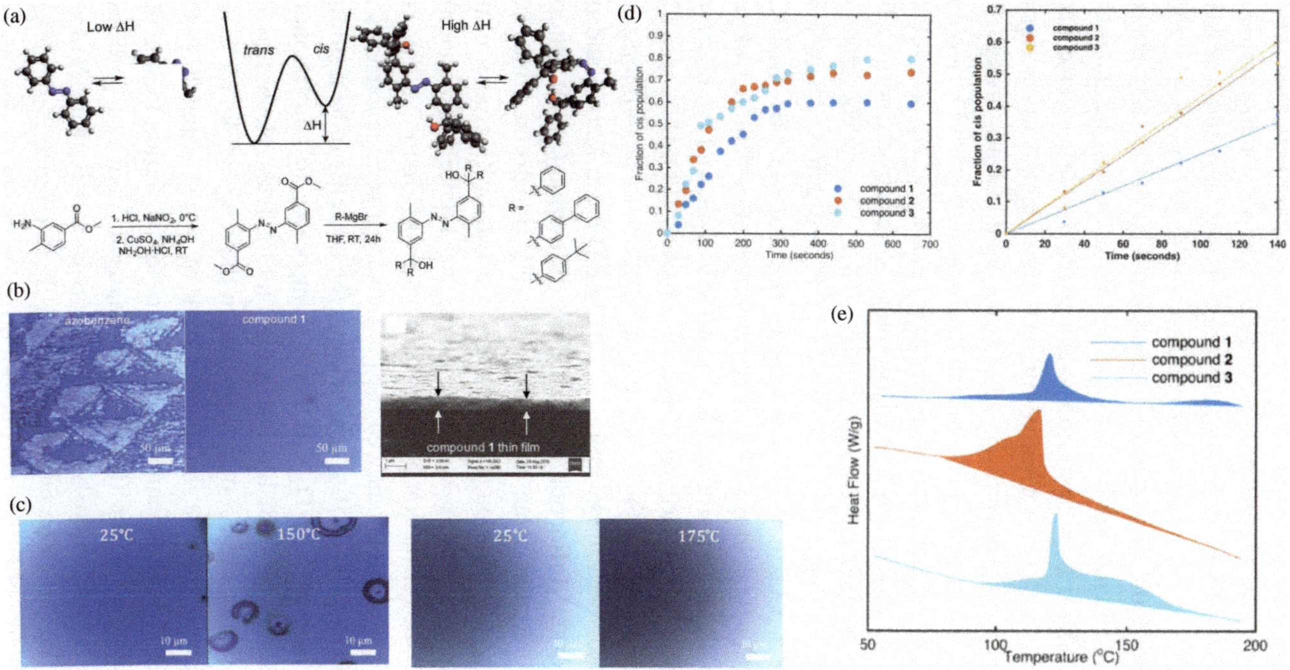

Figure 2.7. (a) The synthesized rote of the azobenzene derivatives. (b) The optical microscope and SEM images of the compound 1 solid-state film. (c) The optical microscope images of the compounds 2 and 3. (d) Chargeability and fraction of *cis*-isomer linear fit of compounds 1, 2, and 3. (e) DSC traces for compounds 1, 2, and 3 from top to bottom. Reprinted with permission from Ref. [21]. Copyright 2011, American Chemical Society. Reprinted with permission from Ref. [22]. Copyright 2013, American Institute of Physics. Reprinted with permission from Ref. [23]. Copyright 2014, Macmillan Publishers Limited.

Room-temperature ionic liquids (ILs) generally refer to the salts composing entirely of ionic states that appeared liquid at room-temperature or nearly room-temperature, which has attracted extensive attention due to its unique physical and chemical properties. Kimizuka's [25] research group takes advantage of the fact that ionic crystals (ICs) of the azobenzene derivatives converted into *cis*-isomer azobenzene derivatives IL under UV irradiation, and that *cis*-isomer azobenzene derivatives IL reverse back into ICs of the azobenzene derivatives under visible light. And a series of cationic azobenzene derivatives (*n*, *m*)-X were synthesized through the design of molecular structure of azobenzene, where X was chloride ion, bromide ion and bis(trifluoromethyl sulfonyl)amide (Tf_2N), as well as flexible and bulky oligo (ethylene oxide) units were introduced into the ammonium salts to reduce the melting point of ICs. The molecular structure of azobenzene derivatives was shown in Figure 2.8(a). There into, (1,2)-Cl, (1,4)-Cl, and (6,4)-Tf_2N belong to ILs and (4,6)-Br, (6,4)-Cl, (6,4)-Br, and (8,2)-Br belong to ICs at room-temperature. The ICs of yellow (6,4)-Br gradually became red ILs under UV light, and the observed birefringence of ICs in polarizing optical microscopy (POM) images disappeared, accompanied by the enhancement of the n–π^* absorption peak in the UV–Vis absorption spectrum $\lambda = 450$ nm (Figure 2.8(b)) and the disappearance of the diffraction peak in X-ray diffraction (XRD) patterns (Figure 2.8(d)). At the same time, ILs reversely transformed into ICs under visible light irradiation, of which the n–π^* absorption peak in the UV–Vis absorption spectrum $\lambda = 450$ nm weakens (Figure 2.8(c)) and the reappearance of the diffraction peak in XRD patterns (Figure 2.8(d)). The DSC curves (Figure 2.8(e)) showed the photothermal storage capacity of (6,4)-Br ICs and (6,4)-Tf_2N ILs, in which that the enthalpy of heat storage (ΔH_{iso}) of *cis*-isomer (6,4)-Tf_2N ILs was comparable to chromophore, while the *cis*-isomer (6,4)-Br has two peaks that the heat released by *cis* to *trans* is almost twice ΔH_{iso} of azobenzene in the range of 34–65°C and the sharp peak near 87°C is the melting peak of *trans*-isomer ICs. The *cis*-isomer (6,4)-Tf_2N ILs changed from red to yellow during the heating process, and the *cis*-isomer (6,4)-Br ILs changed from red to brown (*trans*-isomer ICs) continuously to yellow (*trans*-isomer ILs), as shown in Figure 2.8(f). The ionic conductivity of (6,4)-Br ICs and (6,4)-Tf_2N ILs can be adjusted reversibly through photo-induced isomerization, as shown

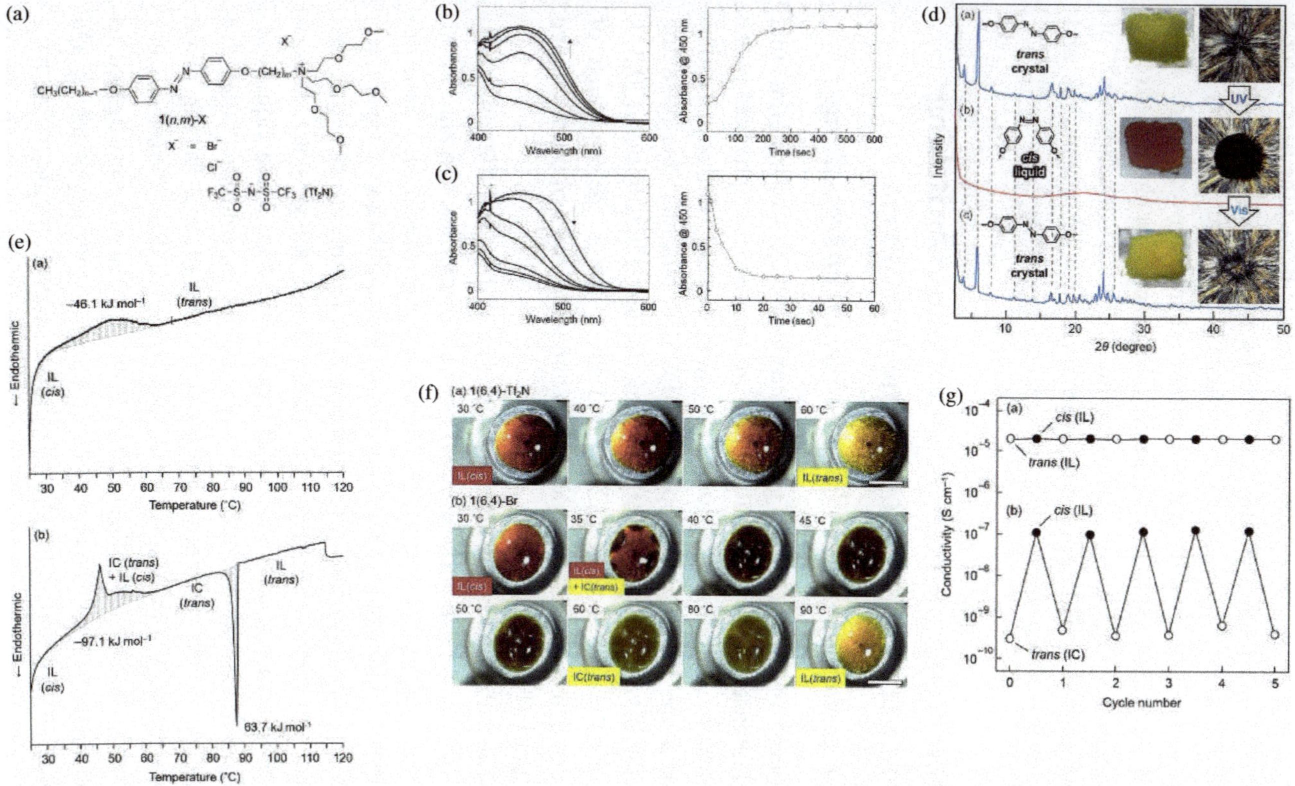

Figure 2.8. (a) The molecular structure of azobenzene derivatives. (b) and (c) The n–π* absorption peak of UV–Vis absorption spectra $\lambda = 450$ nm. (d) XRD patterns of (6,4)-Br ICs under UV light and visible light. (e) DSC curves of the azobenzene derivatives. (f) Picture of ILs during the heating process. (g) Changes of ionic conductivity upon repeated UV and Vis irradiation. Reprinted with permission from Ref. [25]. Copyright 2015, Wiley-VCH.

in Figure 2.8(g). This study demonstrated that the ICs of the azobenzene derivatives can be converted into ILs by the photoliquefaction process and significantly improve the photothermal storage capacity. The integration of self-assembly and photoinduced interconversion of ionic liquids (IC-IL) addresses the limitations of current solar thermal storage technology, offering a design principle for the development of multifunctional ILs. This approach overcomes existing constraints and opens up new possibilities for solar thermal energy storage.

2.2 Solar thermal fuels based on photo-switching phase-transition azobenzene derivatives

2.2.1 *Materials of phase change thermal energy storage*

Phase change materials (PCMs) generally refer to a class of materials that store and release energy (thermal energy) in the process of physical phase change, which can realize the storage of external energy and adjust ambient temperature. It is widely applied for PCM in building energy-saving [26,27], intelligent fabric [28,29], air conditioning refrigeration [30,31], solar thermal energy storage [32,33] and so on due to the advantage of high energy density, small volume change and wide range of phase change temperature [34–36].

PCM are classified into different types according to different standards. PCMs can be categorized based on their properties. Organic PCM, inorganic PCMs, and composite PCMs are distinguished based on material composition. Furthermore, considering the application temperature, they can be classified into low-temperature, medium-temperature, and high-temperature PCMs. Lastly, based on the phase change states, materials are grouped as solid–solid, solid–liquid, solid–gas, and liquid–gas PCMs. However, solid–gas and liquid–gas PCMs are seldom studied due to significant volume changes and gas storage challenges during phase transitions. Inorganic PCMs, such as inorganic crystalline hydrated salts, molten salts, metals, and alloys, have been extensively investigated as the initial focus of phase change energy storage technology research. On the subjects of inorganic crystalline hydrated salts, inorganic PCMs have shown to be optimal candidates due to their low cost and high energy storage density. These materials undergo a process where they dissolve crystalline water and absorb heat, resulting in a temperature increase. Conversely, when the temperature decreases, the crystalline water is reabsorbed, releasing heat. This reversible phase change behavior contributes to their effectiveness in energy storage applications. However, there are also the problems of phase separation and a high degree of supercooling during phase change, which lead to poor cycle stability and limit the range of practical applications. Molten salts, metals, and alloys are rarely used as energy storage materials in practical applications due to their high temperature of phase transition. Organic PCMs mainly include alkanes, paraffin, fatty acids, and polyols, which are widely considered as research objects phase change energy storage materials because of their excellent performance such as high energy storage density, good thermal stability, no phase separation and low degree of supercooling [37].

There are more than 20000 PCMs, appropriate phase change energy storage materials that can be selected according to different application scenarios to achieve the purpose of temperature and energy storage. In general, the selection of PCM should meet the following performance requirements [38,39]: (1) Phase temperature matching the operating temperature. (2) High energy storage density. (3) Good chemical stability, non-toxic and non-corrosive. (4) Excellent cycle stability. (5) The crystallization rate is fast and the degree of supercooling is low. (6) Extraordinary thermal conductivity. (7) Low cost and wide source of raw materials.

2.2.2 *Phase change materials applied in solar thermal energy conversion and storage*

PCMs enable the efficient utilization of solar thermal energy by converting light energy into heat energy and storing it within the material through the process of phase change. This mechanism allows for effective capture and storage of solar energy. Due to the low solar thermal energy absorption efficiency of PCM, the photothermal conversion efficiency and energy storage efficiency of solar thermal energy can be improved by introducing materials with outstanding photothermal conversion performance into PCMs, which mainly include carbon materials, metal as well as metal oxides and dyes.

Organic PCMs exhibit significant potential for practical applications in solar thermal energy storage and waste heat recovery. This is attributed to their promising phase change enthalpies, wide range of melting points, and chemical stability. However, challenges persist in the actual energy storage of organic PCMs that need to be addressed, such as low thermal conductivity resulting in low energy storage and conversion efficiency, leakage problems caused by the fluidity after phase change during cycle use as well as electrical insulation and poor light absorption ability prevent the utilization of electrical energy and solar power. Zou *et al.* [40] prepared polyurethane@graphite foam (PU@GF) composites PCMs by *in-situ* polymerization and impregnation, and the morphology is shown in Figure 2.9(a). The phase change enthalpy and melting-point of PU@GF decrease with the increase of polyethylene glycol (PEG) content. PU@GF composites exhibit significant improvements in terms of supercooling degree compared to PEG PCMs. Furthermore, Figure 2.9(b) demonstrates that the DSC curves of PU@GF composites remain largely unchanged even after 200 cycles, indicating their high thermal cycling stabilities. Additionally, the thermal conductivity of PU@GF composites has been substantially enhanced, reaching a value that is 12 times higher than that of PEG, as shown in Figure 2.9(c). The electro-thermal conversion and storage of PU@GF composites can be realized at low potential (1.0 V, 1.2 V, and 1.4 V), of which the relationship curve of PU@GF composites between temperature and applied voltage time is shown in Figure 2.9(d). When the temperature reached 39°C, 41°C, and 42°C respectively, PU@GF composites began to undergo phase change, and the temperature continued to rise to reach the equilibrium state and the electrothermal conversion efficiency could run up to 80%. Because GF has excellent light absorption and photothermal conversion ability, it could undergo phase change and store heat energy under simulated solar illumination, and corresponding energy storage efficiency can come up to 44–66.9%, which can realize the photothermal conversion and storage for solar energy.

PCMs that absorb and release significant amounts of latent heat at a constant temperature have been widely employed in the thermal management of electronic devices. However, the PCMs currently used for thermal management are primarily rigid solids or leaking liquids, lacking flexibility. This limitation hinders their application in wearable thermal

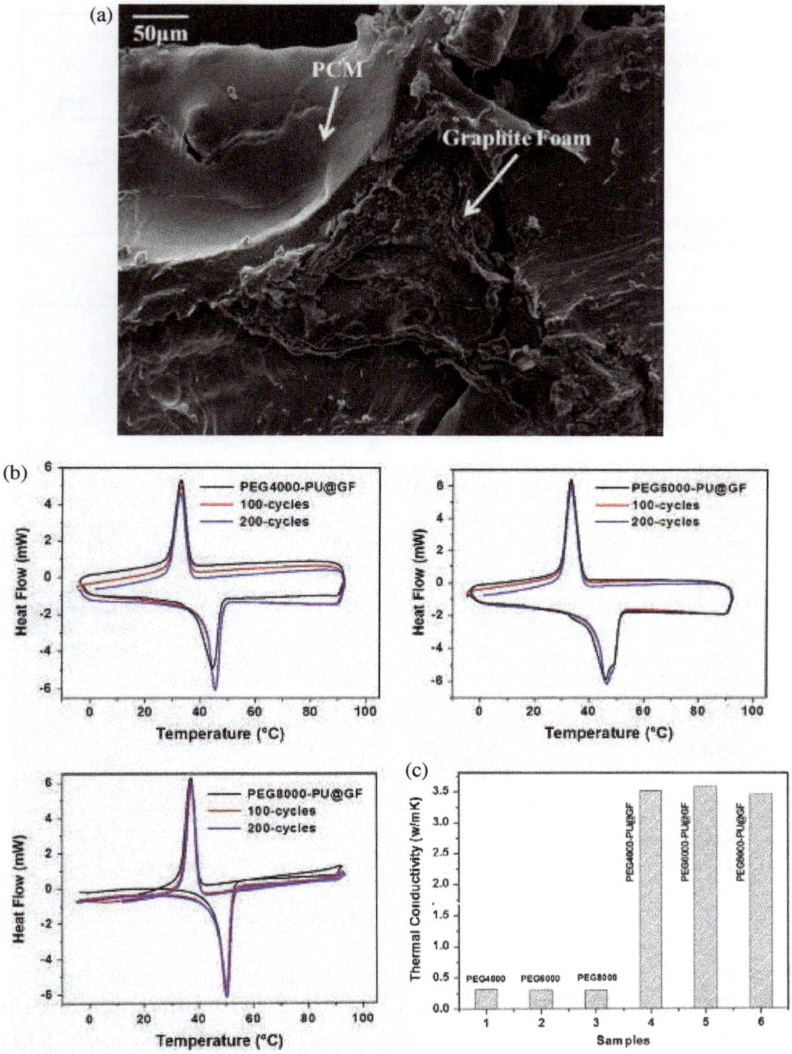

Figure 2.9. (a) SEM of PU@GF composites. (b) DSC curves of three kind of PU@GF composites. (c) Thermal conductivities of pure PEG and PU@GF composites. (d) Electra-thermal conversion and storage characterization. (e) Photo-thermal conversion and storage characterization. Reprinted with permission from Ref. [40]. Copyright 2015, Elsevier Ltd.

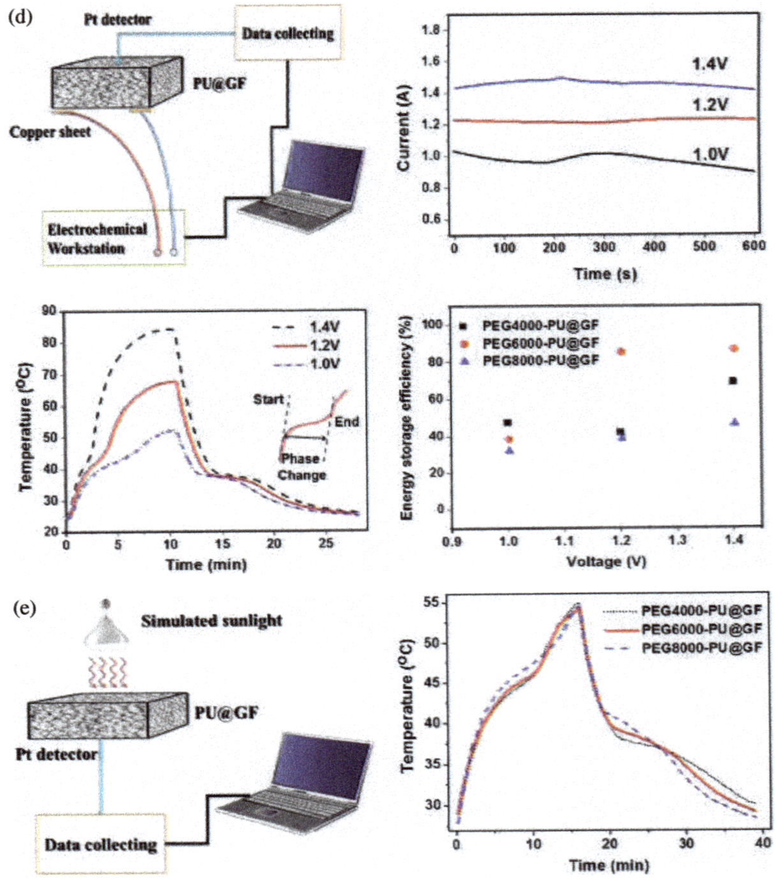

Figure 2.9. (*Continued*)

management applications. Although flexible PCM composites have been constructed by physical blending and chemical modifying with different flexible polymers, designing and fabricating intrinsically flexible PCM with stable phase transition properties, shape tailorability, and foldability remains a significant challenge for wearable thermal management applications. Achieving all these desired characteristics in a single material poses difficulties but is crucial for enabling efficient and adaptable thermal management in wearable devices. Shi *et al.* [41] synthesized the intrinsic PCM film (MTPEG) involved two step polymerization of toluene diisocyanate (TDI), polyethylene glycol (PEG), and melamine, which is remarkable self-supporting, ultra-flexible, and shape-conformable properties, as shown in Figure 2.10(a). The melting range and crystallization range of MTPEG film with different molecular weights were adjustable in the temperature range of 36.7–59.8°C and 5.1–33.9°C respectively, a broad and overlapping transition temperature range encompassing both crystallization and melting can be achieved spanning approximately 5°C to 60°C. This wide temperature range allows for versatile and efficient thermal energy storage and release in various applications. (Figure 2.10(b)). The phase change enthalpy (ΔH_c = 118.7 J/g) of MTPEG film with PEG 10000 is the highest. MTPEG 10,000 film maintained a constant temperature at 12 and 20s respectively during the heating and cooling processes when the temperature reaches 48°C, indicating that exceptional thermal behaviors of absorbing and releasing heat energy is attributed to the reversible phase transition between an ordered crystalline phase and a disordered amorphous state of the PEG chain (Figure 2.10(c)). A sandwich-like structure temperature control and thermal energy storage device was prepared via layer-by-layer assembly of MTPEG/GF/MTPEG film, as shown in Figure 2.10(d). As the current increased from 0.5 to 1.5 A, the phase change time (thermal energy storage time) decreased gradually and thermal heat release can be kept about 300s during the cooling process (Figure 2.10(e)). The electrothermal conversion efficiency of MTPEG/GF/MTPEG film increased with the applied current increasing and can reach up to 94%. The MTPEG/GF/MTPEG film was thermally charged within a controlled temperature range using a 2000 mAh battery. It demonstrated the ability to maintain a surface temperature between 25°C and 50°C for 6 h, even in a −10°C temperature environment, across more than 15 cycles (Figure 2.10(f)). The excellent temperature control performance provided a comfortable temperature environment for the human body or wearable electronics in extremely cold conditions. Figure 2.10(g) showed the wearability and temperature control stability of the device. Meanwhile, The MTPEG/GF/MTPEG film exhibited exceptional photothermal conversion ability due to the excellent light absorption ability of GF (Figure 2.10(h)). The PCM composites display promising potential in the field of electro/photo-thermal management, providing a new direction for the development of wearable smart fabrics (Figure 2.10(i)).

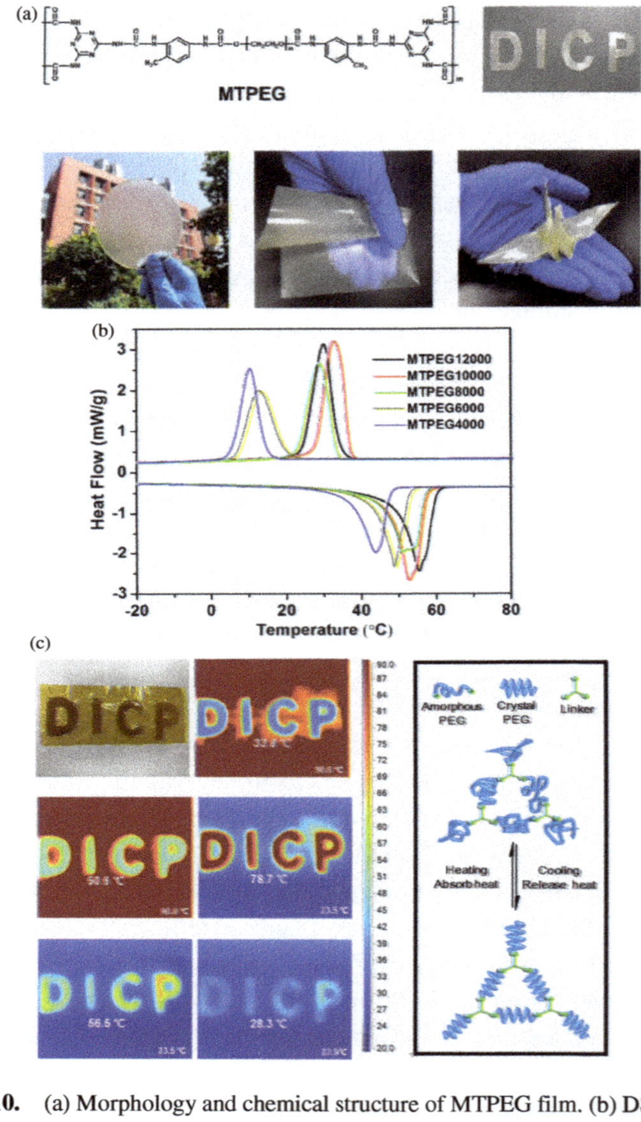

Figure 2.10. (a) Morphology and chemical structure of MTPEG film. (b) DSC curves of MTPEG films representing the phase change temperature. (c) Thermal storage of MTPEG films. (d) Electrothermal conversion device and MTPEG/GF/MTPEG film. (e) The temperature curves of thermal control device obtained at different currents. (f) Electrothermal cycles at different environments and temperatures with a battery. (g) The flexible electrothermal device sewed into a lab coat. (h) The temperature-time curves under simulated sunlight irradiation. (i) Photograph of smart temperature control cloth. Reprinted with permission from Ref. [41]. Copyright 2020, Elsevier B.V.

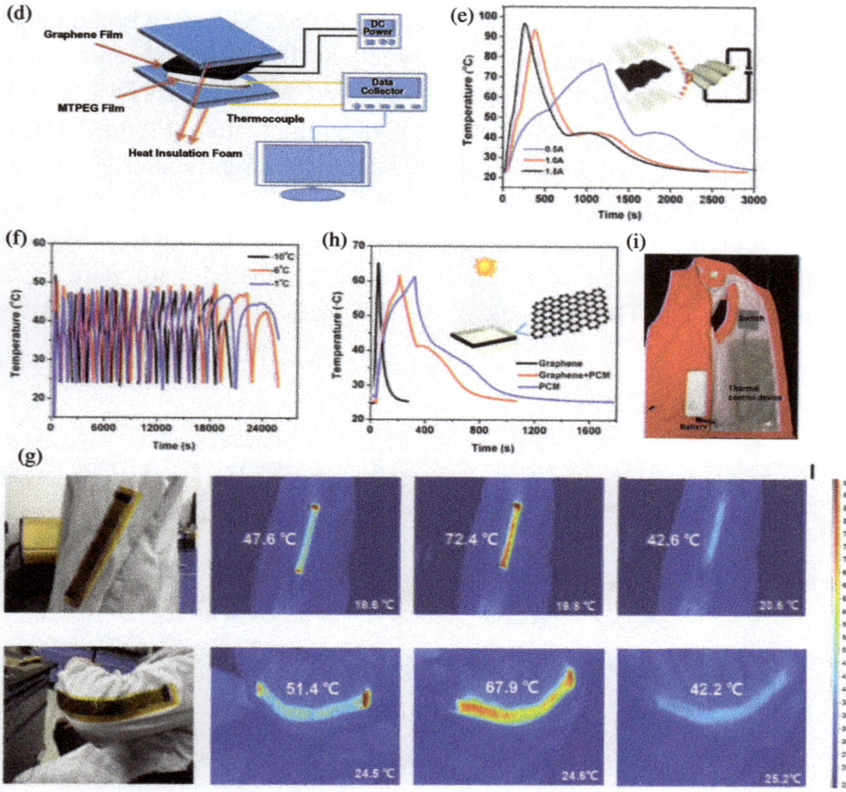

Figure 2.10. (*Continued*)

Generally, PCM cannot directly and efficiently utilize ultraviolet and visible light which account for about 45% of the solar spectrum due to their low thermal efficiency, as well as the mismatching of solar irradiation in time and space. Both these factors lead to the significant reduction of solar energy for PCM. Therefore, the development of full-band solar absorption and photothermal conversion and storage PCM is of great significance for efficient utilization of clean and renewable solar energy. Zhang *et al.* [42] prepared MWCNT/PEG/SiO_2 form-stable PCM composites by ultrasound-assisted sol-gel method, and the preparation process is shown in Figure 2.11(a). The schematic diagram of full-band light driven reversible phase transition for MWCNT/PEG/SiO_2 form-stable PCM composites is shown in Figure 2.11(b). PEG and SiO_2 were considered as PCM and supporting material, MWCNT have a broad range of UV–Vis light absorbance and excellent light-thermal conversion capacity. Compared with PCM without MWCNT, MWCNT/PEG/SiO_2 form-stable PCM composites showed a higher temperature (Figure 2.11(c)) and significantly improved thermal conductivity, greatly shortened heat storage and release time, and improved energy utilization efficiency (Figure 2.11(d)). During the heat process, MWCNT/PEG/SiO_2 PCM composites exhibited solid–solid phase change compared with pure PEG (Figure 2.11(e)). The prepared MWCNT/PEG/SiO_2 PCM composites

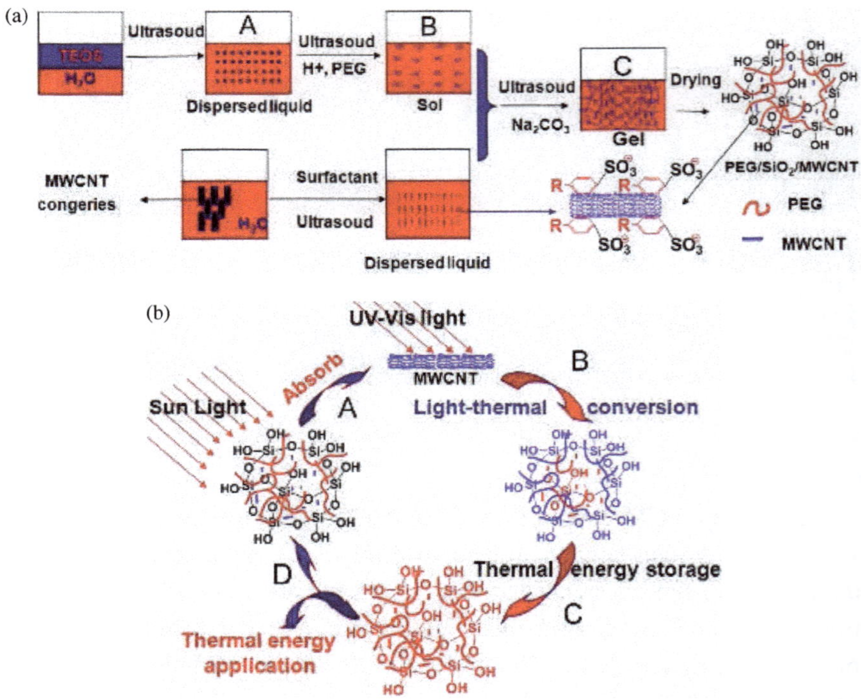

Figure 2.11. (a) Schematic of the synthesis of PCM assisted by ultrasound. (b) Schematic of light to heat conversion and storage. (c) Photo-induced exothermic properties of PCM. (d) The heating and freezing temperature curves of pure PEG, PEG/SiO_2 and MWCNT/PEG/SiO_2 composites. (e) Photograph of MWCNT/PEG/SiO_2 composites and pure PEG taken in a hot stage. Reprinted with permission from Ref. [42]. Copyright 2014, Elsevier B.V.

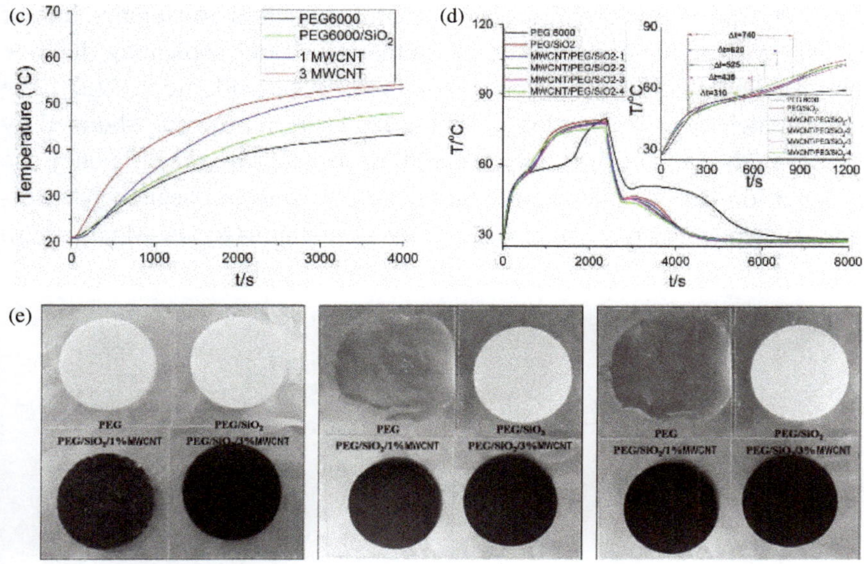

Figure 2.11. (*Continued*)

offer a novel platform for enhancing solar irradiation utilization efficiency. These composites have broad applications in various fields related to energy conversion and storage. Their unique properties make them promising materials for improving energy efficiency and advancing renewable energy technologies.

Dye is one of the most extraordinary materials of light absorption and photothermal conversion, which can be selected for absorption of visible light and molar extinction coefficient can reach tens of thousands. Tang *et al.* [43] first synthesized the PCM containing terminal isocyanate group (–NCO) by step-condensation reaction process between TDI and polyethylene glycol ether (MPEG). Then, the three dyes of red, yellow and blue were further chemical graft reacted with the pre-prepared PCM and polyethylene polyamine to obtain the organic solid–liquid phase change material (OSPCM), which was further cross-linked by TDI to obtain organic form-stable phase change material (OCSPCM), and the specific synthesis route is shown in Figure 2.12(a). The UV–Vis absorption spectra of OCSPCM showed obvious visible light absorption characteristic peak compared with MPEG, indicating that the dye has obvious light absorption capacity (Figure 2.12(b)). Meanwhile, the temperature of OCSPCM rose rapidly compared with that of MPEG, indicating that dye has excellent photothermal conversion performance (Figure 2.12(c)). The color-matching with broadband visible light absorption was obtained by adjusting the ratio of dyes, which could further improve the efficiency use of solar irradiation, and its photothermal conversion and heat energy storage efficiency was more than 74%. At the same time, OCSPCM has a high

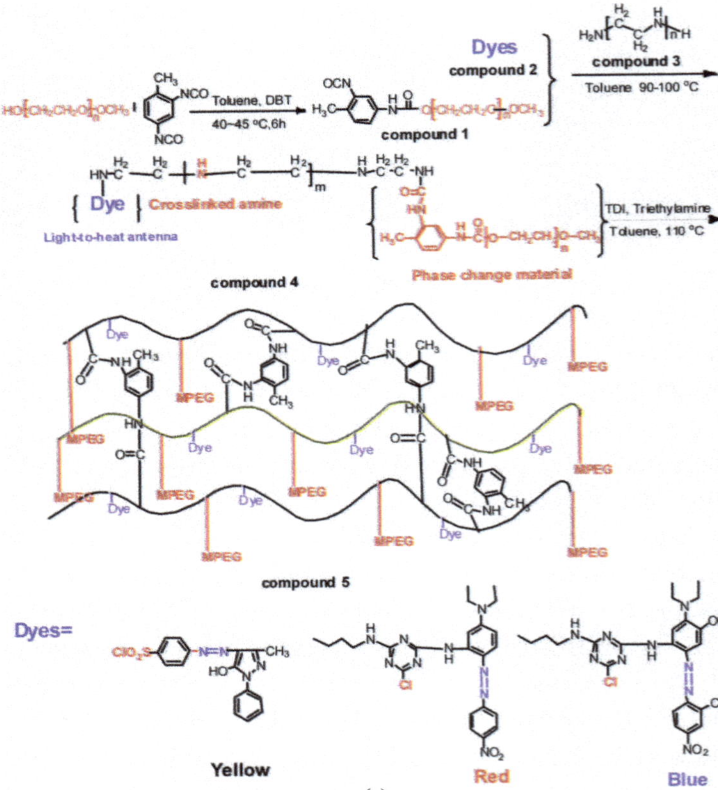

(a)

Figure 2.12. (a) Chemical structure and synthetic scheme of solar thermal conversion materials. (b) UV–Vis absorption spectra and color image of MPEG and OCSPCM. (c) Light-driven efficiencies of MPEG and OCSPCM. (d) Shape-stabilized effects as the temperature increases. Reprinted with permission from Ref. [43]. Copyright 2013, Published by Elsevier Ltd.

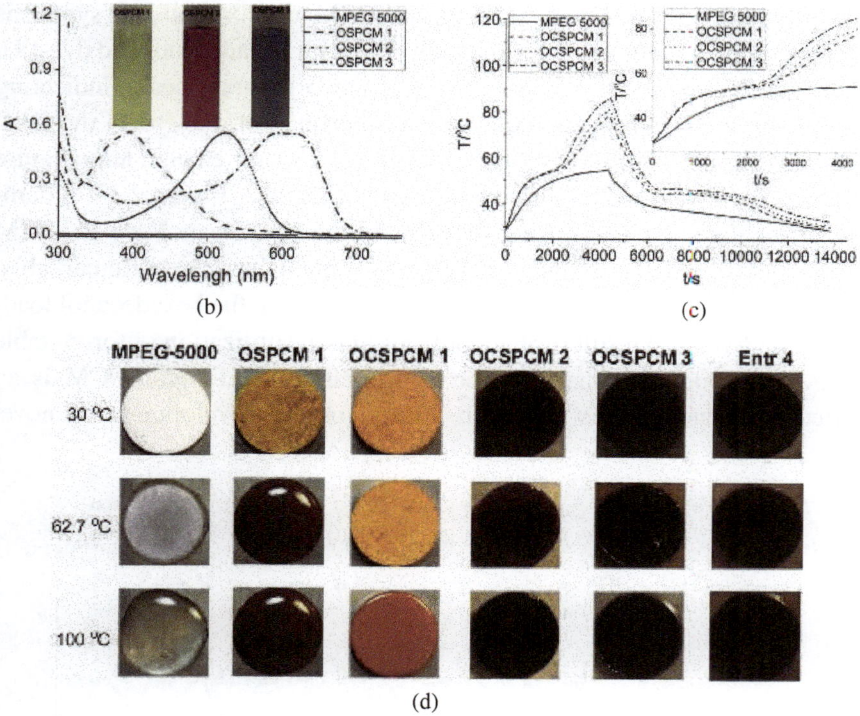

Figure 2.12. (*Continued*)

energy density ($\Delta H > 120$ J/g). OCSPCM and their color-matching PCM remained solid and no liquid was observed even when the temperature exceeded 100°C compared with MPEG during the heating process from 30°C to 100°C, which showed obvious solid–solid phase transition behavior, as shown in Figure 2.12(d). Color-matching, with its simplicity and low cost, not only offers a new platform for enhancing the efficiency of solar irradiation utilization but also expands the color options of materials to meet people's diverse aesthetic preferences. This dual advantage allows for improved solar energy utilization while simultaneously offering a wide range of visually appealing color choices in various applications.

Tang *et al.* [44] synthesized dye-polyurethane (dye-PU) with TDI and PEG as characteristic raw materials and azobenzene dye diol as chain extender, and then the hexadecanol/dye-PU composite by integrating hexadecanol into a visible light-driven PCM matrix, and the synthesis route is shown in Figure 2.13(a). Compared with hexadecanol and dye-PU, hexadecanol/dye-PU PCM composite has two separate peaks indicating the melting of hexadecanol and the phase change of dye-PU in the DSC curves (Figure 2.13(b)), and the corresponding phase change temperature range can be adjusted by altering the ratio of the two. Figure 2.13(c) demonstrates that the phase change enthalpy of the hexadecanol/dye-PU PCM composite increases as the hexadecanol content increases. Remarkably, the phase change enthalpy can reach 229.5 J/g when the hexadecanol loading is 63.8%. This result further confirms that the utilization of form-stable PCMs as supportive matrices for incorporating solid–liquid PCM is an effective strategy. This approach allows for the development of novel

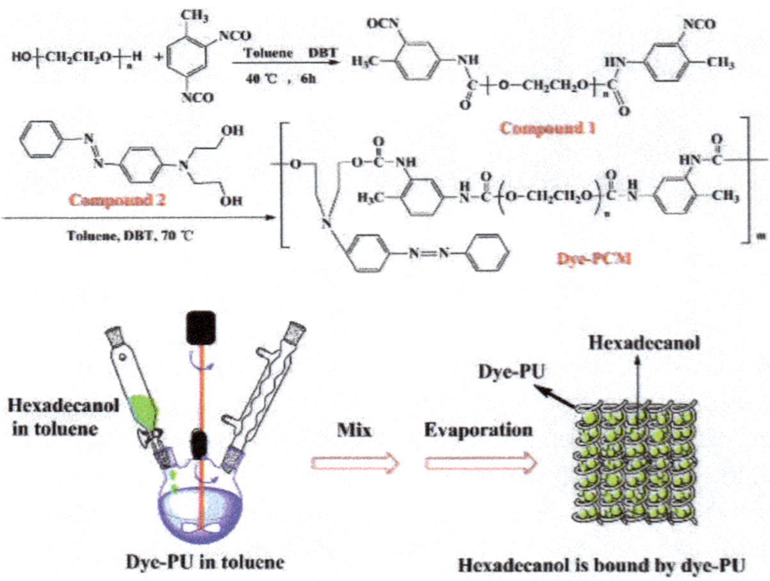

(a)

Figure 2.13. (a) The synthesis route of the PCM composite. (b) DSC curves of dye-PU, hexadecanol and hexadecanol/dye-PU composite. (c) DSC curves of hexadecanol/dye-PU composites with different weight percentage of hexadecanol. (d) Light-thermal conversional curves of sample under solar simulator. (e) The form-stable photographs of (1) hexadecanol/dye-PU composite (2) hexadecanol (3) dye-PU. Reprinted with permission from Ref. [44]. Copyright 2015, Elsevier B.V.

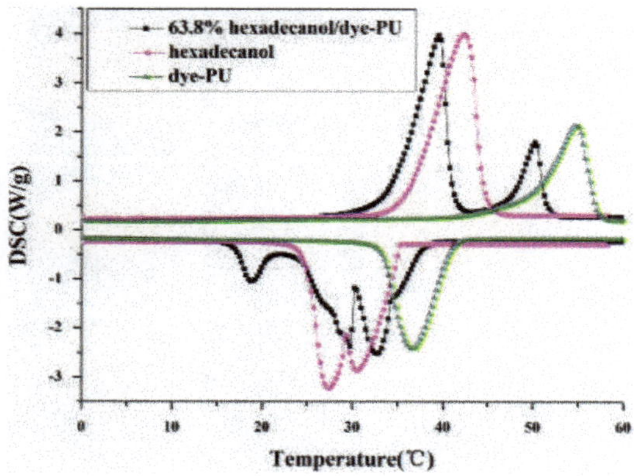

(b)

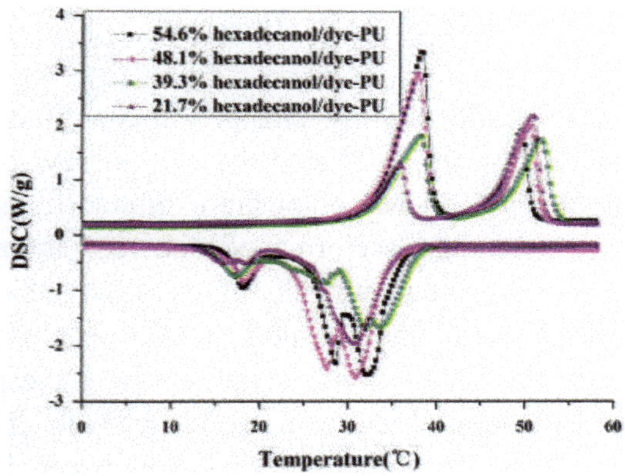

(c)

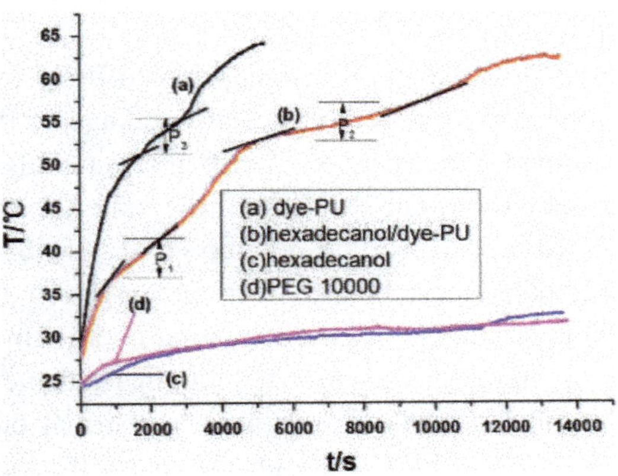

(d)

Figure 2.13. (*Continued*)

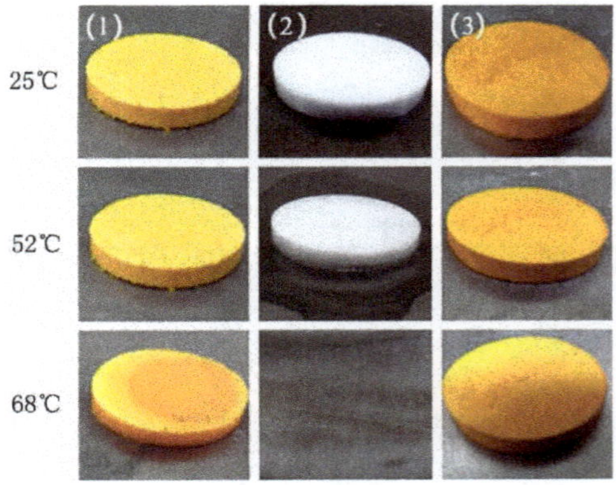

(e)

Figure 2.13. (*Continued*)

form-stable PCMs with high phase change enthalpy and suitable phase change temperature, offering enhanced thermal energy storage capabilities. Hexadecanol/dye-PU PCM composite exhibited superior solar thermal conversion and storage performance due to light absorption and photothermal conversion characteristic of dyes, in which temperature increased with increasing radiation time under the solar simulator irradiation and there are two phase transition platforms corresponding to the phase change of hexadecanol and dye-PU, respectively (Figure 2.13(d)). Hexadecanol/dye-PU PCM composite maintained a stable shape even heating to 68°C above the phase change temperature and no liquid leakage was observed (Figure 2.13(e)).

The core-shell structure in microencapsulated PCM (MPCM), that is PCM as the core and shell structure prevents leakage of PCM during the process of solid–liquid phase change. The shell is generally polymer material, which separates the inner PCM from the external environment and reduces the volatilization of certain organic PCM [45,46]. The combination of photothermal materials and microencapsulated technology can realize the efficient storage and release of solar energy. In typical scenarios, solar energy is captured and converted into thermal energy through the use of photothermal materials. The generated thermal energy is subsequently transferred to the PCM through thermal diffusion. This process allows for the efficient storage and utilization of the captured solar energy in the PCM. However, the thermal resistance between PCM and photothermal materials is enhanced owing to the existence of shell structure, which further impedes the transfer of thermal energy to the inner PCM [47,48]. Yu *et al.* [49]. developed a MPCM as part of a highly efficient solar energy storage system, the MPCM structure consisted of a shell made of polymethyl methacrylate (PMMA), while the core was composed of an organic PCM (eicosane) integrated with PMMA-modified black phosphorus sheets (BPs). BPs with excellent photothermal conversion performance reduce heat loss during solar thermal transfer and accelerate energy storage. The purpose of PMMA modification of BPs is to enable BPs to be stably dispersed in organic solvents, which can be achieved through a one-step tip sonication process, as shown in Figure 2.14(a). PMMA, eicosane, and mBPs were dissolved in CH_2Cl_2 as the oil-phase solution and polyvinyl alcohol (PVA) was dissolved in hot water to form a water-phase solution. Next, the oil-phase solution is introduced into the water-phase solution while vigorously stirring, resulting in the formation of a stable emulsion. After evaporation of CH_2Cl_2, microencapsulated PCM (mBPs-MPCM) was formed, and the preparation process is shown in Figure 2.14(b). The MPCM exhibit a spherical shape and clean surface. When the mass ratio of the core/shell is increased to 13.3, the microcapsules collapse indicating lower encapsulation efficiency of the core. Meanwhile, the minimum core/shell structure of microcapsules is retained when the ratio of the core/shell is decreased to 4.4, as shown in Figure 2.14(c). The average size of microcapsules was 20 μm, and the melting temperature and crystallization temperature of mBPs-MPCM decrease due to the small internal space in the microcapsules compared with eicosane and pure MPCM [50]. Moreover, both the melting latent heat and crystallization latent heat of the mBPs-MPCM decreased with decreasing the content of eicosane, but it still showed a high thermal energy storage capacity of energy density (ΔH > 180 J/g), as shown in Figure 2.14(d). The absence of any leakage between mBPs-MPCM and MPCM by heating to 70°C for 10 min confirmed that the PMMA shell structure provides sufficient mechanical strength for the microencapsulated composites and prevents leakage of the melted eicosane, resulting in good thermal reliability. The diffuse reflectance spectra (DRS) analysis revealed that MPCM samples without modified black phosphorus (mBPs) exhibited weak light absorption and limited

(a)

(b)

(c)

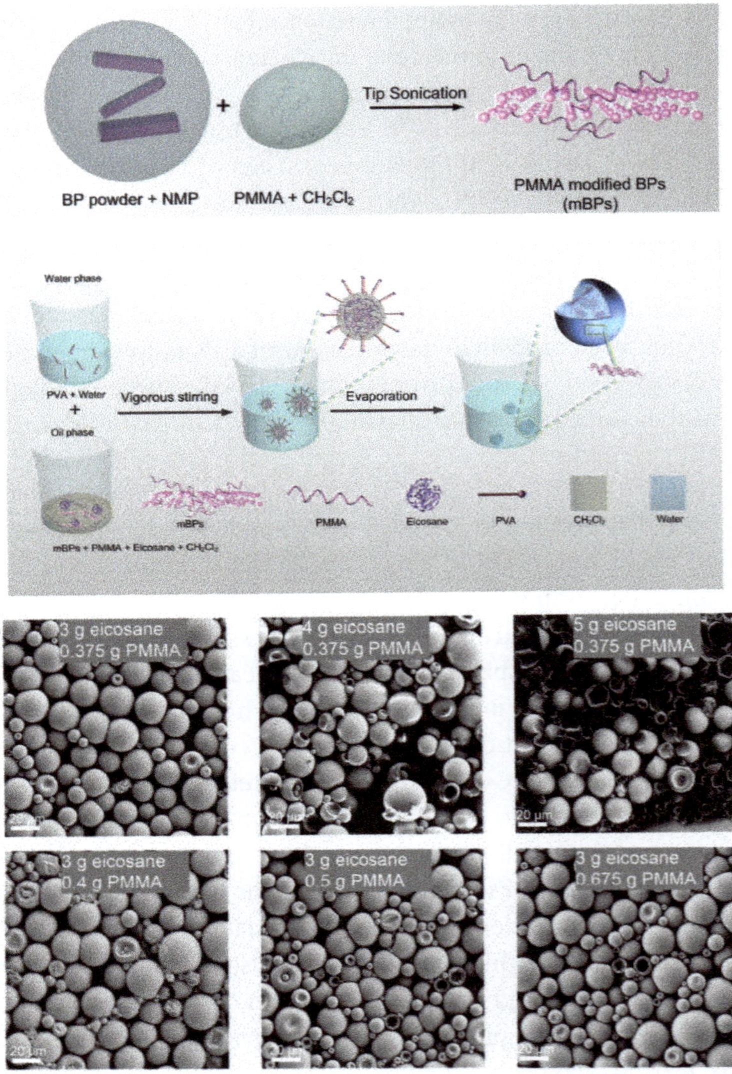

Figure 2.14. (a) Synthesis of mBPs. (b) Schematic illustration of synthesis of the mBPs–MPCM. (c) SEM images of the MPCM composites with different core/shell ratios. (d) DSC curves of (1) eicosane (2) MPCM composites (3) mBPs–MPCM composites with the inset showing the thermal reliability. (e) Absorbance spectra of the MPCM and mBPs– MPCM composites. (f) SEM images of the mBPs decorated MPCM composite and the mBPs–MPCM composites. (g) Photothermal characteristics of decorated MPCM, mBPs–MPCM and MPCM. (h) Solar energy storage mechanism of the mBPs decorated MPCM composite and the mBPs–MPCM composites. Reprinted with permission from Ref. [49]. Copyright 2020, the authors. Reprinted with permission from Ref. [50]. Copyright 2018, Elsevier Ltd.

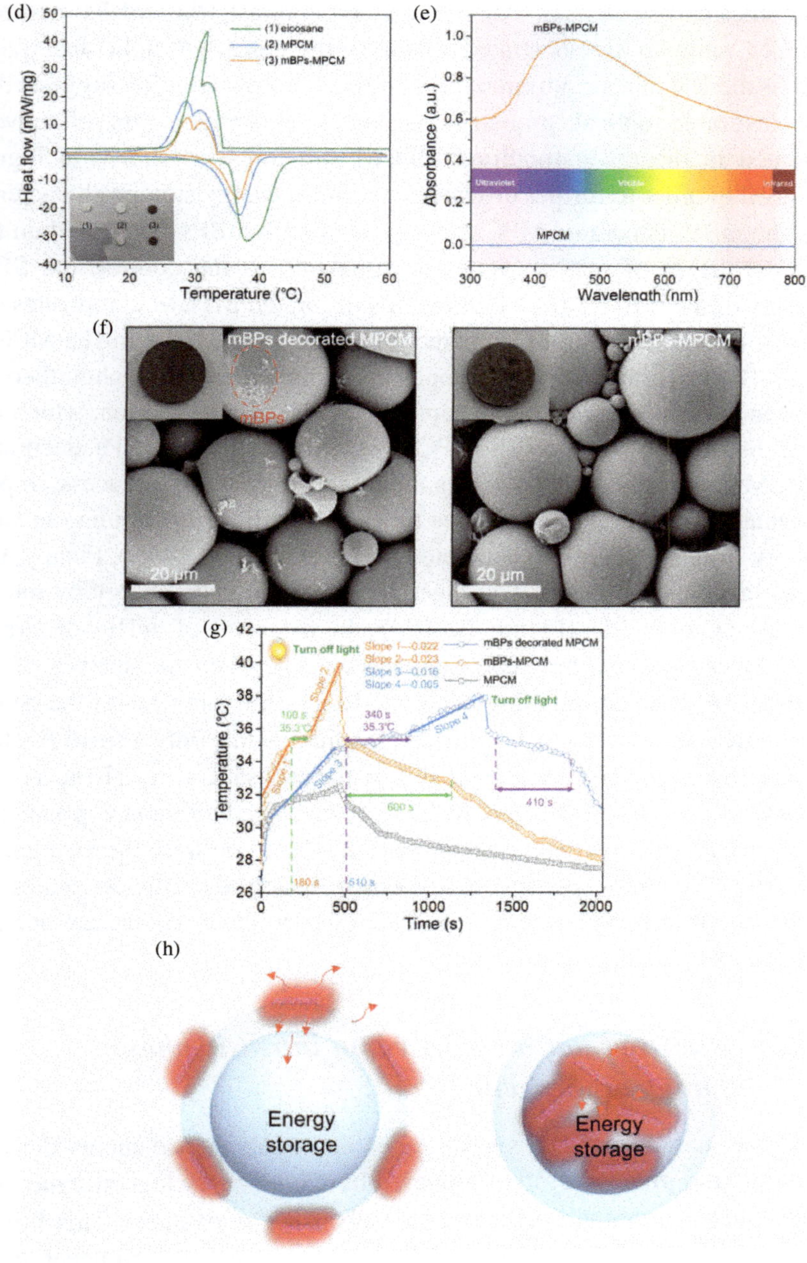

Figure 2.14. (*Continued*)

photothermal conversion capabilities. In contrast, the mBPs-modified MPCM samples demonstrated significantly enhanced light absorption across the UV to near- infrared (NIR) range. This observation suggests that the favorable optical properties of black phosphorus are effectively retained in the mBPs-modified MPCM structures, as shown in Figure 2.14(e). In order to further investigate the solar energy storage characteristic, the mBPs decorated MPCM composites with mBPs distributed on the shell structure of microcapsules are prepared as the control, the SEM images of the mBPs-MPCM and mBPs decorated MPCM composites are shown in Figure 2.14(f). Compared with pure MPCM, the mBPs-MPCM and mBPs decorated MPCM composites exhibited excellent photothermal conversion and energy storage capacity under solar irradiation, which the time required by the mBPs-MPCM is 280s and the mBPs decorated MPCM composites is 850s during the total energy storage process, implying that the mBPs-MPCM is three times more efficient pertaining to solar energy storage than mBPs decorated MPCM (Figure 2.14(g)). Finally, this investigation illustrated the storage mechanism of solar energy by microencapsulated PCMs (Figure 2.14(h)). In the case of mBPs-decorated MPCM composites, heat convection occurs between the surface of the structure and the surrounding medium. As a result, a portion of the generated heat is transferred to the adjacent medium, while only a small fraction of the heat is ultimately transferred to the eicosane core. However, the presence of mBPs in the MPCM leads to enhanced efficiency in harnessing solar energy and rapid energy storage. This is because the heat generated by the black phosphorus, located inside the microcapsules, is promptly transferred to the eicosane core without significant energy dissipation.

2.2.3 *Solar thermal fuels based on intrinsic phase change azobenzene*

Although nanohybrid STFs have the potential to enhance energy density, the charging process is often hindered by the close-packed structure and high molar extinction coefficient of azobenzene derivatives. This limitation restricts the depth of incident light during charging and necessitates the use of solvents. However, the use of organic solvents exacerbates environmental concerns and adds to the overall cost. Photoliquefiable azobenzene (PLAZ) derivatives have received extensive attention for their excellent performance in changing from solid to liquid at room temperature during the process of *cis–trans* isomerization without solvent assistance [51,52]. The photo-induced *cis*-isomers helped decrease the melting-point of mixture, and the liquid-state PLAZ provided more free volume for the *cis-trans* isomerization, which is more favorable for heat release. Nevertheless, the fluidity of liquid-state PLAZ breaks the morphology of STFs. Based on the above-mentioned properties, Yu *et al.* [53] prepared the solid-state STFs which combines azobenzene derivatives (4-methoxyazobenzene) with one kind of easily obtained cloth fabric by hot-melting method, and the preparation process is shown in Figure 2.15(a). The melting-pointing of *cis*-isomer and *trans*-isomer of 4-methoxyazobenzene are 25°C and 53°C, respectively. Thus, photoliquefaction can occur below room temperature. Figure 2.15(b) shows the SEM images of the composites of azobenzene derivatives and fabric with different filling ratios. The energy density of pure PLAZ STFs increases along with more incident energy, but at the expense of reducing solar energy conversion efficiency (Figure 2.15(c)). The energy density did not follow a normally linear relationship with the *cis* ratio, which is different from that report for liquid azobenzene STFs. Meanwhile, the energy density-*cis* ratio and the energy density-incident energy showed a similar nonlinear growth trend (Figure 2.15(d)). This is because the PLAZ energy storage enthalpy ($\Delta H_{storage}$) consists of two parts: *cis–trans* isomerization energy (ΔH_{iso}) and phase change enthalpy ($\Delta H_{ph\text{-}trans}$) as shown in Figure 2.15(e). The theoretical energy density limit of PLAZ STFs is more than 300 J/g (83.3 Wh/g) through theoretical calculation and experimental fitting results. The energy density of the composite STFs is higher than that of pure PLAZ STFs as well as the energy density increases with increasing the filling ratio of azobenzene derivatives. The solar energy conversion efficiency of the composite STFs attenuated slower than the pure azobenzene derivatives STFs (Figure 2.15(f)), which demonstrated the superiority of the flexible fabric template. The energy storage life of the composite STFs (6d) is longer than corresponding PLAZ STFs (4d), as shown in Figure 2.15(g). The energy storage heat release of the composites STFs can be accelerated under high temperature and blue light irradiation

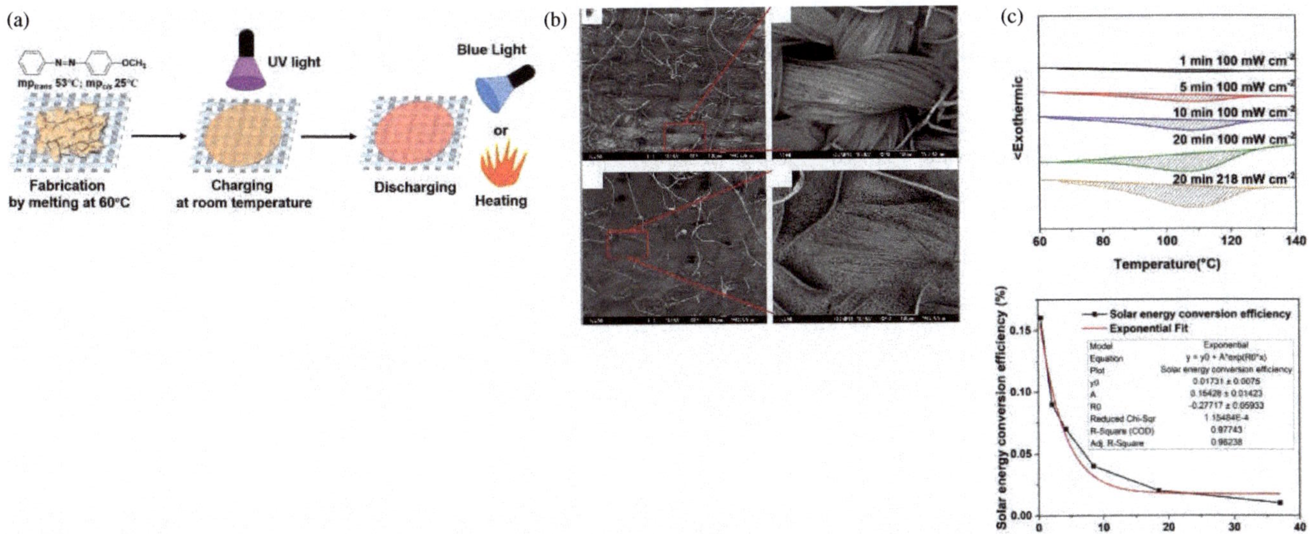

Figure 2.15. (a) Schematic of the composites STF fabrication, charging and discharging. (b) The SEM images of the composites with different filling ratios. (c) DCS curves of the composites STFs charged for different time and the relationship between solar energy conversion efficiency and incident energy. (d) Relationship between the incident energy and the *cis* ratio. (e) Relationship between the *cis* ratio and the energy density as well as relationship between the energy density and incident energy. (f) The energy-storage mechanism of the STFs. (g) Charging and discharging of the composites STFs and the pure azobenzene derivatives STFs. (h) Discharging of the composites STFs stimulated by blue light irradiation. Reprinted with permission from Ref. [53]. Copyright 2019, WILEY-VCH.

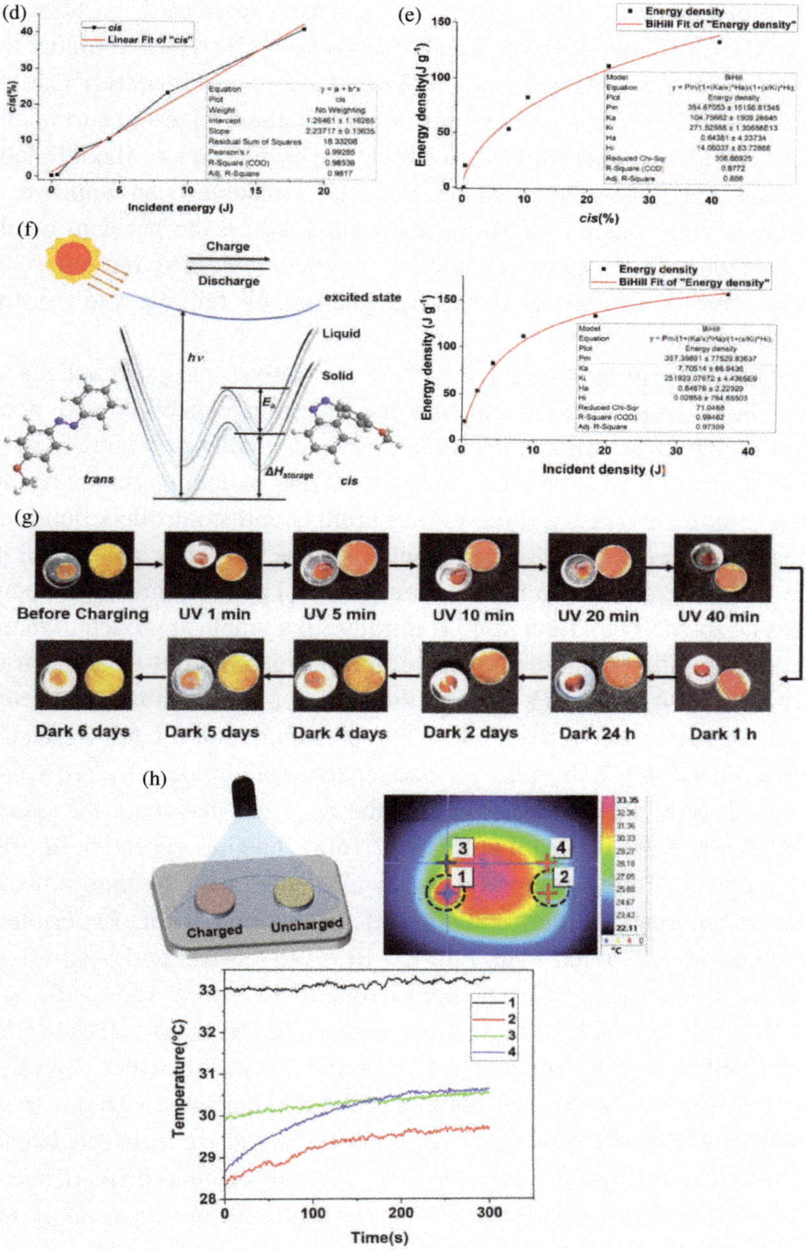

Figure 2.15. (*Continued*)

conditions. When the temperature is increased from 60°C to 80°C, the surface temperature of the charged composites STFs was 2°C higher than that of the uncharged composites STFs and was approximately 4°C higher than that of the uncharged composites STFs under blue light irradiation (Figure 2.15(h)). This study showed that the composites of flexible fabric templates and photoliquefied azobenzene derivatives can improve the energy storage density and storage life, and solved the problem of photoliquefaction after charging, which supplied reference for other high energy density azobenzene derivatives and flexible templates to construct a STF system.

The exploration of liquid and PLAZ derivatives has generated significant interest due to their solvent-free charging capability and photo-controllable heat release. However, several challenges remain in the development of these azobenzene derivatives, including achieving high energy storage density, long-term stability, and controlled rapid heat release. It is crucial to investigate and optimize the performance of liquid and PLAZ derivatives to realize their practical applications as molecular STFs (Azo-MSTFs). Li *et al.* [54] introduced a viable approach to manipulate the melting point and achieve light-triggered solid-to-liquid transformation in Azo-MSTFs through the modification of substituent groups on azobenzene derivatives. A series of liquid and PLAZ derivatives (4-bromo-alkoxy-2′,6′-diethyl azobenzene abbreviated Azo-Cn-Br, n = 2, 4, 6, 8, 10, and 12) were synthesized and designed, the specific synthesis process is shown in Figure 2.16(a). The incorporation of ortho- and para-substituted groups synergistically influenced the intermolecular interaction, enabling the regulation of the melting point. Experimental findings demonstrated that Azo-C4-Br, Azo-C6-Br, and Azo-C10-Br exhibited liquid properties at room temperature, while Azo-C2.Br, Azo-C8-Br, and Azo-C12.Br underwent photo-induced solid-to-liquid phase transitions among the three solid compounds. All azobenzene derivatives showed similar UV–Vis absorption spectra and reach their photo-stationary state (PSS) within 30 s under UV irradiation due to *trans–cis* isomerization. Meanwhile, the intensity of π–π* absorption peak decreased gradually and the corresponding wavelength of the absorption peak shifted from 325 nm to 306 nm. While the intensity of n–π* absorption peak increased gradually and the corresponding wavelength of the

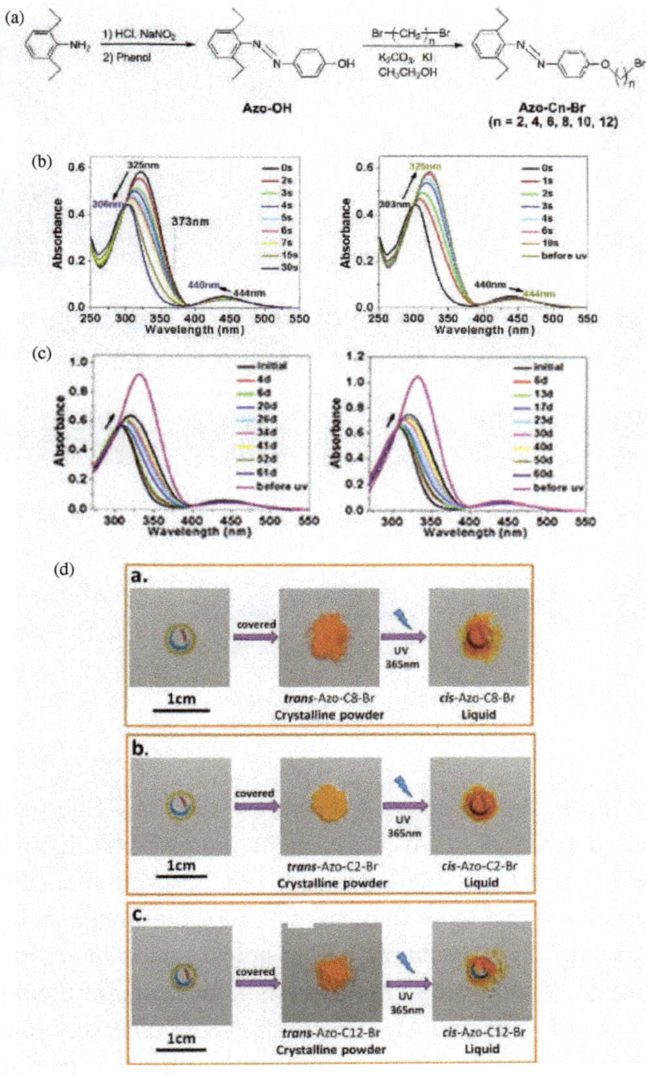

Figure 2.16. (a) The synthesized schematic of Azo-Cn-Br. (b) The UV–Vis absorption spectra of Azo-C2.Br. (c) The UV–Vis absorption spectra of Azo-C4-Br and Azo-C6-Br, respectively. (d) Pictures of Azo-C2.Br, Azo-C8-Br, and Azo-C12.Br before and after UV light irradiation. (e) POM images of Azo-C2.Br, Azo-C8-Br and Azo-C12.Br before and after UV light irradiation. (f) IR thermal imaging of "uncharged" and "charged" of Azo-C4-Br and Azo-C8-Br under blue light irradiation at room temperature. Reprinted with permission from Ref. [54]. Copyright 2022, American Chemical Society

(e)

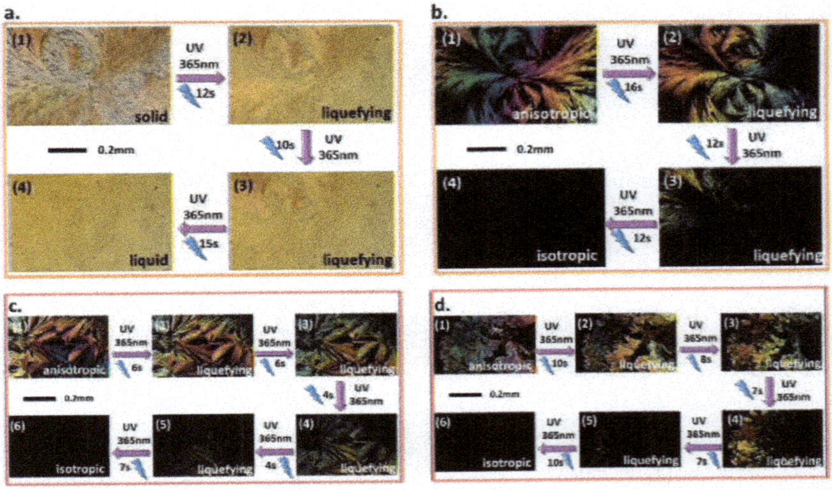

(f)

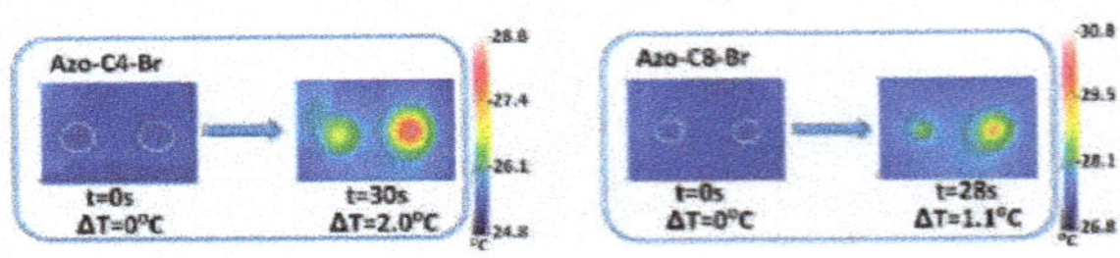

Figure 2.16. (*Continued*)

absorption peak shifted from 444 nm to 439 nm. Subsequently, the azobenzene derivatives recover within 13 s under blue light irradiation due to the *cis–trans* reversion. The UV–Vis absorption spectra of Azo-C2.Br was shown in Figure 2.16(b). The rapid response to light allows them to "charge" and "discharge" more quickly upon alternating irradiation of ultraviolet and visible light, which is the basis for them to be used as MSTFs. Azo-C4-Br and Azo-C6-Br showed a relatively long half-life (about 60d), as shown in Figure 2.16(c), indicating that the introduction of the bromo-alkyl chain changes the van der Waals interactions and steric hindrances, improving the *cis–trans* isomerization activation energy. The UV–Vis absorption spectra of Azo-C4-Br and Azo-C6-Br neat-liquid thin films were almost identical to those in dilute organic solvents, which indicates that there are no appreciable intermolecular interactions in the condensed azobenzene derivatives to suppress the photoisomerization. At the same time, Azo-C4-Br and Azo-C6-Br neat-liquid thin films showed relatively fast photoisomerization, and the rate constant $k_{trans\text{-}cis}$ was 0.19 s^{-1} and 0.18 s^{-1}, respectively as well as the rate constant $k_{cis\text{-}trans}$ was 0.6 s^{-1} and 0.23 s^{-1}, respectively. Azo-C2.Br, Azo-C8-Br, and Azo-C12.Br changed from solid–state to liquid state under UV light irradiation (Figure 2.16(d)), and the birefringence observed in POM gradually disappeared and finally transformed to a dark field (Figure 2.16(e)), which indicates the photoliquefication under UV light irradiation. The photoliquefied Azo-C8-Br showed the highest energy density ($\Delta H_{\text{storage}} = \Delta H_{\text{iso}} + \Delta H_{\text{ph-}trans} = 145.95$ J/g) and was able to release heat at relatively low temperature due to their low crystalline temperature. The ΔT of the heat released by liquid Azo-C4-Br and photoliquefied Azo-C8-Br under blue light irradiation was 2.0°C and 1.1°C, respectively (Figure 2.16(f)), which indicates that the prepared azobenzene derivatives could quickly release the storage photothermal energy under visible light irradiation. This study presented an effective molecular design approach to achieve solvent-free liquid and photoliquefiable STFs. By manipulating the molecular structure, it was demonstrated that the performance of azobenzene derivatives can be optimized, offering great potential for the development of highly efficient MSTFs operating at low temperatures.

The energy density of azobenzene derivatives can be improved by introducing phase transition in the process of photo-induced *cis–trans* isomerization, that is say that realizing photo-induced liquefaction of *trans*-isomer crystal and crystallization of *cis*-isomer in the course of *cis–trans* back isomerization, which solved the problems of low energy density and solvent assistance in the operation process of azobenzene derivatives STFs. The remaining issue is developing methods to enhance the gravimetric energy density and the mole fraction of *cis*-isomers at the PSS. Kimizuka *et al.* [55] reported the development of two liquid bisazobenzene derivatives, referred to as compounds 1 and 2 (Figure 2.17(a)). These derivatives exhibited excellent properties as solvent-free molecular switches (MSTFs). Compound 1 had a bent meta-bisazobenzene structure, which effectively lowered its melting point. In comparison, compound 2 possessed a more symmetric structure due to the introduction of a methoxy group in the 4-position of the azobenzene moiety. The researchers analyzed the UV–Vis is absorption spectra of the compounds

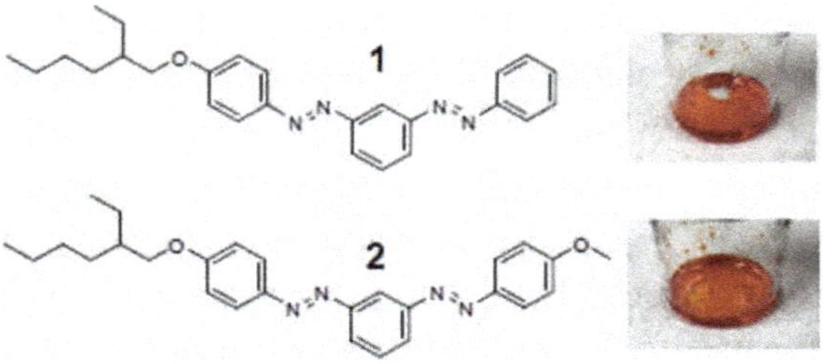

(a)

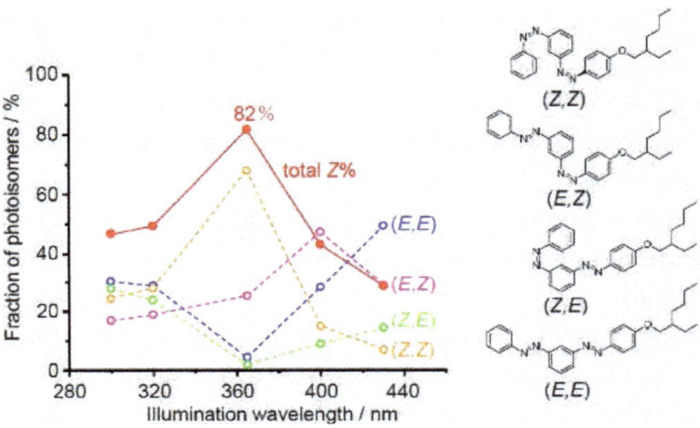

(b)

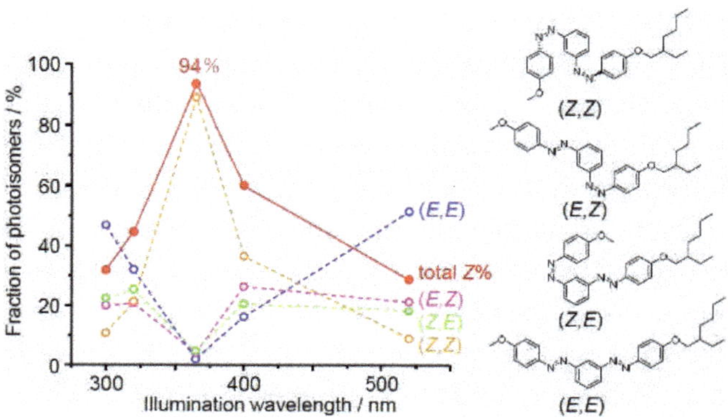

(c)

Figure 2.17. (a) Chemical structures of liquid bisazobenzene derivatives (1 and 2) and their photographs at room temperature. (b) Fraction of photo-stationary isomers of 1 as a function of the illumination wavelength. (c) Fraction of photo-stationary isomers of 2 as a function of the illumination wavelength. Reprinted with permission from Ref. [55]. Copyright 2022, The Chemical Society of Japan.

and found that the two azobenzene units in the meta-bisazobenzene were electronically decoupled from each other. This decoupling indicated that the azobenzene units behaved independently in terms of their photochemical properties. When compound (*E*, *E*)-1 was subjected to UV irradiation, it underwent photoisomerization and converted into a mixture of four possible geometric isomers: (E, E), (E, Z), (Z, E), and (Z, Z). The relative fractions of these photo-stationary isomers varied depending on the excitation wavelengths used during the irradiation process. The relationship between the excitation wavelength and the fraction of each isomer at PSS of compounds 1 and 2 is shown in Figures 2.17(b) and 2.17(c), it can be seen that the total fraction of *cis*-isomers of compounds 1 and 2 at the PSS under 365 nm UV irradiation was 82% and 94%, respectively. The efficient *trans–cis* isomerization of (*E*, *E*)-2 compared to (*E*, *E*)-1 would be attributed to the sharper S_0–S_2 transition band with a higher molecular extinction coefficient that derives from the symmetrical structure. According to the DSC results and theoretical calculation, the specific heat values for bisazobenzene derivatives 1 and 2 were 108.7 kJ/mol (262 J/g) and 102.2 kJ/mol (230 J/g), respectively, which were almost twice as large as the monoazobenzene derivatives [14,19] and was significantly improved as compared to the value of the existing liquid azobenzene STFs. This study elucidated the basic characteristics of condensed-phase bisazobenzene STFs for the first time and laid an important foundation for further improving the energy density by introducing the photo-induced phase transition in the bisazobenzene derivatives system.

The integration of photo-induced solid-to-liquid phase transition and photo-induced *trans–cis* isomerization can improve the energy density of the azobenzene derivatives and reduce photothermal energy excitation temperature. If the melting-point (T_m) and the crystalline temperature (T_c) of azobenzene derivatives in *trans* and *cis* forms are above and below room temperature respectively, the photo-induced reversible solid-to-liquid phase transition can be achieved and the opportunity for low-temperature exothermic can be provided. Feng *et al.* [56] designed two azobenzene derivatives containing different alkyl chains (T-Azo and F-Azo), as shown in Figure 2.18(a). The time required for T-Azo and F-Azo to reach the PSS in the charging process under UV irradiation is 26 and 35 min, respectively. Subsequently, the time required to recover to

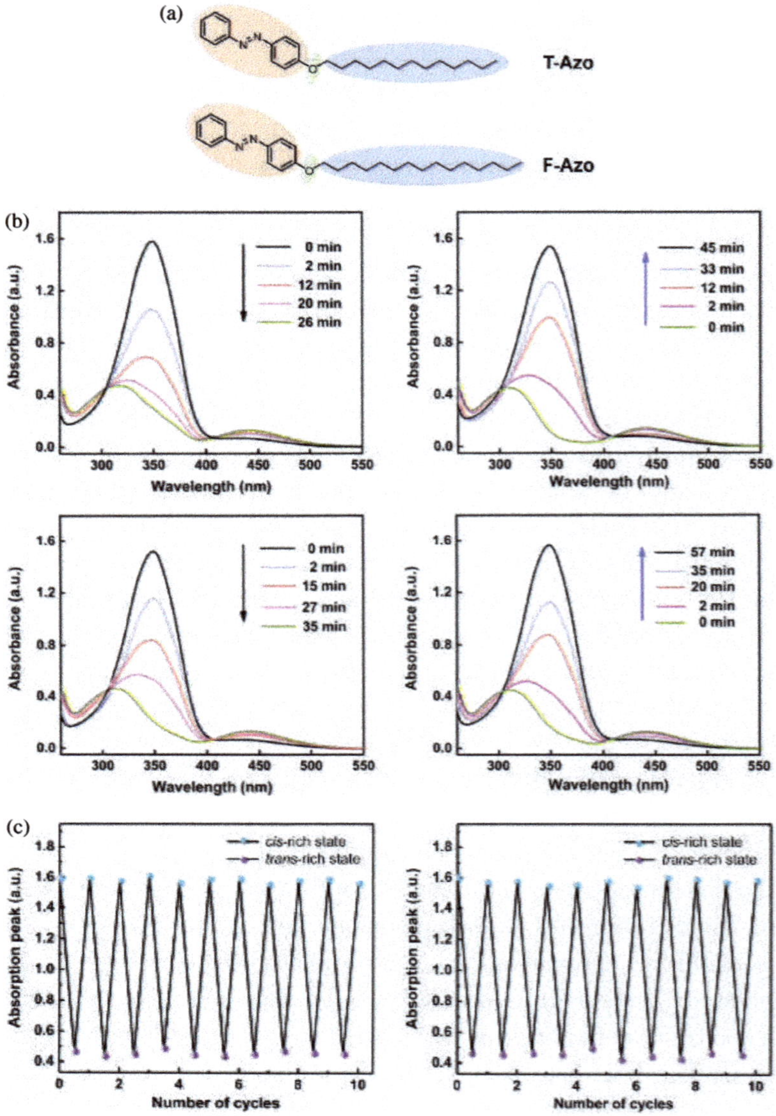

Figure 2.18. (a) Chemical structure of T-Azo and F-Azo. (b) UV–Vis absorption spectra of T-Azo and F-Azo under UV irradiation and blue light irradiation. (c) Cycling stability of T-Azo and F-Azo under alternating UV light and blue light irradiation. (d) DSC traces of charged T-Azo and F-Azo. (e) Photographs of changing with F-Azo under UV irradiation and blue light irradiation. (f) IR thermal-imaging of charged T-Azo and F-Azo. Reprinted with permission from Ref. [56]. Copyright 2020, Elsevier Ltd.

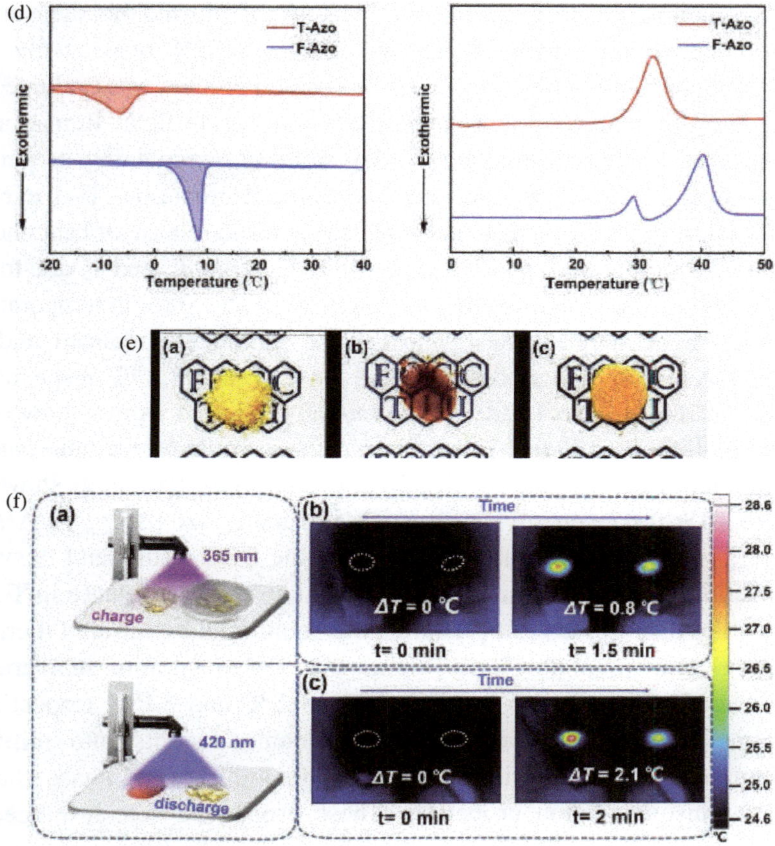

Figure 2.18. (*Continued*)

the initial stable state during the discharging process under 420 nm blue light irradiation was 45 and 57 min, respectively, as shown in Figure 2.18(b). This is because the introduction of a longer alkyl chain in F-Azo molecules leads to increased entanglement between the molecules, resulting in greater steric hindrance. This steric hindrance, in turn, hinders the structural transformation of the molecules. The longer alkyl chain effectively restricts the movement and flexibility of the molecules, impeding their ability to undergo the desired structural changes. Consequently, the presence of the longer alkyl chain in F-Azo molecules prevents or inhibits the expected structural transformation [23,57]. ^{1}H NMR results showed that the isomerization degree of T-Azo and F-Azo were 74% and 72%, respectively. The repetitive photo-induced *trans–cis* isomerization process was evaluated by subjecting the material to alternating UV and visible light. Remarkably, the material exhibited excellent cycling stability, maintaining its performance over 10 cycles without any noticeable degradation. This cycling stability was particularly crucial for the storage and release of heat energy (Figure 2.18(c)). T-Azo and F-Azo exhibited similar T_m and T_c due to the similar molecular structure, both of which were above room temperature. After UV light charging, the T_m and T_c were significantly reduced, and the T_c of T-Azo and F-Azo decreased to −6.3°C and 8.2°C, respectively (Figure 2.18(d)). T-Azo and F-Azo gradually changed from yellow solid to red liquid during the UV charging process at room temperature, and it could return to solid state by blue light (420 nm) excitation, as shown in Figure 2.18(e). The energy density of T-Azo and F-Azo ($\Delta H_{\text{storage}} = \Delta H_{\text{iso}} + \Delta H_{\text{ph-}trans}$) were 170.5 J/g and 198.4 J/g by the DSC results and theorical calculation. The temperature change caused by the heat release in T-Azo and F-Azo under blue light (420 nm) was monitored by infrared thermal-imaging camera, and it was found that the maximum temperature difference between T-Azo and F-Azo was 0.8°C and 2.1°C, respectively (Figure 2.18(f)). This study offered valuable insights into utilizing azobenzene derivatives with photo-induced solid-to-liquid phase change and photoisomerization properties. These properties were leveraged to address the challenge of achieving high heat release temperatures in solid-to-liquid PCMs. By employing these unique azobenzene derivatives, the researchers successfully overcame the limitation of high heat release temperatures commonly encountered in STFs. The findings open up new possibilities for the development of advanced materials capable of efficiently storing and releasing heat energy at more desirable and practical temperatures.

Designing and synthesizing a simple azobenzene based on the photothermal conversion molecule with high energy density and long energy storage half-life to prepare solid-state azobenzene STFs without any solvent-assistant pretreatment process is more practical and convenient for daily use. Liu *et al.* [58] designed and synthesized a novel azobenzene derivative 1-(4-(isopentyloxy)phenyl)-2.(p-tolyl)diazene (IPDAZ) with photoliquefiable ability at room temperature utilizing the melting point difference between the *trans*-isomer and the *cis*-isomer of IPDAZ, as shown in Figure 2.19(a). The photo-induced *trans–cis* isomerization process of IPDAZ was investigated by the UV–Vis absorption spectra of IPDAZ CH_2Cl_2 solution, as shown in Figure 2.19(b). It can be seen that only 5s was needed for the *trans*-isomer to reach the PSS, the photoisomerization efficiency was 95%, and the charging rate is nearly linearly proportional to the charging time. The time required for discharging under 570 nm visible light irradiation was 570 min, and the half-time was 110 min. Meanwhile, the photo-induced isomerization of IPDAZ CH_2Cl_2 solution performed good reversibility with little attenuation in five cycles as shown in Figure 2.19(c). The photothermal energy stored by IPDAZ during the charging process under UV irradiation increases with the prolongation of irradiation time. As the irradiation time increased, it was evident that the broad exothermic peaks in the temperature range of 70–110°C became more pronounced. Notably, after 150 min of UV irradiation, the exothermic peak reached its maximum intensity, measuring 181.6 J/g. This observation suggests a direct correlation between the duration of UV irradiation and the enhancement of the exothermic behavior in the material (Figure 2.19(d)). The energy storage life of charging IPDAZ in dark environment is 48 h according to the relationship between the exothermic peak of DSC curves and the irradiation time, as shown in Figure 2.19(e). The photo-induced liquidation and solidification process of IPDAZ was shown in Figure 2.19(f), and solid IPDAZ began to melt with 355 nm UV light irradiation after 5 min and completely melted after 80 min. This photo-induced liquefaction process can also be further verified by the fact that the endothermic peak of DSC curves in Figure 2.19(d) decreased with the prolongation of charging time, which is the reason for the content of *trans*-isomer increases with the prolongation of charging time, and the melting point of *cis*-isomer is low than room temperature. The energy density of IPDAZ ($\Delta H_{storage} = \Delta H_{iso} + \Delta H_{ph\text{-}trans}$) was 259.1 J/g by the DSC results and theoretical calculation. The solid-state STF device was fabricated by photo-liquation and thermal-liquation methods without any solvent assistance due to the IPDAZ exhibited excellent film-processing ability on commercial fabric cloth. The prepared composite solid STFs changed from yellow to red under UV light irradiation

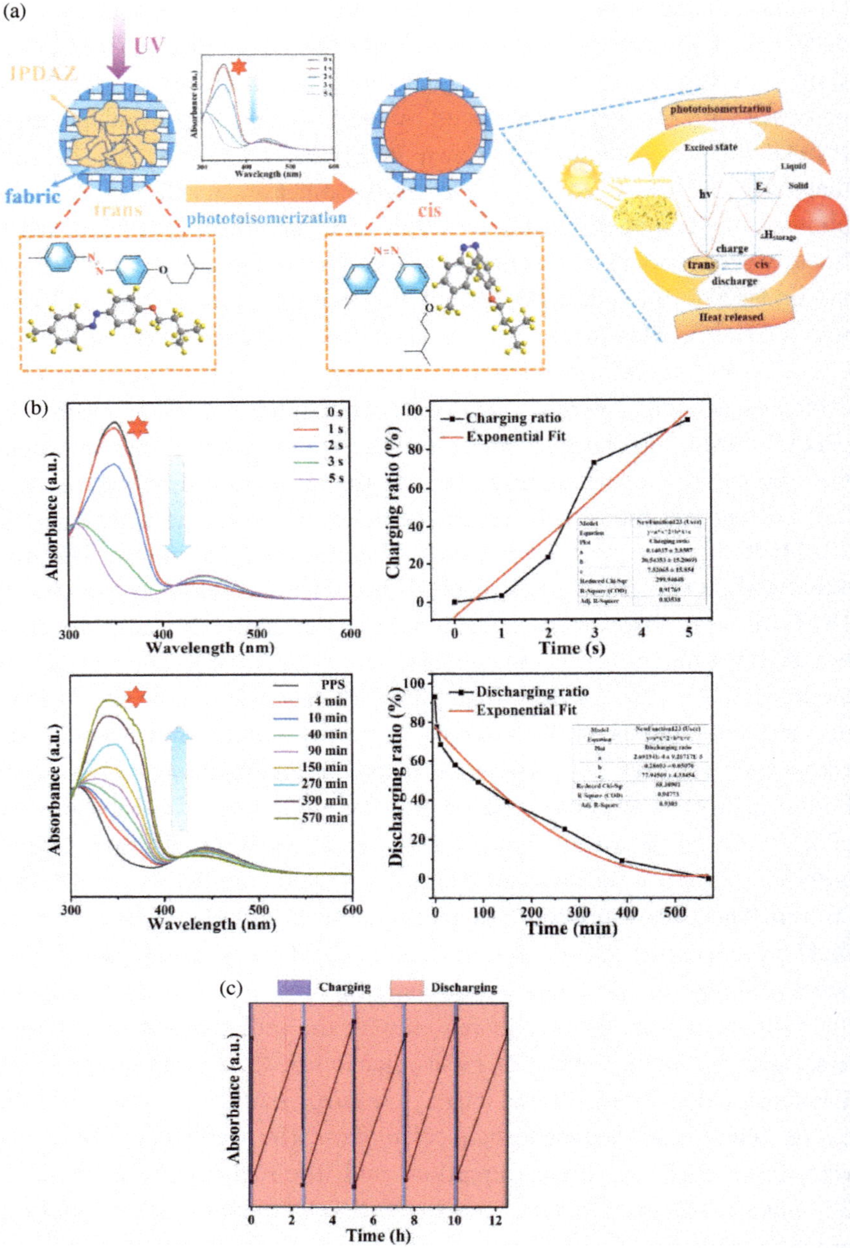
(a)
UV
IPDAZ
fabric
trans
phototoisomerization
cis
phototoisomerization
Excited state
Liquid
Solid
hv
charge
trans
cis
discharge
Heat released
(b)
Absorbance (a.u.)
Wavelength (nm)
0 s
1 s
2 s
3 s
5 s
Charging ratio
Exponential Fit
Charging ratio (%)
Time (s)
PPS
4 min
10 min
40 min
90 min
150 min
270 min
390 min
570 min
Discharging ratio
Exponential Fit
Discharging ratio (%)
Time (min)
(c)
Charging
Discharging
Absorbance (a.u.)
Time (h)

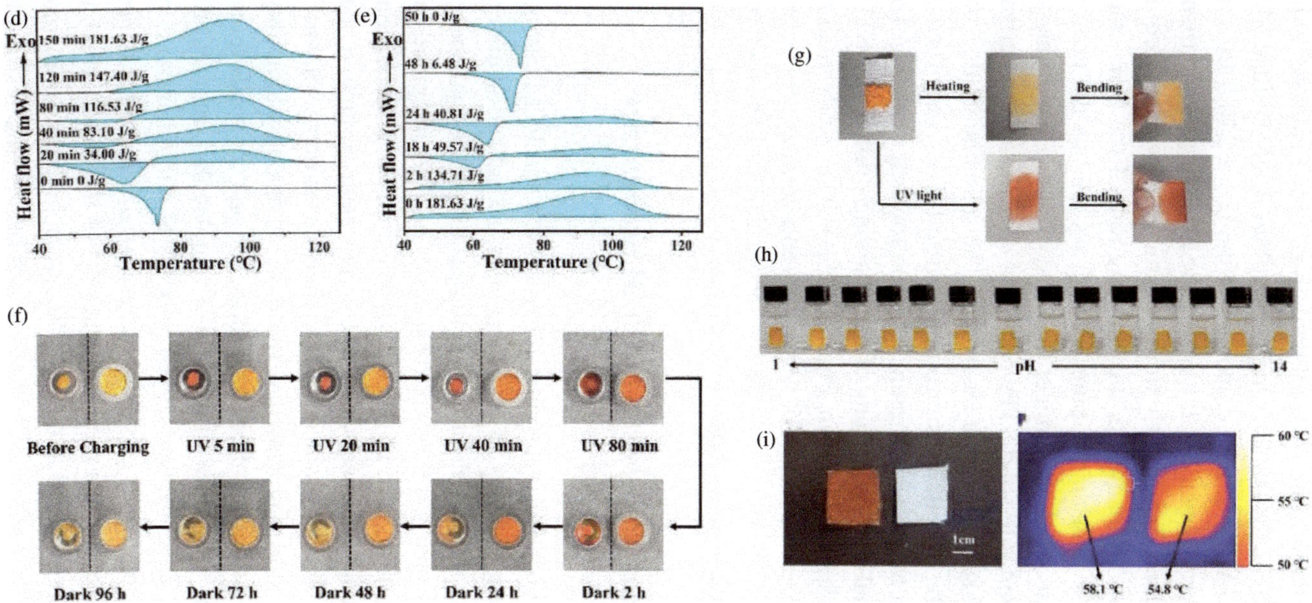

Figure 2.19. (a) The schematic diagram of the IPDAZ STF film preparation process and its possible energy-storage mechanism. (b) The UV–Vis absorption spectra of the IPDAZ STF. (c) The charging and discharging reversibility curves. (d) DSC curves of pure IPDAZ for different charging time. (e) DSC curves of charging IPDAZ for different discharging time in the dark. (f) The digital photo of photo-induced phase changing process of pure IPDAZ (left) and IPDAZ STF film (right). (g) The composite STF film fabricated by photo-induced and thermal-induced method with IPDAZ and commercial fabric. (h) The digital photo of IPDAZ STF film immersed in water solution with pH 1–14 for 24 h. (i) IR thermal-imaging of charged IPDAZ STF film and commercial fabric. Reprinted with permission from Ref. [58]. Copyright 2022, Elsevier Ltd.

(Figure 2.19(g)) and did not dissolve and precipitate in the aqueous solution of PH 1–14 (Figure 2.19(h)), showing good solution pH stability. The isomerization energy (ΔH_{iso}) of the composite solid STFs is 210.5 J/g and the corresponding energy density is 288 J/g by the DSC results and theoretical calculation, which significantly improved the energy density compared with pure IPDAZ. Meanwhile, the charging rate of the composite solid STFs was higher than that of pure IPDAZ. As well as the energy storage life (96 h) in the dark was longer than that of pure IPDAZ due to the H-bond in the composite solid STFs restricted *cis*-to-*trans* isomerization and improved the stability of the *cis*-isomer. The solar energy thermal storage and controllable release of the composite solid STFs were further verified through the temperature difference caused for heat releasing that was excited by thermal-induced means of the charging composites STFs and commercial fabric, as shown in Figure 2.19(i). This study mainly designed and synthesized a new PLAZ derivatives and fabricated the composite solid STFs based on Yu's [53] research, which further expanded the types of STFs prepared from azobenzene derivatives.

2.2.4 *Solar thermal fuels based on PCM/azobenzene derivative composites*

The integration of azobenzene molecular photo-switch and PCM can not only improve the energy density of STFs, but also the activation energy barrier introduced, providing stability to the phase storing thermal energy and trigger-ability for energy release. Grossman *et al.* [59]. prepared the composite consisting of organic PCM (tridecanoic acid) and azobenzene derivative, which is functionalized with a tridecanoic ester group to render high miscibility with the PCM molecules, the chemical structure of which is shown in Figure 2.20(a). The lamellar structure is formed by long alkyl chain packing, the π–π stacking of benzene rings, and van der waals forces between alkyl chains and ester linkers affect molecular interactions. Figures 2.20(b) and 2.20(c) showed a cycle process of photothermal conversion, storage, and controlled release of the prepared composites. Upon heating the composite above the melting point (T_m) of the PCM, which is 42°C in this case, a transformation occurs. The composite transitions into a mixture consisting of molten PCM and crystalline aggregates of the

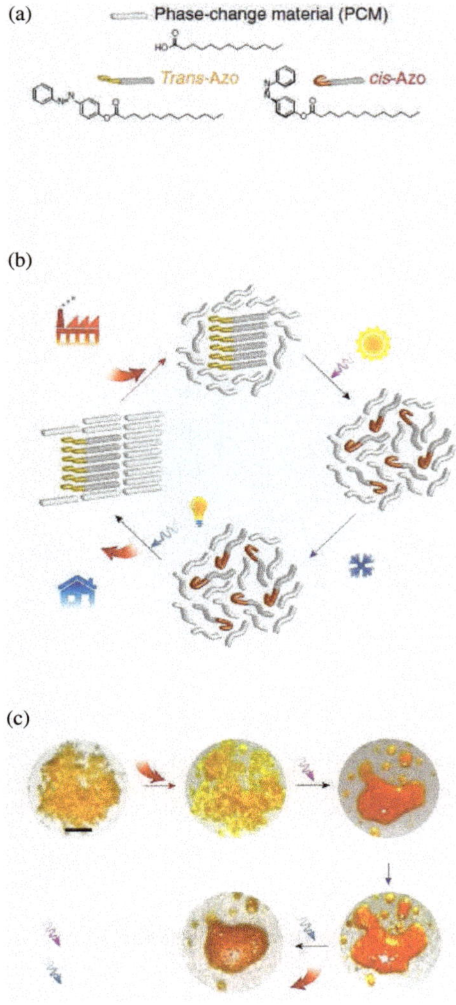

Figure 2.20. (a) The chemical structure of azobenzene derivative. (b) Schematic of photothermal conversion, storage and release. (c) Photographs of the PCM composites during photothermal conversion, storage and release process. (d) DSC curves of charged and uncharged composites obtained while cooling from 70°C at a rate of 5°C/min, illustrating different crystallization points (T_1 and T_2) and the gap (ΔT_c). (e) T_c and ΔT_c as well as ΔH with varying additive ratios. (f) Stability of heat storage in a charged liquid composite in the dark at 36°C. (g) Schematic energy diagrams of azobenzene derivative PCM composite without thermal insulation and conventional PCM with thermal insulation. Reprinted with permission from Ref. [59]. Copyright 2017, the authors.

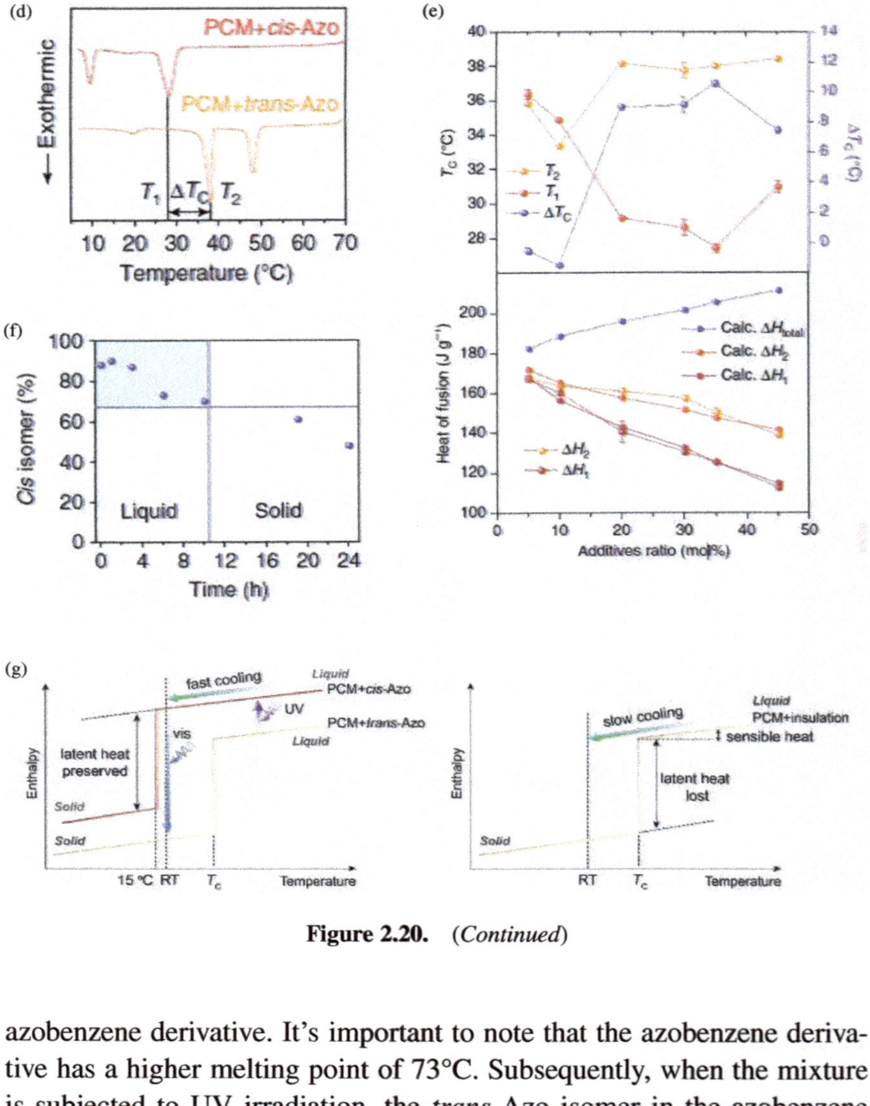

Figure 2.20. (*Continued*)

azobenzene derivative. It's important to note that the azobenzene derivative has a higher melting point of 73°C. Subsequently, when the mixture is subjected to UV irradiation, the *trans*-Azo isomer in the azobenzene derivative undergoes a photochemical reaction, converting into the *cis*-Azo isomer. The resulting *cis*-Azo isomer adopts a twisted conformation, allowing it to disperse well within the liquid PCM. By cooling the composite to a temperature below the original transition temperature (T_c), which is 38°C in this scenario, the liquid state of the composite can be conserved. Importantly, throughout this cooling process, the latent thermal energy stored within the composite is fully maintained. Then under the visible light irradiation, the *cis*-Azo is transformed into *trans*-Azo and allowed the PCM composites to crystallize and release the stored latent heat, recovering the original state of the composite. The crystallization temperature T_c of the uncharged and charged composites was measured by DSC, as shown in Figure 2.20(d). The exothermic peaks of charged composite appear at 28°C and 9°C, corresponding to the crystallization of PCM and *cis*-Azo respectively, and the exothermic peaks of the uncharged composite appear at 48°C and 38°C, corresponding to the crystallization of *trans*-Azo and PCM respectively. In the PCM composites, the presence of the liquid PCM molecules affects the crystallization temperature of *trans*-Azo. The crystallization temperature of *trans*-Azo, which is measured as 60°C in its pristine form, is notably reduced when it is incorporated into the PCM composites. This reduction occurs due to the solvation effect caused by the liquid PCM molecules at temperatures above the crystallization point of the PCM. It is worth noting that the extent of changes in phase transition temperatures can be adjusted by modifying the doping level in the composite and controlling the cooling rate. By altering the amount of azobenzene derivative incorporated into the composite (doping level), or by changing the rate at which the composite is cooled, the resulting composite material can exhibit different phase transition temperatures. These adjustments provide a means to fine-tune the properties of the composite material to suit specific application requirements. The T_c changes of the composites with different doping levels are shown in Figure 2.20(e). The heat released (ΔH_{total}) from liquid-state *cis*-Azo/PCM to solid-state *trans*-Azo/PCM included ΔH_{PCM}, $\Delta H_{trans\text{-Azo}}$ and ΔH_{iso}. If the pure azobenzene is charged, it must be heated to above the melting point of *trans*-Azo ($T_m = 73$°C), but *cis*–*trans* isomerization transition will occur at this temperature, preventing the control over thermal storage. Also, in the solid state, UV illumination of azobenzene without heating proves ineffective due to steric confinement within the crystalline lattice caused by the aromatic groups. As a result, the design of PCM and azobenzene composites presents a unique challenge in achieving thermal storage at temperatures higher than the relatively low melting point (T_m) of the PCM. The chemical structure of azobenzene derivatives also affected T_c of PCM. The decyl-oxy and *tert*-butyl groups were introduced to the pristine azobenzene derivative to form compounds 2 and 3, respectively. Compound 2 exhibited the highest T_c of 68°C, indicating strong binding forces between the azobenzene derivative through π–π stacking of benzene rings and side packing between alkyl chains. The *tert*-butyl group of compound 3 significantly diminishes the π–π stacking of benzene rings, which helps to prevent the formation of nucleating agents in the composites (nucleating agents help to crystallize PCM and reduce the degree of supercooling) and effectively disrupt the PCM alignment. The stability of thermal energy storage in the charged PCM composites was measured by ^{1}H NMR analysis of the *cis* fraction and by monitoring solidification of the composite over a period of 24 h in the dark at the temperature of 36°C (below the T_m). The half-life of *cis* in such conditions was about 24 h and the liquid phase was conserved for at least 10 h. Finally, a unique approach that uses the photo energy as a way to control the phase transition properties of traditional thermal storage PCM was demonstrated in this study. The photothermal conversion efficiency of the PCM composite containing 30% of azobenzene derivative is about 5–6%, that is, UV photo energy of 3.6 kJ per 1 gram of the composite was required for photoinduced phase fixation, while the total heat storage in 1 gram of the composite is about 0.2 kJ, which is higher than most conventional STFs for photothermal conversion efficiency. For example, the energy storage capacity of pristine azobenzene STF is 0.23 kJ/g [60], while the input UV photon energy density is 18 kJ/g (1.3% efficiency). For traditional PCM, thermal insulation is used to reduce the cooling rate to maintain latent thermal storage, but this method cannot stop the spontaneous heat transfer between thermally charged PCM and the cooling surroundings (Figure 2.20(g)). In contrast, photo switched PCM systems can hold on to the latent thermal until it is released even in the absence of insulation by changing the intrinsic properties (crystalline temperature) of the PCM, while triggering the latent thermal release through visible light.

Latent heat storage, as compared to sensible heat storage, offers higher storage capacity and efficiency. It allows for heat release at a specific temperature, determined by the phase transition temperature of the PCM. However, a drawback of latent heat storage in PCM is its dependence solely on the ambient temperature for crystallization of the PCM. This reliance on ambient temperature limits the control over the release of stored energy. In other words, the phase transition of the PCM occurs based on the surrounding temperature, which means the release of stored energy is not easily controllable or adjustable. This lack of control over the release of stored energy poses a challenge in applications where precise and targeted heat release is required. Overcoming this limitation and developing methods to regulate the release of energy in latent heat storage systems is an ongoing area of research and development. Grossman *et al.* [61]. explored the optical regulation of thermal energy storage in a variety of aliphatic PCMs. Notably, the research demonstrated that T_c of each PCM could be effectively reduced to varying degrees by manipulating their chain lengths and functional groups. The results of the investigation revealed specific parameters and conditions that could be controlled to influence the nucleation and crystallization processes of these thermal storage materials. This finding is particularly significant as it allows for fine-tuning the properties of aliphatic PCMs, making them more adaptable and efficient for thermal energy storage applications. Also the strategy to make the T_c of organic PCMs optically triggerable is described in Figure 2.21(a). The structure of aliphatic PCM is shown in Figure 2.21(b), and the melting point T_m of PCM generally increased as the chain become longer. The maximum supercooling (ΔT_c = 10.5°C) of PCM composite with polar C_{13}-acid and azobenzene derivative is found at 35% Azo molar fraction, while the maximum supercooling (ΔT_c = 3.5°C) of PCM composite with nonpolar C_{28}-alkane and azobenzene derivative occurs at 50–60% Azo molar fraction, as shown in Figure 2.21(c). The ΔT_c of these diverse PCMs increased first and then decreased with an increase in Azo content, which was caused by the decrease of supercooling degree of *trans*-Azo nucleating agent formed via incomplete charging with Azo content increasing. The ΔT_c of these diverse PCMs increased as alkyl chain become longer (Figure 2.21(d)), which is the reason for T_m of PCM increasing with the increase of the length of the alkyl chain and heat storage must be carried out above the T_m. Azo will undergo isomerization competition as the temperature increases, resulting in the decrease in the content of *cis*-Azo and *trans*-Azo as nucleating seeds reduced ΔT_c of PCM composite as shown in Figure 2.21(e). The degree of Azo isomerization varies in the range of 28–90% (Figure 2.21(f)) when different PCMs were charged under UV

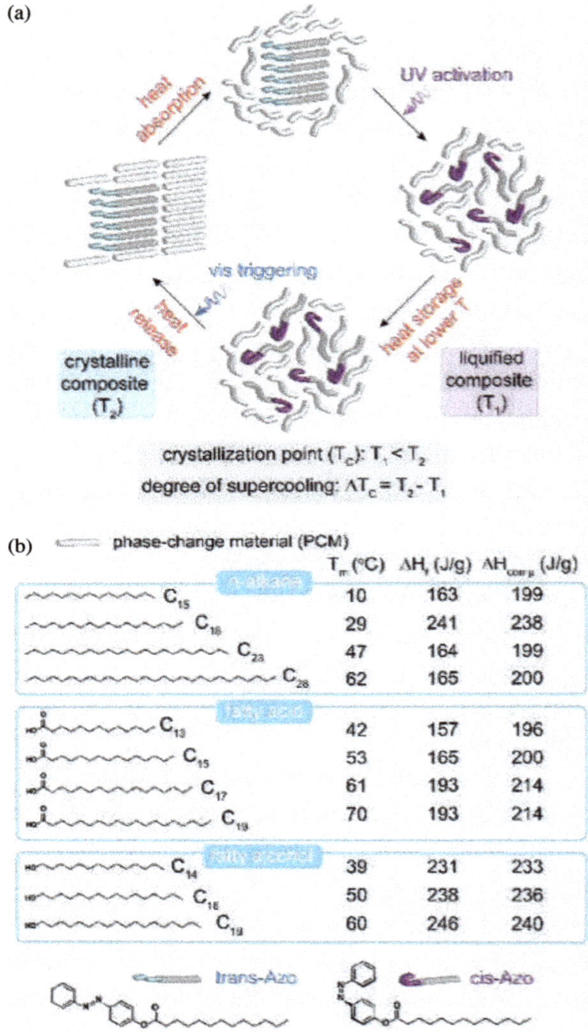

Figure 2.21. (a) Schematic illustration of heat storage and release cycle in the PCM composite. (b) Chemical structure of diverse PCMs. (c) ΔT_c measured for both C_{13}-acid and C_{28}-alkane composites while varying the molar additive ratio in composite. (d) ΔT_c measured for each composite and plotted against the chain length of each PCM. (e) ΔT_c plotted against the melting point of each PCM. (f) Charging amount of azobenzene in PCMs and the corresponding ΔT_c value. (g) ΔT_c plotted against charging amount. Reprinted with permission from Ref. [61]. Copyright 2018, The Royal Society of Chemistry.

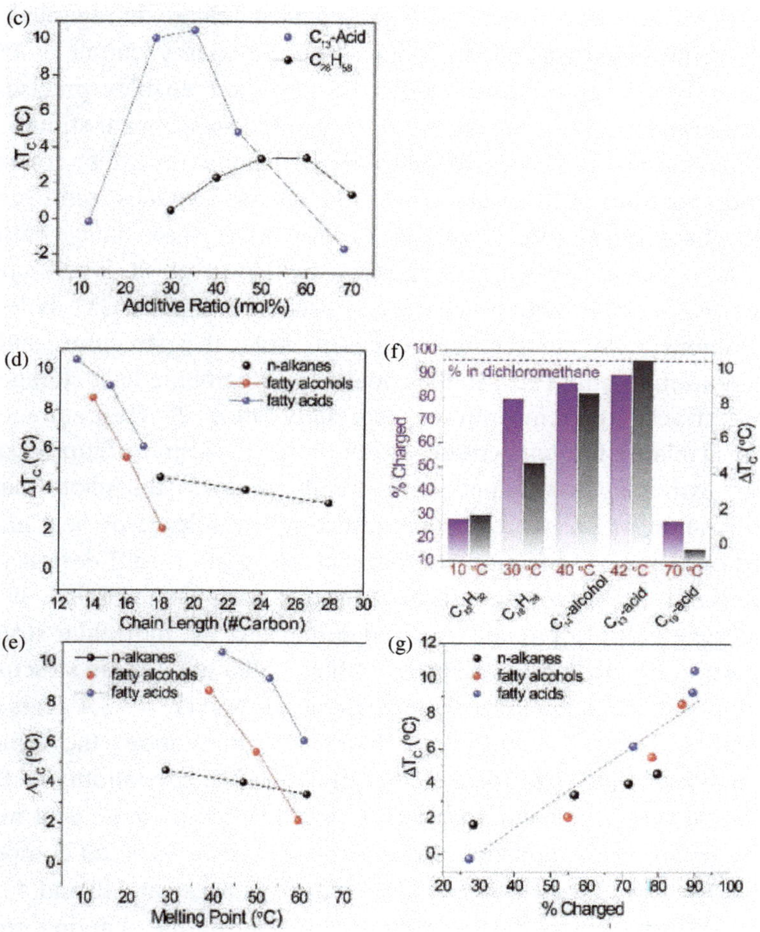

Figure 2.21. (*Continued*)

irradiation for the same time, and the degree of Azo isomerization also affected ΔT_c of PCM (Figure 2.21(g)). It can be seen that the comparison between non-polar PCMs and polar PCMs reveals that non-polar PCMs generally exhibit lower levels of supercooling compared to their polar counterparts. This can be attributed to the weaker interaction between the non-polar PCMs and the *cis*-Azo isomer. This study showed that PCM selection is critical for heat storage in a suitable temperature range.

The utilization of optically triggered heat release is regarded as a highly promising approach for controlling the energy output of PCMs across a broad temperature range. This method enables precise and selective activation of heat release from PCMs using optical stimuli. The phase transition process of PCM at specific temperatures can be controlled by incorporating azobenzene molecular photo-switches into organic PCM, which provides the possibility to release the phase change enthalpy and photo-thermal energy simultaneously. However, it is difficult to induce the phase transition behavior of azobenzene and PCM molecules synchronously due to the difference in molecular structure, and the limitation of organic PCM is their inability to combine heat storage and release at very low temperatures, especially below T_c. Heat storage and release at relative low temperature by improving the degree of supercooling, but an extremely low temperature usually impedes the photo-induced isomerization of azobenzene, thus affecting the release of heat energy. Based on the above-mentioned properties, Feng *et al.* [62] designed and synthesized an alkyl-grafted azobenzene and incorporated it into tetradecane (Ted) to prepare Azo/Ted composite, the molecular structure is shown in Figure 2.22(a). Because of the similar alkyl chain structure of Azo and Ted, the crystallization temperature of Azo ($T_c = -3.4$°C) is close to that of Ted ($T_c = 0.11$°C). Under UV irradiation, the structure conversion of *trans*-Azo to *cis*-Azo will reduce the crystallinity of PCM, that is, increasing the supercooling, so that low temperature heat release can be achieved by adjusting the content of Azo in Azo/Ted composite. Azo showed π–π* transition and *n*–π* transition peaks at 349 and 437 nm of UV–Vis absorption spectra, respectively. During the charging process under UV irradiation, the peak intensity of the π–π* transition showed a decrease, while the peak of the *n*–π* transition exhibited a slight increase. This trend continued until reaching the PSS, indicating the transformation and equilibrium between different electronic states of the azobenzene derivatives, as shown in Figure 2.22(b). The degree of isomerization of Azo reached 81% under UV irradiation for 30 min, and it took 156 h for *cis*-Azo to completely transform *trans*-Azo in the dark, which could be accelerated by blue light (420 nm) irradiation. The melting point (T_m = 47.7°C) and crystallization point ($T_c = -3.4$°C) of *cis*-Azo decreased significantly compared with that of *trans*-Azo ($T_m = 80.5$°C, $T_c = 63.8$°C).

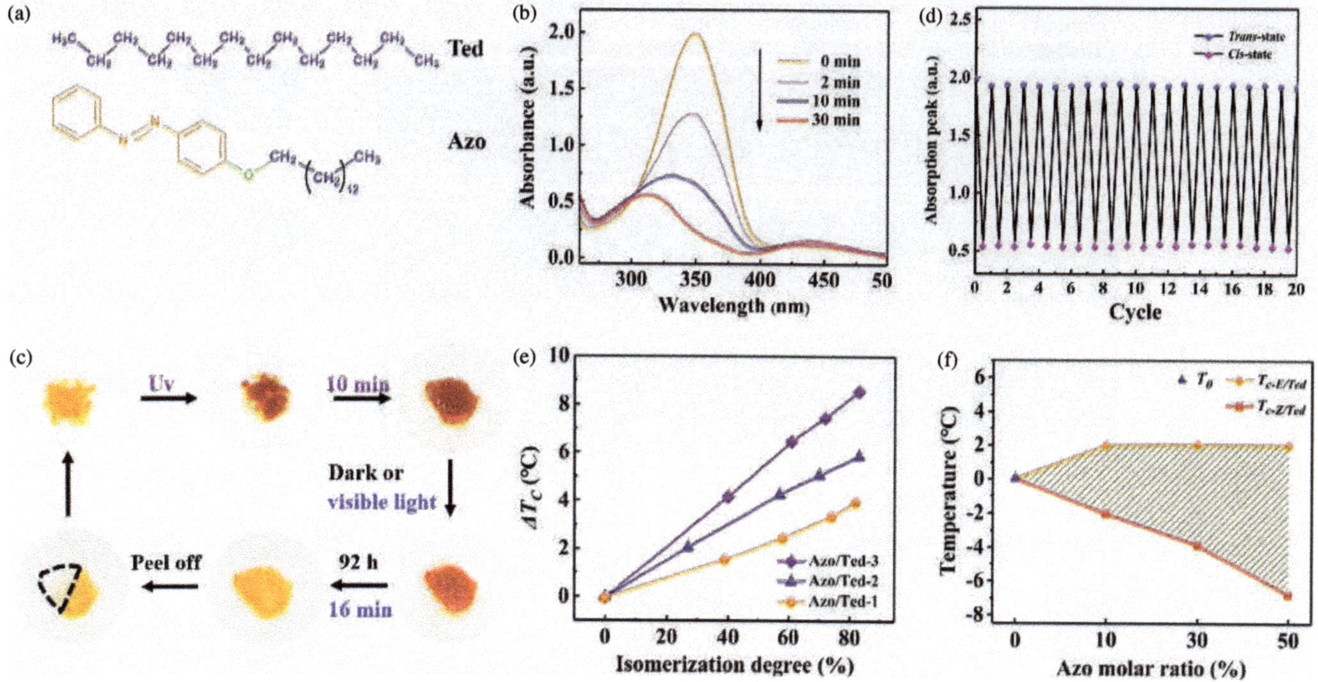

Figure 2.22. (a) Chemical structure of Azo and Ted. (b) The UV–Vis absorption spectra of Azo. (c) Photographs of Azo/Ted during UV light illumination/dark/blue-light-induced control heat release and storage. (d) Cycling stability of Azo/Ted induced by alternatively irradiating UV light and blue light. (e) ΔT_c of the Azo/Ted composites under UV irradiation for different times. (f) $T_{c\text{-}E/Ted}$ and $T_{c\text{-}z/Ted}$. (g) Annular device for distributed energy utilization realize the cycle of latent heat storage and release via optical trigger. Reprinted with permission from Ref. [62]. Copyright 2020, Wiley-VCH GmbH.

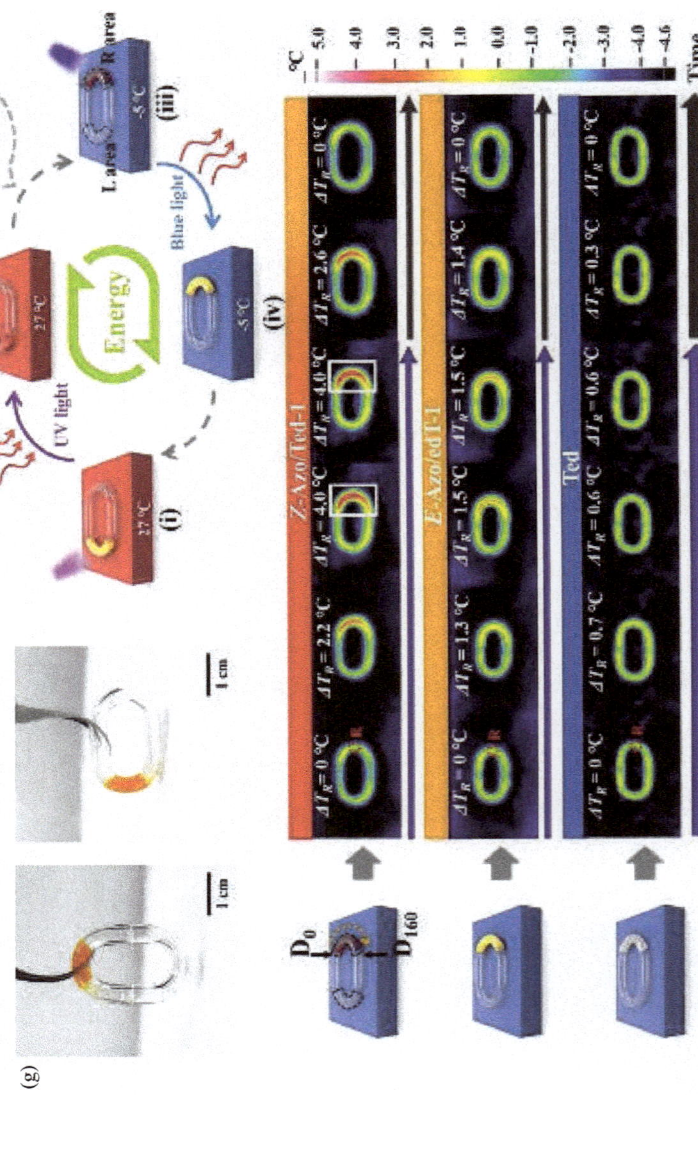

Figure 2.22. *(Continued)*

Azo/Ted composite underwent a reversible solid–liquid phase transition through photo-induced isomerization, transforming from a yellow solid power to a red liquid under UV irradiation, and then from a red liquid to a yellow solid under visible light or in the dark, as shown in Figure 2.22(c). The time required for Azo/Ted to reach the PSS increased with the increasing of Azo content owing to weakening of the π–π stacking effect of adjacent azobenzene molecules when *trans*-Azo successfully intercalated into the lamellar alignment of Ted, which promotes the isomerization transition. Azo/Ted exhibited a high degree of isomerization (83%) and excellent cyclic stability (Figure 2.22(d)). *Trans*-Azo/Ted exhibited two exothermic peaks in the DSC curves during the crystallization process, corresponding to the phase change enthalpy of Ted at 2°C and *trans*-Azo at 30–65°C, respectively. The T_c of *trans*-Azo decreased with an increase in Azo content. There is only an exothermic peak in the DSC curve of *cis*-Azo/Ted during the crystalline process due to T_c of *cis*-Azo overlapped with T_c of Ted to become T_c of *cis*-Azo/Ted, which could be adjusted within the range of −6.71–0.11°C by changing the content of Azo. The T_c of *cis*-Azo/Ted can realize the simultaneous release of phase change enthalpy and photo-thermal energy as well as the storage and release of latent heat at T_c of Ted. The supercooling degree of Azo/Ted ($\Delta T_c = T_{c\text{-}E/\mathrm{Ted}} - T_{c\text{-}z/Ted}$) increased with increasing of Azo content and degree of isomerization as shown in Figure 2.22(e), of which the ability of photo-controllable supercooling allows for optically triggered heat release across a broad temperature range (−6.71–2.13°C) as shown in Figure 2.22(f). Azo/Ted was placed into annular device for distributed energy utilization to realize the cycle of latent heat storage and release via optically triggered, as shown in Figure 2.22(g). The high fluidity of the charged liquid enables energy recycling and distribution throughout the system. This fluid nature also contributes to the recyclability and controllability of heat storage and release, which are crucial for efficient utilization of distributed energy resources.

Traditional PCMs face limitations in achieving long-term storage and controllable release of solar thermal energy. This is primarily due to the reliance on ambient temperature for the crystallization process, which seriously limits the large-scale application of PCM in the field of solar thermal storage. It is an efficient strategy of enhancing the supercooling degree of PCM to realize solar thermal long-term storage [63,64]. Feng *et al.* [65]. designed and synthesized a long-chain alkyl azobenzene with low cost tetradecanoate, and integrated it with tetradecyl alcohol (TA) to prepare PCM composite for solar thermal storage and controlled release (Figure 2.23(a)). The long-chain alkyl azobenzene showed π–π* transition and n–π* transition peaks at 322 and 435 nm of UV–Vis absorption spectra, respectively. The peak intensity of π–π* transition decreased while the n–π* transition peak slightly increased until reaching PSS during charging process under UV irradiation. Inversely, the peak intensity of π–π* transition increased and the n–π* transition peak slightly decreased under 450 nm blue light irradiation or in the dark, as shown in Figure 2.23(b). The half-life of long-chain alkyl azobenzene was 118 h, and the proportion of *cis*-isomers achieved 79% when reaching PSS (Figure 2.23(c)), indicating that the long-chain alkyl azobenzene remained an excellent photo-induced *trans–cis* isomerization performance. The long-chain alkyl azobenzene completed 50 cycles without photodegradation during repeated charging and discharging, demonstrating outstanding cycling stability, as shown in Figure 2.23(d). The isomerization degree of PCM composites with different contents of long-chain alkyl azobenzene was consistent with the long-chain alkyl azobenzene, indicating that the long-chain alkyl azobenzene in PCM composite still remains an outstanding photo-isomerization performance. All the uncharged PCM composites showed similar phase change behavior, and with increasing long-chain alkyl azobenzene content, there are negligible changes in the PCM section, while the crystallization peak of *trans*-Azo shifts towards a higher temperature, as shown in Figure 2.23(e). Due to the enhanced intermolecular interactions between TA and long-chain alkyl azobenzene, the T_c of the PCM composite decreases after charging. This phenomenon introduces supercooling, allowing the thermal energy to be retained at cooler temperatures compared to the original PCM [59]. The ΔT_c between charged and discharged PCM composites increased with the increasing long-chain alkyl azobenzene content due to the polar *cis*-Azo with steric bulk but decreased after 55% as shown in Figure 2.23(f). This is because the incomplete charging of *trans*-Azo and the residual *trans*-Azo promoted crystallization and reduced the supercooling degree by further increasing long-chain alkyl azobenzene content. The charged PCM composite can remain liquid for 10 h at 33°C (lower

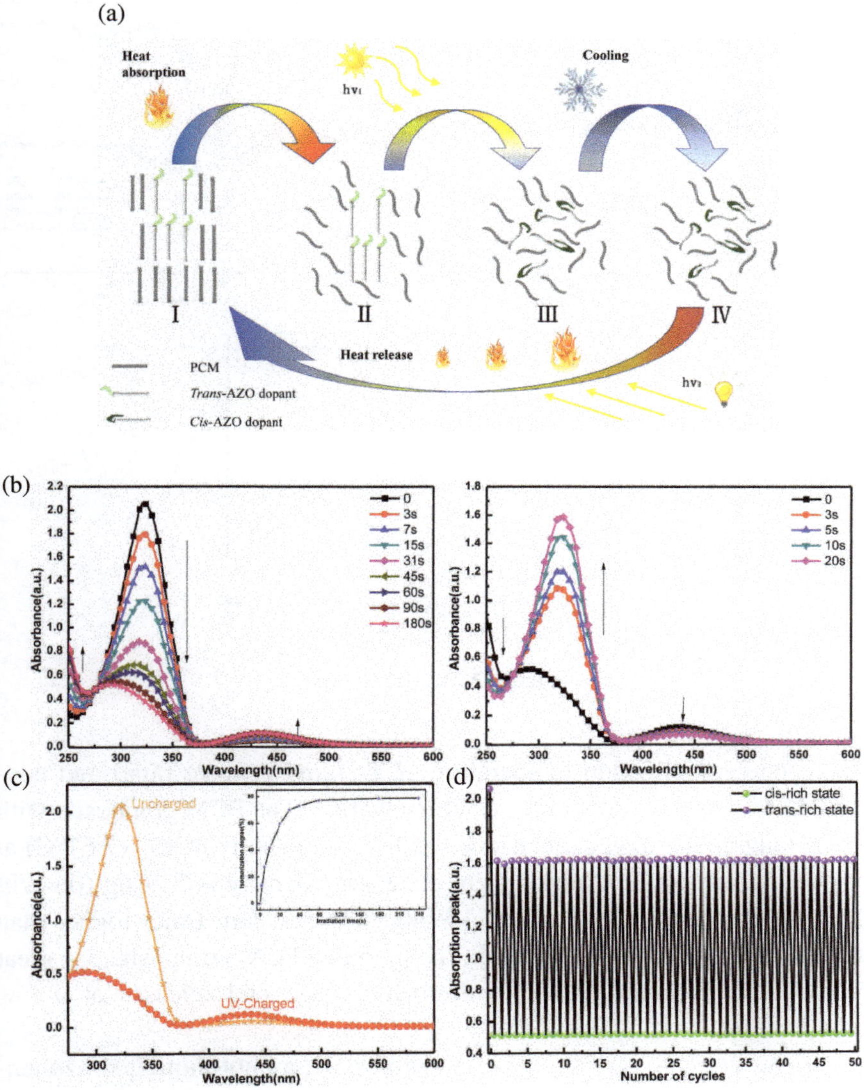

Figure 2.23. (a) Solar energy storage and controllable release of light-actuated PCM composite by introducing long-chain alkyl azobenzene. (b) Time-dependent UV–Vis spectra of long-chain alkyl azobenzene in DCM solution at 25°C. (c) Absorption spectrum before and after UV charging with the isomerization degree show in the inset. (d) Cycling stability by alternative charging and discharging process for 50 cycles. (e) DSC curves of uncharged and charged PCM composite. (f) The changes of ΔT_c with the increasing long-chain alkyl azobenzene content in TA. Reprinted with permission from Ref. [65]. Copyright 2021, Elsevier B.V.

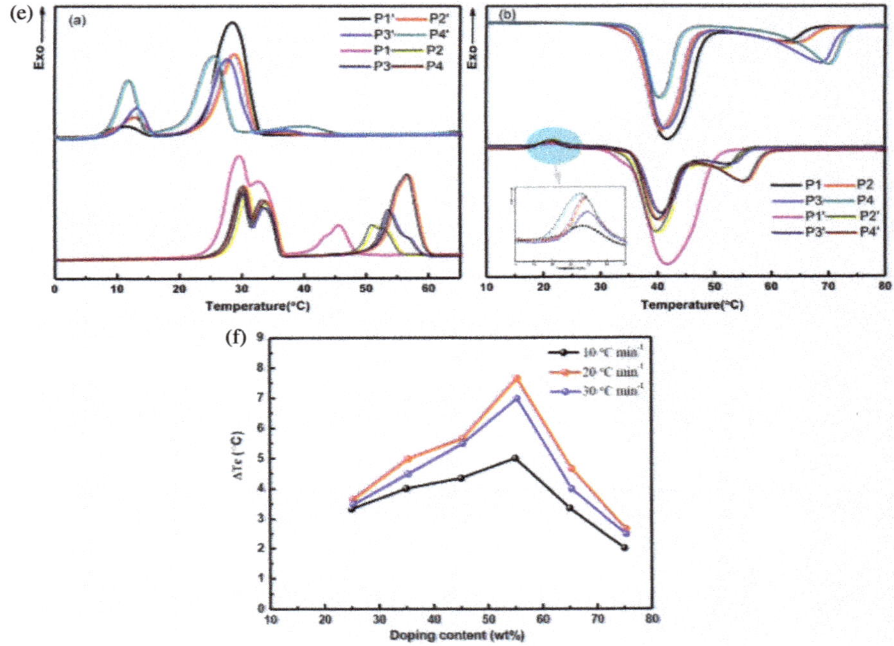

Figure 2.23. (*Continued*)

than the crystallization temperature of TA) and achieve rapid and controlled heat release under blue light irradiation. The PCM composite with 55% long-chain alkyl azobenzene reached the maximum ΔT_c (7.67°C) at a cooling rate of 20°C/min, where the energy density is 239 J/g (66 Wh/kg) and the power density is 594 W/kg, which is four times higher than that of pure TA (114 W/kg) [66]. The prepared PCM composite has great application potential for solar thermal storage and releases at a low temperature.

Feng *et al.* [67] designed and synthesized a photoliquefied azobenzene derivative (PAzo) and aluminum nitride aerogel (AlNA) to prepare phase change azobenzene/ aluminum nitride aerogel composite (PAzo/AlNA) as shown in Figure 2.24(a). The interlayer porous structures of the aerogel provide suitable light transmittance and ensure good interface compatibility with azobenzene, and the SEM images of the surface and cross-section of AlNA and PAzo/AlNA are shown in Figure 2.24(b).

Figure 2.24. (a) Schematic illustration of energy storage and release of PAzo/AlNA. (b) SEM images of surface (1,4) and cross-section (2,3,5,6) of AlNA (1–3) and PAzo/AlNA (4–6). (c) DCS curves of charged PAzo and PAzo/AlNA. (d) Morphology of charged and discharged PAzo and PAzo/AlNA at 25°C. (e) IR thermal imaging of charged and uncharged PAzo/AlNA under blue light at 25°C. Reprinted with permission from Ref. [67]. Copyright 2020, Elsevier Ltd.

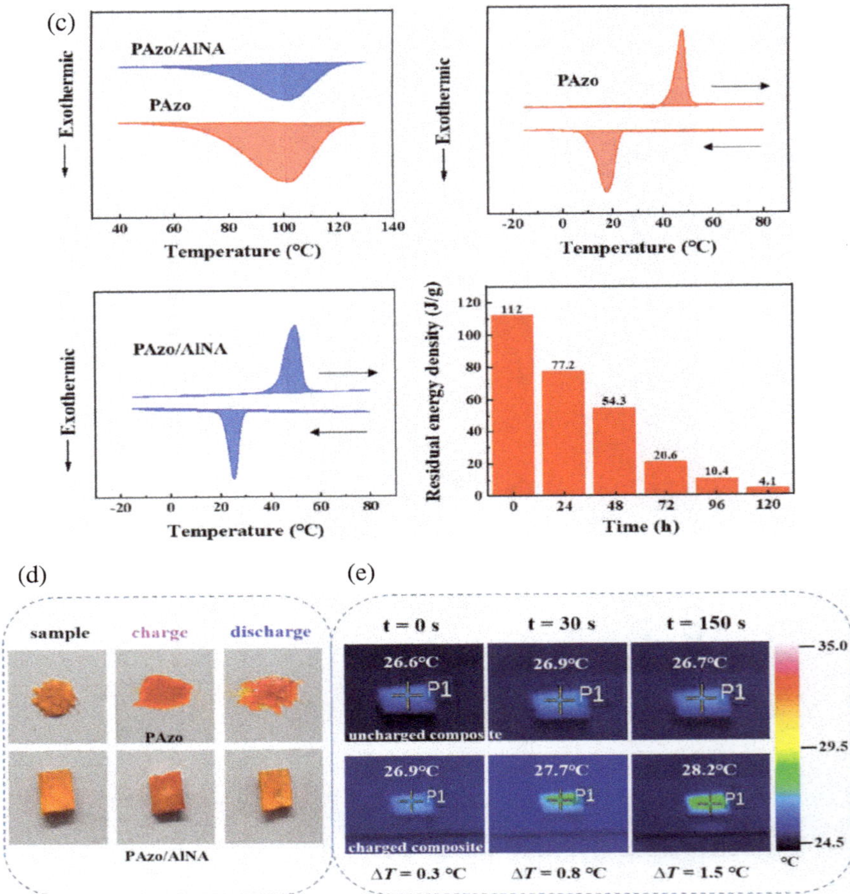

Figure 2.24. (*Continued*)

PAzo showed π–π* transition and n–π* transition peaks at 350 and 450 nm of UV–Vis absorption spectra, respectively. After 25 min of UV irradiation, PAzo/AlNA reached a metastable state, which took longer than the irradiation time of PAzo (19 min) due to the partial weakening of UV light transmittance. The time required for PAzo/AlNA to return to stable *cis*-isomer under blue light or in the dark is shorter than that of PAzo, because the presence of a porous lamellar structure in the composite facilitates the dispersion of azobenzene molecules and reduces their interaction with each other. The degree of isomerization of PAzo can be as high as 81%, which is attributed to the presence of a nonlinear alkyl chain, which disrupts the π–π stacking and introduces steric hindrance to the PAzo molecule. Additionally, the *cis*-isomer of PAzo lowers the melting temperature of the mixture and creates a larger free volume, promoting subsequent isomerization. This positive feedback loop enhances the overall performance of the system. The ΔH_{iso} of PAzo and PAzo/AlNA are 112 J/g and 65 J/g, ΔH_{phas} are 69 J/g and 41 J/g, $\Delta H_{total} = \Delta H_{iso} + \Delta H_{phas}$ are 181 J/g and 106 J/g, respectively (Figure 2.24(c)). The T_m and T_c of PAzo were 47°C and 18°C respectively, which is lower than T_m (50°C) and T_c (25°C) of PAzo/AlNA due to hydrogen bond and van der Waals interaction in aerogel network. At room temperature, PAzo turned into a red liquid with good fluidity under UV (365 nm) irradiation, and then returned to the original yellow crystalline state under blue light (420 nm), while PAzo/AlNA always maintained a shape-stabilized in the process, as shown in Figure 2.24(d). The heat released by charged PAzo/AlNA when UV irradiation for 150 s, ΔT achieved the maximum value of 1.5°C as shown in Figure 2.24(e). These results indicated PAzo/AlNA can achieve light-controlled rapid heat release at room temperature, which is expected to be used for solid-state self-heating devices in the future.

Azobenzene PCMs are distinct from traditional PCMs due to the occurrence of a photo-induced isomerization process accompanying the phase transition. This process allows for the simultaneous storage of latent heat and photothermal energy, which can be released within a specific temperature range. The crystallinity and crystallization temperature of the *trans* and *cis* isomers of azobenzene respectively determine the heat output capacity and temperature range. However, due to the variations in crystallinity and isomerization capability among different azobenzene molecules, the combination of phase-transition enthalpy and photothermal energy release at low temperatures remains challenging. Thus, optimizing the thermal release and energy transfer of azobenzene molecules through chemical means still poses significant challenges. Feng *et al.* [68] designed and synthesized two types of asymmetric alkyl-linked grafted azobenzenes, namely A-Azo-g and B-Azo-g, where the length of the alkyl chain was increased from 6 carbon atoms to 10 carbon atoms. The specific molecular structures and photoinduced isomerization transitions are illustrated in Figure 2.25(a). Figure 2.25(b) showed the UV–Vis absorption spectra of A-Azo-g and B-Azo-g, demonstrating alternating π–π* and n–π* transitions under alternate irradiation with ultraviolet and blue light (420 nm). As the alkyl chain length increases, the degree and rate of isomerization for all Azo-g derivatives decrease (Figure 2.25(c)), indicating that the spatial steric hindrance effect of the alkyl chain limits the structural transition during molecular self-assembly via van der Waals forces to some extent. Additionally, the introduction of methyl groups in B-Azo-g increases the free volume, making the photoinduced isomerization transition easier compared to B-Azo-g. The *trans*-isomers of Azo-g exhibit stronger crystallization ability and a higher crystallization temperature T_c compared to the *cis*-isomers. However, Z-A6-Azo-g and Z-B7–10Azo-g cannot crystallize at low temperatures (T_g ranges from −78°C to −73°C). This indicates that the methyl substituent further reduces the crystallization ability of Azo-g, which is further supported by the fact that E-A-Azo-g has a smaller supercooling degree compared to E-B-Azo-g (Figure 2.25(d)). This weaker crystallization ability of E-Azo-g allows for the synchronous release of crystallization enthalpy ($\Delta H_{c)}$ and photothermal energy (ΔH_p) at low temperatures through the isomerization-induced phase-transition process. To elucidate the three different exothermic modes, the authors categorized Azo-g into three classes: (1) A7–10-Azo-g: both *trans* and *cis* isomers can crystallize at specific temperatures; (2) A6-Azo-g and B7–10-Azo-g: the *trans* isomer can crystallize, while the *cis*-isomer remains in an amorphous state with an extremely low temperature (T_g); (3) B6-Azo-g: both *trans* and *cis* isomers are in an amorphous state. Correspondingly, there are three exothermic modes: (1) A7–10-Azo-g can release $\Delta H_{E\text{–}c}$, $\Delta H_{Z\text{–}c}$, and ΔH_p by controlling temperature and isomerization conditions; (2) A6-Azo-g and B7–10-Azo-g release $\Delta H_{E\text{–}c}$ and ΔH_p; (3) B6-Azo-g solely stores and releases photothermal energy under alternate irradiation with ultraviolet and blue light, independent of temperature variations, as shown in Figure 2.25(e). Pattern (1) allows for selective heat release under different temperature conditions. However, for A7–10-Azo-g, there is no phase transition below T_c, and it can only release ΔH_p. The limited temperature range hinders heat output at low temperatures. On the other hand, pattern (2) enables

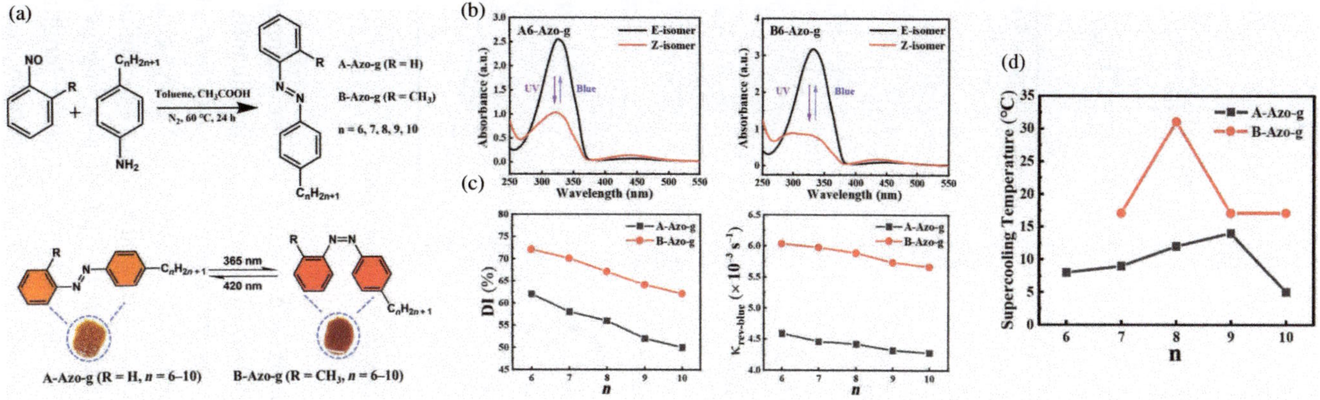

Figure 2.25. (a) Chemical structure and scheme Azo-g photo-induced reversible *trans-cis* isomerization process. (b) UV–Vis absorption spectra of A- Azo-g and B- Azo-g (c) Degree of isomerization and Krev-blue of Azo-g as a function of n at room temperature (d) Supercooling temperature of all Azo-g except for A6-Azo-g (e) Schematic illustration of three exothermic modes to utilizes the latent heat and photothermal energy in a close cycle based on reversible phase-change and isomerization (f) ΔH_c of Azo-g (g) Schematic illustration of a distributed energy utilization system. Reprinted with permission from Ref. [68]. Copyright 2023, the authors. Energy & Environmental Materials published byJohn Wiley & Sons Australia, Ltd on behalf of Zhengzhou University.

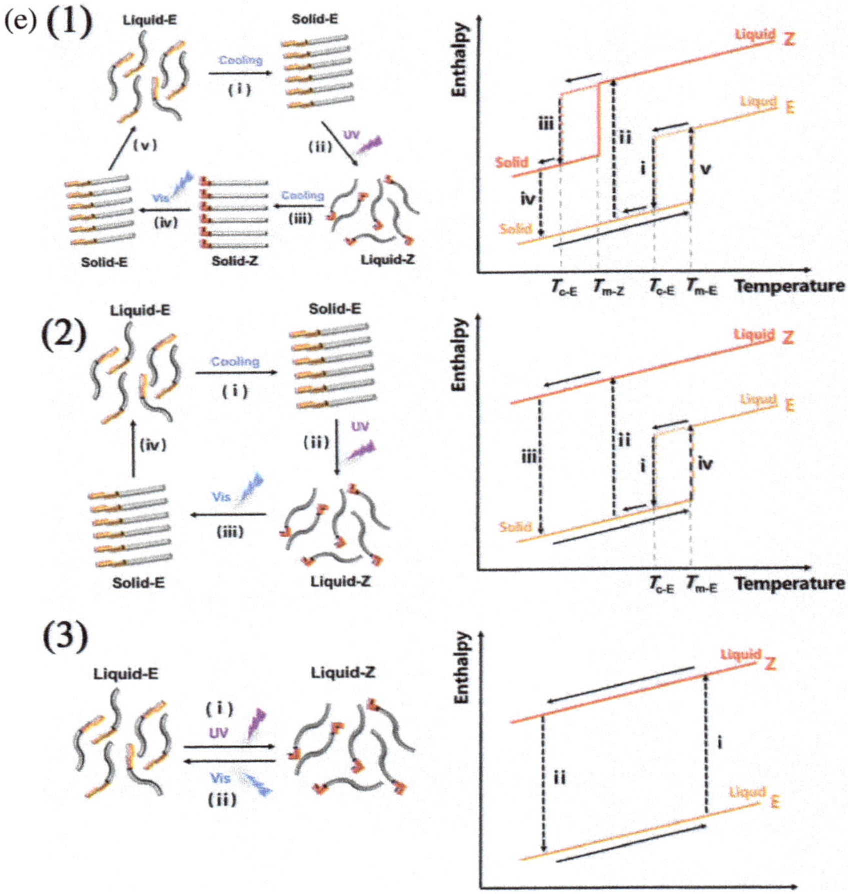

Figure 2.25. (*Continued*)

photothermally-induced heat release at extremely low temperatures since the *cis*-isomer cannot crystallize. This system significantly widens the temperature range for heat output compared to existing photothermal materials. The study investigated the influence of alkyl chain length on ΔH_c of Azo-g and found that A-Azo-g and B-Azo-g exhibited different dependencies on the alkyl chain length. ΔH_c of A-Azo-g increased from 86.6 J/g to 106.6 J/g with an increasing alkyl chain length due to the favorable van der Waals forces-driven self-assembly, enhancing the

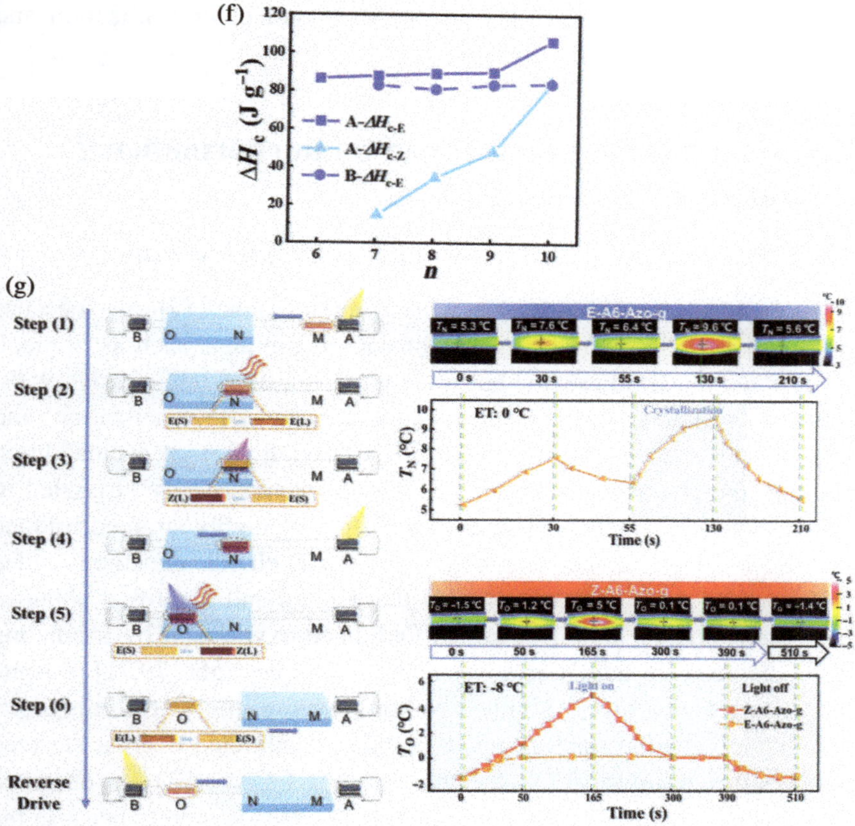

Figure 2.25. (*Continued*)

regularity of *trans*-isomer crystallization. In contrast, ΔH_c of B-Azo-g remained unchanged with varying alkyl chain length, as the introduction of methyl groups weakened the intermolecular interactions, and the alkyl chain could not enhance the regularity of *trans*-isomer crystallization (Figure 2.25(f)). Finally, the authors developed a novel distributed energy utilization system based on light-driven and heat-releasing processes, as depicted in Figure 2.25(g). This device achieved two-step controllable heat output, raising the surface temperature by 4.3°C and 6.6°C in low-temperature environments at 0°C and −8°C, respectively. This research provides valuable insights into utilizing latent heat and photothermal energy through molecular structure optimization of crystallization and isomerization processes.

2.3 Solar thermal fuels based on norbornadiene/quadricyclane derivatives

2.3.1 *Introduction of norbornadiene/quadricyclane*

Norbornadiene (NBD) derived its name from the island of Borneo because of its resemblance to the structure of natural products such as borneol, NBD is not a natural product but was first synthesized in the early 1950s. NBD is characterized by photo-induced valence isomerization and can be converted to quadricyclane (QC) through [2 + 2] cycloaddition reaction under certain conditions. However, chemists have struggled to synthesize QC which is the special structure of highly strained bridged polycyclic compound and its preparation by pure chemical methods ended in failure [69]. It was not until 1957 that Cristol and Snell first obtained the corresponding photoisomerized QC dicarboxylic acid by irradiating NBD dicarboxylic acid derivatives under ultraviolet light [70,71]. Despite the presence of a highly strained carbon skeleton in QC, the compound exhibits remarkable stability under ambient conditions. Even at elevated temperatures of up to 140°C, the conversion of QC to NBD was observed to occur at a relatively slow rate ($\tau_{1/2}$ = 14 h). This is mainly because the back-conversion of QC to NBD belongs to [2 + 2] ring-opening reaction, according to the famous rules on the *the conservation of orbital symmetry* proposed by Woodward and Hoffmann in1969, which is formally forbidden by the aforementioned restrictions, resulting in a high energy barrier for thermal-induced back-conversion of QC to NBD.

The reversible photo-induced valence isomerization transformation of NBD-QC is considered as one of the most promising systems for solar energy conversion and storage. NBD converts absorbed solar energy into chemical energy and stores it in QC through valence isomerization reaction, which has a higher isomerized enthalpy of 96 kJ/mol and can be released in the form of heat under the specific catalyst. It was noted that NBD is prone to polymerization, and may form explosive peroxides in the presence of light and air. However, the drawback of NBD is that the maximum absorption wavelength in the gas phase is not more than 226 nm and in the ethanol solution is not more than 220 nm [72], while the minimum wavelength of the solar spectrum is 300 nm. As a result, the absorption spectrum of NBD is not matched with solar spectrum, and sunlight cannot be directly utilized. This unsatisfactory absorption characteristic is associated with the very low quantum yield ($\Phi = 0.05$) of NBD [73]. Therefore, in order to make NBD-QC applied to molecular solar thermal storage (MOST) system, there are usually two strategies to solve this problem. One is the absorption must be redshifted by introducing substituents on NBD. The other is to sensitize NBD photo-induced isomerization with a photosensitizer.

In 1977, Turro *et al.* [74]. proposed the mechanism of photo-induced valence isomerization between NBD and QC. First, NBD and QC were photo-dissociated into precursor molecules NBD_{azo} and QC_{azo} under the light irradiation, respectively, and norbornadienyl and quadricyclenyl diradicals were generated by the loss of N_2. The two diradicals D_N and D_Q can be formed in an excited single and triplet state, which then recombine to either form NBD or QC, as shown in Figure 2.26(a). The experimental results show that QC is mainly formed in the process of NBD_{azo} and QC_{azo}

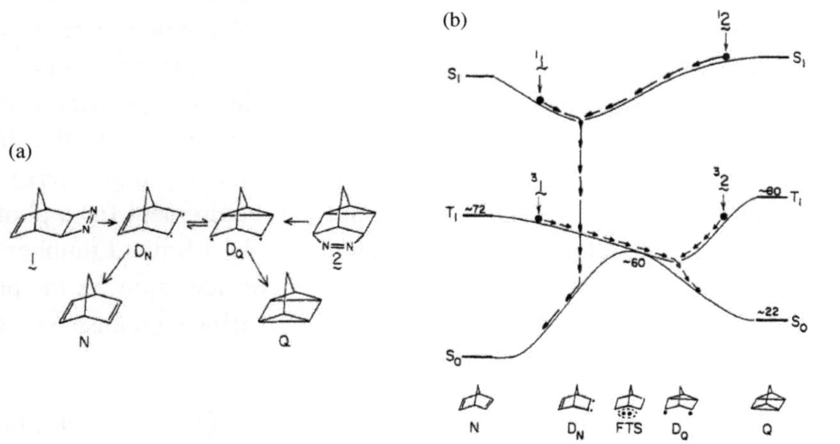

Figure 2.26. (a) The reaction process of NBD and QC to from D_N and D_Q. (b) Postulated energy surfaces for the interconversion of NBD and QC. Reprinted with permission from Ref. [25]. Copyright 1977, American Chemical Society.

from $S_0 \rightarrow T_1$ and NBD_{azo} from $S_0 \rightarrow S_1$. In the process of $S_0 \rightarrow S_1$, QC_{azo} mainly forms NBD. Based on the above experimental results, the photo-induced valence isomerization process between NBD and QC is further explained by introducing the concept of the potential energy surface. When NBD is excited to its triplet state, it will rearrange and move in a radiationless transition to the S_0 surface of QC, and when NBD is excited to its single state will recombine into D_N and form NBD by non-radiative transition to S_0, as shown in Figure 2.26(b). The proposed mechanism offers a theoretical foundation for overcoming the limited absorption properties of NBD and devising a pathway for the light-induced isomerization to QC via triplet transition.

The photosensitizers used in the photo-induced valance sensitized isomerization of NBD and its derivatives include triplet energy transfer photosensitizers, electron transfer photosensitizers, transition metal compound photosensitizers, and semiconductor photosensitizers. Among them, triplet photosensitizer was the earliest one taken up for the most in-depth study and its sensitized NBD photo-induced isomerization reaction mechanism is shown in Figure 2.27 [75]. The photosensitizer is first brought to its excited single state S_1 under light irradiation, and then from S_1 to T_1 through intersystem crossing. The occurring triplet-triplet energy transfer between photosensitizer and NBD converts NBD to its excited triplet state while the photosensitizer triplet state returns to the ground state. Subsequently, NBD triplet state is transformed into QC triplet state by photo-induced valence isomerization and QC triplet state will quickly return to the QC ground state ($t < 1$ns) since QC triplet state is so unstable [76]. It is worth noting that the relatively high energy of the triplet state of NBD (257 kJ/mol) and the requirement for a photosensitizer triplet state even higher in energy, only a limited number of molecules are deemed suitable and worthy of consideration for the process [77]. In 1964, Hammond *et al.* [78] used triplet photosensitizers

$$\mathrm{S} \xrightarrow{h\nu} {}^{1}\mathrm{S} \longrightarrow {}^{3}\mathrm{S} \xrightarrow{\mathrm{NBD}} \mathrm{S} + {}^{3}\mathrm{NBD} \longrightarrow \mathrm{S} + {}^{3}\mathrm{QC} \longrightarrow \mathrm{S} + \mathrm{QC}$$

Figure 2.27. The triplet-triplet energy transfer scheme provides a classical explanation for the pathway of NBD isomerization sensitized by triplets. Reprinted with permission from Ref. [75]. Copyright 2022, dissertation from the University of Erlangen-Nuremberg.

acetone, benzophenone, and acetophenone in isopentane solution to investigate the photo-isomerization of NBD, which was the pioneer of NBD-QC in the field of energy storage. Since then, a large number of researchers have carried out extensive studies. Smith *et al.* [79] used acetophenone as photosensitizers and found that the yield of QC obtained by NBD reacting in diethyl ether solvent for 36–48 h was 70–80%. Cahill *et al.* [80] use 4,4′-diethyl (diethylamino) dibenzophenone (tetraethyl mihalone) and mihalone as triplet photosensitizers to study the photo isomerization reaction of NBD in solve-free conditions. The results showed that tetraethyl mihalone reduced the induction time of the reaction and increased the yield of QC by 40% and the yield of the reaction was about 99.5%. Unfortunately, the presence of photosensitizers can promote side reactions and polymerization of NBD, leading to fouling of the storage media. While measures like excluding oxygen or adding radical scavengers can partially suppress the formation of unwanted by products, complete prevention of light-induced decomposition and side reactions is difficult to achieve [81].

Electron transfer photosensitizer is the electron transfer with NBD under photo excitation, so as to undergo valence isomerization reaction to produce QC. The selection of electron transfer photosensitizers is not limited by the triplet energy. As long as they match the REDOX potential of the substrate, the photo isomerization reaction of NBD can be achieved through the electron transfer process. Grutsch and Kuta [82] used iridium as photosensitizer to study the photoisomerization of NBD in acetonitrile solvent with nitrogen bubbles. The results showed that the excited photosensitizer diffused with NBD to form a NBD triplet state or charge transfer stable complex, and then released energy to form ground state photosensitizer and QC. The charge transfer between the photosensitizer and NBD was demonstrated. Transition metal compound photosensitizers and semiconductor photosensitizers are not elaborated here.

2.3.2 *Research advances of NBD-QC applied in molecular solar thermal system*

Since the main drawback of NBD is that its onset absorption wavelength is less than 300 nm, sunlight cannot directly drive the valence isomerization reaction of NBD. Consequently efforts are required to shift the absorption spectrum of NBD closer to the solar spectrum range. In recent years, numerous substituted NBD have been synthesized via Diels–Alder reaction between substituted cyclopentadiene and alkyne [83,84]. However, a common problem is that the introduction of groups that red-shifting auxochrome in NBD leads to the reverse isomerization of photo-isomer, which reduced the thermal stability. Meanwhile, the Diels–Alder reaction imposes several limitations on the types of substituted NBDs that can be synthesized. These restrictions may impact the range of potential applications and the efficiency of solar thermal energy storage systems. Tranmer *et al.* [85] developed a new method to synthesize aryl-substituted NBD by Suzuki-Mayaura cross-coupling reaction. Based on the above reactions, Kaper Moth-Poulsen *et al.* [86] synthesized a series of five diaryl-substituted NBD (3a–e) to create a novel push–pull conjugated system in NBD. These compounds feature one aryl group that acts as an electron donor and another aryl group that acts as an electron acceptor, the structure of which is shown in Figure 2.28(a). The onset absorption wavelength of all substituted NBD showed a significant redshift compared with that of non-substituted NBD, and the most significant shift was observed for NBD 3e with an absorption onset of 462 nm, corresponding to a red-shift of approximately 195 nm, as shown in Figure 2.28(b). The quantum yield of NBD (3a–e) was 45.8–75.5%. Aryl-substituted compounds are known to increase the thermal lability of QC. Analogous to this QC,QC (4a–e) are readily converted to corresponding NBD above 80°C. The stability of the synthesized compounds was evaluated, revealing that compounds 4a, 4b, and 4d exhibit sufficient stability with half-lives ranging from weeks to months at 25°C. On the other hand, compounds 4c and 4e, which contain a trifluoromethyl electron-withdrawing group (EWG), display higher thermal lability with half-lives of 2.9 days and 1.9 h at 25°C, respectively. To demonstrate the concept of energy storage, a parabolic solar collector was constructed and equipped with a glass pipe serving as a continuous flow reactor, as shown in Figure 2.28(c). NBD 3c was dissolved in toluene and injected into a transparent glass pipe to quantitatively observe the change of photo-induced valence isomerization transformation with residence time. It was found that the conversion of NBD 3c reached 100% when the residence time was 60 min (Figure 2.28(d)).

The heating capabilities of liquid MOST system shows promising applications, such as the preparation of functional windows for temperature regulation and de-icing that can reduce the energy requirements of traditional heating and cooling [57,87]. However, this could be as short as 4–8 h for a lamination on windows that regulates the interior temperature on a daily basis. On the other hand, metastable liquid photo-isomers can control the release of heat with the catalyst for domestic heating in extreme seasonal climates. The incorporation of molecular photo-switches into polymers can achieve MOST applications in solid-state devices. In previous studies, NBD derivatives have been introduced into polymethyl methacrylate [88] or copolymerized with other monomers to form polymers

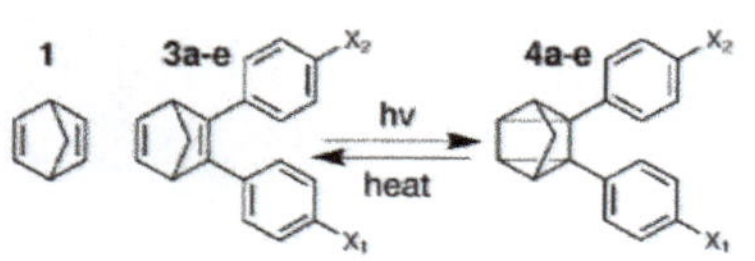

Entry	X_1	X_2
3a	-H	-H
3b	-H	-OMe
3c	$-CF_3$	-OMe
3d	-CN	-OMe
3e	-CF3	$-NMe_2$

(a)

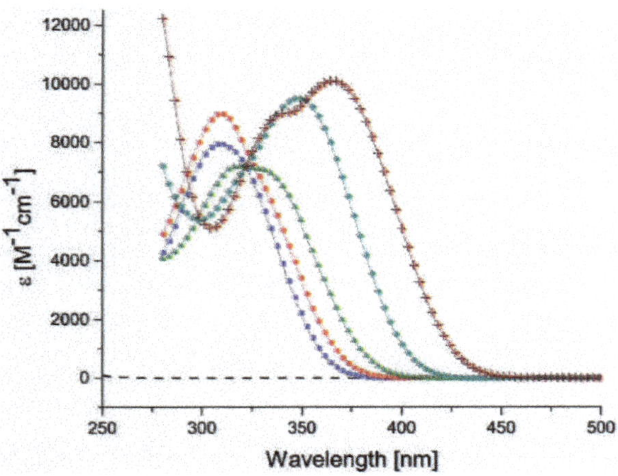

(b)

Figure 2.28. (a) Chemical structure of the synthesized NBD and corresponding QC. (b) Absorption spectrum of the unsubstituted NBD in acetonitrile and substituted NBD in toluene. (c) Isometric drawing of the solar collector. (d) Percentage of NBD to QC conversion with varying residence time. Reprinted with permission from Ref. [86]. Copyright 2014, The Royal Society of Chemistry.

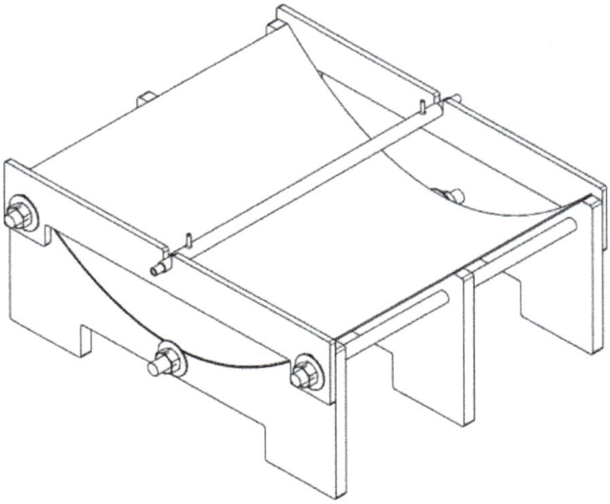

(c)

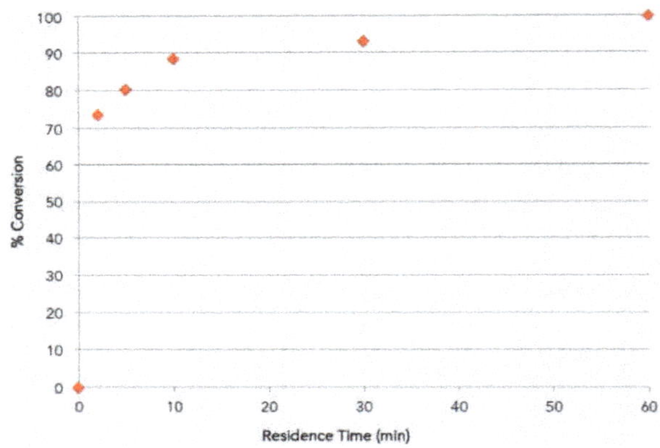

(d)

Figure 2.28. (*Continued*)

[89–91]. In addition, since most NBD and its corresponding QC are liquid, it is easy to form a solid NBD polymer composite by blending with the polymer, whose mechanical properties are determined by the polymer matrix. Kaper Moth-Poulsen *et al.* [92]. developed an NBD-QC polymer composite system for applications in windows and coatings where solid-state functionality is required. To achieve this, a series of NBD-based molecules were designed and synthesized with a focus on achieving good solar spectrum matching and target half-lives in the range of 4–8 h as shown in Figure 2.29(a). Compared with NBDs N5b–d bearing a cyano acceptor group, N4b–d with trifluoroacetyl acceptor group improved overlap with the most intense region of the solar spectrum and the onset absorption wavelength was redshifted by 70–100 nm. The solar absorption efficiency of N4b, N4d, and NN4e were 2.9%, 1.5%, and 3.8%, respectively. The larger the dipole moment through C2-C3 olefin for the NBD, the lower the lifetime for the corresponding QC. The half-life of N4a–c gradually decreases, and N5a–c has a similar trend, as shown in Figure 2.29(b). The reverse isomerization of QQ4e required two processes, namely QQ4e→QN4e and QN4e→NN4e, and the first conversion was significantly faster than the second step (Figure 2.29(c)). When the N4b was incorporated into four different polymer matrices (PS, PMMA, PC, and PVDC), all composites of N4b@polymer resulted in the formation of free-standing films with a yellow color. However, upon exposure to irradiation, these materials underwent a color change and turned colorless due to the formation of Q4b. N4b@PS demonstrated excellent cyclability with almost no attenuation during the four cycles as shown in Figure 2.29(d). The heat released by charging Q4c@PS could not be detected as the back-converted too rapidly, and the storage energy enthalpy of Q4b@PS, Q4d@PS and QQ4e@PS were 30.2 kJ/kg, 50.4 kJ/kg, and 51.8 kJ/kg, respectively. The NBD@PS composite was cast on the glass surface treated with hexamethylenedisilazane to form a 70 μm thin film, and the film was converted within a few seconds under sunlight irradiation, and the indoor temperature could be adjusted in real life, as shown in Figure 2.28(e). This study presented here showcases the ability of NBD dispersed within a polystyrene matrix to effectively absorb sunlight and subsequently release the captured energy as heat. This development marks an initial advancement towards the creation of window tinting materials that incorporate NBD MOST systems, offering substantial potential for energy-saving applications. By harnessing the sun's energy and converting it into heat, this technology holds great promise for contributing to energy efficiency efforts.

Mild oxidizing agents [93,94] can be used to initiate the formation of QC^+ in the NBD-QC system, allowing for controlled release of the stored thermal energy. This process triggers a radical chain reaction that leads to the reconversion of QC^+. Heterogeneous catalysts are generally selected

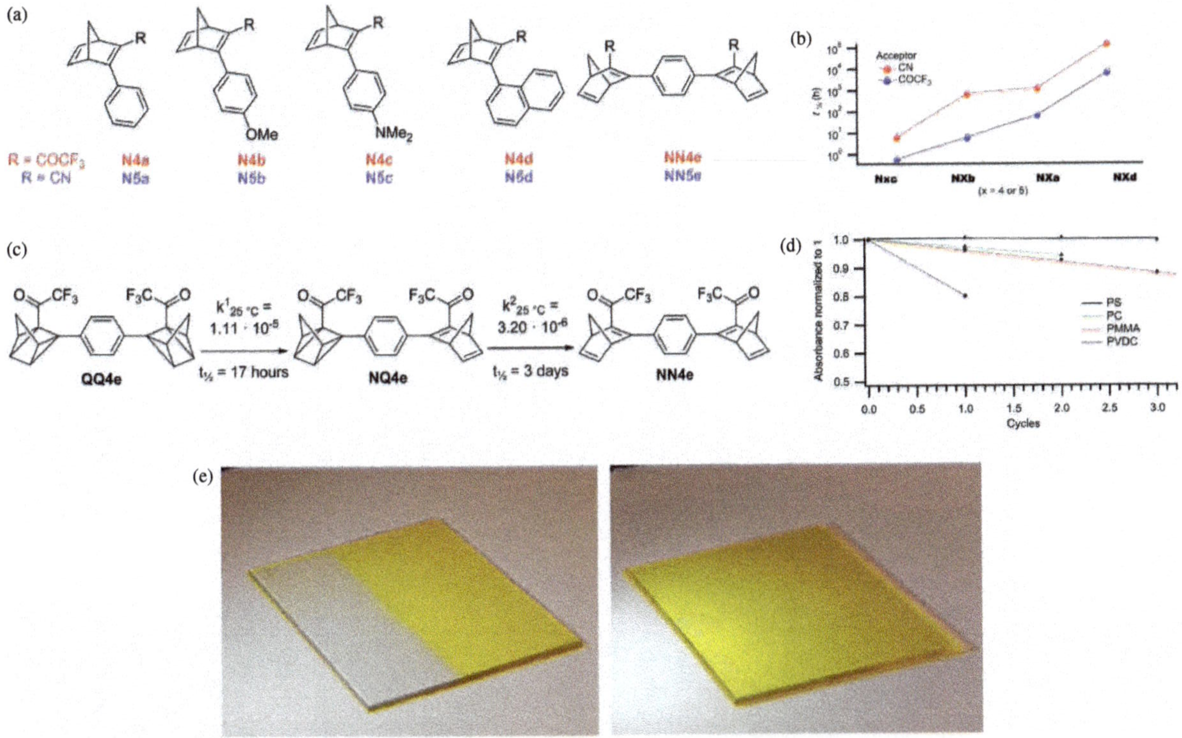

Figure 2.29. (a) Chemical structure of NBDs N4a–d and NN4e as well as N5a–d and NN5e. (b) The evolution curve of half-life and acceptor strength between N4a–d and N5a–d. (c) A schematic diagram illustrating the sequential back-reaction steps observed in the UV–Vis absorption spectrum from QQ4e to NN4e. (d) Cyclability of N4b in different polymers. (e) Hexamethylenedisilazane treated glass coated with 70 μm thick layer of N4b@PS upon exposure of half of the sample to light and after back-conversion. Reprinted with permission from Ref. [92]. Copyright 2019, the authors.

due to the low control degree of homogenous catalysts. Kaper Moth-Poulsen *et al.* [95]. employed a carbon-supported cobalt phthalocyanine catalyst for controlled heat release of MOST devices based on NBD-QC. This catalyst exhibited a remarkable reversibility of over 99.8% per storage cycle. Recently, some researchers have developed the electrochemical controlled thermal release of NBD-QC system, that is, QC^+ is formed through an electrochemical process, which converts into NBD by autocatalytic reaction [96–98]. Bachmann *et al.* [97]. have successfully demonstrated the electrochemical triggering of energy release from NBD through the application of a potential. Brummel *et al.* [99]. synthesized a NBD derivative 2 cyano-3-(3,4-dimethoxy-phenyl) by designing the molecular structure, the structure of which is shown in Figure 2.30(a). The unique pull–push electron couple present in the NBD derivative enables a significant redshift in the maximum absorption wavelength, extending it to 340 nm. This redshift enables direct light absorption in the near UV region within the molecule of the NBD derivative. The electrochemically controlled heat released was studied by the photoelectrochemical infrared reflection absorption spectroscopy (PEC-IRRAS) device (Figure 2.30(b)) in the preset process of photochemical charging and electrochemical discharging cycle (Figure 2.30(c)), and the valence isomerization was judged by the difference between the –CN of NBD and QC at the IR peak position. When the electrode potential was ramped from –0.88 V_{fc} to 0.62 V_{fc} in steps of 0.1 V, the measured IR spectrum analysis (Figure 2.30(d)) showed that ν(CN)-QC at 2200 cm^{-1} and ν(CN)-NBD at 2195 cm^{-1} gradually disappeared with the increase of potential, indicating QC transformed into NBD via valence isomerization. From the quantitative analysis derived from the band intensities, QC completely transformed into NBD at a potential of 0.32 V_{fc}, as shown in Figure 2.30(e). Figure 2.30(e) also showed that the transformed rate of QC→NBD increased with the potential increase in the range of –0.88 to –0.08 V_{fc}, which is mainly due to the catalytic activation of Pt electrode. At the same time, it can be seen in Figure 2.30(d) that when the potential rose to 0.92 V_{fc}, ν (CN)-NBD reappeared, indicating the consumption of electric oxidation of NBD. Therefore, the electrochemical controlled conversion of QC to NBD is suitable for the potential range of 0.32–0.82 V_{fc}. In the test of photochemical charging and electrochemical heat release cycles, the NBD-QC system has a reversibility of more than 99% per cycle, as shown in Figure 2.30(f).

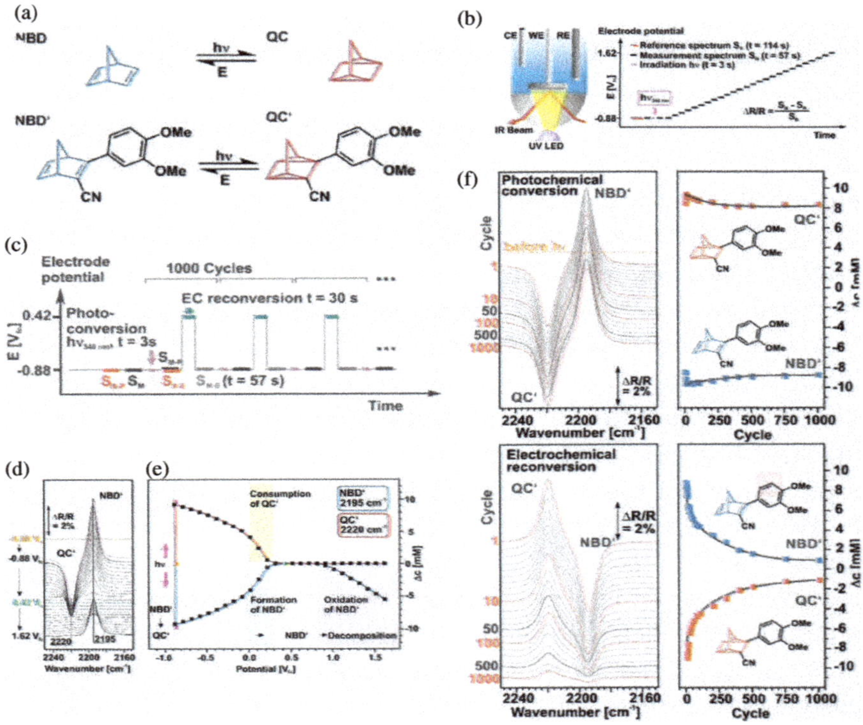

Figure 2.30. (a) Chemical structure of NBD derivative and corresponding QC. (b) Schematic representation of the PEC-IRRAS setup. (c) Experimental procedure of covering 1000 conversion cycles monitored by PEC-IRRAS. (d) ν(CN) region as a function of the electrode potential. (e) The concentration of the NBD derivative and its corresponding QC can be determined as a function of the potential by analyzing the band intensities in the ν(CN) region. (f) IR spectra as well as concentrations after photochemical conversion and electrochemically triggered reconversion. Reprinted with permission from Ref. [99]. Copyright 2019, Elsevier Ltd.

In order to solve the mismatch between NBD-QC system and solar spectrum, photosensitizing valence isomerization of NBD by photosensitizer is an effective method. Developing photosensitizers that can effectively optimize MOST systems has been a complex task. One of the key challenges lies in finding photosensitizers that possess electronic transitions similar to solar photons and a triplet state energy higher than that of NBD. In order to enhance the efficiency of the system, a potential solution is to chemically link the photosensitizer with NBD, enabling intramolecular energy transfer. Chou *et al.* [100]. introduced thermally activated delayed fluorescence (TADF) molecule into NBD-QC system and strategically designed a novel MOST-relevant compound using phenoxazine–triphenyltriazine (PXZ–TRZ) [101] as a core. The PXZ–TRZ molecule was discovered to successfully achieve the separation of HOMO and LUMO within a single molecule, resulting in a nearly zero-energy gap between the S_1 and T_1 states. Three PXZ–TRZ–NBD compounds, PZDN, PZTN, and PZQN were synthesized by anchoring PXZ–TRZ to NBD molecule, as shown in Figure 2.31(a). PZDN and PZTN exhibited strong TADF quenching depending on the reaction time of NBD→QC, PZQN showed an almost constant TADF rate where slow energy transfer hindered the NBD→QC reaction. It can be seen from the absorption and emission fluorescence spectra of PZDN, PZTN, and PZQN (Figure 2.31(b)) that the maximum emission wavelength of PZQN (530 nm) is redshifted compared with that of PZDN and PZTN (490 nm). Under the irradiation of 405 nm light, the decay of delayed fluorescence gradually increased (Figure 2.31(c)), while the intensity of steady-state TADF gradually increased (Figure 2.31(b)). By fitting the delayed fluorescence, the lifetime was determined to be 10.8 μs, and the area enclosed by the kinetic curve increased by 3.6 times, concluding that following triplet–triplet energy transfer, the photoisomerization reaction occurred for NBD→QC. A photolysis experiment using a mixture of PZDMe and NBD-OH that the molar ratio is 1:2 was used to simulate each PZDN having two NBD units was carried out, the kinetic curve and steady-state fluorescence spectra did not change significantly over time (Figure 2.31(d)). The above results indicated that the photoisomerization of NBD→QC was controlled by intramolecular energy transfer, which is further confirmed by the NMR analysis of PZDN before and after illumination. The emission intensity of PZQN in the steady-state fluorescence spectra and TADF decay dynamic were almost invariable after continuous irradiation, indicating that there is no triplet-triplet energy transfer and hence a lack of NBD→QC isomerization. This is mainly because PZDN was constituted from NBD units connected with a 1,3,5-triazine of PXZ–TRZ core by electron-donating ether bond, which increased LUMO energy, leading to an enhanced energy level difference between the S_1 and

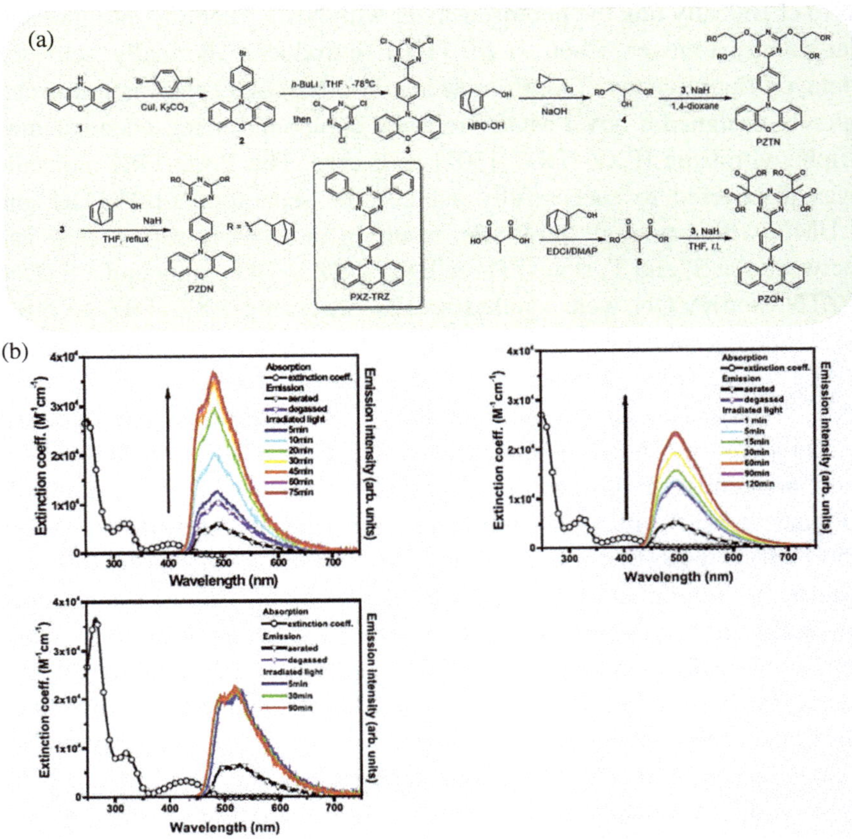

Figure 2.31. (a) The synthesized rote and chemical structure of PZDN, PZTN and PTQN. (b) Irradiation time-dependent fluorescence spectra PZDN, PZTN and PTQN. (c) The emission kinetic profiles of PZDN and PZTN exhibit time-dependent decay dynamics in their TADF emission. These dynamics are characterized by different conditions: "aer" refers to aerated conditions, "deg" refers to degassed conditions, and "irr" refers to light irradiation conditions. (d) The chemical structure of PZDMe and NBD-OH as well as steady state measurements and kinetic trace as a function of irradiation time. Reprinted with permission from Ref. [100]. Copyright 2022, the authors.Reprinted with permission from Ref. [101]. Copyright 2012, The Royal Society of Chemistry.

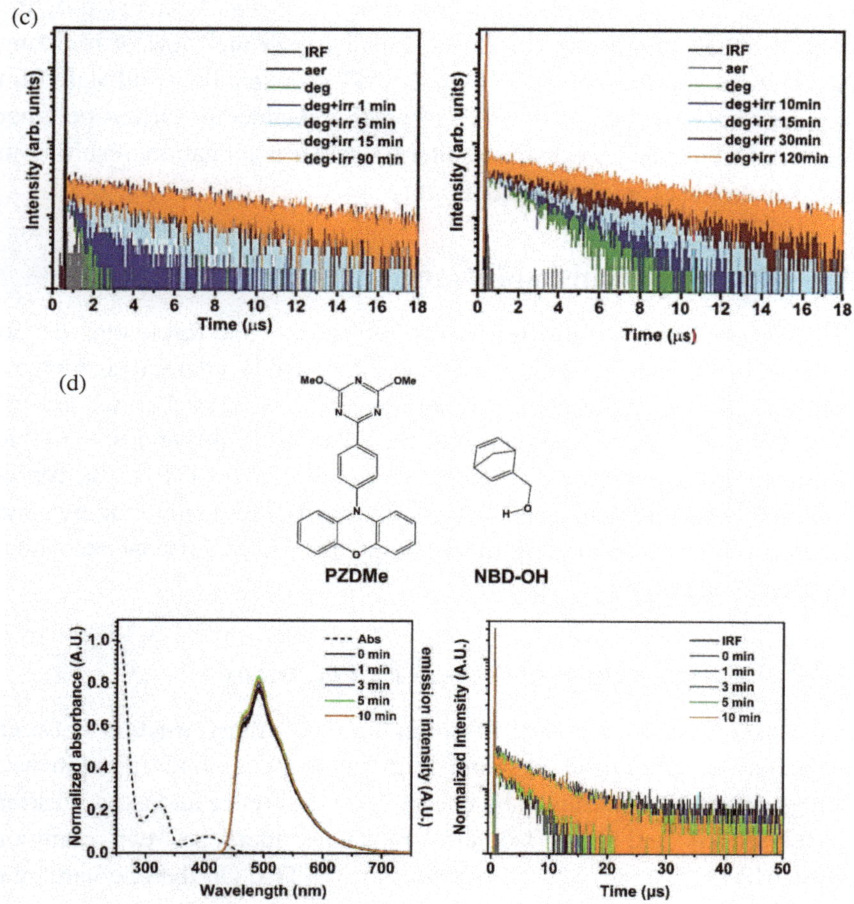

Figure 2.31. (*Continued*)

S_0 states, thereby increasing the energy level difference between the S_1 and T_1 states. Conversely, PZQN incorporates an alkyl bridge link that acts as a weak EWG. This leads to a decrease in the LUMO energy, resulting in a reduced energy level difference between S_1 and S_0. Consequently, the energy level of T_1 in PZQN is lower than that of NBD. The thermal-induced isomerization of QC→NBD was confirmed by the NMR analysis of PZDQC, and the heat released was 161.8 kJ/mol by DSC test.

Meanwhile, the rapid release of stored heat energy can be realized by the catalyst cobalt tetraphenylporphyrin. Finally, with the view of realizing the practical application of PZDN in MOST systems, the solid-state film of pure PZDN was prepared and PZDN was embedded in various polymer films. All films underwent severe intermolecular aggregation, resulting in low conversion efficiency and durability.

2.4 Challenges and solutions

STFs have achieved a large number of remarkable and fascinating developments in the past few decades, whereas they still faced great challenges, such as low energy density, short storage life, low photothermal conversion efficiency, narrow overlap with the solar spectrum, and irreversible photodegradation. Taking STF-based azobenzene (Azo-STF) as the representative, researchers have provided solutions to these shortcomings and developed STFs with extraordinary comprehensive performance to trade-off their properties.

2.4.1 *Energy density of Azo-STFs enhances*

Azobenzene takes advantage of configurational transformation between molecules to store absorbed solar energy in chemical bonds, and intrinsic energy density is determined by energy level difference and isomerization degree of *cis* and *trans* isomer. Until now, there are two common approaches to increase the energy density of STFs. On the one hand, the intrinsic energy density is improved by molecular engineering strategy and template method. On the other hand, the introduction of phase transformation system and the addition of the phase change on the basis of intrinsic energy density can improve the energy density of Azo-STFs. In this regard, the detailed introduction and discussion have been made in the previous part, and will not be circumstantial here.

The higher the fraction of *cis*-isomers in the PSS, the greater energy of the azobenzene molecules. There are plenty of factors affecting the isomerization degree and rate, including molecular structure, aggregation state [102], solvent properties [103,104], irradiation wavelength [105–107], temperature [108–110], and pressure [111,112], etc. Most investigations on the quantum yield of photoisomerization are concentrated in liquid or solution. This is because the absorption intensity of different parts of the solid film is unconformity, and it is different to control the uniform distribution of light in the space of the solid film. As a result, the research in the solid state exhibits a more complex problem than that in liquid or solvent.

The *trans–cis* transition of azobenzene underwent two exciting processes, $S_0 \rightarrow S_1$ and $S_0 \rightarrow S_2$, and the *cis–trans* transition also underwent a similar process [106]. Nevertheless, the combined quantum yields of $\Phi_{trans-cis}$ and $\Phi_{cis-trans}$ do not add up to unity, suggesting the presence of multiple pathways for isomerization. So far, the photo-induced *trans–cis* isomerization mechanism of azobenzene mainly includes four mechanisms-rotation, inversion, concerted inversion, and inversion-assisted rotation, as shown in Figure 2.32. The rotational pathway involves the rupture of the N=N π-bond to allow free rotation about the N–N bond. Rotation changes the C–N–N–C dihedral angle while the N–N–C angle remains fixed at 120°. The inversion mechanism is that one N=N–C angle increases

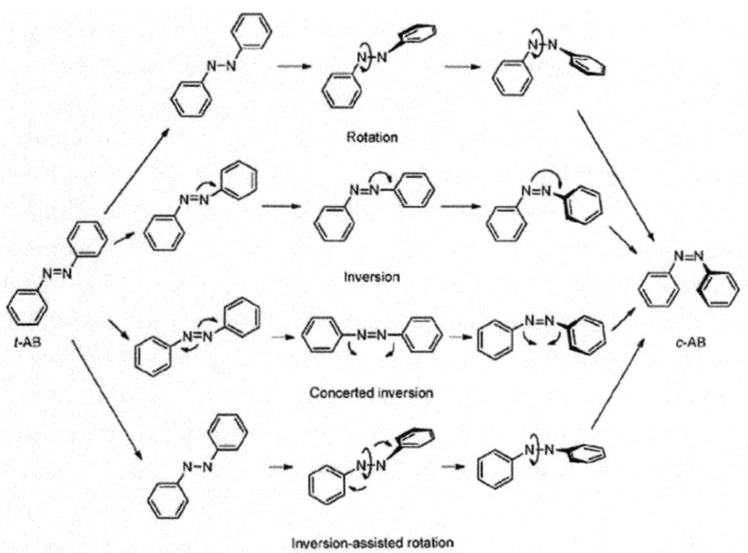

Figure 2.32. The isomerization mechanism of azobenzene. Reprinted with permission from Ref. [106]. Copyright 1979, American Chemical Society.

to 180° while the C–N=N–C dihedral angle is unchanged at 0°. For the concerted inversion mechanism, both N=N–C bond angle increase to 180° generating a linear transition state. The inversion-assisted rotation mechanism is to rotate C–N=N–C dihedral angle while inversing N=N–C angle until it is transformed into *cis*-isomer. Ultrafast time-resolved spectroscopy can detect the short lifetime of excited states, and combine with *ab initio* calculation method and semi-empirical calculation to better explain the isomerization mechanism, which lays a theoretical foundation for guiding the design of azobenzene molecules. The electronic and steric effects of position and type of azobenzene substituents affect the degree and rate of isomerization, and in essence, the pathway of isomerization. Roitberg *et al.* [113] studied a series of azobenzene by *ab initio* calculation method and determined the influence of substituents on the isomerization pathway and the energy barrier from the three-dimensional potential energy surface of the ground state and the electron excited state. The results show that the electron-donating group (EDG) increases the ground state inversion energy barrier, while the EWG is just the opposite.

Solvent polarity and viscosity are also important factors affecting the quantum yield of isomerization of azobenzene. Burdette *et al.* [114] summarized the influence of regular solvent polarity and viscosity on the quantum yield of isomerization according to the literature. The quantum yield ($\Phi_{trans\text{–}cis}$) of *trans–cis* isomerization after n–π^* excitation demonstrates an increase with higher solvent polarity, while it decreases with increasing solvent polarity. However, solvent polarity does not impact the quantum yields of photo-isomerization following π–π^* excitation. The viscosity and polarity of the solvent can influence the isomerization mechanism. Non-viscous polar solvents tend to promote rotation, whereas viscous non-polar solvents favor inversion. Isomerization quantum yields experience a significant decrease when embedded in a solid matrix. The temperature has no significant effect on the quantum yield of azobenzene *trans–cis* isomerization, while the quantum yield of *cis-trans* isomerization decreases as the temperature increases.

2.4.2 *Prolonging the energy storage life of Azo-STFs*

The energy storage life of Azo-STFs strongly depends on the energy barrier of *cis–trans* isomerization. The stability of the *cis*-isomer in the high energy state means that the more difficult the isomerization is, and the half-life is longer. The shorter the half-life of azobenzene, the faster the dissipation of energy stored in the *cis*-isomer, which is not conducive to the storage of solar thermal energy. On the contrary, the longer the half-life of azobenzene, the slower the release of energy stored in the *cis*-isomer, which is also unfavorable for special situations requiring rapid discharge. Therefore, it is critical to develop Azo-STFs with suitable half-life according to specific application scenarios. The half-life of azobenzene is related to the substituent group and solvent polarity. As a result, improving the half-life of azobenzene can be done from the perspective of molecular structure design. In the past few decades, researchers have designed and synthesized a series of novel azobenzene derivatives to address the drawback of short half-life.

Pozhidaeva *et al.* [115] adjusted the half-life and absorption spectrum of azobenzene derivatives by changing the electronic effect of the para-substituent of azobenzene molecules. It was found that the half-life gradually decreased from 43 h to 11s with the enhanced electron-donating effects of the substituent and the absorption spectrum is redshifted from 342 nm to 382 nm, as shown in Figure 2.33(a). Azobenzene bridging compound refers to the ring structure constituted from bridging groups between adjacent positions where N=N bond. Renth and Temps *et al.* [116] studied the photo-switching behavior of ethylene bridging azobenzene molecules. Although this type of molecule has been reported synthesized before [117–119], its photophysical and photochemical properties have not previously been recognized and studied in detail. The results show that the n–π^* absorption peaks of the *trans*-isomer and *cis*-isomer of these azobenzene derivatives exhibit a distinct separation behavior, and the *cis*-isomer becomes more thermodynamically stable with a half-life of 4.5 h. On this basis, the azobenzene molecular photo-switches with a longer half-life were prepared by adjusting the azobenzene molecular substituents in ethylene bridging, as shown in Figure 2.33(b). Hammerich *et al.* [120] introduced oxygen and sulfur atoms into the bridging azobenzene compounds, the half-life is longer than ethylene bridging azobenzene. Woolley *et al.* synthesized a series of molecular photo-switches of azobenzene derivatives by introducing amino substituents of EDG into the ortho-position of 4,4′-diamino-substituted azobenzene molecules, as shown in Figure 2.33(c). It was found that the half-life of ring structure

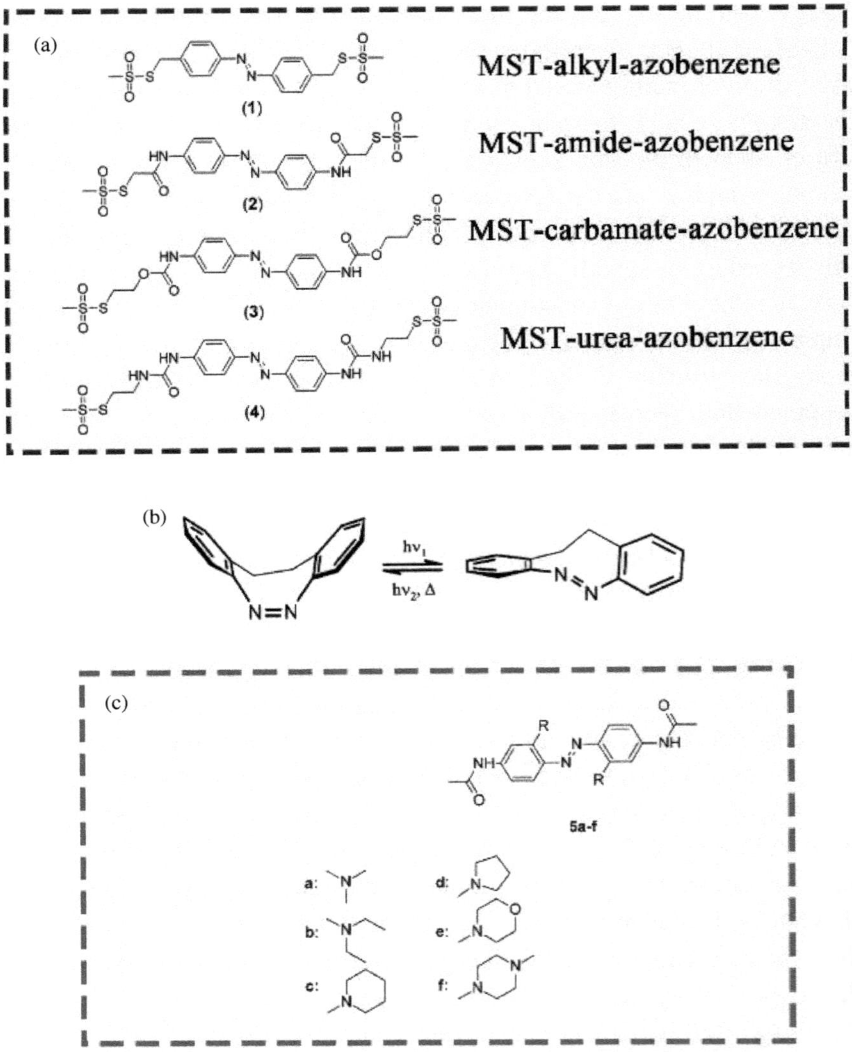

Figure 2.33. (a)–(h) The chemical structure of the azobenzene derivatives. Reprinted with permission from Ref. [115]. Copyright 2004, American Chemical Society. Reprinted with permission from Ref. [116]. Copyright 2009, American Chemical Society. Reprinted with permission from Ref. [120]. Copyright 2016, American Chemical Society. Reprinted with permission from Ref. [121]. Copyright 2007, The Royal Society of Chemistry. Reprinted with permission from Ref. [125]. Copyright 2010, The Royal Society of Chemistry. Reprinted with permission from Ref. [126]. Copyright 2012, American Chemical Society. Reprinted with permission from Ref. [127]. Copyright 2013, The Royal Society of Chemistry and Owner Societies.

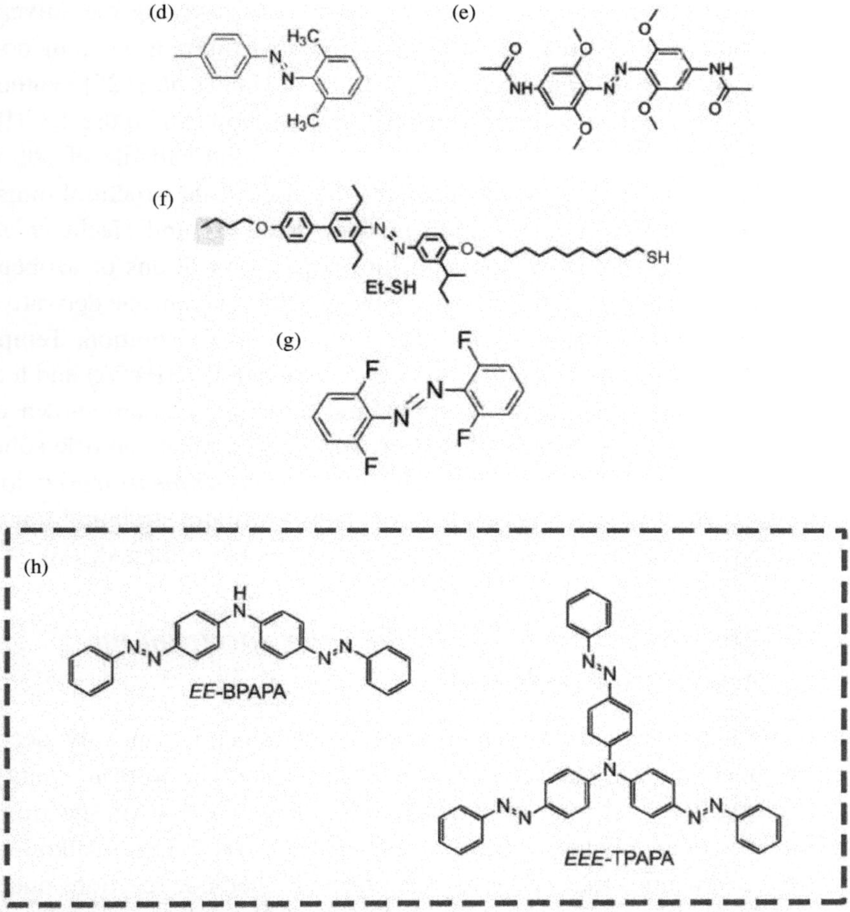

Figure 2.33. (*Continued*)

substituent is longer than that of non-ring structure substituent, and the half-life of six-membered ring substituent is longer than that of five-membered ring substituent, which is due to the steric hindrance effect increased by the introduction of ring structure in the molecular ortho-position of azobenzene derivatives, resulting in slower isomerization recovery rate of *cis–trans*. Asanuma *et al.* [121] introduced methyl group into 2′,6′-ortho-position of azobenzene (Figure 2.33(d)), the half-life was 25 h at 60°C. Woolley *et al.* [122] introduced methoxy groups into four ortho-position of 4,4′-diamino-substituted azobenzene (Figure 2.33(e)), and the half-life was as long as 53 h (2.4 days). Lowering the solvent polarity further prolonged the lifetime of the *cis*-isomer, a phenomenon attributed to a dipolar transition state [123,124]. Han *et al.* [125] synthesized an ortho-diethylated azobenzene (Et-SH), as shown in Figure 2.33(f). The azobenzene derivative exhibits an extremely long half-life of 380 h due to the bulky ethyl group that prevents the large-scale structural transformation during the isomerization process. Brouwer and Hecht *et al.* [126] introduced fluorine atoms into four adjacent positions of azobenzene molecules (Figure 2.33(g)), and found that the azobenzene derivative showed an amazing half-life (700 days) in acetonitrile solution. Temps *et al.* [127] synthesized binary azobenzene compounds (BPAPA) and ternary azobenzene compounds (TPAPA), and the structures are shown in Figure 2.33(h). The half-lives of BPAPA and TPAPA in acetonitrile solution were 2.4 h and 22.4 h, respectively. Since the *cis–trans* isomerization process of azobenzene compounds involves two or more structural transformation processes, the half-life of the *cis*-isomer is prolonged.

2.4.3 *The absorption range of the solar spectrum for Azo-STFs broadening*

According to the optical properties of azobenzene, it can only occur photo-induced *trans–cis* isomerization by absorbing UV to store energy. However, ultraviolet light only accounts for 5% of the solar spectrum, resulting in relatively low utilization efficiency for Azo-STFs. In addition, organic compounds may cause degradation and other side reactions under ultraviolet irradiation for a long time. The most effective way to solve this problem is to enable the absorption spectrum to redshift and make full use of visible and near infrared light (NIR). Meanwhile, another important problem will be caused, that is, visible light will also induce *cis–trans* isomerization transformation, which will lead to the reduction of isomerization degree and energy density. Thus, the efficient development of Azo-STFs that can effectively function under complete solar spectrum irradiation poses a significant challenge.

Over the past few decades, researchers have successfully designed and synthesized a series of novel azobenzene molecules that respond to visible and near-infrared light [128–136]. These azobenzene compounds are mainly applied in the field of life science and biology, which has guiding and instructive significance for solar energy conversion and storage.

The introduction of EDGs or push-pull electron structure at the para-position of azobenzene can make the π–π* transition absorption band redshift, but at the same time, it will decrease the isomerization degree at the cost of rapid thermal *cis-trans* isomerization [136–138].

The introduction of methoxy groups in all four ortho-positions of azobenzene significantly retards the thermal relaxation of *cis–trans* isomerization, thereby promoting the generation of a substantial proportion of *cis*-isomers [136,139]. Additionally, this particular substitution pattern induces a differentiation in the n–π* transitions between the *cis* and *trans* isomers, enabling the potential for bidirectional photo-switching under visible light. Siewertsen *et al.* [116] discovered a system consisting of an azobenzene molecule connected through ortho-positions by an ethylene linker. This system exhibited a significant 100 nm separation between the n–π* transitions of the *cis* and *trans* isomers, enabling bidirectional switching using visible light. Notably, this system achieved nearly complete photoconversions, with 100% *trans–cis* conversion and 92% *cis–trans* conversion. Woolley's group [122] initially reported an interesting azobenzene that four methoxy groups at the ortho-positions, which separated the n–π* transitions *cis* and *trans* isomers by 36 nm, capable of visible light switching behavior (82% *trans–cis*, 85% *cis–trans*). These two kinds of azobenzene photo-switches distorted the planar of the *cis*-isomers due to the ortho-position substituents to the N=N bond, and thus affected the energy of π and π* molecular orbitals, which may be the main reason for the splitting of the n–π* bands of both isomers. Subsequently, Woolley's group [139,140] replaced the oxygen atoms of the methoxy group with chlorine atoms, bromine atoms, and sulfur atoms. A series of tetra-ortho-substituted azobenzene derivatives all showed the separation of n–π* absorption bands for *cis* and *trans* isomers, and could carry out reversible photo-induced isomerization under the irradiation of red-light wavelength range. Brouwer and Hecht *et al.* [126] synthesized azobenzene derivatives with tetra-ortho-fluoro substitutions. These substitutions resulted in the separation of the n–π* transitions between the *cis* and *trans* isomers, while maintaining the trans geometry without any twisting. Herges's group [120] successfully synthesized oxygen and sulfur-bridged azobenzene on the basis of the ethylene-bridged azobenzene, these two azobenzene photo-switches can respond to reversible isomerization under red light (650 nm).

Upconversion nanomaterials can emit high energy light under low energy light excitation, that is, materials emit short wave and high frequency light after long wave and low frequency light excitation. Up-conversion nanoparticles doped with a certain amount of lanthanide elements convert near-infrared light (NIR) to ultraviolet or visible light by absorbing two or more photons consecutively. Yu *et al.* [141] demonstrated that the utilization of up-conversion nanoparticles (UCNPs) enables the induction of *trans–cis* photoisomerization in azobenzene compounds through exposure to NIR light. This finding represents the first instance of such a phenomenon being reported. Crosslinked liquid crystal polymer which is composed of NaYF4: 20%mol Yb^{3+}, 1%mol Tm^{3+}, as UCNPs combined with azobenzene can initiate photoisomerization under 980 nm near-infrared light irradiation, which was due to the overlap of absorption wavelength between UCNPs and azobenzene crosslinked liquid crystal polymer in the wavelength range of emission light (450–475 nm), as shown in Figure 2.34(a). Li *et al.* [128] presented a novel self-organized helical superstructure. This structure was created by incorporating a newly developed light-driven chiral molecular switch and UCNPs into a liquid crystal composed of readily available components. The system exhibited remarkable reversible tunable behaviors solely by adjusting the excitation power density of NIR light at 980 nm, as shown in Figure 2.34(b). The reversible light regulation process is caused by the reversible photoisomerization of chiral azobenzene photo-switches induced by ultraviolet light and visible light emitted by UCNPs under different excitation power density NIR. Under the excitation of high-power density NIR, ultraviolet light emitted by UCNPs induced *trans–cis* isomerization of chiral azobenzene molecules. Visible light emitted by UCNPs can induce *cis–trans* isomerization of chiral azobenzene molecules under low power density NIR excitation.

To sum up, increasing the absorption wavelength of azobenzene photo-switches can be realized by separating the n–π^* transition band in the visible range and redshifting the π–π^* transition absorption band. Corresponding implementation methods include (I) the introduction of EDGs, EWGs, or push–pull electronic structures in the para or ortho position of azobenzene molecules. (II) Design and synthesis of bridged

(a)

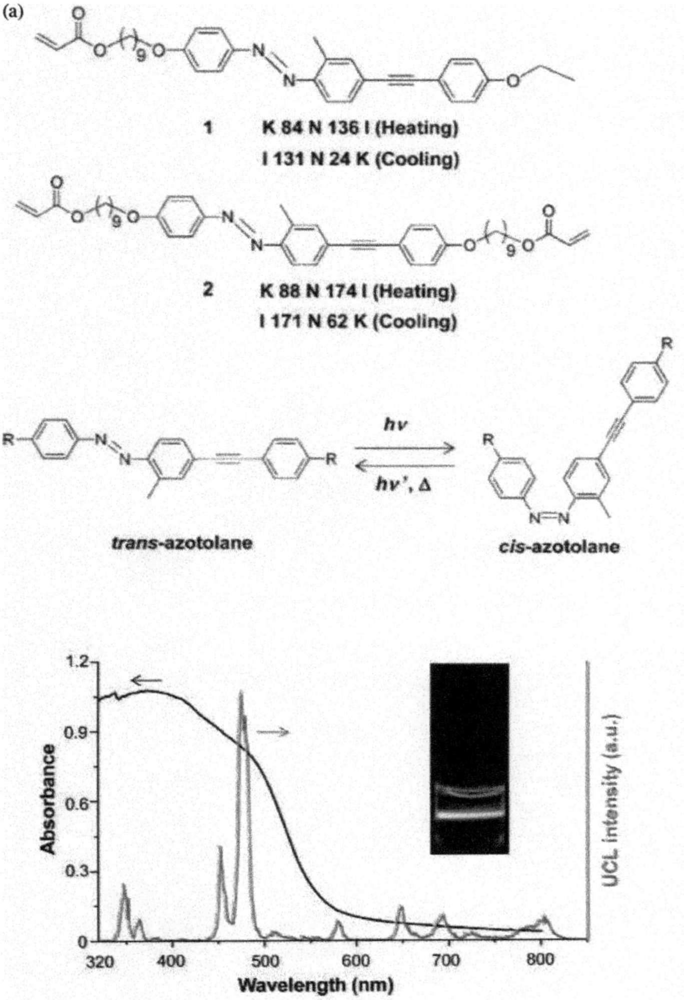

Figure 2.34. (a) Chemical structure of Azo-containing liquid crystal compounds 1 and 2 as well as UV–Vis absorption spectra of azobenzene units and Upconversion luminescence emission spectra of UCNPs excited with a 980 nm NIR laser. (b) The molecular structure of the light-driven chiral switch and a schematic representation of the photoisomerization process in this chiral azobenzene are illustrated. The remote control process involves the generation of *in-situ* UV light under high excitation power density NIR light and the generation of *in-situ* visible light under low excitation power density NIR light at the same wavelength. Furthermore, the schematic mechanism highlights the reversible tuning of the self-organized helical superstructure using the chiral switch and UCNPs upon irradiation with NIR laser at various power densities. Reprinted with permission from Ref. [128]. Copyright 2014, American Chemical Society. Reprinted with permission from Ref. [141]. Copyright 2011, American Chemical Society

(b)

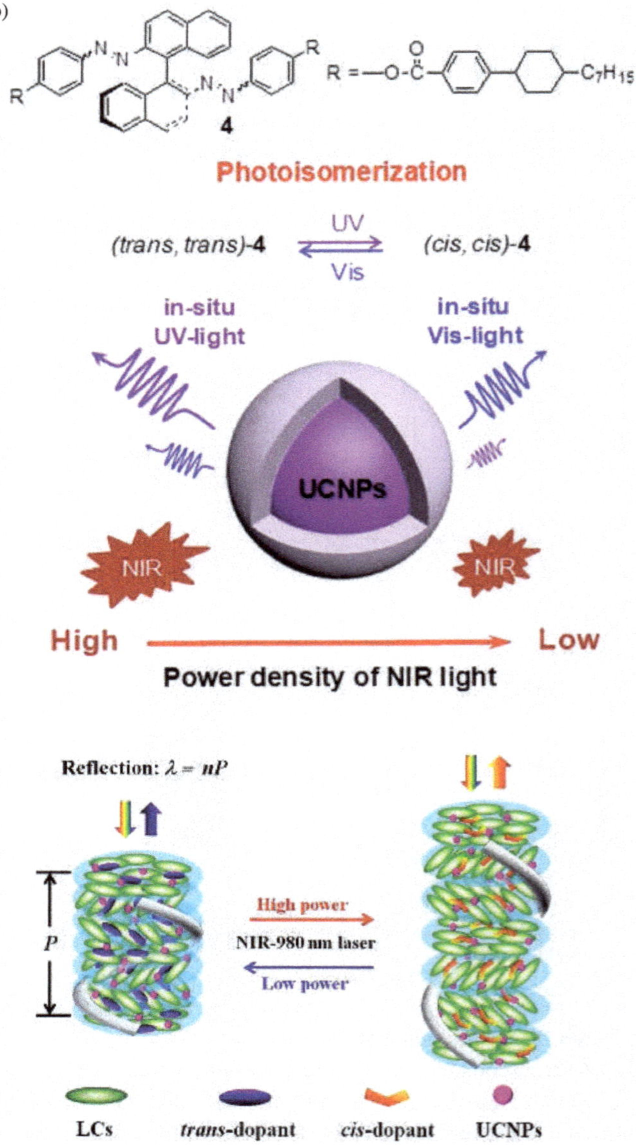

Figure 2.34. (*Continued*)

azobenzene or tetra-ortho-substituted azobenzene. (III) The upconversion nanoparticles were introduced into the azobenzene system, and the unique photophysical properties of UCNPs were utilized to achieve visible light *trans–cis* isomerization response of azobenzene photo-switches. Generally speaking, the absorption wavelength redshift of azobenzene photo-switches will lead to the acceleration of thermal recovery *cis–trans* isomers, which is not conducive to the long-term storage of solar energy by Azo-STFs. As a result, it is of vital significance to develop Azo-STFs that can achieve redshift of absorption wavelength without significant change of *cis–trans* isomerization rate in future practical applications.

2.5 Direction in future

Azo-STFs have been proven to be an efficient and promising solar energy utilization system. After decades of research and development, from the initial perspective that azobenzene cannot store photo-thermal energy for a long term to the emergence and development of various azobenzene derivatives, azobenzene phase-transition systems, and azobenzene-template systems, Azo-STFs undoubtedly shows the potential of efficient utilization of solar energy and broad application prospect in the future. However, the rapid development of azobenzene solar energy storage materials is still limited by some challenges, including low energy density, short energy storage-life, low solar spectral utilization, single utilization mode, and incomplete heat release. Herein, we summarize the obstacles to be solved in the future development of azobenzene STF and the main research directions to overcome.

(I) Azo-STFs are usually applied to practical solar photothermal storage and controlled release in the form of solid-state film, but the preparation of solid-state film mainly relies on spin-coating technology. Therefore, research direction in the future can focus on the development of other preparation methods (electrostatic spinning, 3D printing, etc.) to realize the production and preparation of large-scale thin films, and solid thin films can be integrated in the equipment of photothermal energy utilization (solar water heater, solar photothermal blanket, solar fabric, solar thermal insulation devices, solar photothermal tactile sensing devices, information encryption and anti-counterfeiting, intelligent temperature control windows and coatings, etc.), so that Azo-STFs to diversification and functional and intelligent direction of development.

(II) The energy stored in Azo-STFs needs to be released under the conditions of external heating or visible light irradiation. Photo-triggered heat release usually leads to incomplete heat release of Azo-STFs, and thermal-trigger is bound to require new energy consumption. If the external new energy consumption is greater than the energy released by azobenzene, it will cause energy loss in azobenzene photothermal storage system, and this is not advisable. If the new external energy consumption is less than the energy released by azobenzene, although there is energy surplus, it still depends on the external energy input to stimulate the system to release heat, which limits the application of Azo-STFs to a certain extent. Therefore, it is an attractive research direction in the future to develop new stimulus response methods (chemical catalysis [141–143], electrical stimulation [99,144–147], mechanical stimulation [148–150], etc.) instead of traditional light and heat stimulation.

References

[1] Hartley GS. (1937). The Cis-form of azobenzene. *Nature*, **140**, pp. 281.

[2] Hosono N, Kajitani T, Fukushima T, Ito K, Sasaki S, Takata M, Aida T. (2010). Large-area three-dimensional molecular ordering of a polymer brush by one-step processing. *Science*, **330**, pp. 808–811.

[3] Bisoyi HK, Li Q. (2016). Light-driven liquid crystalline materials: from photo-induced phase transitions and property modulations to applications. *Chem Rev*, **116**, pp. 15089–15166.

[4] Delaire JA, Nakatani K. (2000). Linear and nonlinear optical properties of photochromic molecules and materials. *Chem Rev*, **100**, pp. 1817–1846.

[5] Matharu AS, Jeeva S, Ramanujam PS. (2007). Liquid crystals for holographic optical data storage. *Chem. Soc Rev*, **36**, pp. 1868–1880.

[6] Ikeda T, Tsutsumi O. (1995). Optical switching and image storage by means of azobenzene liquid-crystal films. *Science*, **268**, pp. 1873–1875.

[7] Taoda H, Hayakawa K, Kawase K, Yamakita H. (1987). Photochemical conversion and storage of solar energy by azobenzene. *J Chem Eng. Jpn*, **20**, pp. 265–270.

[8] Beharry AA, Woolley GA. (2011). Azobenzene photoswitches for biomolecules. *Chem Soc Rev,* **40**, pp. 4422–4437.

[9] Sekkat Z, Knoll W. (2002). *Photoreactive Organic Thin Films*. Elsevier.

[10] Pang X, Lv J, Zhu C, Qin L, Yu Y. (2019). Photodeformable azobenzene — containing liquid crystal polymers and soft actuators. *Adv Mater,* **31**, pp. 1904224.

[11] Natansohn A, Rochon P. (2002). Photoinduced motions in azo-containing polymers. *Chem Rev,* **102**, pp. 4139–4176.

[12] Research Gate. (2022) Polymerazobenzene complexes from supramolecular concepts to efficient photoresponsive. *Polymer.*

[13] Dong L, Feng Y, Wang L, Feng W. (2018). Azobenzene-based solar thermal fuels: design, properties, and applications. *Chem Soc Rev*, **47**, pp. 7339–7368.

[14] Kucharski TJ, Tian Y, Akbulatov S, Boulatov R. (2011). Chemical solutions for the closed-cycle storage of solar energy. *Energy Environ Sci*, **4**, pp. 4449–4472.

[15] Bren' VA, Dubonosov AD, Minkin VI, Chernoivanov VA. (1991). Norbornadiene quadricyclane an effective molecular system for the storage of solar energy. *Russ Chem Rev*, **60**, pp. 451.

[16] CE Center. (2023). *What's So Cool About Cool Roofs*. CEU_WhatsSoCool. Pdf. https://coolroofs.org/documents/CEU_WhatsSoCool.pdf (accessed 2023-12-18).

[17] Adamson AW, Vogler A, Kunkely H, Wachter R. (1978). Photocalorimetry. Enthalpies of photolysis of trans-azobenzene, ferrioxalate and cobaltioxalate ions, chromium hexacarbonyl, and dirhenium decarbonyl. *J Am Chem Soc*, **100**, pp. 1298–1300.

[18] Olmsted J, Lawrence J, Yee GG. (1983). Photochemical storage potential of azobenzenes. *Solar Energy*, **30**, pp. 271–274.

[19] Masutani K, Morikawa M, Kimizuka N. (2014). A liquid azobenzene derivative as a solvent-free solar thermal fuel. *Chem Commun*, **50**, pp. 15803–15806.

[20] Zhitomirsky D, Cho E, Grossman JC. (2016). Solid-state solar thermal fuels for heat release applications. *Adv Energy Mater*, **6**, pp. 1502006.

[21] Kolpak AM, Grossman JC. (2011). Azobenzene-functionalized carbon nanotubes as high-energy density solar thermal fuels. *Nano Lett*, pp. 113156–3162.

[22] Kolpak AM, Grossman JC. (2013). Hybrid chromophore/template nanostructures: a customizable platform material for solar energy storage and conversion. *J Chem Phys*, **138**, pp. 034303.

[23] Kucharski TJ, Ferralis N, Kolpak AM, Zheng JO, Nocera DG, Grossman JC. (2014). Templated assembly of photoswitches significantly increases the energy-storage capacity of solar thermal fuels. *Nat Chem*, **6**, pp. 441–447.

[24] Zhitomirsky D, Grossman JC. (2016). Conformal electroplating of azobenzene-based solar thermal fuels onto large-area and fiber geometries. *ACS Appl Mater Interfaces*, **8**, pp. 26319–26325.
[25] Ishiba K, Morikawa M, Chikara C, Yamada T, Iwase K, Kawakita M, Kimizuka N. (2015). Photoliquefiable ionic crystals: a phase crossover approach for photon energy storage materials with functional multiplicity. *Angew Chem Int Ed*, **54**, pp. 1532–1536.
[26] Al-Yasiri Q, Szabó M. (2021). Incorporation of phase change materials into building envelope for thermal comfort and energy saving: a comprehensive analysis. *J Build Eng*, **36**, pp. 102122.
[27] da Cunha SRL, de Aguiar JLB. (2020). Phase change materials and energy efficiency of buildings: a review of knowledge. *J Energy Sto*, **27**, pp. 101083.
[28] Mondal S. (2008). Phase change materials for smart textiles — An overview. *Appl Thermal Eng,* **28**, pp. 1536–1550.
[29] Niu Z, Yuan W. (2021). Smart nanocomposite nonwoven wearable fabrics embedding phase change materials for highly efficient energy conversion–storage and use as a stretchable conductor. *ACS Appl Mater Interfaces*, **13**, pp. 4508–4518.
[30] Omara AAM, Abuelnour AAA. (2019). Improving the performance of air conditioning systems by using phase change materials: a review. *Int J Energy Res*, **43**(10), pp. 5175–5198.
[31] Takudzwa Muzhanje A, Hassan MA, Hassan H. (2022). Phase change material based thermal energy storage applications for air conditioning: review. *Appl Ther Eng*, **214**, pp. 118832.
[32] Kahwaji S, Johnson MB, Kheirabadi AC, Groulx D, White MA. (2018). A comprehensive study of properties of paraffin phase change materials for solar thermal energy storage and thermal management applications. *Energy*, **162**, pp. 1169–1182.
[33] Qiu L, Ouyang Y, Feng Y, Zhang X. (2019). Review on micro/nano phase change materials for solar thermal applications. *Renew Energy*, **140**, pp. 513–538.
[34] Liu M, Saman W, Bruno F. (2012). Review on storage materials and thermal performance enhancement techniques for high temperature phase change thermal storage systems. *Renew Sust Energy Rev*, **16**, pp. 2118–2132.
[35] Farid MM, Khudhair AM, Razack SAK, Al-Hallaj S. (2004). A review on phase change energy storage: materials and applications. *Energy Convers Manage*, **45**, pp. 1597–1615.
[36] Xiao M, Feng B, Gong K. (2002). Preparation and performance of shape stabilized phase change thermal storage materials with high thermal conductivity. *Energy Convers Manage*, **43**, pp. 103–108.

[37] Magendran SS, Khan FSA, Mubarak NM, Vaka M, Walvekar R, Khalid M, Abdullah EC, Nizamuddin S, Karri RR. (2019). Synthesis of organic phase change materials (PCM) for energy storage applications: a review. *Nano-Struct Nano-Objects*, **20**, pp. 100399.

[38] Behzadi S, Farid MM. (2014). Long term thermal stability of organic PCMs. *Appl Energy*, **122**, pp. 11–16.

[39] Nazir H, Batool M, Bolivar Osorio FJ, Isaza-Ruiz M, Xu X, Vignarooban K, Phelan P, Inamuddin, Kannan AM. (2019). Recent developments in phase change materials for energy storage applications: a review. Inter. *J Heat Mass Tran*, **129**, pp. 491–523.

[40] Chen R, Yao R, Xia W, Zou R. (2015). Electro/photo to heat conversion system based on polyurethane embedded graphite foam. *Appl Energy*, **152**, pp. 183–188.

[41] Kou Y, Sun K, Luo J, Zhou F, Huang H, Wu Z-S, Shi Q. (2021). An intrinsically flexible phase change film for wearable thermal managements. *Energy Storage Mater*, **34**, pp. 508–514.

[42] Tang B, Wang Y, Qiu M, Zhang S. (2014). A full-band sunlight-driven carbon nanotube/PEG/SiO_2 composites for solar energy storage. *Sol. Energy Mater Sol C*, **123**, pp. 7–12.

[43] Wang Y, Tang B, Zhang S. (2014). Organic, cross-linking, and shape-stabilized solar thermal energy storage materials: a reversible phase transition driven by broadband visible light. *Appl Energy*, **113**, pp. 59–66.

[44] Tang B, Wang L, Xu Y, Xiu J, Zhang S. (2016). Hexadecanol/phase change polyurethane composite as form-stable phase change material for thermal energy storage. *Sol. Energy Mater. Sol. C*, **144**, pp. 1–6.

[45] Liu C, Rao Z, Zhao J, Huo Y, Li Y. (2015). Review on nanoencapsulated phase change materials: preparation, characterization and heat transfer enhancement. *Nano Energy*, **13**, pp. 814–826.

[46] Giro-Paloma J, Martínez M, Cabeza LF, Fernández AI. (2016). Types, methods, techniques, and applications for microencapsulated phase change materials (MPCM): a review. *Renew Sust Energy Rev*, **53**, pp. 1059–1075.

[47] Jiang Z, Yang W, He F, Xie C, Fan J, Wu J, Zhang K. (2018). Microencapsulated paraffin phase-change material with calcium carbonate shell for thermal energy storage and solar-thermal conversion. *Langmuir*, **34**, pp. 14254–14264.

[48] Liu H, Wang X, Wu D. (2017). Fabrication of graphene/TiO2/paraffin composite phase change materials for enhancement of solar energy efficiency in photocatalysis and latent heat storage. *ACS Sust Chem Eng*, **5**, pp. 4906–4915.

[49] Huang H, Shi T, He R, Wang J, Chu PK, Yu X-F. (2020). Phase-changing microcapsules incorporated with black phosphorus for efficient solar energy storage. *Adv Sci*, **7**, pp. 2000602.

[50] Zhao Y, Kong W, Jin Z, Fu Y, Wang W, Zhang Y, Liu J, Zhang B. (2018). Storing solar energy within Ag-Paraffin@Halloysite microspheres as a novel self-heating catalyst. *Appl Energy*, **222**, pp. 180–188.

[51] Norikane Y, Uchida E, Tanaka S, Fujiwara K, Koyama E, Azumi R, Akiyama H, Kihara H, Yoshida M. (2014). Photoinduced crystal-to-liquid phase transitions of azobenzene derivatives and their application in photolithography processes through a solid–liquid patterning. *Org Lett*, **16**, pp. 5012–5015.

[52] Hu J, Li X, Ni Y, Ma S, Yu H. (2018). A programmable and biomimetic photo-actuator: a composite of a photo-liquefiable azobenzene derivative and commercial plastic film. *J Mater Chem C*, **6**, pp. 10815–10821.

[53] Hu J, Huang S, Yu M, Yu H. (2019). Flexible solar thermal fuel devices: composites of fabric and a photoliquefiable azobenzene derivative. *Adv Energy Mater*, **9**, pp. 1901363.

[54] Yang Y, Huang S, Ma Y, Yi J, Jiang Y, Chang X, Li Q. (2022). Liquid and photoliquefiable azobenzene derivatives for solvent-free molecular solar thermal fuels. *ACS Appl Mater Inter*, **14**, pp. 35623–35634.

[55] Morikawa M, Yamanaka Y, Kimizuka. (2022). Liquid bisazobenzenes as molecular solar thermal fuel with enhanced energy density. *Chem Lett*, **51**, pp. 402–406.

[56] Liu H, Feng Y, Feng W. (2020). Alkyl-grafted azobenzene molecules for photo-induced heat storage and release via integration function of phase change and photoisomerization. *Compos Commun*, **21**, pp. 100402.

[57] Yang W, Feng Y, Si Q, Yan Q, Long P, Dong L, Fu L, Feng W. (2018). Efficient cycling utilization of solar-thermal energy for thermochromic displays with controllable heat output. *J Mater Chem A*, **7**, pp. 97–106.

[58] Zhao R, Li Y, Bai J, Mu J, Chen L, Zhang N, Han J, Liu F, Yan S. (2022). Preparation of flexible photo-responsive film based on novel photo-liquefiable azobenzene derivative for solar thermal fuel application. *Dyes Pigm*, **202**, pp. 110277.

[59] Han GGD, Li H, Grossman JC. (2017). Optically-controlled long-term storage and release of thermal energy in phase-change materials. *Nat Commun*, **8**, pp. 1446.

[60] Corruccini RJ, Gilbert EC. (1939). The heat of combustion of cis- and trans-Azobenzene. *J Am Chem Soc*, **61**, pp. 2925–2927.

[61] Han GGD, Deru JH, Cho EN, Grossman JC. (2018). Optically-regulated thermal energy storage in diverse organic phase-change materials. *Chem Commun*, **54**, pp. 10722–10725.

[62] Liu H, Tang J, Dong L, Wang H, Xu T, Gao W, Zhai F, Feng Y, Feng W. (2021). Optically triggered synchronous heat release of phase-change enthalpy and photo-thermal energy in phase-change materials at low temperatures. *Adv Funct Mater*, **31**, pp. 2008496.

[63] Zhang N, Yuan Y, Cao X, Du Y, Zhang Z, Gui Y. (2018). Latent heat thermal energy storage systems with solid–liquid phase change materials: A review. *Adv Eng Mater*, **20**, pp. 1700753.

[64] Shen J, Zhang P, Song L, Li J, Ji B, Li J, Chen L. (2019). Polyethylene glycol supported by phosphorylated polyvinyl alcohol/graphene aerogel as a high thermal stability phase change material. *Compos Part B: Eng*, **179**, pp. 107545.

[65] Jiang Y, Liu J, Luo W, Quan X, Li H, Huang J, Feng W. (2021). High-energy and light-actuated phase change composite for solar energy storage and heat release. *Surf Interfaces*, **24**, pp. 101071.

[66] Cai W, Sun Z-G, Ma H-K. (2014). An experimental study of the property of phase transition temperature for organic composites phase change materials. *Atlantis Press*, pp. 309–313.

[67] Tang J, Feng Y, Feng W. (2021). Photothermal storage and controllable release of a phase-change azobenzene/aluminum nitride aerogel composite. *Compos Commun*, **23**, pp. 100575.

[68] Gao J, Feng Y, Fang W, Wang H, Ge J, Yang X, Yu H, Qin M, Feng W. (2023). Co-harvest phase-change enthalpy and isomerization energy for high-energy heat output by controlling crystallization of alkyl-grafted azobenzene molecules. *Energy Environ Mater*, pp. e12607.

[69] Cristol SJ. (1986). The synthesis of quadricyclenes: a retrospective look. *Tetrahedron*, **42**, pp. 1617–1619.

[70] Dauben WG, Cargill RL. (1961). Photochemical transformations — VIII: the isomerization of Δ2,5-bicyclo[2.2.1]heptadiene to quadricyclo [2.2.1.02,6.03,5]heptane (quadricyclene). *Tetrahedron*, **15**, pp. 197–201.

[71] Cristol SJ, Snell RL. (1958). Bridged polycyclic compounds. VI. The photoisomerization of Bicyclo [2, 2, 1] hepta-2, 5-diene-2, 3-dicarboxylic acid to Quadricyclo [2, 2, 1, 02, 6, 03, 5] heptane-2, 3-dicarboxylic Acid1, 2. *J Am Chem Soc*, **80**, pp. 1950–1952.

[72] Dilling WL. (2002). *Intramolecular Photochemical Cycloaddition of Nonconjugated Olefins*. ACS Publications.

[73] Miki S, Asako Y, Yoshida Z. (1987). Photochromic solid films prepared by doping with donor–acceptor norbornadienes. *Chem Lett*, **16**, pp. 195–198.

[74] Turro NJ, Cherry WR, Mirbach MF, Mirbach MJ. (1977). Energy acquisition, storage, and release. Photochemistry of cyclic azoalkanes as alternate entries to the energy surfaces interconnecting norbornadiene and quadricyclene. *J Am Chem Soc*, **99**, pp. 7388–7390.

[75] Weiß C. (2022). *Novel Concepts of Closed Cycle Energy Storage and Release in Molecular Solar Thermal Energy Storage Systems Based on the Valence Isomers Norbornadiene and Quadricyclane*. Friedrich-Alexander-Universität Erlangen-Nürnberg (FAU).

[76] Cuppoletti A, Dinnocenzo JP, Goodman JL, Gould IR. (1999). Bond-coupled electron transfer reactions: Photoisomerization of norbornadiene to quadricyclane. *J Phys Chem A*, **103**, pp. 11253–11256.

[77] Helms AM, Caldwell RA, Triplet Species from Norbornadiene. (1995). Time-resolved photoacoustic calorimetry and ab initio studies of energy, geometry, and spin-orbit coupling. *J Am Chem Soc*, **117**, pp. 358.

[78] Hammond GS, Wyatt P, DeBoer CD, Turro NJ. (2002). *Photosensitized Isomerization Involving Saturated Centers*. ACS Publications.

[79] Smith CD. (2003). *Quadricyclane, in: Organic Syntheses*. John Wiley & Sons, Ltd., pp. 133–133.

[80] Cahill P, Steppel R. (2004). Process of quadricyclane production. US20040054244A1.

[81] Taoda H, Hayakawa K, Kawase K. (1987). Photosensitized isomerization of imorbornadiene for solar energy storage and suppression of side reaction. *J Chem Eng Jpn*, **20**, pp. 335–338.

[82] Grutsch PA, Kutal C. (1986). Charge-transfer sensitization of the valence photoisomerization of norbornadiene to quadricyclene by an orthometalated transition-metal complex. *J Am Chem Soc*, **108**, pp. 3108–3110.

[83] Yoshida Z. (1985). New molecular energy storage systems. *J Photochem*, **29**, pp. 27–40.

[84] Ikezawa H, Kutal C, Yasufuku K, Yamazaki H. (1986). Direct and sensitized valence photoisomerization of a substituted norbornadiene. Examination of the disparity between singlet- and triplet-state reactivities. *J Am Chem Soc*, **108**, pp. 1589–1594.

[85] Tranmer GK, Yip C, Handerson S, Jordan RW, Tam W. (2000). Synthesis of 2,3-disubstituted norbornadienes. *Can J Chem*, **78**, pp. 527–535.

[86] Gray V, Lennartson A, Ratanalert P, Börjesson K, Moth-Poulsen K. (2014). Diaryl-substituted norbornadienes with red-shifted absorption for molecular solar thermal energy storage. *Chem Commun*, **50**, pp. 5330–5332.

[87] Saydjari AK, Weis P, Wu S. (2017). Spanning the solar spectrum: azopolymer solar thermal fuels for simultaneous UV and visible light storage. *Adv Energy Mater*, **7**, pp. 1601622.

[88] Miki S, Asako Y, Yoshida Z. (1987). Photochromic solid films prepared by doping with donor–acceptor norbornadienes. *Chem Lett*, **16**, pp. 195–198.

[89] Iizawa T, Hijikata C, Nishikubo T. (1992). Synthesis and solar energy storage property of polymers containing norbornadiene moieties. *Macromolecules*, **25**, pp. 21–26.

[90] Nishikubo T, Kameyama A, Kishi K, Kawashima T, Fujiwara T, Hijikata C. (2002). *Synthesis of New Photoresponsive Polymers Bearing Norbornadiene Moieties by Selective Cationic Polymerization of 2 [[(3-Phenyl-2,5-Norbornadienyl)-2.Carbonyl]Oxy]Ethyl Vinyl Ether and Photochemical Reaction of the Resulting Polymers*. ACS Publications.

[91] Oueslati A, Abdelhedi I, Mercier R, Drockenmuller E, Ben Romdhane H. (2013). Photoresponsive polyamides containing pentamethylated norbornadiene moieties: synthesis and photochemical properties under sunlight irradiation. *J Poly Sci Part A: Poly Chem*, **51**, pp. 4650–4656.

[92] Petersen AU, Hofmann AI, Fillols M, Mansø M, Jevric M, Wang Z, Sumby CJ, Müller C, Moth-Poulsen K. (2019). Solar energy storage by molecular norbornadiene–quadricyclane photoswitches: polymer film devices. *Adv Sci*, **6**, pp. 1900367.

[93] Koser GF, Faircloth JN. (1976). Silver(I)-promoted reactions of strained hydrocarbons. Oxidation vs. rearrangement. *J Org Chem*, **41**, pp. 583–585.

[94] Maruyama K, Tamiaki H, Kawabata S. (1985). Development of a solar energy storage process. Photoisomerization of a norbornadiene derivative to a quadricyclane derivative in an aqueous alkaline solution. *J Org Chem*, **50**, pp. 4742–4749.

[95] Wang Z, Roffey A, Losantos R, Lennartson A, Jevric M, Petersen AU, Quant M, Dreos A, Wen X, Sampedro D, Börjesson K, Moth-Poulsen K. (2019). Macroscopic heat release in a molecular solar thermal energy storage system. *Energy Environ Sci*, **12**, pp. 187–193.

[96] Gassman PG,Yamaguchi R, Koser GF. (1978). The ease of oxidation of highly strained polycyclic molecules. *J Org Chem*, **43**, pp. 4392–4393.

[97] Brummel O, Besold D, Döpper T, Wu Y, Bochmann S, Lazzari F, Waidhas F, Bauer U, Bachmann P, Papp C, Steinrück H-P, Görling A, Libuda J, Bachmann J. (2016). Energy storage in strained organic molecules: (spectro) electrochemical characterization of norbornadiene and quadricyclane. *Chem Sus Chem*, **9**, pp. 1424–1432.

[98] Brummel O, Waidhas F, Bauer U, Wu Y, Bochmann S, Steinrück H-P, Papp C, Bachmann J, Libuda J. (2017). Photochemical energy storage and electrochemically triggered energy release in the norbornadiene–quadricyclane system: UV photochemistry and IR spectroelectrochemistry in a combined experiment. *J Phys Chem Lett*, **8**, pp. 2819–2825.

[99] Waidhas F, Jevric M, Fromm L, Bertram M, Görling A, Moth-Poulsen K, Brummel O, Libuda J. (2019). Electrochemically controlled energy storage in a norbornadiene-based solar fuel with 99% reversibility. *Nano Energy*, **63**, pp. 103872.

[100] Meng F-Y, Chen I-H, Shen J-Y, Chang K-H, Chou T-C, Chen Y-A, Chen Y-T, Chen C-L, Chou P-T. (2022). A new approach exploiting thermally activated delayed fluorescence molecules to optimize solar thermal energy storage. *Nat Commun*, **13**, pp. 797.

[101] Tanaka H, Shizu K, Miyazaki H, Adachi C. (2012). Efficient green thermally activated delayed fluorescence (TADF) from a phenoxazine–triphenyltriazine (PXZ–TRZ) derivative. *Chem Commun*, **48**, pp. 11392–11394.

[102] Stranius K, Börjesson K. (2017). Determining the photoisomerization quantum yield of photoswitchable molecules in solution and in the solid state. *Sci Rep*, **7**, pp. 41145.

[103] Siampiringue N, Guyot G, Monti S, Bortolus P. (1987). The cistrans photoisomerization of azobenzene: an experimental re-examination. *J Photochem*, **37**, pp. 185–188.

[104] Gegiou D, Muszkat KA, Fischer E. (1968). Temperature dependence of photoisomerization. VI. Viscosity effect. *J Am Chem Soc*, **90**, pp. 12–18.

[105] Lednev IK, Ye T-Q, Matousek P, Towrie M, Foggi P, Neuwahl FVR, Umapathy S, Hester RE, Moore JN. (1998). Femtosecond time-resolved UV-visible absorption spectroscopy of trans-azobenzene: dependence on excitation wavelength. *Chem Phys Lett*, **290**, pp. 68–74.

[106] Bortolus P, Monti S. (1979). Cis-trans photoisomerization of azobenzene. Solvent and triplet donors effects. *J Phys Chem*, **83**, pp. 648–652.

[107] Fischer E, Frankel M, Wolovsky R. (1955). Wavelength dependence of photoisomerization equilibria in azocompounds. *J Chem Phys*, **23**, pp. 1367–1367.

[108] Fischer R. (1960). Temperature dependence of photoisomerization equilibria. Part I. Azobenzene and the Azonaphthalenes. *J Am Chem Soc*, **82**, pp. 3249–3252.

[109] Malkin S, Fischer E. (1962). Temperature dependence of photoisomerization. Part II. 1 quantum yields of cis trans isomerizations in azocompounds. *J Phys Chem*, **66**, pp. 2482–2486.

[110] Yoshii K, Machida S, Horie K. (2000). Local free volume and structural relaxation studied with photoisomerization of azobenzene and persistent spectral hole burning in poly(alkyl methacrylate)s at low temperatures. *J Poly Sci Part B: Poly Phys*, **38**, pp. 3098–3105.

[111] Bassotti E, Carbone P, Credi A, Di Stefano M, Masiero S, Negri F, Orlandi G, Spada GP. (2006). Effect of strain on the photoisomerization and stability of a congested Azobenzenophane: a combined experimental and computational study. *J Phys Chem A*, **110**, pp. 12385–12394.

[112] Asano T, Yano T, Okada T. (1982). Mechanistic study of thermal Z-E isomerization of azobenzenes by high-pressure kinetics. *J Am Chem Soc*, **104**, pp. 4900–4904.

[113] Crecca CR, Roitberg AE. (2006). Theoretical study of the isomerization mechanism of azobenzene and disubstituted azobenzene derivatives. *J Phys Chem A*, **110**, pp. 8188–8203.

[114] Bandara HMD, Burdette SC. (2012). Photoisomerization in different classes of azobenzene. *Chem Soc Rev*, **41**, pp. 1809–1825.

[115] Pozhidaeva N, Cormier M-E, Chaudhari A, Woolley GA. (2004). Reversible photocontrol of peptide helix content: Adjusting thermal stability of the cis state. *Bioconjugate Chem*, **15**, pp. 1297–1303.

[116] Siewertsen R, Neumann H, Buchheim-Stehn B, Herges R, Näther C, Renth F, Temps F. (2009). Highly efficient reversible Z–E photoisomerization of a bridged azobenzene with visible light through resolved $S1(n\pi^*)$ absorption bands. *J. Am Chem Soc*, **131**, pp. 15594–15595.

[117] Duval H. (1910). Recherches sur la benzidination. *Bull Soc Chim Fr,* **7**, pp. 727–732.

[118] Kellogg RM, Schaap AP, Wynberg H. (1969). Bromination, deuteration, and lithiation of the dithienyls. *J Org Chem*, **34**, pp. 343–346.

[119] Tauer E, Machinek R. (1996). ortho-Azobenzenophanes: photochemical and thermal reactions of [2,2][2,2′]azobenzenophane. *Liebigs Annalen*, **1996**(7), pp. 1213–1216.

[120] Hammerich M, Schütt C, Stähler C, Lentes P, Röhricht F, Höppner R, Herges R. (2016). Heterodiazocines: synthesis and photochromic properties, trans to cis switching within the bio-optical window. *J Am Chem Soc*, **138**, pp. 13111–13114.

[121] NishiokaH,LiangX,KashidaH,AsanumaH.(2007).2′,6′-Dimethylazobenzene as an efficient and thermo-stable photo-regulator for the photoregulation of DNA hybridization. *Chem Commun*, pp. 4354–4356.

[122] Beharry AA, Sadovski O, Woolley GA. (2011). Azobenzene photoswitching without ultraviolet light. *J Am Chem Soc*, **133**, pp. 19684–19687.

[123] Dokić J, Gothe M, Wirth J, Peters MV, Schwarz J, Hecht S, Saalfrank P. (2009). Quantum chemical investigation of thermal cis-to-trans isomerization of azobenzene derivatives: substituent effects, solvent effects, and comparison to experimental data. *J Phys Chem A*, **113**, pp. 6763–6773.

[124] Nishimura N, Sueyoshi T, Yamanaka H, Imai E, Yamamoto S, Hasegawa S. (1976). Thermal cis-to-trans isomerization of substituted azobenzenes II. Substituent and solvent effects. *BCSJ*, **49**, pp. 1381–1387.

[125] Han M, Ishikawa D, Honda T, Ito E, Hara M. (2010). Light-driven molecular switches in azobenzene self-assembled monolayers: effect of molecular structure on reversible photoisomerization and stable cis state. *Chem Commun*, **46**, pp. 3598–3600.

[126] Bléger D, Schwarz J, Brouwer AM, Hecht S. (2012). O-Fluoroazobenzenes as readily synthesized photoswitches offering nearly quantitative two-way isomerization with visible light. *J Am Chem Soc*, **134**, pp. 20597–20600.

[127] Bahrenburg J, Sievers CM, Schönborn JB, Hartke B, Renth F, Temps F, Näther C, Sönnichsen FD. (2013). Photochemical properties of multi-azobenzene compounds. *Photochem Photobiol Sci*, **12**, pp. 511–518.

[128] Wang L, Dong H, Li Y, Xue C, Sun L-D, Yan C-H, Li Q. (2014). Reversible near-infrared light directed reflection in a self-organized helical superstructure loaded with upconversion nanoparticles. *J Am Chem Soc*, **136**, pp. 4480–4483.

[129] Wang L, Dong H, Li Y, Liu R, Wang Y-F, Bisoyi HK, Sun L-D, Yan C-H, Li Q. (2015). Luminescence-driven reversible handedness inversion of self-organized helical superstructures enabled by a novel near-infrared light nanotransducer. *Adv Mater*, **27**, pp. 2065–2069.

[130] Bléger D, Hecht S. (2015). Visible-light-activated molecular switches. *Angew Chem Int Ed*, **54**, pp. 11338–11349.

[131] Dong M, Babalhavaeji A, Samanta S, Beharry AA, Woolley GA. (2015). Red-shifting azobenzene photoswitches for in vivo use. *Acc Chem Res*, **48**, pp. 2662–2670.

[132] Weis P, Wu S. (2018). Light-switchable azobenzene-containing macromolecules: from UV to near infrared. *Macromol Rapid Commun*, **39**, pp. 1700220.

[133] Wu S, Butt H-J. (2016). Near-infrared-sensitive materials based on upconverting nanoparticles. *Adv Mater*, **28**, pp. 1208–1226.

[134] Wang L, Gutierrez-Cuevas KG, Urbas A, Li Q. (2016). Near-infrared light-directed handedness inversion in plasmonic nanorod-embedded helical superstructure. *Adv Opt Mater*, **4**, pp. 247–251.

[135] Gutierrez-Cuevas KG, Wang L, Xue C, Singh G, Kumar S, Urbas A, Li Q. (2015). Near infrared light- driven liquid crystal phase transition enabled

by hydrophobic mesogen grafted plasmonic gold nanorods. *Chem Commun*, **51**, pp. 9845–9848.

[136] Sadovski O, Beharry AA, Zhang F, Woolley GA. (2009). Spectral tuning of azobenzene photoswitches for biological applications. *Angew. Chem Int Ed*, **48**, pp. 1484–1486.

[137] Mourot A, Kienzler MA, Banghart MR, Fehrentz T, Huber FME, Stein M, Kramer RH, Trauner D. (2011). Tuning photochromic ion channel blockers. *ACS Chem Neurosci.*, **2**, pp. 536–543.

[138] Chi L, Sadovski O, Woolley GA. (2006). A blue-green absorbing cross-linker for rapid photoswitching of peptide helix content. *Bioconjugate Chem*, **17**, pp. 670–676.

[139] Samanta S, Beharry AA, Sadovski O, McCormick TM, Babalhavaeji A, Tropepe V, Woolley GA. (2013). Photoswitching azo compounds in vivo with red light. *J Am Chem Soc*, **135**, pp. 9777–9784.

[140] Samanta S, McCormick TM, Schmidt SK, Seferos DS, Woolley GA. (2013). Robust visible light photoswitching with ortho-thiol substituted azobenzenes. *Chem Commun*, **49**, pp. 10314–10316.

[141] Wu W, Yao L, Yang T, Yin R, Li F, Yu Y. (2011). NIR-light-induced deformation of cross-linked liquid-crystal polymers using upconversion nanophosphors. *J Am Chem Soc*, **133**, pp. 15810–15813.

[142] Titov E, Lysyakova L, Lomadze N, Kabashin AV, Saalfrank P, Santer S. (2015). Thermal cis-to-trans isomerization of azobenzene-containing molecules enhanced by gold nanoparticles: an experimental and theoretical study. *J Phys Chem C*, **119**, pp. 17369–17377.

[143] Chu Z, Han Y, Bian T, De S, Král P, Klajn R. (2019). Supramolecular control of azobenzene switching on nanoparticles. *J Am Chem Soc*, **141**, pp. 1949–1960.

[144] Moldt T, Przyrembel D, Schulze M, Bronsch W, Boie L, Brete D, Gahl C, Klajn R, Tegeder P, Weinelt M. (2016). Differing isomerization kinetics of azobenzene-functionalized self-assembled monolayers in ambient air and in vacuum. *Langmuir*, **32**, pp. 10795–10801.

[145] Williams DE, Martin CR, Dolgopolova EA, Swifton A, Godfrey DC, Ejegbavwo OA, Pellechia PJ, Smith MD, Shustova NB. (2018). Flipping the switch: fast photoisomerization in a confined environment. *J Am Chem Soc*, **140**, pp. 7611–7622.

[146] Greenfield JL, Gerkman MA, Gibson RSL, Han GGD, Fuchter MJ. (2021). Efficient electrocatalytic switching of azoheteroarenes in the condensed phases. *J Am Chem Soc*, **143**, pp. 15250–15257.

[147] Waidhas F, Jevric M, Bosch M, Yang T, Franz E, Liu Z, Bachmann J, Moth-Poulsen K, Brummel O, Libuda J. (2020). Electrochemically

controlled energy release from a norbornadiene-based solar thermal fuel: increasing the reversibility to 99.8% using HOPG as the electrode material. *J Mater Chem A*, **8**, pp. 15658–15664.

[148] Surampudi SK, Patel HR, Nagarjuna G, Venkataraman D. (2013). Mechanoisomerization of azobenzene. *Chem Commun*, **49**, pp. 7519–7521.

[149] Zheng X, Liu X, Liu L, Li X, Jiang S, Niu C, Xie P, Liu G, Cao Z, Ren Y, Qin Y, Wang J. (2022). Multi-stimuli-induced mechanical bending and reversible fluorescence switching in a single organic crystal. *Angew Chem Int Ed*, **61**, pp. e202113073.

[150] Berson J, Moosmann M, Walheim S, Schimmel T. (2019). Mechanically induced switching of molecular layers. *Nano Lett*, **19**, pp. 816–822.

Chapter 3

Azobenzene-Containing Polymers for Solar Thermal Fuels: Design and Synthesis

Abstract

The pursuit of sustainable global society has led scientists to extensively research and develop renewable energy technology. Solar thermal fuels (STFs) have attracted enormous interest for their ability to harness solar energy. By capturing photon energy, they can generate high-energy metastable isomers and release the stored energy as heat on demand through the reversible photoisomerization of molecular photo-switches. Recently, numerous azobenzene-based STFs have been fabricated by various methods including molecular engineering and template self-assembly, among which azobenzene-containing polymers (azopolymers) are considered the most promising materials for the purpose of developing STFs due to the advantages of their ease of forming uniform and flat films of controlled thickness. This chapter first introduces the fundamentals of designing and synthesizing azopolymers, including the properties of azobenzene, the advantages of azopolymers, and the key parameters for designing azopolymers. Then, recent research advances in azopolymers for solar energy storage and conversion are described including azobenzene compound/polymer composites, linear azopolymers, dendrimer azopolymers, and other types of azopolymers. Finally, several ways of triggering energy release about azopolymers are summarized, the most prospective advantages and present limitations of azopolymers for STFs are analyzed, and future research and application prospects are suggested.

3.1 Fundamentals of azopolymer STFs

The depletion of fossil fuels, global warming, and the potential dangers of international conflict have created an unsettling shadow over global energy prospects [1,2]. Addressing these crises requires developing renewable and sustainable energy technologies and strategies to reduce the reliance on conventional fossil fuels, which are now a major driver of the global economy [3,5]. Frequently mentioned alternatives encompass nuclear energy, clean coal, as well as various renewable energy sources like hydroelectric power, biomass/biofuels, geothermal energy, ocean energy, wind power, and solar energy [6,7]. Solar energy is the richest and infinite natural resource in the solar system: in 1 h, the Sun supplies what humanity needs in a year [8,9]. Over the last few years, there has been a great deal of work on the development of innovative technologies, particularly for the capture, conversion, and storage of solar energy, such as photovoltaics, solar thermal collectors, photovoltaic solar batteries, and STFs [10–12]. Especially, STFs have attracted great attention for solar energy conversion and storage, which can harvest, convert, store photon energy, and release the stored energy as heat [13–16]. In recent years, many azobenzene-based STFs have been developed such as azobenzene small molecule, azobenzene–nanocarbon hybrids and azobenzene-containing polymers [17–20]. As one of the superior candidates for the development of STFs, azobenzene-containing polymers (azopolymers) or polymer composites not only can form uniformly flat films, but also display the feasibility of large-area exotherm [21–24]. The uniform azopolymer films aid in the absorption of photons, and heat can be released both in the solid-state and by regulating their *cis*-to-*trans* conversion, which offers the possibility to incorporate STF materials into a number of current heaters [25,26]. Over the last couple of decades, scientists have been very enthusiastic about the properties of azopolymers, such as photoinduced phase change, photoinduced birefringence and so on [27–31]. Therefore, a great number of photoactive azopolymers have been developed. In this chapter, we summarize systematically the key parameters and state-of-the-art progress on azobenzene-containing polymers from the perspectives of molecular design and synthesis. Various kinds of azobenzene-containing polymers for STFs such as azobenzene compound/polymer composites, linear azopolymers, dendrimer azopolymers, and other types of azopolymers are addressed. Several approaches to induce energy release from azo polymers are described, and the most promising advantages and challenges of azobenzene-containing polymers for STFs are analyzed.

3.1.1 *Properties of azobenzene*

STFs, a kind of latent thermal storage materials, can save the harvested solar energy in chemical bonds and release it as heat energy under certain conditions. Their properties offer the possibility of stable solar energy output. For optimum utilization of solar energy, the spectrum of STFs should approach the solar spectrum [32–34]. It is also necessary to focus on the amount of specific sunlight wavelengths that meet the conditions for STFs to be excited, and to make the STFs redshift to the visible region [35,36]. The STFs are required to have highly photochemically active chemical bonds, for example, a conjugated double bond and N=N. Various photo-switchable molecules have been suggested as STF candidates, including anthracene [37], stilbene [38,39], and fulvalene-tetracarbonyl-diruthenium [40,41]. So far, three systems have received the highest level of focus and are being subjected to extensive studies: the norbornadiene/quadricyclane (NBD/QC) couple [42–44], the dihydroazulene/vinylheptafulvene (DHA/VHF) couple [45–48] and the azobenzene [49–52]. The photoreactions of Norbornadiene (NBD) to quadricyclane (QC) were initially documented in 1958 for a dicarboxylic derivative and in 1961 for its unsubstituted form [53,54]. The corresponding photoreaction involves of a [2+2] cycloaddition between the alkene groups in NBD, resulting in the formation of QC. The reaction can proceed through direct irradiation or under sensitized conditions. NBD, however, only absorbs in the ultraviolet (UV) region, making it unsuitable for solar radiation [55–57]. Photoisomerization of DHA to VHF takes place via breaking of a carbon–carbon bond. Recent theoretical studies on the switching mechanism have demonstrated that planarization takes place after DHA photoexcitation, leading to the opening of the ring and subsequent degradation of the molecule to a VHF ground state via an internal transformation of S_1/S_0 [45]. VHF exhibits photochemical inactivity when exposed to irradiation. This unique property can greatly enhance the photo-stationary states within STF systems by utilizing a spectral overlap distinct from that of azobenzene, thereby preventing photoinduced back conversion under solar light [58,59]. However, there is a high degree of absorption overlap between VHF and DHA, and therefore there is potential competition for absorption [60].

Azobenzene, the isomerization process occurs from *trans*-to-*cis* form, exhibiting a considerable quantum yield and distinct characteristics between the *trans* and *cis* states [61–63]. Azobenzene photoisomerization is a special process in which the excess energy is dissipated by controlling the molecular configuration [64,65]. This transformation is frequently associated with other processes such as vibration relaxation, radiationless/radiative transitions, and energy transfer quenching [66,67]. Up to now, there have been four possible pathways for azobenzene photoisomerization, including inversion, rotation, coordination inversion, and inversion-assisted rotation. Most azobenzene molecules can be isomerized from the *trans* to *cis* form through UV irradiation, while a small fraction of molecules can be reduced to the *trans*-isomers [68]. The *cis*-isomer has the ability to revert to the *trans*-isomer by emitting heat when heated or exposed to visible light irradiation [69–71]. Hence, in contrast to other STF energy storage systems, researchers can explore the energy storage mechanism of azobenzene-based STFs more effectively without the need for additional chromophores. Furthermore, under light irradiation, azobenzene molecules undergo isomerization, leading to a slight alteration in molecular size where the molecular axis transitions from 9 Å to 5.5 Å [72,73]. The changes in molecular size resulting from the isomerization of azobenzene molecules can accumulate, leading to noticeable macroscopic deformations. Light-driven flexible actuators based on azobenzene photoresponsive polymers have fast, flexible and reversible light-induced deformations, making them promising for a variety of applications. These actuators find applications ranging from solar thermal storage to molecular motors, memory devices, and manipulators [74,75]. Various polymer structures such as main chain polymers, cross-linked networks, side-chain polymers have been designed and synthesized to achieve light-induced transformation by the forces of the azobenzene unit and the polymer [21,76]. This indicates that the controlled conversion of stored energy in azobenzene into mechanical energy is achievable. While it is true that the energy storage density of azobenzene is not as high as that of the NBD/QC derivative system (which can be up to 0.97 MJ kg^{-1}), the unique isomerization mechanisms involved in the processes of heat storage and release make the azobenzene energy storage system suitable for a broader range of potential applications. These applications can be utilized for solar thermal storage and conversion.

3.1.2 *Advantages of azopolymer STFs*

In 1978, Adamson *et al.* initially measured the storage enthalpy (49 kJ mol^{-1} or 269 J g^{-1}) of azobenzene and reported that the *cis*-isomer has a half-life of approximately 4.2 days [10]. Olmsted and colleagues have investigated the possibility for solar energy storage with substituted azobenzene [14]. Yet, they concluded that azobenzene is not suitable for photochemical solar storage due to its rapid thermal reduction rate and limited solubility in polar solvents. If the azobenzene solution is not stored in a cool and dark environment, the *cis*-isomer will release heat at high temperatures. Implementing these precautions enables the attainment of long-term storage and repeated utilization of azobenzene STFs. Recently, in order to improve the storage performances of azobenzene, a variety of functional azobenzene STFs have been proposed by researchers, including pure azobenzene derivatives, nano-carbon-templated azobenzene, and polymer-templated azobenzene [77–80]. In the beginning, azobenzene derivatives were deemed unsuitable for energy storage due to their short half-lives [81–83]. Nevertheless, as the demand for solid-state integration of small molecules in functional devices increased, molecular thin films became highly attractive owing to their versatility, ease of processing, and cost-effectiveness [84–86]. Subsequently, substantial computational and experimental efforts have been directed toward enhancing the performance of azobenzene-based STFs. These findings suggest that specifically targeted molecular engineering of azobenzene holds great potential as an approach for the design of solid-state STFs

[87–89]. While molecular engineering of azobenzene photo-switches has shown promise in improving their performance, challenges such as short half-life and limited exothermicity still need to be addressed [90]. In order to overcome these challenges, the template self-assembly strategy was created. It is expected that not only the energy density and storage half-life of the resultant STFs can not only be improved by templated azobenzene using a carbon-based material or a polymer, but also by thermal and chemical durability [91,92]. Many findings suggest that although nanocarbon templated azobenzene can significantly enhance the energy density and half-life of STFs, the introduction of nanocarbon materials may affect the light absorption of the material [93,94]. Furthermore, these studies focusing on carbon-based templated azobenzene suggest that similar principles can be employed in the design of polymer-templated STFs.

Polymer-templated azobenzene has the ability to generate uniform and flat films with controlled thicknesses, making them suitable for large-area heat release applications [14]. These uniform films, made of azopolymers, can efficiently capture photons and release heat through the controlled *cis*-to-*trans* isomerization, allowing for integration into various heating devices. Fortunately, in recent decades, azobenzene-containing polymers have been developed as photosensitive materials that combine the photo-switching properties of azobenzene with excellent mechanical properties and processability. Azopolymers have garnered significant attention in fundamental research due to their advanced and intricate structures compared to small-molecule azobenzenes, which opens up possibilities for innovative material design [21]. Some azopolymers exhibit collective, cooperative, and synergistic motions induced by light, resulting in intriguing photosensitive properties. Additionally, photosensitive azopolymers have found applications in information storage, photopharmacology, photolithography, photonics, solar energy storage, and actuation [25,75]. Consequently, a wide range of photoactive azopolymers has been developed, facilitating further exploration in this rapidly evolving field. While azopolymers may have lower storage energy densities and half-lives compared to carbon-templated azobenzene STFs, their ease of integration into current heating equipment presents tremendous potential for the advancement of novel applications in energy storage and heat release. This is the most important advantage of azopolymer STFs. Not only that, the studies of azopolymers with photo-induced reversible solid–liquid phase change properties in recent years not only enables the introduction of latent heat of phase change to enhance storage energy density, but also further expands their application potential as STFs. Therefore, azobenzene-containing polymers have attracted significant research interest as STF materials.

3.1.3 *Requisites for azopolymer STFs*

The exploitation of azopolymer STFs has become increasingly prominent over the past few decades due to the high expectations associated with this technology. However, the ideal azopolymer STFs remains elusive, prompting the employment of various strategies that combine experimental and computational approaches to assist in molecular design [95–97]. Achieving the optimal efficiency of STFs for nitrogen-containing polymers has gained significant attention, which requires many objectives to be satisfied in the design of suitable nitrogen-containing benzene molecules. Such design criteria are governed by several parameters, including engineering and other challenges [86,98]. The initial crucial step in the azopolymer STF involves light absorption by the parent molecule. This reversible photoisomerization reaction leads to the formation of its corresponding metastable isomer. This azobenzene molecule must possess sufficient stability to retain the stored chemical potential for an appropriate duration, dependent on the intended application [99–101]. Subsequently, the stored energy should be released as heat, precisely when and where it is needed [92,102]. To ensure the success of the initial photoisomerization stage in the STF cycle, the azobenzene pair must fulfill a comprehensive list of requirements, which may at times be contradictory:

- A significant disparity in free energies between the azobenzene molecule and its isomer, while minimizing the increase in molecular weight to maximize energy density.
- A reasonably high kinetic barrier for reverse conversion.
- An absorption spectrum of the azobenzene molecule that aligns with the solar spectrum.

- A high quantum yield for the efficient generation of the metastable isomer through photoisomerization.
- Minimal degradation of both the azobenzene molecule and its isomer during repeated cycles, particularly at elevated temperatures.

These are the primary criteria that need to be optimized in order to enhance the performance of any potential candidate [85,103]. Therefore, a more thorough elucidation of certain requirements pertaining to the photochemistry of these compounds will be provided: the absorption spectrum (matching with solar radiation), photoreaction efficiency (quantum yield), and energy storage capacity.

Solar Match: The primary emphasis of the photochemical aspect of the STF cycle lies in the conversion from a low-energy configuration to a high-energy configuration through a photochemical reaction. To enable this transformation, the primary requirement is the absorption of energy from sunlight. Therefore, ideal azopolymer STFs should absorb as many photons as possible in the UV light and visible light range of the solar spectrum, specifically between 300 nm and 700 nm [104]. Preferably, the absorption spectrum of the lower-energy isomer should align with the most intense region of the solar energy window (solar match), ideally within the range of 340–540 nm, where the solar radiation is relatively high. This would ensure efficient absorption of photons from sunlight for the photochemical process [105,106]. To prevent unwanted competition for photon absorption between the initial isomer and the metastable isomer, it is preferable to have no overlap in their absorption spectra. However, most of the azobenzene cores reported so far do not exhibit significant absorption beyond 350 nm, which is a drawback since the flux of solar photons at wavelengths below 330 nm is quite low [107]. To address this limitation, progress has been made both experimentally and computationally in developing functionalized azobenzene molecules that absorb at longer wavelengths. One successful chemical strategy involves incorporating donor–acceptor substituents to enhance the conjugated π-system of the molecule and creating a push–pull effect [108–113]. Ideally, when targeting the generation of relevant azobenzene molecules, it is preferable to focus on low molecular weight electron donor and acceptor groups.

This choice minimizes the impact on energy density, which is influenced by both the energy difference between isomers and the overall molecular weight [114,115]. Indeed, optimizing a cluster of molecules for STF applications is an ambitious task due to the potential adverse effects on other relevant properties beyond energy storage resulting from chemical modifications of azobenzene.

Quantum Yield: When the *trans*-isomer absorbs a photon from the sun, the transition from the ground state to the excited state occurs. Then some of the molecules will undergo photo-conversion, but some of them could undergo relaxation and return to their original state [96,116]. In order to measure and quantify the effectiveness of the photoconversion process from the azobenzene molecule to the *cis*-isomer, it is necessary to determine the quantum yield. This dimensionless parameter indicates the likelihood of the *trans*-isomer generating a transferrable high-energy *cis*-isomer for each absorbed photon. From an efficiency perspective, the photo-conversion leading to the high isomer should be maximized, ideally approaching unity (a value of 1). This should allow solar power to be converted into chemical energy in an efficient manner, thereby preventing other competing processes like radiation, non-radiation, or quenching [117–119].

Storage Energy Density: Energy storage is another key issue in azopolymer STFs. The purpose of the STF technique is to achieve the maximum temperature rise when the stored chemical energy is released as heat in the *cis*-isomer. Thus, the key to achieving this objective is the difference in enthalpy between isomers. That is, the greater the difference in energy between the *trans*-isomer and *cis*-isomer, the greater the accumulation of energy in the system [120–122]. As a general guideline, azopolymer STFs should have a minimum energy storage capacity of 0.3 MJ kg^{-1} in order to be practical for applications. This energy can be subsequently released through the application of an external stimulus, such as a temperature increase or with the help of a catalyst. Therefore, it is not desirable for an azobenzene molecule to undergo rapid reverse transformation at room temperature so that energy can be stored over a period of hours, days, or months (storage period), depending on the intended use. Although this chapter deals with the photochemical aspects of azopolymer STFs, it is important to note that there are alternative ways of cooling and heating. For example, the commercial utilization of phase change materials in zeolites and water adsorption has actually been achieved.

3.2 Research progress of azopolymer STFs

Azobenzene-based STFs have exhibited great potential for solar energy storage, stimulated heat release, and stable multi-cycling. However, this rapidly evolving research field is still constrained by low storage energy density and short half-life [9,33]. Azobenzene-containing polymer provide the potential to improve the storage performances of STFs by tuning the monomer structure and the polymer backbone and have therefore attracted great research enthusiasm. Not only that, azopolymers or polymer composites can also form uniformly flat, controlled thickness films that are feasible for large-area heat release, and allow heat release to occur in solid state. Herein, the research progress of azopolymer STFs are introduced such as azobenzene compound/polymer composites, linear azopolymers, dendrimer azopolymers, and other types of azopolymers.

3.2.1 *Azobenzene compound/polymer composites*

Azobenzene compound/polymer composites are a class of materials that allow for applications in solar energy storage and conversion by composing azobenzene compounds with polymer templates such as fabrics and aerogels. These composite azobenzene compounds usually have photoinduced reversible solid-liquid phase change properties, and this can enable solvent-free energy storage of azobenzene/polymer composites and increase the storage energy densities. Meanwhile, the introduction of porous polymer templates not only can address the problem of disrupting the STF device morphology due to the flow of liquid azobenzene compounds, but also provide an opportunity for the wide application of azobenzene-based STFs.

In 2019, Yu *et al.* fabricated a flexible device for solar thermal storage by compositing azobenzene compounds (4-methoxyazobenzene) with a polymer fabric, as depicted in Figure 3.1(a) [213]. Fabrication of easily accessible cloth fabrics through thermal-melting rather than immersion in

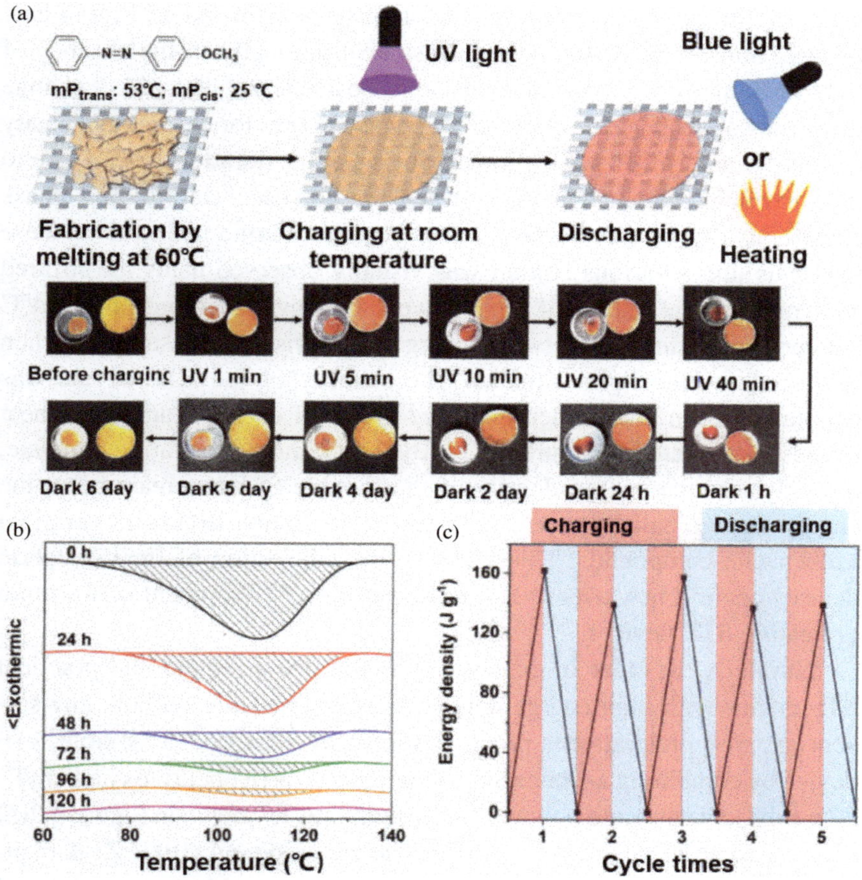

Figure 3.1. (a) Schematic diagram of energy storage and heat release processes of flexible device. Energy storage and heat release of the azobenzene compound and flexible device. (b) DSC results of samples discharged in dark for different durations. (c) The charging and discharging cycles were repeated five times to assess the performance. Reproduced with permission [123]. Copyright 2019, Wiley-VCH.

a solution for solid STF device preparation. The fabric template with well-dispersed azobenzene compounds demonstrates a favorable energy-storage capacity that is comparable to other solid-state STFs. The azobenzene compounds could be transformed into liquid from solid under dark or light-induced conditions. Subsequently, they can release heat gradually in the absence of light or demonstrate light-induced rapid heat release. Subsequently, they can release heat gradually in the absence of light or demonstrate light-induced rapid heat release. The phase change properties of azobenzene compounds enhanced the storage energy density to approximately 201 J g^{-1}, allowing for a storage duration of 40 h due to the intermolecular interaction between the polyester (fabric) and phase change, as depicted in Figure 3.1(b). The exothermic behavior of these materials under various stimuli was visually observed using an infrared camera. The heat release of flexible wearable device demonstrated a 4°C temperature difference between charged and discharged samples when subjected to blue light stimulus. The stability of the STF device was maintained even after undergoing five cycles of energy storage and heat release, as depicted in Figure 3.1(c). By incorporating the fabric template, not only was the energy density of the flexible STFs increased, but the energy storage half-life was also extended in comparison to the original azobenzene compounds. This work on the fabrication of flexible fabric devices opens a new avenue for the subsequent development of warming or heating STF devices.

Similarly, in order to design and synthesize a simple and practical STF device with high energy density and long half-life without any solvent-assisted pretreatment process, Liu *et al.* designed a flexible STF device by combining azobenzene compound (1-(4-(isopentyloxy)phenyl)-2-(p-tolyl)diazene) and polymer fabric template, as shown in Figure 3.2(a) [51]. At room temperature, the azobenzene compound (IPDAZ) demonstrated a photo-controlled reversible phase transition between solid and liquid states. This transition was achieved by utilizing the melting point variation between the trans-configuration and *cis*-configuration, as depicted in Figure 3.2(b). The azobenzene compound demonstrated exceptional film-processing capability on polymer fabric cloth using either the photo-liquation or thermal-liquation method. Notably, this fabrication process did not require any solvent assistance. The aim was to create a solid-state STF device. The storage energy density of flexible STF device could reach up to 288 J g^{-1} (80 W h kg^{-1}) including the phase change energy (ΔH_{pc}=77 J g^{-1}) and photoisomerization enthalpy (ΔH_{iso}=210 J g^{-1}), which was higher than most of other organic azobenzene compounds, as shown in Figure 3.2(c). Compared with azobenzene

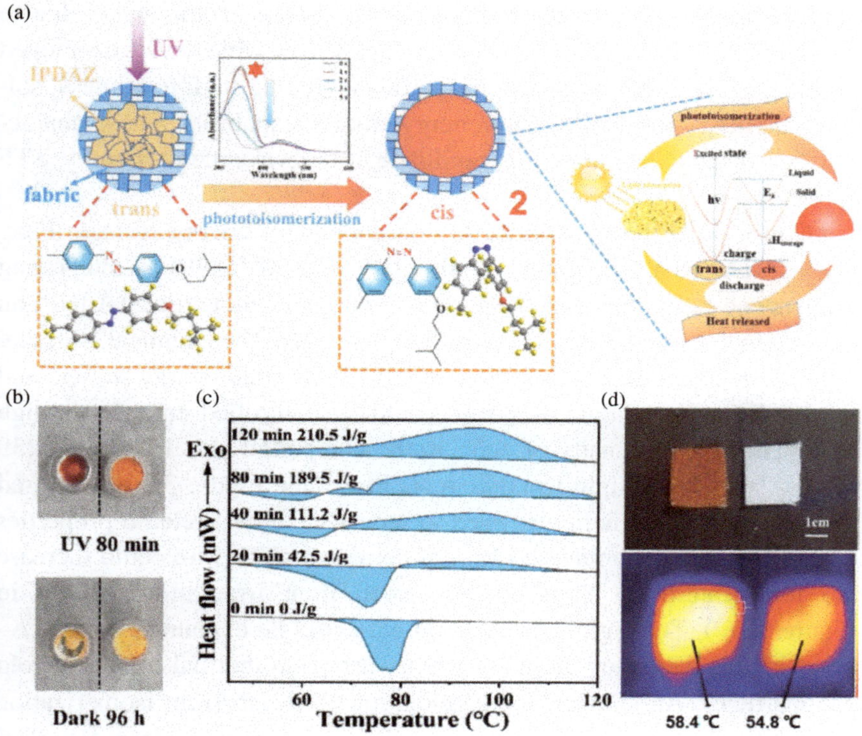

Figure 3.2. (a) The schematic diagram illustrates the process of preparing the STF device and its potential energy-storage mechanism. (b) The digital photos of photo induced phase changing process of azobenzene powder (left) and STF device (right). (c) DSC curves of STF device for various durations of phase changing. (d) A digital photo captures the STF device after UV light irradiation, along with a blank fabric cloth. The infrared thermal imager photo of the two samples on 55°C. Reproduced with permission [51]. Copyright 2022, Elsevier B.V.

compound, both the energy density and storage half-life of the composite STF device were increased because of the intermolecular H-bond interaction. The released heat of STF device after UV light irradiation was higher 3.3°C than that of unirradiated device, reflecting the fast heat release caused by the stimulated *cis*-to-*trans* isomerization of azobenzene compound, as shown in Figure 3.2(d). The composite STF apparatus exhibits a good recycling performance over five cycles, since the flexible cloth template prevents the loss of energy storage components due to the fluidity of liquid IPDAZ after charging. The flexible device developed in this study exhibits excellent properties including high toughness, solvent stability, recyclability and energy storage capability. These characteristics make it a promising candidate for environment friendly STF applications.

Considering the superiority of composites for energy storage, Feng *et al.* developed a composite STF film (PVA/Azo@AuNP) by integrating azobenzene compound (*tri*-azobenzene) and gold nanoparticle (AuNP) on a polyvinyl alcohol (PVA) polymer template [89]. The chemical structure of *tri*-azobenzene is shown in Figure 3.3(a). The solar energy storage and release behavior of the composite could be controlled entirely through light. The energy density of azobenzene compound was 139.4 J g^{-1} (39 Wh kg^{-1}) after UV light irradiation, as illustrated Figure 3.3(b). Infrared thermal imaging camera was used to monitor the heat release properties of the STF film. The released heat of composite STF film could increase the temperature by 8.8°C under green light irradiation, shown in Figure 3.3(c). The significant heat output could be explained by the *cis–trans* isomerization of the azobenzene, the photothermal effect of gold nanoparticles (AuNPs) and the acceleration of the *cis–trans* isomerization by AuNPs. The catalytic effect of AuNPs primarily arises from the electron transfer that takes place between tri-Azo and the surface of the AuNPs. This electron transfer reduces the activation barrier of the *cis*-Azo molecule, facilitating its easy conversion back to its thermodynamically stable state. The heat could be released rapidly from the charged STF film through these multiple stimulation modes, as shown in Figure 3.3(d). The findings strongly suggest that the same principles can be applied to design efficient STFs using various azobenzene compounds and polymers. This finding provides a facile and widely-applicable approach for developing customizable and economical STF films.

Besides flexible templates, Feng *et al.* recently also investigated a composite device (PAzo/AlNA) using a shape-stabilized aluminum nitride aerogel matrix compounded with phase change azobenzene compound for solar thermal energy storage, shown in Figure 3.4(a) [124]. Through molecular design, a novel azobenzene molecule exhibiting photo-induced solid-liquid transitions has been synthesized. This molecular design

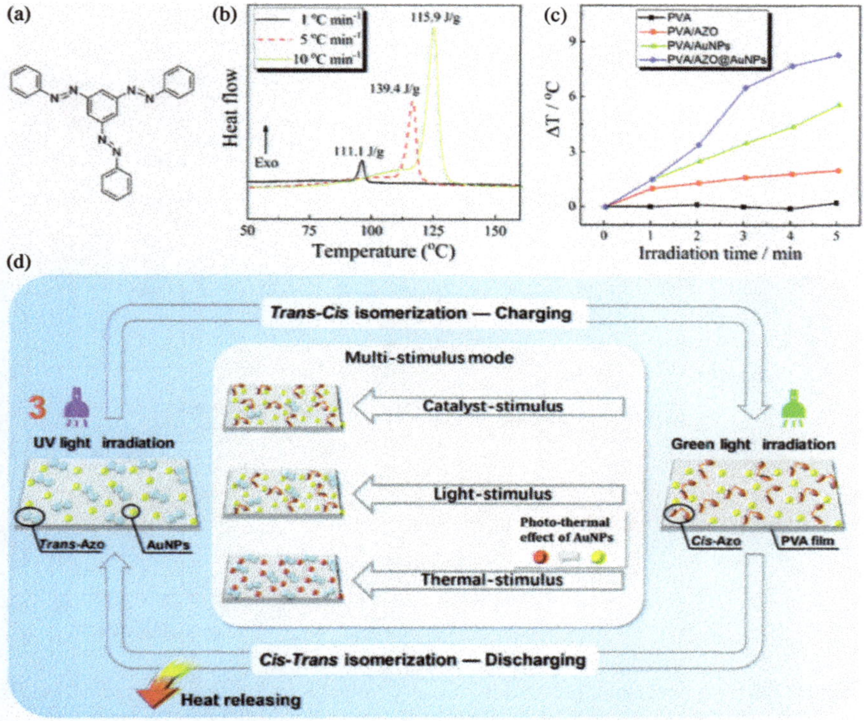

Figure 3.3. (a) Chemical structure of tri-azobenzene. (b) DSC curves of tri-Azo at a heating rate of 1°C min^{-1} (black solid line), 5°C min^{-1} (red dashed line), and 10°C min^{-1} (green dotted line). (c) Temperature difference (ΔT) versus time curves of the four different films. (d) Schematic illustration of the charging-discharging procedure of PVA/Azo@AuNP STF. Reproduced with permission [89]. Copyright 2020, Royal Society of Chemistry.

enables efficient storage of solar and environmental thermal energy from multiple sources. The interlayer porous structures of the aerogel not only could enable proper light transmission and excellent interfacial compatibility with azobenzene, but also benefit the reversion rate of metastable isomers. The composite demonstrated responsiveness to both UV light and blue light, shown in Figure 3.4(b). This composite device possesses the ability to achieve multi-source storage of solar energy and environmental heat. After exposure to UV light irradiation, it exhibits a high

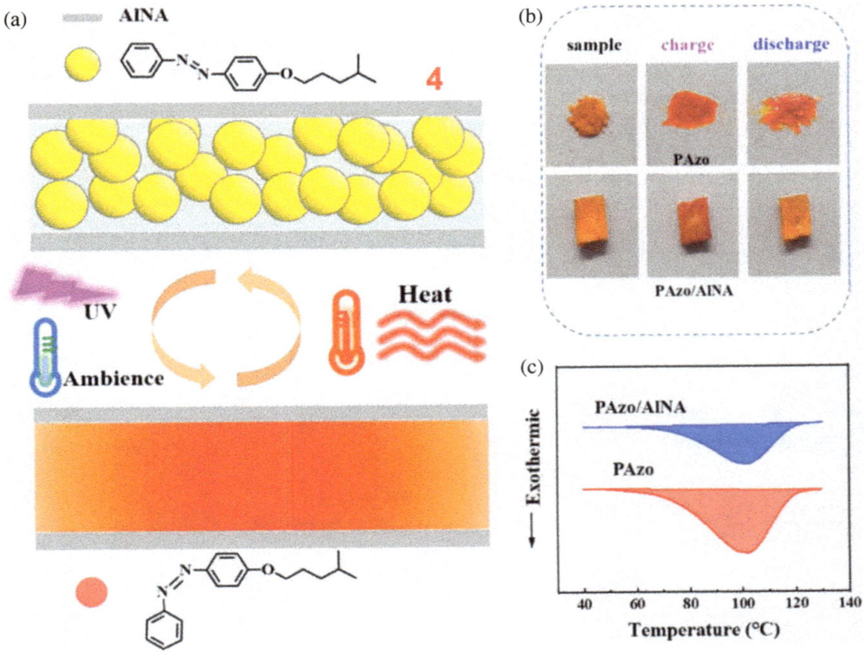

Figure 3.4. (a) Schematic illustration of energy storage and release of PAzo/AlNA. (b) Morphology of charged and discharged PAzo and PAzo/AlNA at 25°C. (c) DSC curves of charged PAzo (red) and PAzo/AlNA (blue) measured while heating from 30°C to 130°C. Reproduced with permission [124]. Copyright 2021, Elsevier B.V.

degree of isomerization (81%) and energy storage capacity (181 J g^{-1}) after UV light irradiation, as seen in Figure 3.4(c). The heat release of PAzo/AlNA composite induced by blue light could be visually monitored to investigate the feasibility of utilizing this composite as a solid-state self-heating device. After UV irradiation, the temperature difference of 1.5°C can be generated in the composites, which shows that PAzo/AlNA is capable of rapidly releasing heat under room temperature conditions through light control. This work presents a novel strategy for investigating self-heating solid-state thin film devices.

The above-mentioned investigations on azobenzene compounds/polymer composites have proven that the novel strategy of constructing composite devices not only facilitates the solar energy absorption, storage and conversion of azobenzene compounds, but also provides the possibility to develop economical and practical STFs. It is still, however, a nascent stage in the development of composite units. Many studies on composite devices fabricated from easily film-forming azobenzene polymers such as linear polymers and dendrimers are also in urgent demand. Therefore, it is particularly important to investigate the solar energy storage and heat release properties of these azopolymers, which will be discussed below.

3.2.2 *Linear azopolymer STFs*

Linear azopolymers commonly feature an alkane chain as the backbone and azobenzene group as the side chain, which have been extensively investigated over the past decades [24,125]. The azobenzene units can interact with each other and can be influenced by the backbone, exhibiting abundant photochemical properties that enable linear azopolymers to be applied for solar energy conversion and storage. Moreover, recent studies have shown that the incorporation of azobenzene units can also regulate the glass transition temperature of the polymer, resulting in solid–liquid phase transition properties [21,76]. These persistent efforts and positive explorations not only further enhance the storage performances of azopolymer STFs, but also allow a deeper and more convenient exploration of this rapidly developing field.

In 2016, Grossman *et al.* designed a linear azopolymer as STFs. Figure 3.5(a) displays the main chain of the polymer, which consists of an alkyl chain. The pendant group in this polymer is the azobenzene moiety. The materials demonstrated long half-lives at room temperature, with the monomer exhibiting a half-life of 92 h and the polymer having a half-life of 55 h [126]. The storage energy density of the monomer and polymer after UV light irradiation were 151 J g^{-1} (42 W h kg^{-1}) and 104 J g^{-1} (29 W h kg^{-1}), separately, shown in Figure 3.5(b). The disparity in energy density between the monomer and polymer might result from variances in light absorption efficiency and photoisomerization quantum yield. Steric hindrance, caused by the coexistence of rotational (steric-sensitive) and inversion mechanisms for isomerization, could impede photo-switching in the solid state, thereby restricting the performance of the STFs. Large freestanding polymer films obtained by a simple solution spin-coating

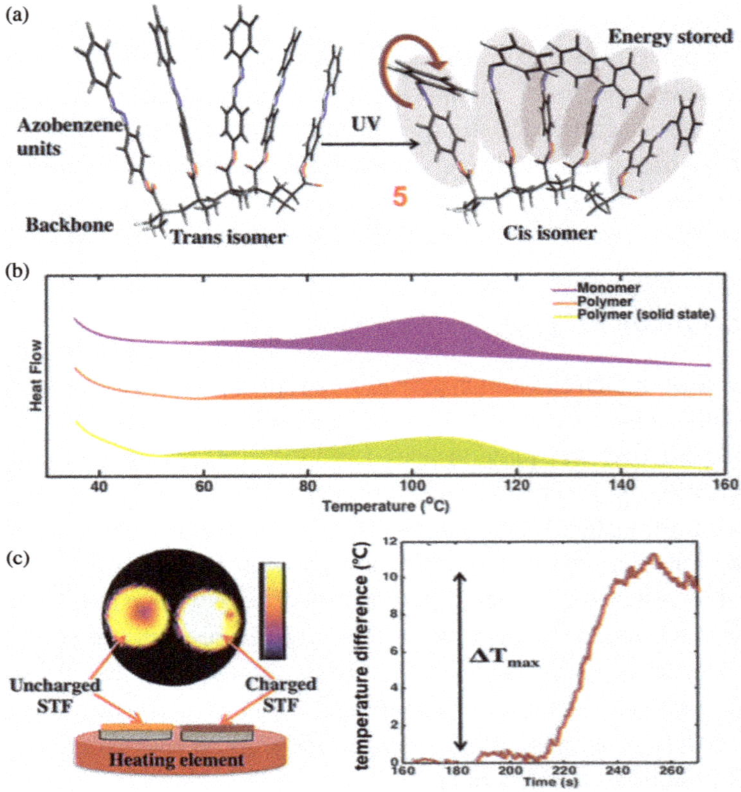

Figure 3.5. (a) Schematic of linear azopolymers consisting of four monomers in the *trans*-state being converted to the *cis*-state upon UV illumination. (b) Differential scanning calorimetry traces for monomer and polymer charged under various conditions. (c) Top-view IR heat map of *trans*-isomers and *cis*-isomers placed on a heating element, with side-view illustration below. Heat map depicted for maximal temperature difference between samples with color bar indicating relative heat magnitude. Average temperature differences between the two STFs. Reproduced with permission [126]. Copyright 2016, Wiley-VCH.

process, where a liquid cross-linked polymer was applied to balance the dissolution of the underlying film during the spin-coating process. Importantly, this approach enabled fabricating uniform films that could withstand temperature up to 180°C. An infrared camera was utilized to measure the spatial temperature profile, and the heat released could raise the temperature by up to 10°C, which paves the avenue for the application of STFs in the solid state, as illustrated in Figure 3.5(c). Moreover, the rapid temperature variation resulting from heat release within a few seconds plays a critical role in certain applications that require rapid heat release. This research utilizing linear azopolymers in the solid state offers tremendous attraction for new solid-state applications for renewable energy storage and heat release.

Later, the same group developed a new linear azopolymer that could be easily synthesized by electrodepositing azobenzene and a carboxylic acid-based monomer on various kinds of substrates via copolymerization to further integrate azopolymer films into some solid-state devices, shown in Figure 3.6(a). By controlling the deposition conditions, such as thickness and uniformity, the team achieved highly uniform films over a large surface area [20]. Interestingly, the azopolymer allowed for conformal deposition on various conducting substrates, which could be either flat or structured, depending on the specific application, shown in Figure 3.6(b). As a result, the integration of low-thermal conductivity STF devices with high-thermal-conductivity metals enables efficient heat transfer to other components or regions within the fabric or device. The STFs in these devices store solar energy when exposed to UV irradiation. The energy density of the azopolymer fuel was about 90 J g^{-1} along with a storage half-life of 75 h. The stored energy can be released as heat through the process of *cis*-to-*trans* isomerization, then directionally transported along the wire to specific locations. This directional heat transport is achieved by utilizing the wire as a conduit, allowing the released heat to be guided to desired regions, as shown in Figure 3.6(c). This novel work proves that electrodeposition is a valid method for integrating azobenzene STFs into various solid-state applications. Furthermore, the copolymerization process allows for an easy introduction of additional functional groups or molecules into the azopolymer STF materials without significantly compromising their energy density. However, it is important to note a potential issue. When the film thickness increases, there is a risk of cracks and deformities caused by the significant shrinkage that occurs during drying. To address this challenge, future research on fiber-based STFs could focus on designing high-molecular-weight polymers that contain azobenzene or incorporating binders while minimizing the impact on the energy densities

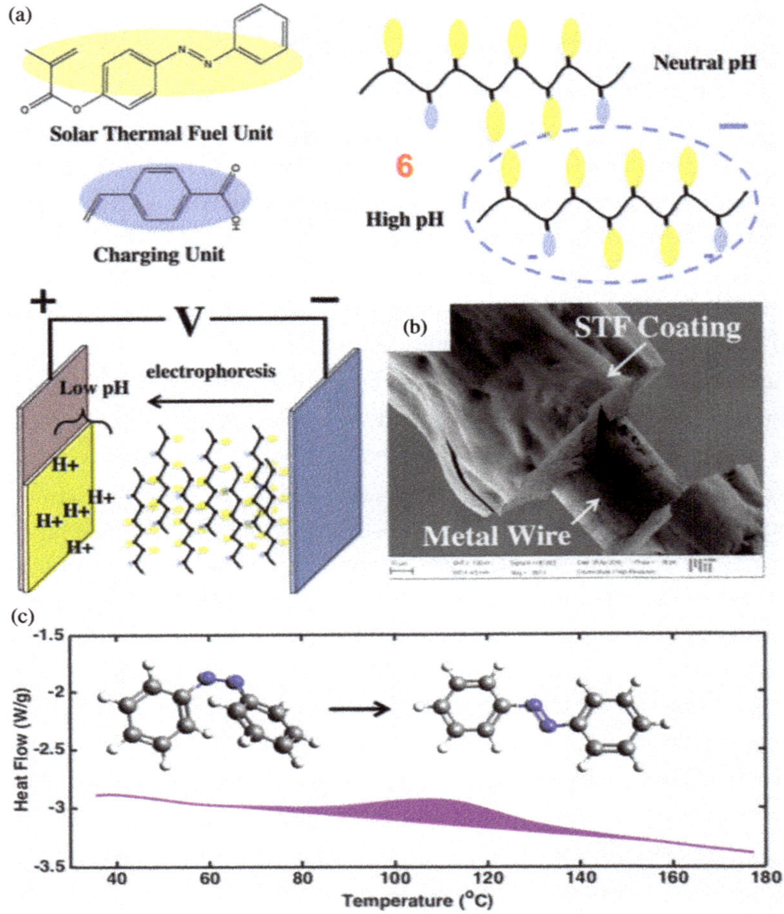

Figure 3.6. (a) Chemical structure of linear azopolymers and concept of the electrodeposition scheme. (b) Proof-of-concept using a 50 μm fiber coated with an electrodeposited thick STF coating, showing both the wire and the copolymer coating. (c) DSC heating curve of azopolymers after UV light irradiation and the *cis*-to-*trans* reversion process. Reproduced with permission [20]. Copyright 2016, American Chemical Society.

of the STFs. This would help enhance the structural integrity and overall performance of the STF films.

Previous investigations suggested that different solvent processing could alter the azopolymer's physical characteristics to regulate storage energy densities and the activation energy of reversion process of STFs.

In 2017, Venkataraman *et al.* designed three poly(methacrylate) syndiotactic-rich polymers containing azobenzene groups to investigate the effect of processing solvent and film structure on storage energy density,

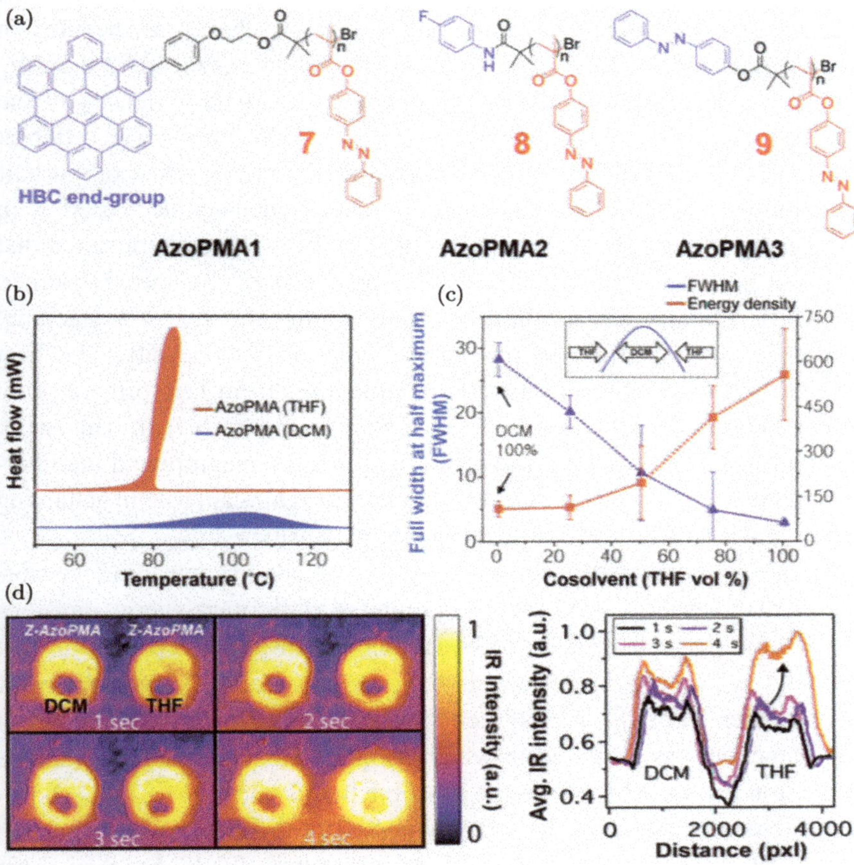

Figure 3.7. (a) Chemical structures of azopolymers with varying end groups. (b) DSC curves of a *cis*-rich AzoPMA3 dried from either THF or DCM. (c) Diagram of FWHM and energy density of AzoPMA3 based on the different co-solvent ratios of THF and DCM. The inset is a representation of DSC curve change based on different solvent ratio (d) IR imaging of *cis*-rich AzoPMA3 dried from THF or DCM. (d) A plot of average IR intensity as a function of distance across the frame shows the average IR intensity profiles of the AzoPMA3 samples over time. Reproduced with permission [127]. Copyright 2017, Springer Nature.

as shown in Figure 3.7(a) [127]. In a favorable solvent, polymer chains undergo expansion, creating an environment that fosters strong interactions among the polymer molecules. The dichloromethane (DCM) and tetrahydrofuran (THF) were selected to investigate the interactions between the polymer and the solvent. The results displayed that both THF and DCM were effective solvents for azopolymers and different solvent treatments could affect the shape of the DSC exotherm, as depicted in Figure 3.7(b). The energy density of the azopolymer AzoPMA3 dried by DCM was 110 J g^{-1} (30.6 W h kg^{-1}), whereas the energy density of the AzoPMA3 obtained from THF was 510 J g^{-1} (142 W h kg^{-1}). The different mixing ratios of THF and DCM demonstrated that the full width at half maximum (FWHM) increased and the energy density decreased with increasing DCM ratio, as shown in Figure 3.7(c). The authors stated that the solvent effectively dissolves the dipole formed by the *cis*-isomer, reducing aggregation in solution and enabling effective π-π stacking, which is the critical to acquire high storage energy density. The heat release performance reveals that the sample dried from THF offered superior exothermic effect after 4 s, as shown in Figure 3.7(d). This work demonstrated that the optimization of polymer architecture and the interactions between polymers and solvents play a significant role in achieving azopolymer STFs with high storage energy densities.

Although much advancements have been made in controlling the storage performances of solid-state azopolymers, the storage energy densities of polymer systems remain to be further improved. There is an urgent demand to propose some new strategies to advance the development of azopolymer STFs. In a study conducted by Wu *et al.* in 2017, a novel phenomenon was observed in which azopolymers exhibited a reversible transition from a solid state to an isotropic liquid state due to T_g, as shown in Figure 3.8(a) [128]. The azopolymers with a high *trans* content are solid materials with a glass transition temperature (T_g) higher than room temperature, while the azopolymers with a high *cis* content are liquid materials with a T_g lower than room temperature, as shown in Figure 3.8(b). These polymers incorporating azobenzene, which display photoinduced reversible phase change characteristics, exhibited excellent solar energy storage and conversion properties, as shown in Figure 3.8(c). This work provides a novel strategy for understanding and designing phase-change

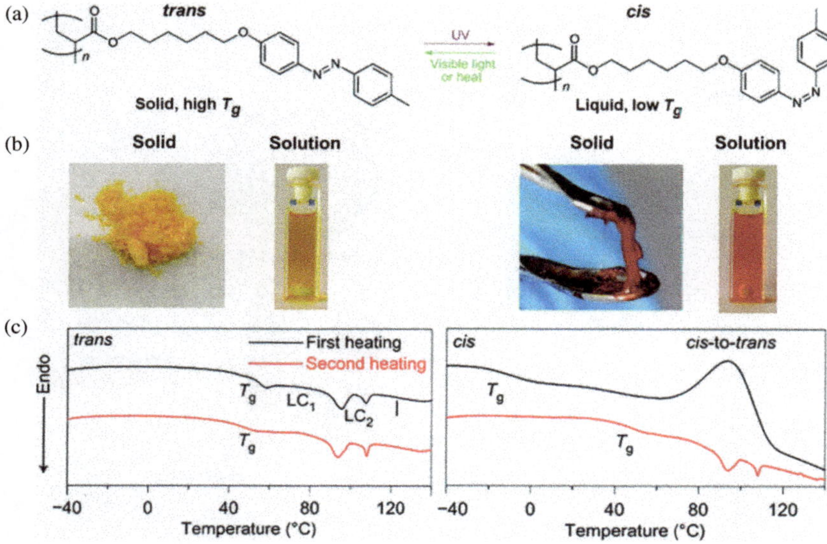

Figure 3.8. (a) Chemical structure and reversible photoisomerization of azopolymer with a flexible linker and an azobenzene moiety attached to the side chain of the polymer. (b) Photographs of azopolymer in solid and solution. The *trans*-isomer is solid and the *cis*-isomer is liquid. (c) DSC heating curves of *trans*-rich and *cis*-rich azopolymers. Reproduced with permission [128]. Copyright 2017, Springer Nature.

azopolymer materials, which not only can simultaneously capture the enthalpy of isomerization and phase change energy between solids and liquids to remarkably increase the storage energy densities of azobenzene-based STFs but also offer a fascinating view for the investigation of solar energy storage technology.

Inspired by phase change strategy of azopolymers, Yu *et al.* recently reported the microphase separation and nanoconfinement of liquid-crystalline block copolymer (LCBC) (PEO-*b*-PAzo) to improve the storage performances of STFs. In this system, the separated phase of poly(ethylene oxide) (PEO) was one light-inactive phase change material. The continuous phase of azobenzene-containing liquid-crystalline polymers (azobenzene-containing LCP) could store both solar energy by the isomerization of azobenzene groups and latent heat by phase change, as shown in Figure 3.9(a) [129]. The sPEO was capable of preferentially dispersing into the PEO

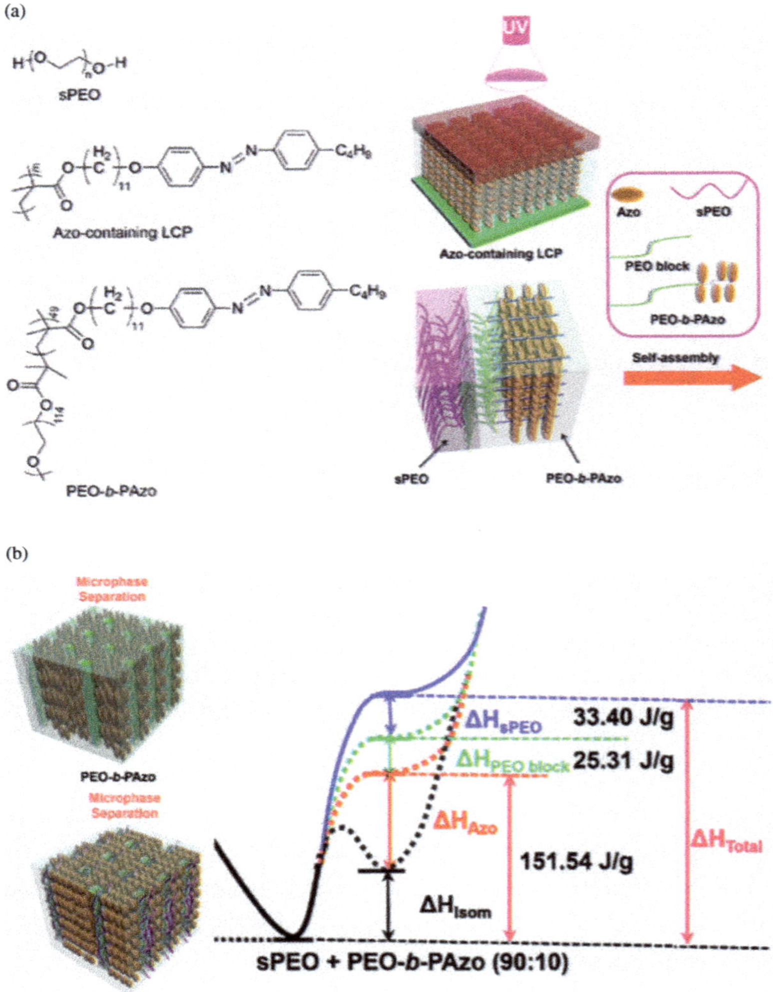

Figure 3.9. (a) Chemical structures of sPEO, azo-containing LCP, and the block copolymer PEO-b-PAzo. Illustrations depict the behavior of azo-containing LCP under UV irradiation, the microphase separation of PEO-b-PAzo, and the self-assembly of sPEO and PEO-b-PAzo composites. (b) Solar thermal energy storage within sPEO and PEO-*b*-PAzo composites. (c) Photographs of *trans*-rich (left) and *cis*-rich composite film (right). IR thermal images of heat release from sPEO and PEO-*b*-PAzo composites films with the addition ratio of 90:10. (d) Schematic illustration of the sPEO and PEO-*b*-PAzo composites (90:10) within cloth for human body thermal application. IR thermal images of *cis*-rich sPEO and PEO-bPAzo composites (90:10) within cloth on the wrist under 460 nm irradiation. Reproduced with permission [129]. Copyright 2021, American Chemical Society.

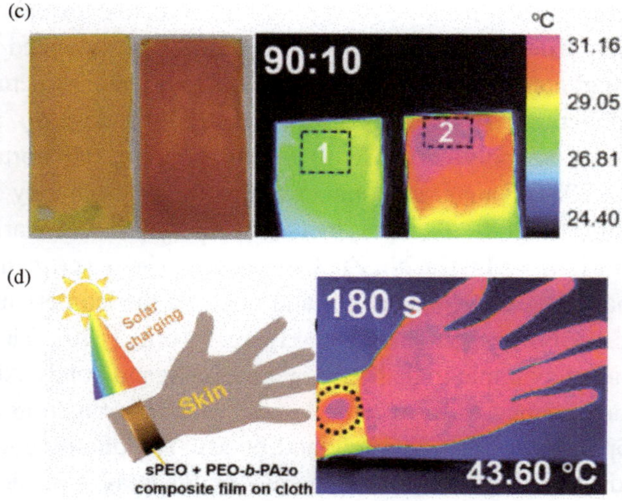

Figure 3.9. (*Continued*)

nanodomains of PEO-*b*-PAzo to form composite films (sPEO + PEO-*b*-PAzo). At room temperature, the *trans*-to-*cis* isomerization of azobenzene moieties allows for the storage of photon energy and phase change energy. Additionally, the LC-to-isotropic phase transition of the liquid crystal polymer (LCP) block enables the storage of phase change energy. Figure 3.9(b) illustrates that crystalline-to-isotropic transitions of semi-crystalline poly(ethylene oxide) (sPEO) and the PEO block above the melting temperature (T_m) of PEO also contribute to the storage of phase change energy. The total storage energy density of sPEO and PEO-*b*-PAzo composite film with the addition ratio of 90:10 could up to 210.25 kJ kg^{-1} (77.14 kJ mol^{-1}), accompanied by a storage half-life of 9.3 h. The stored energy could be released upon irradiation with 460 nm light, increasing the temperature rapidly by about 6.5°C, shown in Figure 3.9(c). The sPEO and PEO-*b*-PAzo composite could be applied to prepare wearable thermal fabrics, where the surface temperature of the fabric increased from 32.23°C to 43.60°C with increasing blue light irradiation time, shown in Figure 3.9(d). This work on enhancing the storage performances of azopolymer STFs through microphase separation and nanoconfinement paves the way for advances and applications of STFs materials in the field of wearable smart materials, particularly in the development of warm fabrics.

Although phase change strategy can be effective in enhancing the storage energy densities of azopolymer STFs, photoinduced reversible solid-to-liquid conversion is usually difficult as a chain structure hindering crystallization. Thus, it is a strong need to develop alternative approaches for azopolymer STFs to further improve the storage energy densities. In 2022, Feng *et al.* conducted a study in which they designed a series of sulfonic-grafted azopolymers that interacted with various metal ions (PAzo-M, M = Mg, Ca, Ni, Zn, Cu, and Fe). The purpose of this study was to enhance the energy densities and improve the isomerization kinetics of these polymers, as shown in Figure 3.10(a) [130]. The dynamic bonds included dynamic ionic bonds and metal-ligand bonds. Alkali metal ions have a tendency to form ionic bonds, whereas transition metal ions have a propensity to form metal-ligand bonds. The photoinduced formation and dissociation of M–O dynamic bonds play a critical role in enhancing the energy gap (ΔE) between *trans*-isomers and *cis*-isomers, thereby contributing to high-energy storage capabilities and facilitating rapid isomerization rates. This phenomenon is crucial for improving the overall performance of the system, as shown in Figure 3.10(b). The presence of dynamic interactions also reduces the thermal barrier for *cis*-to-*trans* isomerization and enables reversible crosslinking of polymer chains through dynamic bonds. The PAzo-Mg demonstrated the highest energy storage density, reaching 408.6 J g^{-1} (113.5 Wh kg^{-1}). The cyclic utilization of the isomerization energy and the dynamic bond enthalpy makes PAzo-M exhibit a rapid structural transformation rate at low temperatures, allowing for the release of high-power heat. The manufactured flexible textile coated with PAzo-Mg is capable of providing a significant temperature increase of 7.7–12.5°C in low-temperature environment by selectively self-releasing heat both indoors and outdoors, as shown in Figure 3.10(c). This novel strategy of manipulating the formation and dissociation of photodynamic bonds through isomerization not only creates an opportunity to develop azopolymer STFs with high storage energy density, but also provides a way to design self-heating wearable thermal management devices.

Except for the storage capacity, heat release performance is also essential for the application of azopolymer STFs [88,131]. The investigation of the heat release inducing approach of STFs is of great significance for

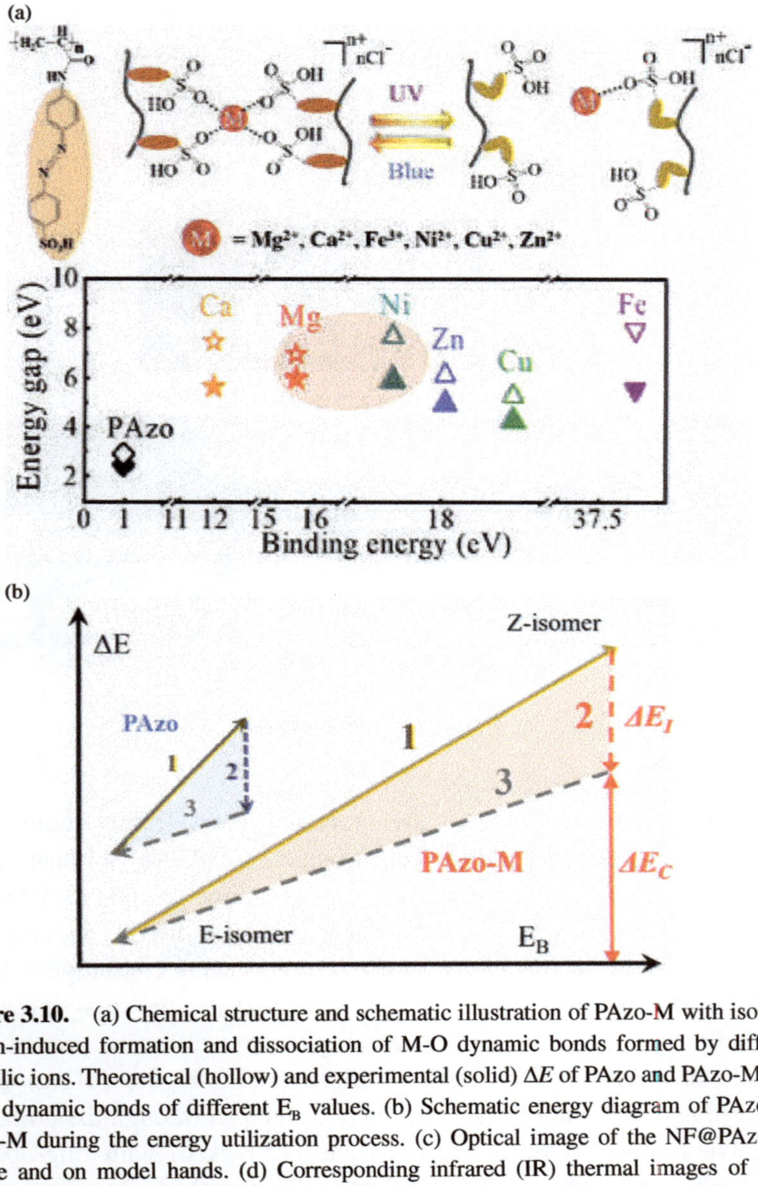

Figure 3.10. (a) Chemical structure and schematic illustration of PAzo-M with isomerization-induced formation and dissociation of M-O dynamic bonds formed by different metallic ions. Theoretical (hollow) and experimental (solid) ΔE of PAzo and PAzo-M with M-O dynamic bonds of different E_B values. (b) Schematic energy diagram of PAzo and PAzo-M during the energy utilization process. (c) Optical image of the NF@PAzo-Mg textile and on model hands. (d) Corresponding infrared (IR) thermal images of NF@PAzo-Mg on model hands in indoor and outdoor, respectively. Reproduced with permission [130]. Copyright 2022, Wiley-VCH.

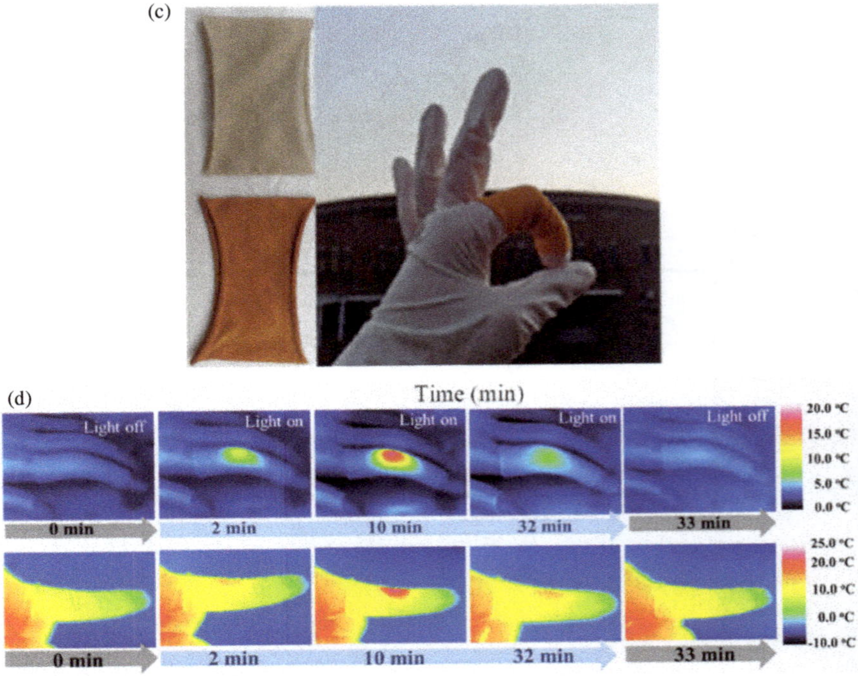

Figure 3.10. (*Continued*)

realizing widespread applications. In recent years, most studies have focused on the heat release by light irradiation and external heat source stimulation. However, the poor penetration ability of light can lead to incomplete heat release, while external heat sources usually require large energy input that can inevitably cause excessive energy consumption [132–134]. To address this issue, Feng *et al.* conducted a recent study investigating the impact of external force stretching on the exothermic properties of flexible azopolymer STF films, as illustrated in Figure 3.11(a) [135]. Uniform and stretchable STF films were developed through the utilization of polynorbornene-templated azobenzene (PNB-Azo) with ring-opening metathesis polymerization and covalent grafting methods. Two homogeneous and flexible PNB-Azo films were prepared by solution processing. The polymer PNB-Azo-1 displayed a significantly high *trans*-to-*cis* isomerization (72%) and stable energy storage, which can be attributed to the intermolecular interactions and steric hindrance present in the system. In

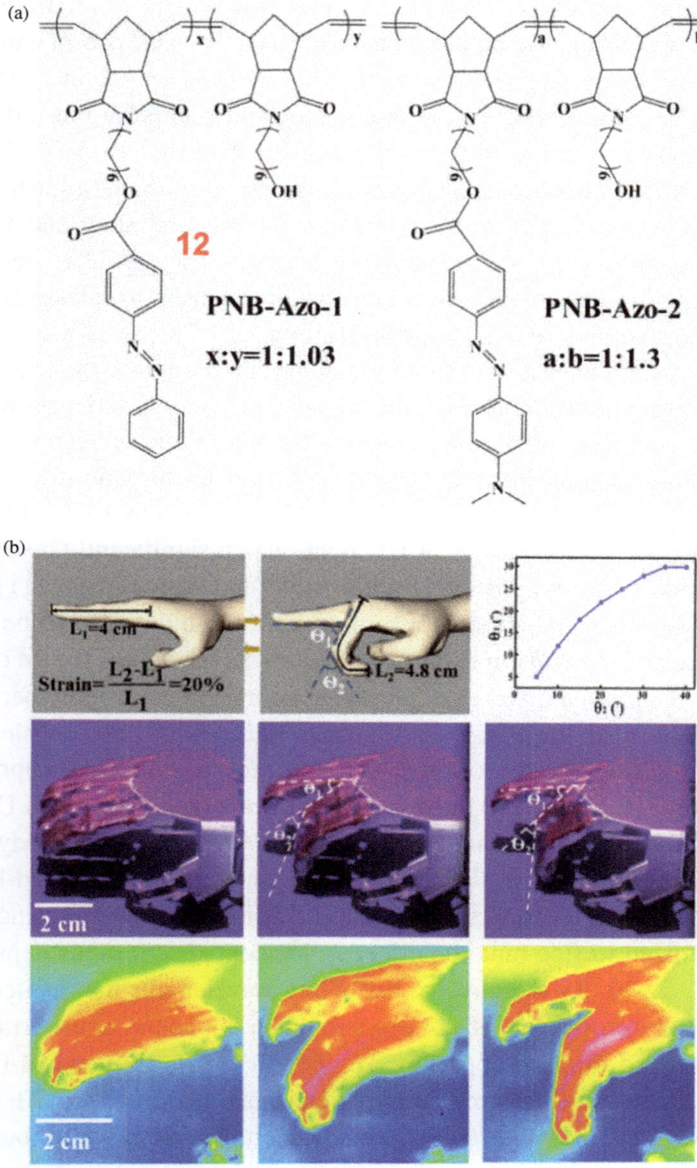

Figure 3.11. (a) Chemical structure of PNB-Azo-1 (left) and PNB-Azo-2 (right). (b) Schematic illustration of the dynamic stretching caused by the bending of the robot's fingers (left) and the motion of $\theta 1$ and $\theta 2$ along with the bending of fingers (right). Optical image of PNBAzo-1 film positioned on four fingers of the robot during the cyclic bending and the corresponding IR thermal images. Reproduced with permission [135]. Copyright 2019, American Chemical Society.

comparison, PNB-Azo-2 exhibited a lower *trans*-to-*cis* isomerization rate (19.2%) and displayed rapid reversion due to the presence of pull–push substituents on the azobenzene groups. The stretched azopolymer film, subjected to a 20% strain, demonstrated excellent photocharging efficiency (85%) and achieved a high energy density of 176.4 kJ kg^{-1}. A high-resolution IR thermal imaging camera was utilized to monitor the temperature changes resulting from the heat released during both static and dynamic stretching processes during charging and discharging. The reversible dynamic stretching of PNB-Azo-1 enabled the fingers to release heat at a rapid rate, resulting in a temperature increase of 1.5°C, as shown in Figure 3.11(b). The findings revealed that stretching could improve the energy storage and exothermic rate through favorable isomerization, and that reversible dynamic stretching induced a faster heat rate than static stretching.

Despite the considerable progress that has been made in enhancing the storage performance of azopolymer STFs, most reports are restricted to storing the photon energy of UV light, which significantly restricts the application of azopolymer STFs in visible and natural light [11]. Over 50% of the solar radiation spectrum on Earth is understood to be within the range of 300–800 nm in terms of photon wavelengths [16]. In order to address the problem, Wu *et al.* presented an integrated azopolymer device that could store both UV and visible light efficiently, as shown in Figure 3.12(a). The device, arranged from top to bottom, comprised of coumarin 314, a visible-light-responsive azopolymer (PmAzo), a UV pass filter, and a UV light-responsive azopolymer (PAzo) [136]. The dye layer was utilized to convert blue light into green light, which could be efficiently captured by PmAzo for storage. Additionally, two filtering layers were utilized to fine-tune the effective incident spectrum, as depicted in Figure 3.12(b). By selectively obstructing wavelengths that induced heat release, the proportion of *cis*-isomers under monochromatic irradiation was observed to reach approximately 60–70%. The spin-coated films of PAzo (35 nm) and PmAzo (20 nm) were integrated into the STF device, along with all the other solid layers. By employing a well-thought-out device design and improving the spectral overlap, the solar efficiency was significantly enhanced to 0.4%. The storage energy densities of PAzo and PmAzo were measured to be 125 kJ kg^{-1} (55 kJ mol^{-1}) and 14 kJ kg^{-1} (12 kJ mol^{-1}), separately. These values were obtained with a storage

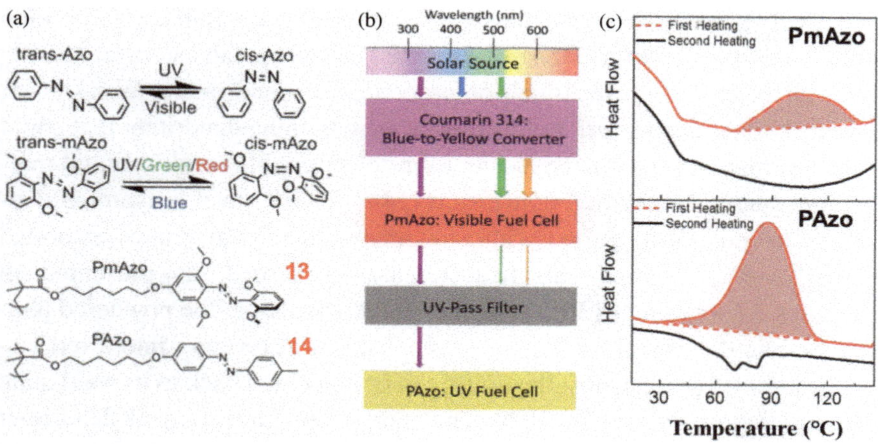

Figure 3.12. (a) Photoisomerization of photoswitches and chemical structures of azo-polymers. (b) Schematic of four-layer UV–Vis solar-thermal cell depicting transmission or absorbance of representative wavelengths. (c) DSC heating curves for photoswitches. Shaded areas were integrated to find the bulk energy density. Reproduced with permission [136]. Copyright 2017, Wiley-VCH.

duration of half a day, as depicted in Figure 3.12(c). Nevertheless, the influence of solid-state phenomena, like polymer crystallinity, has been demonstrated to be significant, underscoring the necessity to precisely define the characteristics of the "solid-state". Not only does this prove that efficient energy storage can be realized by combination, but also strongly promotes STFs toward visible photon energy storage.

3.2.3 *Dendrimer azopolymer STFs*

Dendrimers have unique physicochemical properties such as well-defined nanoarchitectures, a high degree of molecular uniformity, and multiple surface functionalities compared to conventional polymers, which enable dendrimers to attract widespread attention and applications in catalysts, sensors, and biomedical materials [137–141]. In recent times, there has been a growing interest in incorporating photochromic moieties, such as azobenzene chromophores, into dendrimers. This approach holds great potential for the advancement of novel photo-responsive materials, with applications in holographic grating, optical data storage systems, and liquid-crystalline materials

[29,30,141–143]. These previous studies offer the opportunities to develop photochromic dendrimers for solar thermal conversion and storage.

To benefit from the superior properties of dendrimers, Wang *et al.* recently fabricated a series of azobenzene-containing dendrimer that could show remarkable characteristics in terms of reversible solid-to-liquid transitions and solar thermal conversion [144]. These novel polyamide dendritic azopolymers grafted with azobenzene functional groups exhibited solar energy storage and heat release properties, as presented in Figures 3.13(a) and 3.13(b). After UV light irradiation, the powdered photochromic dendrimer undergoes a transition, transforming into a red viscous liquid. Subsequently, the liquid can be converted back to its solid-state by irradiation with visible light. The energy storage density of dendrimer

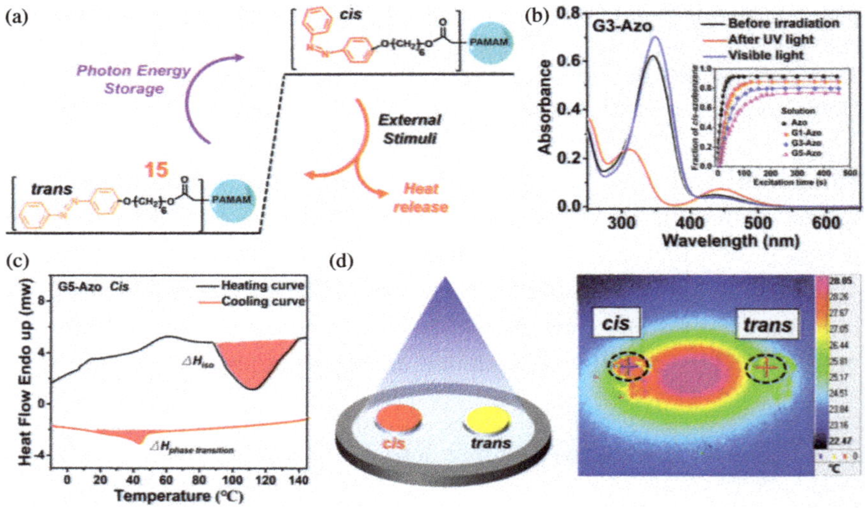

Figure 3.13. (a) Schematic diagram of photon energy storage and heat release of dendrimer photoswitch. (b) UV–Vis absorption spectra of G3.Azo in tetrahydrofuran before irradiation, after UV irradiation, and following subsequent visible light irradiation. The inset presents the fractions of *cis*-azobenzene in the samples after UV irradiation. (c) DSC curve of high-energy dendrimer photoswitches. ΔH_{iso} is the isomerization enthalpy. $\Delta H_{phase\ transition}$ is phase transition energy. (d) Schematic diagram of *cis*-isomer and *trans*-isomer STFs under visible light irradiation. Instantaneous infrared image of G3.Azo before (right) and after (left) UV irradiation under blue light irradiation. The sample locations were circled. Reproduced with permission [144]. Copyright 2020, American Chemical Society.

azopolymer, considering both latent heat energy and isomerization enthalpy, can reach up to 214 J g^{-1}, as seen in Figure 3.13(c). Observations indicate that the energy density of photochromic dendrimers notably rises with an increase in the number of azobenzene groups on the dendrimer's surface. The liquid high-energy dendrimer azopolymer has the ability to release stored energy as heat when exposed to visible light. This leads to a substantial temperature disparity of up to 5.2°C between the *trans*-isomers and the *cis*-isomers. These results offer a novel view and a promising pathway for preparing azobenzene-based STFs with dendrimer structures.

Despite the potential of azobenzene-based STFs in converting solar energy, the majority of current research leans toward utilizing high-energy UV light for energy storage. However, this approach has limitations as it degrades the azobenzene and can potentially disrupt or damage the surrounding environment, posing concerns for material lifetime and environmental impact [101,115]. Visible light, a key component of the solar spectrum, provides infinite possibilities to create practical STF materials in the presence of sunlight. Consequently, there is great interest in exploring molecular STFs that can be stored in the visible light range. To achieve this goal, Wang *et al.* designed an arylazopyrazole (AAP)-containing dendrimer, as shown in Figure 3.14(a); this innovative design enabled stable storage of blue light photon energy and facilitated controlled heat release [145]. After exposure to blue light (405 nm), the proportion of *cis*-isomers in the azopolymer dendrimer increased to around 85%. Subsequently, it decreased to around 6% upon exposure to green light. These results indicate that the dendrimer incorporating AAP demonstrated remarkable capabilities for photon energy storage and controlled heat release. This is illustrated in Figure 3.14(b). The energy density of the dendrimer azopolymer reached a remarkable 140 J g^{-1} (67 kJ mol^{-1}) after effectively capturing blue light, exhibiting a half-life of up to 12.9 days. The dendrimer film is capable of releasing heat in response to various stimuli, including light and heat. Figure 3.14(c) displays that the heat released by dendrimer azopolymer film under green light irradiation could raise the the high-energy isomer temperature by 6.5°C. The fabricated films demonstrated exceptional cycling stability in terms of energy storage and heat release. This characteristic provides access to a broad range of uses, as depicted in Figure 3.14(d). This research, which can

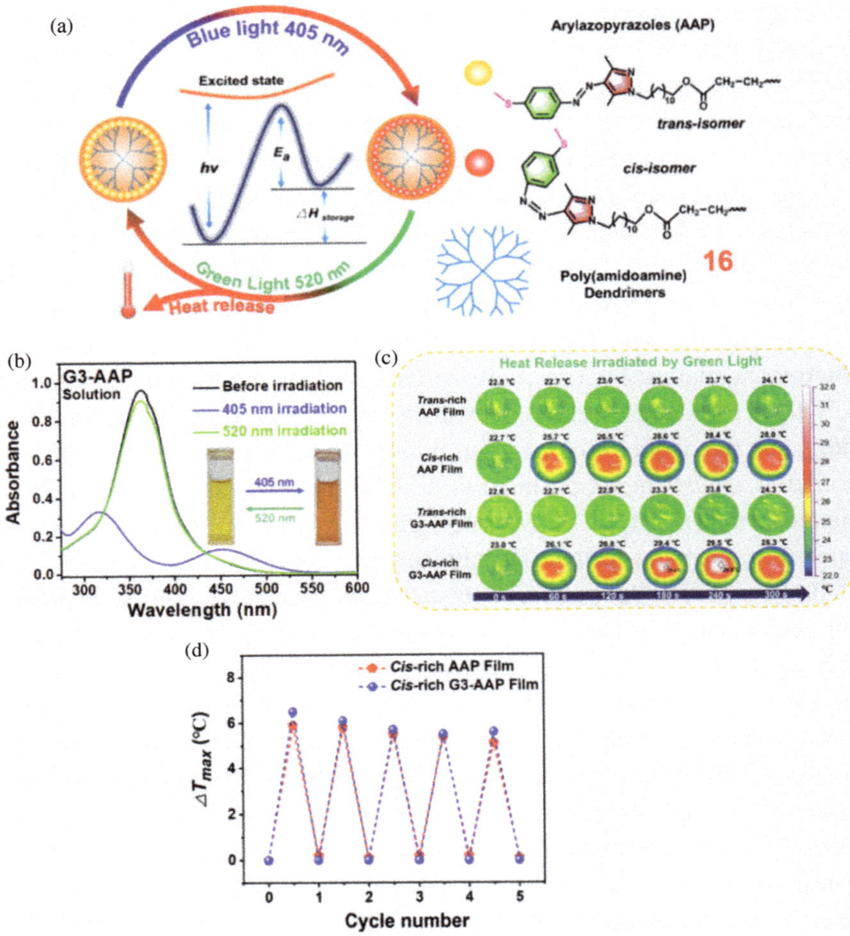

Figure 3.14. (a) Schematic diagram of dendrimer fuel absorbing photon energy of blue light for efficient energy storage and heat release. (b) UV–Vis absorption spectra of dendrimer photoswitch in *N, N*-dimethylformamide. The inset displays a photograph capturing the color change of the dendrimer photoswitch solution. (c) Infrared images of heat release of the *trans*-rich photoswitch film, *cis*-rich photoswitch film, *trans*-rich dendrimer film, and *cis*-rich dendrimer film under green light irradiation. (d) Temperature difference (ΔT_{max}) of *cis*-rich solar thermal films upon five cycles of heat release under green light irradiation. Reproduced with permission [145]. Copyright 2021, American Chemical Society.

significantly improve the energy storage of visible light, offers an opportunity for wider applications in the field of thermal management and thermal control of azobenzene-based STFs.

3.2.4 *Other types of azopolymer STFs*

Apart from azobenzene compound/polymer composites, linear azopolymers and dendrimer azopolymers, there are several other types of azopolymer STFs such as π-conjugated systems and azobenzene-nanocarbon/polymer composites for solar energy storage and conversion. These diverse azopolymers not only can provide theoretical and experimental guidance for further fabrication of azobenzene-based STFs, but also contribute significantly to the widespread applications.

Polydiacetylenes are conjugated polymers synthesized via topochemical polymerization of diacetylene monomers within highly organized structures. Owing to the insightful knowledge of the investigation of poly(diacetylene), Han *et al.* designed a line of azobenzene-functionalized symmetric diacetylenes, the chemical structures and photoisomerization are shown in Figure 3.15(a) [146]. The authors explored the influence of the length of alkyl spacer between the diyne core and terminal azobenzene groups, intermolecular hydrogen bonding, as well as different functional groups on the azobenzene moieties. Diacetylene can be readily functionalized and has a tendency to self-assemble into crystalline-like structures owing to its robust intermolecular interactions. This unique structure resembles a rigid template adorned with densely packed photo-switching units. Figure 3.15(b) shows that the diacetylene monomer could rapidly photopolymerize under UV irradiation into a poly(diacetylene) with an azobenzene-containing side chain on a rigid conjugated backbone. This structure was similar with a rigid template decorated with closely-packed azobenzene units. Both monomer and polymer were crystalline structures before UV light irradiation and after heat release, whereas those after stored energy behaved as viscous liquid states, implying that these azobenzene-containing conjugated polymers could efficiently store phase change energy, as illustrated in Figure 3.15(c). The storage energy density of azobenzene-functionalized diacetylenes (17d) including isomerization enthalpy and phase change

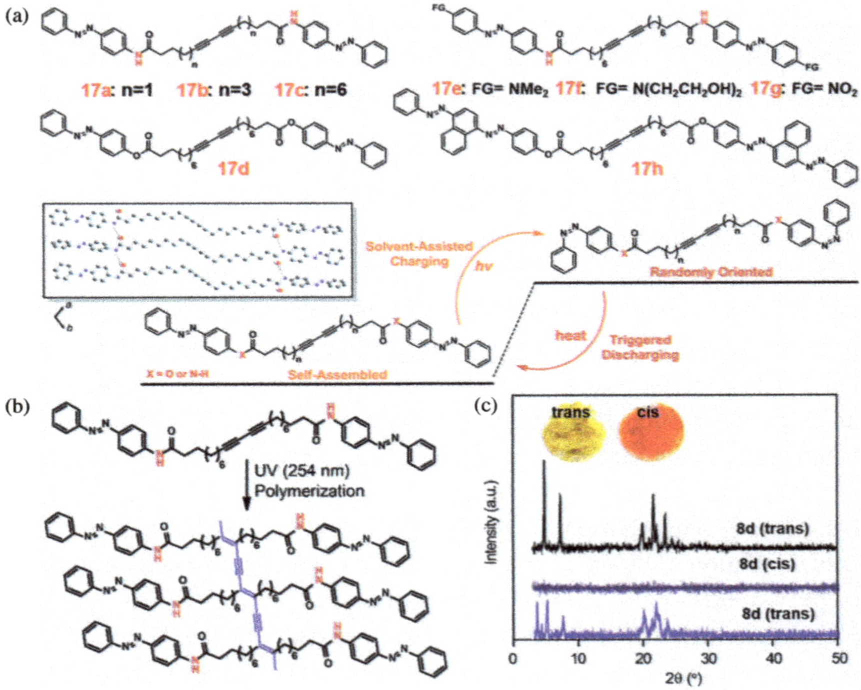

Figure 3.15. (a) Chemical structures of azo-functionalized diacetylenes with different alkyl chain lengths, terminal functional groups, H-bonding units, and an extended p-system along with a schematic of photoisomerization and reverse isomerization of compounds. The inset shows a crystal structure of azobenzene derivative (8c). (b) Scheme of photopolymerization of monomers and the structure of the resulting polymer. (c) Powder X-ray diffraction of monomer 8d. The inset shows photographs of the monomer 8d with different phases (crystalline solid and liquid). Reproduced with permission [146]. Copyright 2016, Royal Society of Chemistry.

energy could reach up to 243.7 kJ kg^{-1} (176.2 kJ mol^{-1}). The enhanced energy density can be attributed to the robust intermolecular interactions found in the well-ordered *trans*-state, as well as the substantial reduction of these interactions in the *cis*-conformation. This study combining conjugated azopolymers with phase change properties presents a comprehensive and insightful design strategy for the further develop, optimize and apply of azobenzene-based STFs.

In recent work, Liu *et al.* developed a ready-to-use azobenzene-based cross-linking polymer film by electrochemical deposition approach,

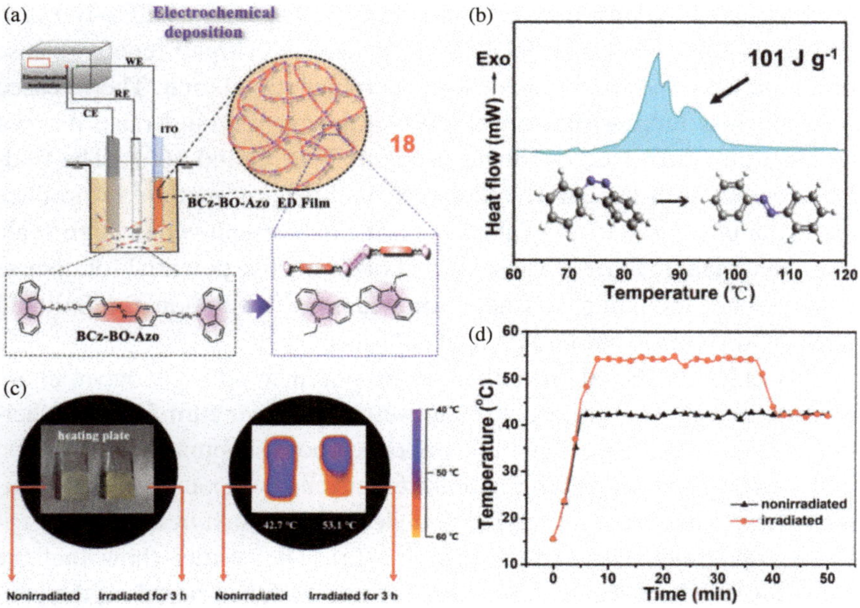

Figure 3.16. (a) The schematic representation for the principle behind the electrochemical deposition technique. (b) DSC heating curve of charged BCz-BO-Azo powder. (c) The thermal release process of BCz-BO-Azo electrochemical deposition film after placing heating plate for 8 min. Normal camera picture and thermal imager. (d) Time resolution heating curves of BCz-BO-Azo ED films determined by thermal imager (black curve) without any irradiation and (red curve) under 365 nm UV-light irradiation for 3 h. Reproduced with permission [146]. Copyright 2020, Wiley-VCH.

shown in Figure 3.16(a). The crosslinked azobenzene film was prepared by electrochemical deposition using a novel designed and synthesized precursor, 1,2-bis(4-(4-(9H-carbazol-9-yl)butoxy)phenyl)-diazene (BCz-BO-Azo), which contained photo-responsive azobenzene moiety and electroactive carbazole moiety. The crosslinked azobenzene film showcased smooth morphology with minimal crack and aggregation, and expected photo-response, making it suitable for application without any pretreatments. After UV light irradiation, the energy density of BCz-BO-Azo reached 101 J g^{-1} as illustrated in Figure 3.16(b). Because of the reduction of molecular motion in some cross-linked structure, the stability of *cis*-configuration BCz-BO-Azo in ED film was greatly improved, which lead to an increase in energy storage life. The half-life of polymer film was 20 h, which was 7.5 times higher than that of BCz-BO-Azo power. The BCz-BO-Azo ED film, which was charged, could release heat, and heat up rapidly under some external stimulation. The released heat of cross-linking film under 40°C atmosphere could raise its own temperature to 53.2°C, as shown in Figure 3.16(c) and 16(d). The findings presented in this study offer a groundbreaking method to produce readily applicable self-template films that can be employed in large-scale industrial manufacturing. These STFs hold immense potential for diverse applications, including self-defrosting glasses, insulation for building walls, and the efficient transport of solar energy.

To enhance the applicability of azopolymer STFs, Cooper *et al.* recently designed a stretchable, lightweight polymer film for solar thermal storage by integrating multiwalled carbon nanotubes (MWCNTs) with carboxylic acid patterning and azobenzene compounds into a flexible matrix composed of a styrene-ethylene-butylene-styrene (SEBS) triblock copolymer, shown in Figure 3.17(a). The results show that the charging and discharging performance could be tuned based on the control of the styrene content of SEBS and the covalent linkage of azobenzene with MWCNTs. The embedding of azobenzene into SEBS can sufficiently stabilize the *cis*-isomers, while linking azobenzene to MWCNTs can enhance absorption for utilization in ambient sunlight, shown in Figure 3.17(b). By increasing the proportion of the rigid, styrene phase in the SEBS block copolymer, the photoisomerization behavior during charging and discharging of the material decreased. This indicates that the storage capacity of the self-template battery (STB) materials is influenced by the available free volume in the polymer substrate. Moreover, the ability to process these devices using drop casting from liquid precursors suggests that they can be manufactured on an industrial scale using high-throughput production methods like slot die or roll-to-roll printing. This research provides valuable insights into the necessary advancements needed to develop scalable and stretchable polymer composites for solar thermal energy storage. The findings contribute to a better understanding of the field and pave the way for further innovations in this area.

Table 3.1 summarizes the energy storage and heat release properties of azopolymer STFs 1–18, which reveals that the design of azobenzene-containing polymers by adjusting the monomer structure and polymer backbone is an excellent strategy to enhance the energy densities and

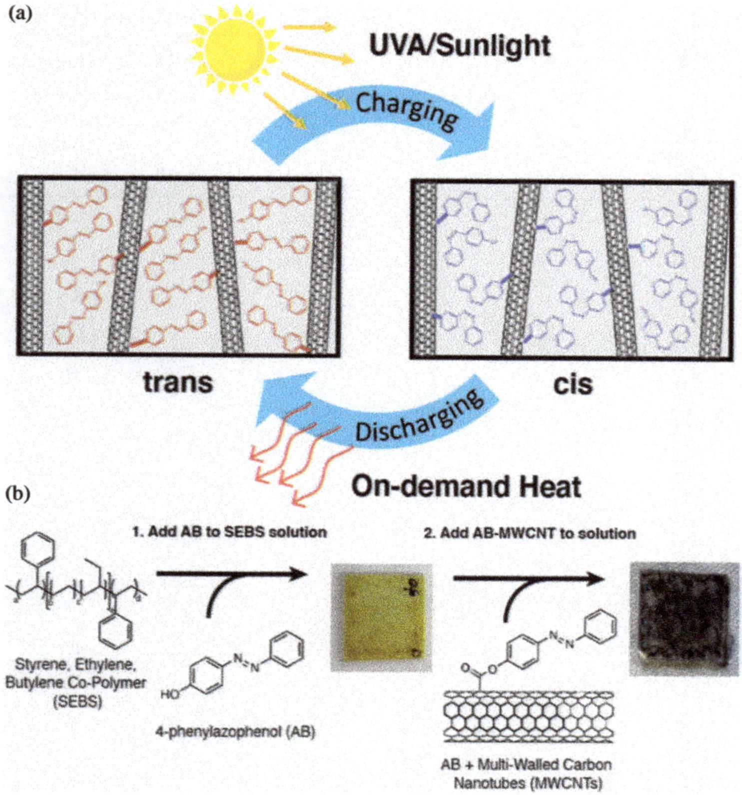

Figure 3.17. (a) Visual representation of the AB transition from the *trans* isomer to the *cis* isomer in the presence of the UVA component of sunlight. (b) Graphical overview of the STB production. The SEBS triblock copolymer is combined first with free AB and then AB linked to MWCNTs to yield a viable, stretchable solar thermal device through drop casting. Reproduced with permission [147]. Copyright 2022, American Chemical Society.

storage half-lives of STFs. Importantly, azopolymers can be readily incorporated into heating devices and hold great potential for the development of new applications in energy storage and heat release, which is the biggest advantage of azopolymer-based STF materials. These findings prove that azobenzene-containing polymers or polymer composites have made considerable advancement in improving storage performances, controlling heat release, and exploring solid-state exothermic applications.

Table 3.1. Summary of the energy storage and heat release properties of azopolymer STFs.

	Storage Energy Density (ΔH)				Half-Life	Heat Release	
STFs	ΔH_{iso} (J g^{-1})	$\Delta H_{phase\ change}$ (J g^{-1})	ΔH_{total} (J g^{-1})	ΔH_{total} (W h kg^{-1})	$\tau_{1/2}$ h	ΔT (°C)	**Refs.**
1	—	—	201	—	40	4	123
2	210	77	287	80	13	3	51
3	139		139	39	—	8.8	89
4	112	69	181	—	—	1.5	124
5	104	—	104	29	55	10	126
6	90	—	90	25	75	—	20
7	373	—	373	103	—	—	127
8	337	—	337	94	—	—	127
9	510	—	510	142	75	—	127
10	151	59	210	77	9	6	129
11	408	—	408	113	181	7	130
12	—	—	—	49	0.25	1.5	135
13	14	—	14	4	16	—	136
14	125	—	125	35	12	—	136
15	181	33	214	59	42	5	144
16	140	—	140	38	312	6	145
17a	154	—	154	—	—	—	146
17b	160	60	220	—	—	—	146
17c	164	81	245	—	27	—	146
17d	238	94	332	—	98	—	146
18	101	—	101	—	20	13.2	148

Note: Unreported.

3.3 Energy release of azopolymer STFs

It is well known that storage energy density is an important parameter for the development of azopolymer STFs, but it is not the only factor affecting STFs. Another important factor that determines the practical application of azopolymer STFs is the controlled heat release [149,150]. The absorption of photons at specific wavelengths by the ground state of the azopolymer STFs can cause it to change from low-energy, stable *trans*-isomers to a high-energy, metastable *cis*-isomers. The metastable isomer mentioned can retain its state for a duration ranging from several hours to years [84,98]. When subjected to external stimulation, the metastable *cis*-isomers can surpass the energy barrier, undergo the process of energy release, and revert to the thermodynamically stable *trans*-isomer. In case of unexpected situations that may arise in daily practice, it should be possible to release the stored heat in a fast and efficient way, so the energy release of azopolymer STF is essential.

The reversible photoisomerization of azobenzene molecules is susceptible to environmental influences such as light, heat, catalyst, solvent, magnetism, and stretching, as shown in Figure 3.18(a). Hence, it is crucial

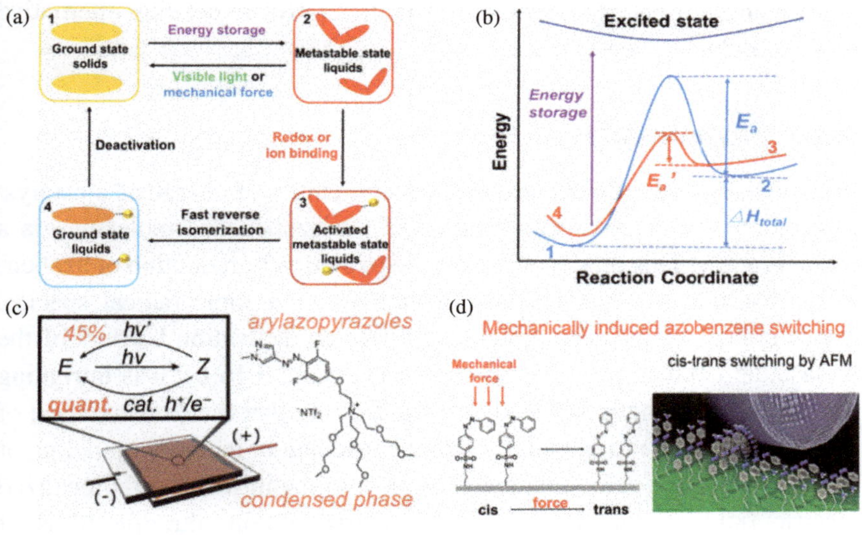

Figure 3.18. (a) Schematic diagram of different induced methods to trigger exotherm after storing energy in the ground state solid molecular solar thermal systems (MOST) materials. (b) Energy diagram illustrating the reduction in activation energy for the reverse isomerization via redox or ion binding. (c) Efficient electrocatalytic switching of azoheteroarenes in the condensed phases. Reproduced with permission [131]. Copyright 2021, American Chemical Society. (D) Mechanically induced azobenzene switching. Reproduced with permission [158]. Copyright 2019, American Chemical Society.

to investigate the isomerization process of azobenzene under various conditions in order to drive the further development of advanced azopolymer STFs. Since azobenzene molecules are the most commercially promising materials for solar thermal energy storage and conversion, there have been many studies on achieving heat release from azobenzene STFs under different stimuli in recent years. Until now, most studies on azobenzene STFs have employed light stimulation and thermal activation to complete the energy release process [130,151]. However, the poor penetration ability of light often only enables the surface STF to complete the heat release, while the internal metastable isomers may find it difficult to realize the heat release. Therefore, the heat release under light conditions is usually incomplete [11]. At the same time, stimulation with an external heat source can usually achieve more complete energy release, but thermal activation typically necessitates a significant input of energy to overcome the activation energy barrier, which inevitably leads to over consumption of energy. Therefore, it is imperative to develop alternative ways of inducing heat release.

3.3.1 *Electrocatalysis*

Electrocatalysis means the use of a small quantity of charge as a catalyst to promote the reverse isomerization of the azobenzene molecule via a redox process. This process involves the oxidation or reduction of azobenzene, resulting in the formation of cationic or anionic radical species. These radical species can effectively lower the activation barriers of the metastable isomers [56,152], as shown in Figure 3.18(b). It is intriguing to note that tests have supplied proof indicating that the isomerization of azobenzene can be triggered through the resonant or inelastic tunneling of electrons. Yogitha *et al.* conducted a study on the influence of a localized electric field on the rate of photoisomerization in ultrathin films of azobenzene-containing side chain LCPs based on PEGIA at the interface between air and water [153]. The generation of a local electric field was accomplished by incorporating polar polyvinylidene fluoride (PVDF) molecules into the PEGIA films. The findings revealed that the exponential effect of PVDF molecule content could be ascribed to the electric field effect arising from the dipole moment of PVDF molecules. The observed influence of the electric field on the isomerization reaction of azobenzene provides justification for exploring the potential utilization of electric fields in conjunction with azo STFs under specific conditions. Recently, Han and colleagues showcased a successful electrocatalytic induction of reverse isomerization in azo-heteroaromatics, both in solution and condensed phases. This method proved to have superior switching integrity compared to photochemical methods [131]. The redox-initiated pathway could decrease the reverse isomerization barrier by 27 kJ mol^{-1}, significantly enhancing the electrochemical conversion efficiency in the condensed phase compared to the solution state, as shown in Figure 3.18(c). Brummel and co-workers successfully proved the recyclability of this process and achieved over 99% reversibility in the release of energy from norbornadiene derivatives through electrochemical triggering [154]. It is anticipated that this electrocatalytic activation process will provide a novel approach for electronically controlled thermal storage of azopolymer STFs with enhanced heat release that is both more comprehensive and faster.

3.3.2 *Mechanical triggering*

In recent years, mechanically induced *cis*-to-*trans* reversion processes of azobenzene functional groups have also received much attention. Singleton *et al.* explored photoinduced isomerization and thermal relaxation of azobenzene-functionalized polymers poly(disperse red 1 acrylate) when the intracellular pressure was increased to 1.5 GPa in diamond-arbor spectroscopy [155]. Using laser-pumped probe spectroscopy, they discovered that the half-life of the isomerization increases as the pressure increases. It is important to note that the pressure is supposed to be low. Dong *et al.* show that compression at 18 GPa leads to irreversible destruction of the molecular structure of azobenzene [156]. Actually, pulling force can also influence azobenzene isomerization. If the polymer films possess an oriented structure or are exposed to linearly polarized light, they show anisotropic photomechanical response. As far back as 2008, Tanaka employed a blend film made up of polycaprolactone and poly(vinyl ether) with azobenzene moiety (Azo-PVE) as a side chain. This innovative approach enabled the production of a prompt and reversible anisotropic photomechanical response within the polymer film [157]. Furthermore, the increase in stretching ratio led to a greater enhancement of anisotropy in the photo-response, indicating the formation of stronger bonds between the amorphous regions via elongated Azo-PVE chains. A recent report on the mechanically scanned inversion of a monomolecular layer of azobenzene by atomic force microscopy (AFM) techniques emphasizes the feasibility of this procedure, which is also confirmed by various mechanochemical isomerization reactions found so far [158]. The activation energy for mechanical switching of azobenzene is determined to be 1.5 eV, which is below the 450 nm (2.7 eV) photon energy needed for inversion, as shown in Figure 3.18(d). There are still challenges in applying mechanical forces to manipulate switches that are randomly distributed. It is possible that mechanical trigger in the shape of shear, compression, and shock waves can also be applied to the reversion process of the azobenzene group by combining a polymer network to efficiently conduct mechanical energy to the switching unit. These findings indicate that mechanically triggered strategies are likely to promote adequate heat release in future azopolymer STF systems.

3.3.3 *Catalyst*

It is possible to efficiently control the heat release from the azopolymer STFs with the use of a catalyst. The thermally catalyzed *cis*-to-*trans* isomerizations of azobenzene can be completed within several hours to several days under ordinary conditions. But the present of AuNPs can speed up the de-isomerization of azobenzene to several minutes. Compared with solution, it is possible to decrease the activation energy needed for the thermal relaxation of azobenzene derivatives from *cis* to *trans* conversion on the gold (111) surface by 75% [159–161]. Sant *et al.* [162] discovered that the inclusion of AuNPs can significantly decrease the thermal lifetime of azobenzene-containing *cis*-isomers, resulting in a reduction of up to three orders of magnitude when compared to nanoparticle-free solutions (Figure 3.19(a)). Angelini *et al.* [163] conducted a study on the impact of these uncoated negatively charged silver nanoparticles on the thermal *cis*-to-*trans* isomerization of azobenzene and discovered that 4-methoxyazobenzene, a resonance-activated derivative of azobenzene, was introduced to the *trans*-Ag nanoparticles in aqueous solution and

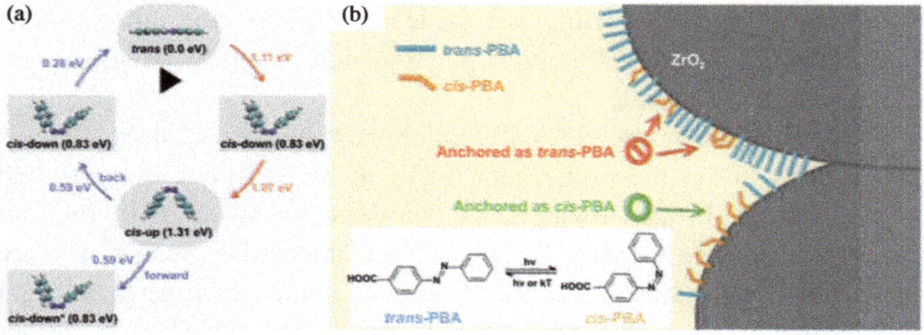

Figure 3.19. (a) Catalytic mechanism of AuNPs. Reproduced with permission [162]. Copyright 2015, American Chemical Society. (b) Catalytic mechanism of ZrO_2 nanoparticle films. Reproduced with permission [164]. Copyright 2018, American Chemical Society.

converted to the *cis* form by irradiation immediately or 120 h after sample preparation. Interestingly, the application of silver nanoparticles as catalysts or reaction media is limited due to their rapid oxidation and aggregation in the absence of any stabilizer.

Besides nanoparticles, thin films of metal oxide nanoparticles can also be used as carriers for azo space energy conversion devices. Achey *et al.* [164] discovered that interfacial attachment of azobenzene derivatives to ZrO_2 nanoparticle films caused a doubling of the rate of thermal and photoisomerization and a threefold reduction in the degree of conversion of azobenzene compared to fluid solutions (Figure 3.19(b)). It is more interesting to note that the extent of these changes can be controlled by selectively anchoring either the *cis* or *trans* isomers of azobenzene onto the ZrO_2 film. The isomerization of azobenzene may also be influenced by metal ions and acid radical ions, such as sodium, bromine, copper and other ions [165]. In general, the catalytic activity of copper ions in azobenzene isomerization is not prominent and it is far less than that of AuNPs. Nevertheless, studies have found that the catalytic activity of divalent copper ions is higher than that of copper ions. Lately, Hossain reported the discovery of two novel macrocyclic azobenzene materials that, besides possessing conventional optical switching properties, are capable of transiently transforming from *cis* to trans isomers in the dark at 273 K and in the presence of Cu^{2+} ions [65]. The Cu^{2+} complex is able to keep the *cis*-isomer in radiation under constant UV irradiation. But when the UV irradiation stops, they change into the *trans*-isomers. A similar phenomenon occurs with silver ions but at a little slower rate.

Overall, most of the relevant literature has chosen to disperse AuNPs as catalysts inside the azobenzene STFs in order to obtain controlled energy release. In this research, we analyzed several kinds of relevant catalysts, including AuNP, ZrO_2 and so on. Compared to AuNPs, they are inexpensive, easily available, and have good catalytic properties, which implies that the relevant applications of azopolymer STFs deserve further investigation to fulfill various environmental demands for the controlled release and use of thermal energy.

Here, we present the key properties of azopolymers as STFs from the perspective of their design and synthesis, presenting the latest research progress on several types of azobenzene-containing polymers such as azobenzene compound/polymer composites, linear azopolymer, dendrimer azopolymer, and other types of azopolymers. All these results show that azopolymer are effective, highly promising, and a versatile tool for solar thermal transfer and energy storage. The unique advantages of azopolymer STFs in terms of ease of forming uniform flat films, feasibility of large surface exotherm, and integration with existing devices will always attract tremendous research enthusiasm. Although azobenzene-containing polymers have made considerable advancement in improving storage performances and controlling heat release, the development of azopolymer STFs remains in its infancy. A number of basic challenges and essential problems are still to be addressed for azopolymer STFs. Therefore, it is urgent to present some novel solutions to address these challenges and create opportunities for developing more advanced azopolymer STFs.

First, the improvement of azopolymer STFs in one aspect of performance usually sacrifices other properties, which makes it very difficult to obtain STFs with excellent performance. For example, developing full-spectrum azopolymer STFs requires a large redshift in the absorption wavelength of the azobenzene molecules, yet this unavoidably reduces the energy density and storage half-life. The rational design of azobenzene molecules with tailored characteristics and positions of substituents has gained increasing attention in order to achieve absorption at longer wavelengths. This approach includes the development of push-pull azobenzene, which exhibits π–π* absorption bands in the visible range. Bridged azobenzene structures have been designed to have thermodynamically stable *cis*-isomers. In addition, tetra-ortho-substituted azobenzene compounds have been developed to exhibit n–π* absorption bands that span from green to near-infrared wavelengths [101,134]. Therefore, on-demand environmental applications of azopolymer STFs need to be designed to balance these parameters depending on the environment of the application.

Secondly, azopolymer STFs have poor control performance. At present, the research on azobenzene STFs mainly focuses on the improvement of energy storage and the prolongation of half-lives. Although the novel strategy of combining azopolymers with covalent bonds, such as dynamic covalent bonds (C–N, C–S, and S–S bonds) or photo-triggered formation and dissociation of coordination bonds (Mg–O and Fe–O) holds promise for the development of azopolymers STFs with ultra-high storage energy densities, but the reported values of energy storage performance are currently theoretical [166]. It is worth noting that photoisomerization of azobenzene under light is a balanced reaction and the degree of reversible isomerization is difficult to reach the desired value. The degree of isomerization may be even lower with changes in climate or light wavelength. In fact, many studies have been carried out on the reasonable control of azobenzene photoisomerization by electrocatalysis [56,154], catalyst catalysis [149,167], acid stimulation [168], and mechanical triggering [150]. Unfortunately, not many studies have combined energy storage studies with studies on the regulation of photoisomerization, and this area deserves further research.

Finally, the low cycling performance of azopolymer STFs also needs to be improved. Currently, the cycle number of azopolymer STFs is about several hundred cycles, while the cycle stability of azobenzene derivatives is only a dozen cycles. Moreover, due to their energy storage mechanism, azopolymer STFs are commonly utilized as thin films. During solar radiation and exothermic cycling, the films are irreversibly damaged in a short time. The material's instability not only raises the cost of its utilization but also lacks an efficient remedy. Studies have demonstrated the potential of specific azobenzene materials in repairing damaged or ruptured materials caused by temperature differences between *cis* and azobenzene derivatives. This capability can effectively prolong the lifespan of azo short-chain films. The poor recovery performance of azopolymer storage systems may be due to frequent isomerization during the recovery process. While the theoretical recovery of the azobenzene groups reaches 100%, the frequent isomerization reactions may lead to irreversible damage to the chemical bonds or structure of the azobenzene groups.

Apart from that, the integration of the STF system into the device structure will be considered when selecting the application phase of the azopolymer STFs. In a device employing an inorganic catalyst or electrochemical trigger method to assist in the heat release process, the cycling of azopolymer STFs on the immobilized catalyst or electrode surface plays a vital role in enabling the charging and discharging of large amounts of material. Photo-active film coating materials for windows, windscreens or fabrics need to have two key properties. Firstly, they should be mostly insoluble in water or common organic solvents. Secondly, these materials should exhibit significant penetration depth, allowing for a substantial coating thickness. Researchers are exploring these specific requirements in order to achieve the greatest number of practical technical applications that are designed to complement traditional solar energy conversion technologies such as photovoltaics and solar fuel power, thereby expanding the possibilities for the efficient use of solar energy. Although the research on azopolymer STFs is still in its infancy, it offers a novel avenue for solar energy harvesting and utilization. The development of existing problems and the presentation of new strategies provide opportunities for further design of advanced azopolymer STFs. Azopolymer STFs are expected to contribute significantly to the energy field.

References

[1] Hu X, Zuo D, Cheng S, Chen S, Liu Y, Bao W, Deng S, Harris SJ, Wan J. (2023). Ultrafast materials synthesis and manufacturing techniques for emerging energy and environmental applications. *Chem Soc Rev*, **52**(3), pp. 1103–1128.

[2] Krevor S, De Coninck H, Gasda SE, Ghaleigh NS, De Gooyert V, Hajibeygi H, Juanes R, Neufeld J, Roberts JJ, Swennenhuis F. (2023). Subsurface carbon dioxide and hydrogen storage for a sustainable energy future. *Nat Rev Earth Environ*, **4**(2), pp. 102–118.

[3] Chen P, Wu Y, Meng J, He P, Li D, Coffman DM, Liang X, Guan D. (2022). The heterogeneous role of energy policies in the energy transition of Asia-pacific emerging economies. *Nat Energy*, **7**(7), pp. 588–596.

[4] Gosens J, Turnbull AB, Jotzo F. (2022). China's decarbonization and energy security plans will reduce seaborne coal imports: results from an installation-level model. *Joule*, **6**(4), pp. 782–815.

[5] Thompson H. (2022). The geopolitics of fossil fuels and renewables reshape the world. *Nature*, **603**(7901), pp. 364.

[6] Lv J, Xie J, Mohamed AGA, Zhang X, Wang Y. (2022). Photoelectrochemical energy storage materials: design principles and functional devices towards direct solar to electrochemical energy storage. *Chem Soc Rev*, **51**(4), pp. 1511–1528.

[7] Tan J, Wang X, Chu W, Fang S, Zheng C, Xue M, Wang X, Hu T, Guo W. (2023). Harvesting energy from atmospheric water: grand challenges in continuous electricity generation. *Adv Mater*, pp. 2211165.

[8] Xu X, Wang G. (2022). Molecular solar thermal systems towards phase change and visible light photon energy storage. *Small*, **18**(16), pp. 2107473.

[9] Zhang B, Feng Y, Feng W. (2022). Azobenzene-based solar thermal fuels: a review. *Nano-Micro Lett*, **14**(1), pp. 138.

[10] Dong L, Feng Y, Wang L, Feng W. (2018). Azobenzene-based solar thermal fuels: design, properties, and applications. *Chem Soc Rev*, **47**(19), pp. 7339–7368.

[11] Gerkman MA, Han GGD. (2020). Toward controlled thermal energy storage and release in organic phase change materials. *Joule*, **4**(8), pp. 1621–1625.

[12] Gonzalez A, Kengmana ES, Fonseca MV, Han GGD. (2020). Solid-state photoswitching molecules: structural design for isomerization in condensed phase. *Mater Today Adv*, **6**, pp. 100058.

[13] Li SW, Wang HQ, Fang J, Liu QB, Wang JJ, Guo SP. (2020). Photoisomerization energy storage using azobenzene and nanoscale templates: a topical review. *J Therm Sci*, **29**(2), pp. 280–297.

[14] Wu S, Butt HJ. (2020). Solar-thermal energy conversion and storage using photoresponsive azobenzene-containing polymers. *Macromol Rapid Commun*, **41**(1), pp. 1900413.

[15] Qiu QF, Shi YR, Han GGD. (2021). Solar energy conversion and storage by photoswitchable organic materials in solution, liquid, solid, and changing phases. *J Mater Chem C*, **9**(35), pp. 11444–11463.

[16] Wang Z, Erhart P, Li T, Zhang Z-Y, Sampedro D, Hu Z, Wegner HA, Brummel O, Libuda J, Nielsen MB, Moth-Poulsen K. (2021). Storing energy with molecular photoisomers. *Joule*, **5**(12), pp. 3116–3136.

[17] Feng W, Li SP, Li M, Qin CQ, Feng YY. (2016). An energy-dense and thermal-stable bis-azobenzene/hybrid templated assembly for solar thermal fuel. *J Mater Chem A*, **4**(21), pp. 8020–8028.

[18] Xu X, Zhang P, Wu B, Xing Y, Shi K, Fang W, Yu H, Wang G. (2019). Molecular solar thermal storage enhanced by hyperbranched structures. *Sol RRL*, **4**(1), pp. 1900422.

[19] Ishiba K, Morikawa MA, Chikara C, Yamada T, Iwase K, Kawakita M, Kimizuka N. (2015). Photoliquefiable ionic crystals: a phase crossover approach for photon energy storage materials with functional multiplicity. *Angew Chem Int Ed*, **54**(5), pp. 1532–1536.

[20] Zhitomirsky D, Grossman JC. (2016). Conformal electroplating of azobenzene-based solar thermal fuels onto large-area and fiber geometries, *ACS Appl Mater Interfaces*, **8**(39), pp. 26319–26325.

[21] Xu WC, Sun S, Wu S. (2019). Photoinduced reversible solid-to-liquid transitions for photoswitchable materials. *Angew Chem Int Ed*, **58**(29), pp. 9712–9740.

[22] Yang B, Yu M, Yu H. (2020). Azopolymer-based nanoimprint lithography: recent developments in methodology and applications. *ChemPlusChem*, **85**(9), pp. 2166–2176.

[23] Ryazantsev MN, Strashkov DM, Nikolaev DM, Shtyrov AA, Panov MS. (2021). Photopharmacological compounds based on azobenzenes and azoheteroarenes: principles of molecular design, molecular modelling, and synthesis. *Russ Chem Rev*, **90**(7), pp. 868–893.

[24] Zhang Z, Xie Z, Nie CSW. (2022). Photo-controlled properties and functions of azobenzene-terminated polymers. *Polymer*, **256**, pp. 125166.

[25] Xu G, Li S, Liu C, Wu S. (2020). Photoswitchable adhesives using azobenzene-containing materials. *Chem — Asian J*, **15**(5), pp. 547–554.

[26] Liang SF, Nie C, Yan J, Zhang QJ, Wu S. (2021). Photoinduced reversible solid-to-liquid transitions and directional photofluidization of azobenzene-containing polymers. *Chin J Polym Sci*, **39**(10), pp. 1225–1234.

[27] Weis P, Wu S. (2018). Light-switchable azobenzene-containing macromolecules: from uv to near infrared. *Macromol Rapid Commun*, **39**(1), pp. 1700220.

[28] Sun SD, Liang SF, Xu WC, Xu GF, Wu S. (2019). Photoresponsive polymers with multi-azobenzene groups. *Polym Chem*, **10**(32), pp. 4389–4401.

[29] Schenning APHJ, Roman Vas C, Weener JW, Baars MWPL, Gaast SJ, van der, Meijer EW. (1998). Amphiphilic dendrimers as building blocks in supramolecular assemblies. *J Am Chem Soc*, **120**, pp. 8199–8208.

[30] Tsuda K, Dol GC, Gensch T, Hofkens J, Latterini L, Weener JW, Meijer EW, De Schryver FC. (2000). Fluorescence from azobenzene functionalized poly(propylene imine) dendrimers in self-assembled supramolecular structures. *J Am Chem Soc*, **122**, pp. 3445–3452.

[31] Jimenez J, Pintre IC, Gascon E, Sanchezsomolinos C, Alcala R, Cavero E, Serrano JL, Oriol L. (2014). Photoresponsive liquid-crystalline dendrimers based on a cyclotriphosphazene core. *Macromol Chem Phys*, **215**(16), pp. 1551–1562.

[32] Gimenez-Gomez A, Magson L, Peñin B, Sanosa N, Soilán J, Losantos R, Sampedro D. (2022). A photochemical overview of molecular solar thermal energy storage. *Photochem*, **2**(3), pp. 694–716.

[33] Le M, Han GGD. (2022). Stimuli-responsive organic phase change materials: molecular designs and applications in energy storage. *Acc Mater Res*, **3**(6), pp. 634–643.

[34] Wang Z, Holzel H, Moth-Poulsen K. (2022). Status and challenges for molecular solar thermal energy storage system based devices. *Chem Soc Rev*, **51**(17), pp. 7313–7326.

[35] Wu S, Li TX, Zhang ZY, Li T, Wang RZ. (2021). Photoswitchable phase change materials for unconventional thermal energy storage and upgrade. *Matter*, **4**(11), pp. 3385–3399.

[36] Baby A, John AM, Balakrishnan SP. (2022). Photoresponsive carbon-azobenzene hybrids: a promising material for energy devices. *ChemPhysChem*, **24**(6), pp. 202200676.

[37] Ganguly G, Sultana M, Paul A. (2018). Designing efficient solar-thermal fuels with [n.N](9,10)anthracene cyclophanes: a theoretical perspective. *J Phys Chem Lett*, **9**(2), pp. 328–334.

[38] Kandinska MI, Kitova SM, Videva VS, Stoyanov SS, Yordanova SB, Baluschev SB, Angelova SE, Vasilev AA. (2019). Precious metal-free molecular machines for solar thermal energy storage. *Beilstein J Org Chem*, **15**, pp. 1096–1106.

[39] Imato K, Sasaki A, Ishii A, Hino T, Kaneda N, Ohira K, Imae I, Ooyama Y. (2022). Sterically hindered stiff-stilbene photoswitch offers large motions, 90% two-way photoisomerization, and high thermal stability. *J Org Chem*, **87**(23), pp. 15762–15770.

[40] Börjesson K, Lennartson A, Moth-Poulsen K. (2014). Fluorinated fulvalene ruthenium compound for molecular solar thermal applications. *J Fluorine Chem*, **161**, pp. 24–28.

[41] Lennartson A, Lundin A, Borjesson K, Gray V, Moth-Poulsen K. (2016). Tuning the photochemical properties of the fulvalene-tetracarbonyl-diruthenium system. *Dalton Trans*, **45**(21), pp. 8740–8744.

[42] Meng FY, Chen IH, Shen JY, Chang KH, Chou TC, Chen YA, Chen YT, Chen CL, Chou PT. (2022). A new approach exploiting thermally activated delayed fluorescence molecules to optimize solar thermal energy storage. *Nat Commun*, **13**(1), pp. 797.

[43] Quant M, Hillers-Bendtsen AE, Ghasemi S, Erdelyi M, Wang Z, Muhammad LM, Kann N, Mikkelsen KV, Moth-Poulsen K. (2022). Synthesis, characterization and computational evaluation of bicyclooctadienes towards molecular solar thermal energy storage. *Chem Sci*, **13**(3), pp. 834–841.

[44] Refaa Z, Hofmann A, Castro MF, Hernandez JO, Wang Z, Hölzel H, Andreasen JW, Moth-Poulsen K, Kalagasidis AS. (2022). Thermo-optical performance of molecular solar thermal energy storage films. *Appl Energy*, **310**, pp. 118541.

[45] Berdnikova DV, Kriesche BM, Paululat T, Hofer TS. (2022). Fused bis(hemi-indigo): broad-range wavelength-independent photoswitches. *Chem — Eur J*, **28**(72), pp. 202202752.

[46] Doronina EP, Jouikov V, Belogolova EF, Sidorkin VF. (2022). Isomer-selective dative bond O→M (M=Si, Ge) for designing new photochromic hemi-indigo systems. *J Organomet Chem*, **958**, pp. 122189.

[47] Kjeldsen ILH, Hovring JF, Von Buchwald TJ, Hillers-Bendtsen AE, Mikkelsen KV. (2022). The effects of solvation on the back reaction and storage capabilities of solar thermal energy storage systems. *Phys Chem Chem Phys*, **24**(9), pp. 5564–5577.

[48] Schøttler C, Vegge SK, Cacciarini M, Nielsen MB. (2022). Long-term energy storage systems based on the dihydroazulene/vinylheptafulvene photo/thermoswitch. *ChemPhotoChem*, **6**(8), pp. 202200037.

[49] Shangguan Z, Sun W, Zhang ZY, Fang D, Wang Z, Wu S, Deng C, Huang X, He Y, Wang R, Li T, Moth-Poulsen K, Li T. (2022). A rechargeable molecular solar thermal system below 0°C. *Chem Sci*, **13**(23), pp. 6950–6958.

[50] Yang Y, Huang S, Ma Y, Yi J, Jiang Y, Chang X, Li Q. (2022). Liquid and photoliquefiable azobenzene derivatives for solvent-free molecular solar thermal fuels. *ACS Appl Mater Interfaces*, **14**(31), pp. 35623–35634.

[51] Zhao RY, Li YC, Bai JY, Mu JH, Chen LX, Zhang N, Han JS, Liu FS, Yan SK. (2022). Preparation of flexible photo-responsive film based on novel photo-liquefiable azobenzene derivative for solar thermal fuel application. *Dyes Pigm*, **202**, pp. 110277.

[52] Kwaria D, Mcgehee K, Liu S, Kikkawa Y, Ito S, Norikane Y. (2023). Visible-light-photomeltable azobenzenes as solar thermal fuels. *ACS Appl Opt Mater*, **1**(2), pp. 633–639.

[53] Wang Z, Roffey A, Losantos R, Lennartson A, Jevric M, Petersen AU, Quant M, Dreos A, Wen X, Sampedro D, Börjesson K, Moth-Poulsen K. (2019). Macroscopic heat release in a molecular solar thermal energy storage system. *Energy Environ Sci*, **12**(1), pp. 187–193.

[54] Orrego-Hernandez J, Dreos A, Moth-Poulsen K. (2020). Engineering of norbornadiene/quadricyclane photoswitches for molecular solar thermal energy storage applications. *Acc Chem Res*, **53**(8), pp. 1478–1487.

[55] Eschenbacher R, Xu T, Franz E, Löw R, Moje T, Fromm L, Görling A, Brummel O, Herges R, Libuda J. (2022). Triggering the energy release in molecular solar thermal systems: norbornadiene-functionalized trioxatriangulen on Au(111). *Nano Energy*, **95**, pp. 107007.

[56] Franz E, Krappmann D, Fromm L, Luchs T, Gorling A, Hirsch A, Brummel O, Libuda J. (2022). Electrocatalytic energy release of norbornadiene-based molecular solar thermal systems: tuning the electrochemical stability by molecular design. *ChemSusChem*, **15**(24), pp. 202201483.

[57] Hillers-Bendtsen AE, Iuel Lunoe Dunweber PG, Olsen LH, Mikkelsen KV. (2022). Prospects of improving molecular solar energy storage of the norbornadiene/quadricyclane system through bridgehead modifications. *J Phys Chem A*, **126**(17), pp. 2670–2676.

[58] Mogensen J, Christensen O, Kilde MD, Abildgaard M, Metz L, Kadziola A, Jevric M, Mikkelsen KV, Nielsen MB. (2019). Molecular solar thermal energy storage systems with long discharge times based on the dihydroazulene/vinylheptafulvene couple. *Eur J Org Chem*, **2019**(10), pp. 1986–1993.

[59] Hillers-Bendtsen AE, Johansen MB, Mikkelsen KV. (2021). Promoting the thermal back reaction of vinylheptafulvene to dihydroazulene by physisorbtion on nanoparticles. *Phys Chem Chem Phys*, **23**(22), pp. 12889–12899.

[60] Vlasceanu A, Koerstz M, Skov AB, Mikkelsen KV, Nielsen MB. (2018). Multistate photoswitches: macrocyclic dihydroazulene/azobenzene conjugates. *Angew Chem Int Ed*, **57**(21), pp. 6069–6072.

[61] Fei L, Yu W, Wu Z, Yin Y, Moth-Poulsen K, Wang C. (2022). Optically controlled thermochromic switching for multi-input molecular logic. *Angew Chem Int Ed*, **61**(44), pp. 202212483.

[62] Nieland E, Voss J, Mix A, Schmidt BM. (2022). Photoresponsive dissipative macrocycles using visible-light-switchable azobenzenes. *Angew Chem Int Ed*, **61**(48), pp. 202212745.

[63] Rickhoff J, Arndt NB, Bockmann M, Doltsinis NL, Ravoo BJ, Kortekaas L. (2022). Reversible, red-shifted photoisomerization in protonated azobenzenes. *J Org Chem*, **87**(16), pp. 10605–10612.

[64] Butler CSG, King JP, Giles LW, Marlow JB, Vidallon MLP, Sokolova A, De Campo L, Tuck KL, Tabor RF. (2021). Design and synthesis of an azobenzene-betaine surfactant for photo-rheological fluids. *J Colloid Interface Sci*, **594**, pp. 669–680.

[65] Hossain MS, Bandyopadhyay S. (2021). Metal ion mediated instant Z→E isomerization of azobenzene macrocycles in the absence of light. *J Org Chem*, **86**(9), pp. 6314–6321.

[66] Baroncini M, D'agostino S, Bergamini G, Ceroni P, Comotti A, Sozzani P, Bassanetti I, Grepioni F, Hernandez TM, Silvi S, Venturi M, Credi A. (2015). Photoinduced reversible switching of porosity in molecular crystals based on star-shaped azobenzene tetramers. *Nat Chem*, **7**(8), pp. 634–640.

[67] Yu JK, Bannwarth C, Liang R, Hohenstein EG, Martinez TJ. (2020). Nonadiabatic dynamics simulation of the wavelength-dependent photochemistry of azobenzene excited to the n-π* and π-π* excited states. *J Am Chem Soc*, **142**(49), pp. 20680–20690.

[68] Jerca FA, Jerca VV, Hoogenboom R. (2021). Advances and opportunities in the exciting world of azobenzenes. *Nat Rev Chem*, **6**(1), pp. 51–69.

[69] Kumar P, Srivastava A, Sah C, Devi S, Venkataramani S. (2019). Arylazo-3,5-dimethylisoxazoles: azoheteroarene photoswitches exhibiting high Z-isomer stability, solid-state photochromism, and reversible light-induced phase transition. *Chem — Eur J*, **25**(51), pp. 11924–11932.

[70] Bhunia S, Dolai A, Samanta S. (2020). Robust bi-directional photoswitching of thiomethyl substituted arylazopyrazoles under visible light. *Chem Commun*, **56**(70), pp. 10247–10250.

[71] He Y, Shangguan Z, Zhang ZY, Xie M, Yu C, Li T. (2021). Azobispyrazole family as photoswitches combining (near-) quantitative bidirectional isomerization and widely tunable thermal half-lives from hours to years. *Angew Chem Int Ed*, **133**(30), pp. 16675–16682.

[72] Soberats B, Uchida E, Yoshio M, Kagimoto J, Ohno H, Kato T. (2014). Macroscopic photocontrol of ion-transporting pathways of a nanostructured imidazolium-based photoresponsive liquid crystal. *J Am Chem Soc*, **136**(27), pp. 9552–9555.

[73] Vapaavuori J, Goulet-Hanssens A, Heikkinen ITS, Barrett CJ, Priimagi A. (2014). Are two azo groups better than one? Investigating the photoresponse of polymer-bisazobenzene complexes. *Chem Mater*, **26**(17), pp. 5089–5096.

[74] Quan J, Guo Y, Ma J, Long D, Wang J, Zhang L, Sun Y, Dhinakaran MK, Li H. (2022). Light-responsive nanochannels based on the supramolecular host-guest system. *Front Chem*, **10**, pp. 986908.

[75] Ren H, Yang P, Yu H. (2022). Recent progress in azopyridine-containing supramolecular assembly: from photoresponsive liquid crystals to light-driven devices. *Molecules*, **27**(13), pp. 3977.

[76] Weis P, Tian W, Wu S. (2018). Photoinduced liquefaction of azobenzene-containing polymers. *Chem — Eur J*, **24**(25), pp. 6494–6505.

[77] Han GG D, Li H, Grossman JC. (2017). Optically-controlled long-term storage and release of thermal energy in phase-change materials. *Nat Commun*, **8**(1), pp. 1446.

[78] Wang Z, Tong Z, Ye Q, Hu H, Nie X, Yan C, Shang W, Song C, Wu J, Wang J, Bao H, Tao P, Deng T. (2017). Dynamic tuning of optical absorbers for accelerated solar-thermal energy storage. *Nat Commun*, **8**(1), pp. 1478.

[79] Feng Y, Liu H, Luo W, Liu E, Zhao N, Yoshino K, Feng W. (2013). Covalent functionalization of graphene by azobenzene with molecular hydrogen bonds for long-term solar thermal storage. *Sci Rep*, **3**, pp. 3260.

[80] Jiang Y, Huang J, Feng W, Zhao X, Wang T, Li CH, Luo W. (2019). Molecular regulation of nano-structured solid-state azo-swcnts assembly film for the high-energy and short-term solar thermal storage. *Sol Energy Mater Sol Cells*, **193**, pp. 198–205.

[81] Olmsted III J, Lawrence J, Yee GG. (1983). Photochemical storage potential of azobenzenes. *Sol Energy*, **30**(3), pp. 271–274.

[82] Yoshida Z-I. (1985). New molecular energy storage systems. *J Photochem*, **29**(1–2), pp. 27–40.

[83] Bren' VA, Dubonosov AD, Minkin VI, Chernoivanov VA. (1991). Norbornadiene-quadricyclane — an effective molecular system for the storage of solar energy. *Russ Chem Rev*, **60**(5), pp. 451–469.

[84] Al-Bataineh QM, Telfah A, Ahlmann N, Tolstik E, Tavares CJ, Hergenröder R. (2022). Photoisomerization kinetics of a novel

photoswitchable film based on methyl red doped with sodium hexachloroplatinate hosted in polyethylene oxide. *J Appl Polym Sci*, **139**(25), pp. 52387.

[85] Gonzalez A, Odaybat M, Le M, Greenfield JL, White AJP, Li X, Fuchter MJ, Han GGD. (2022). Photocontrolled energy storage in azobispyrazoles with exceptionally large light penetration depths. *J Am Chem Soc*, **144**(42), pp. 19430–19436.

[86] Li C, Zhang Z, Zhong TY, Wang G. (2022). Azobenzene modified metal-organic framework: for solar energy storage. *J Energy Storage*, **52**, pp. 104971.

[87] Masutani K, Morikawa MA, Kimizuka N. (2014). A liquid azobenzene derivative as a solvent-free solar thermal fuel. *Chem Commun*, **50**(99), pp. 15803–15806.

[88] Wang Z, Losantos R, Sampedro D, Morikawa M-A, Börjesson K, Kimizuka N, Moth-Poulsen K. (2019). Demonstration of an azobenzene derivative based solar thermal energy storage system. *J Mater Chem A*, **7**(25), pp. 15042–15047.

[89] Dong L, Chen Y, Zhai F, Tang L, Gao W, Tang J, Feng Y, Feng W. (2020). Azobenzene-based solar thermal energy storage enhanced by gold nanoparticles for rapid, optically-triggered heat release at room temperature. *J Mater Chem A*, **8**(36), pp. 18668–18676.

[90] Nucci M, Núñez A, Frutos LM, Marazzi M. (2022). Design of improved molecular solar-thermal systems by mechanochemistry: the case of azobenzene. *Adv Sust Syst*, **6**(7), pp. 2200097.

[91] Yang XY, Li SJ, Li L, Wang XM, Wang YZ, Zhao JG. (2021). A high power density photoactive thermal energy storage material for rapid heat release at low temperatures. *Mater Chem Phys*, **263**, pp. 124336.

[92] Yang XY, Li SJ, Zhao JG, Wang XM, Huang HY, Wang YZ. (2021). Visible light driven low temperature photoactive energy storage materials for high rate thermal output system. *Sol Energy Mater Sol Cells*, **231**, pp. 111330.

[93] Zhao X, Feng Y, Qin C, Yang W, Si Q, Feng W. (2017). Controlling heat release from a close-packed bisazobenzene-reduced-graphene-oxide assembly film for high-energy solid-state photothermal fuels, *ChemSusChem*, **10**(7), pp. 1395–1404.

[94] Pang W, Xue J, Pang H. (2019). A high energy density azobenzene/graphene oxide hybrid with weak nonbonding interactions for solar thermal storage. *Sci Rep*, **9**(1), pp. 5224.

[95] Fei L, Yin Y, Zhang J, Wang C. (2020). A visible energy management by photochromic solar thermal fuel using a color display. *Sol RRL*, **4**(12), pp. 2000499.

[96] Gerkman MA, Gibson RSL, Calbo J, Shi Y, Fuchter MJ, Han GGD. (2020). Arylazopyrazoles for long-term thermal energy storage and optically triggered heat release below 0°C. *J Am Chem Soc*, **142**(19), pp. 8688–8695.

[97] Griffiths K, Halcovitch NR, Griffin JM. (2020). Long-term solar energy storage under ambient conditions in a mof-based solid-solid phase-change material. *Chem Mater*, **32**(23), pp. 9925–9936.

[98] Szabo R, Le KN, Kowalczyk T. (2021). Multifactor theoretical modeling of solar thermal fuels built on azobenzene and norbornadiene scaffolds. *Sust Energy Fuel*, **5**(8), pp. 2335–2346.

[99] Bian Q, Xue Z, Sun P, Shen K, Wang S, Jia J. (2019). Visible-light-triggered supramolecular valves based on β-cyclodextrin-modified mesoporous silica nanoparticles for controlled drug release. *RSC Adv*, **9**(30), pp. 17179–17182.

[100] Zheng X, Bian Q, Ye C, Wang G. (2019). Visible light-, ph-, and cyclodextrin-responsive azobenzene functionalized polymeric nanoparticles. *Dyes Pigm*, **162**, pp. 599–605.

[101] Huang X, Shangguan Z, Zhang Z-Y, Yu C, He Y, Fang D, Sun W, Li Y-C, Yuan C, Wu S, Li T. (2022). Visible-light-induced reversible photochemical crystal-liquid transitions of azo-switches for smart and robust adhesives. *Chem Mater*, **34**(6), pp. 2636–2644.

[102] Fei L, Yin YJ, Yang MF, Zhang SF, Wang CX. (2021). Wearable solar energy management based on visible solar thermal energy storage for full solar spectrum utilization. *Energy Storage Mater*, **42**, pp. 636–644.

[103] Li X, Cho S, Han GGD. (2022). Light-responsive solid-solid phase change materials for photon and thermal energy storage. *ACS Mater Au*, **3**(1), pp. 37–42.

[104] Kashyap V, Sakunkaewkasem S, Jafari P, Nazari M, Eslami B, Nazifi S, Irajizad P, Marquez MD, Lee TR, Ghasemi H. (2019). Full spectrum solar thermal energy harvesting and storage by a molecular and phase-change hybrid material. *Joule*, **3**(12), pp. 3100–3111.

[105] Dreos A, Börjesson K, Wang Z, Roffey A, Norwood Z, Kushnir D, Moth-Poulsen K. (2017). Exploring the potential of a hybrid device combining solar water heating and molecular solar thermal energy storage. *Energy Environ Sci*, **10**(3), pp. 728–734.

[106] Fang J, Liu Q, Guo SP, Lei J, Jin HG. (2019). Spanning solar spectrum: a combined photochemical and thermochemical process for solar energy storage. *Appl. Energy*, **247**, pp. 116–126.

[107] Wang J, Han Z, Liu Y, Zhang X, Cui Z. (2021). Thermodynamic analysis of a combined cooling, heating, and power system integrated with full-spectrum hybrid solar energy device. *Energy Convers Manage*, **228**, pp. 113596.

[108] Beharry AA, Sadovski O, Woolley GA. (2011). Azobenzene photoswitching without ultraviolet light. *J Am Chem Soc*, **133**(49), pp. 19684–19687.

[109] Bleger D, Schwarz J, Brouwer AM, Hecht S. (2012). O-fluoroazobenzenes as readily synthesized photoswitches offering nearly quantitative two-way isomerization with visible light. *J Am Chem Soc*, **134**(51), pp. 20597–20600.

[110] Yang Y, Hughes RP, Aprahamian I. (2012). Visible light switching of a BF_2-coordinated azo compound. *J Am Chem Soc*, **134**(37), pp. 15224–15224.

[111] Samanta S, Beharry AA, Sadovski O, Mccormick TM, Babalhavaeji A, Tropepe V, Woolley GA. (2013). Photoswitching azo compounds in vivo with red light. *J Am Chem Soc*, **135**(26), pp. 9777–9784.

[112] Yang Y, Hughes RP, Aprahamian I. (2014). Near-infrared light activated azo-BF_2 switches. *J Am Chem Soc*, **136**(38), pp. 13190–13193.

[113] Bleger D, Hecht S. (2015). Visible-light-activated molecular switches. *Angew Chem Int Ed*, **54**(39), pp. 11338–11349.

[114] Accardo JV, Mcclure ER, Mosquera MA, Kalow JA. (2020). Using visible light to tune boronic acid-ester equilibria. *J Am Chem Soc*, **142**(47), pp. 19969–19979.

[115] Leistner AL, Kirchner S, Karcher J, Bantle T, Schulte ML, Godtel P, Fengler C, Pianowski ZL. (2021). Fluorinated azobenzenes switchable with red light. *Chem — Eur J*, **27**(31), pp. 8094–8099.

[116] Yan QH, Zhang Y, Dang YF, Feng YY, Feng W. (2020). Solid-state high-power photo heat output of 4-((3,5-dimethoxyaniline)-diazenyl)-2-imidazole/graphene film for thermally controllable dual data encoding/reading. *Energy Storage Mater*, **24**, pp. 662–669.

[117] Hammerich M, Schutt C, Stahler C, Lentes P, Rohricht F, Hoppner R, Herges R. (2016). Heterodiazocines: synthesis and photochromic properties, trans to *cis* switching within the bio-optical window. *J Am Chem Soc*, **138**(40), pp. 13111–13114.

[118] Moreno J, Gerecke M, Grubert L, Kovalenko SA, Hecht S. (2016). Sensitized two-nir-photon Z→E isomerization of a visible-light-addressable bistable azobenzene derivative. *Angew Chem Int Ed*, **55**(4), pp. 1544–1547.

[119] Lameijer LN, Budzak S, Simeth NA, Hansen MJ, Feringa BL, Jacquemin D, Szymanski W. (2020). General principles for the design of visible-light-responsive photoswitches: tetra-ortho-chloro-azobenzenes. *Angew Chem Int Ed*, **59**(48), pp. 21663–21670.

[120] Nishioka H, Liang X, Asanuma H. (2010). Effect of the ortho modification of azobenzene on the photoregulatory efficiency of DNA hybridization and the thermal stability of its *cis* form. *Chem — Eur J*, **16**(7), pp. 2054–2062.

[121] Bahrenburg J, Sievers CM, Schonborn JB, Hartke B, Renth F, Temps F, Nather C, Sonnichsen FD. (2013). Photochemical properties of multi-azobenzene compounds. *Photochem Photobiol Sci*, **12**(3), pp. 511–518.

[122] Yang WX, Feng YY, Si QY, Yan QH, Long P, Dong LQ, Fu LX, Feng W. (2019). Efficient cycling utilization of solar-thermal energy for thermochromic displays with controllable heat output. *J Mater Chem A*, **7**(1), pp. 97–106.

[123] Hu J, Huang S, Yu M, Yu H. (2019). Flexible solar thermal fuel devices: composites of fabric and a photoliquefiable azobenzene derivative. *Adv Energy Mater*, **9**(37), pp. 1901363.

[124] Tang JW, Feng YY, Feng W. (2021). Photothermal storage and controllable release of a phase-change azobenzene/aluminum nitride aerogel composite. *Compos Commun*, **23**, pp. 100575.

[125] Goulet-Hanssens A, Eisenreich F, Hecht S. (2020). Enlightening materials with photoswitches. *Adv Mater*, **32**(20), pp. 1905966.

[126] Zhitomirsky D, Cho E, Grossman JC. (2016). Solid-state solar thermal fuels for heat release applications. *Adv Energy Mater*, **6**(6), pp. 1502006.

[127] Jeong SP, Renna LA, Boyle CJ, Kwak HS, Harder E, Damm W, Venkataraman D. (2017). High energy density in azobenzene-based materials for photo-thermal batteries via controlled polymer architecture and polymer-solvent interactions. *Sci Rep*, **7**(1), pp. 17773.

[128] Zhou H, Xue C, Weis P, Suzuki Y, Huang S, Koynov K, Auernhammer GK, Berger R, Butt HJ, Wu S. (2017). Photoswitching of glass transition temperatures of azobenzene-containing polymers induces reversible solid-to-liquid transitions. *Nat Chem*, **9**(2), pp. 145–151.

[129] Cai F, Song T, Yang B, Lv X, Zhang L, Yu H. (2021). Enhancement of solar thermal fuel by microphase separation and nanoconfinement of a block copolymer. *Chem Mater*, **33**(24), pp. 9750–9759.

[130] Wang H, Feng Y, Gao J, Fang W, Ge J, Yang X, Zhai F, Yu Y, Feng W. (2022). Metallic-ion controlled dynamic bonds to co-harvest isomerization energy and bond enthalpy for high-energy output of flexible self-heated textile. *Adv Sci*, **9**(20), pp. 2201657.

[131] Greenfield JL, Gerkman MA, Gibson RSL, Han GGD, Fuchter MJ. (2021). Efficient electrocatalytic switching of azoheteroarenes in the condensed phases. *J Am Chem Soc*, **143**(37), pp. 15250–15257.
[132] Chu Z, Han Y, Bian T, De S, Kral P, Klajn R. (2019). Supramolecular control of azobenzene switching on nanoparticles. *J Am Chem Soc*, **141**(5), pp. 1949–1960.
[133] Gibson RSL, Calbo J, Fuchter MJ. (2019). Chemical *Z-E* isomer switching of arylazopyrazoles using acid. *ChemPhotoChem*, **3**(6), pp. 372–377.
[134] Waidhas F, Jevric M, Fromm L, Bertram M, Görling A, Moth-Poulsen K, Brummel O, Libuda J. (2019). Electrochemically controlled energy storage in a norbornadiene-based solar fuel with 99% reversibility. *Nano Energy*, **63**, pp. 103872.
[135] Fu L, Yang J, Dong L, Yu H, Yan Q, Zhao F, Zhai F, Xu Y, Dang Y, Hu W, Feng Y, Feng W. (2019). Solar thermal storage and room-temperature fast release using a uniform flexible azobenzene-grafted polynorborene film enhanced by stretching. *Macromolecules*, **52**(11), pp. 4222–4231.
[136] Saydjari AK, Weis P, Wu S. (2017). Spanning the solar spectrum: azopolymer solar thermal fuels for simultaneous uv and visible light storage. *Adv Energy Mater*, **7**(3), pp. 1601622.
[137] Bryszewska BKM. (2001). Dendrimers: properties and applications. *Acta Biochim Pol*, **48**, pp. 199–208.
[138] Didier Astruc EB, Ornelas C. (2010). Dendrimers designed for functions: from physical, photophysical, and supramolecular properties to applications in sensing, catalysis, molecular electronics, photonics, and nanomedicine. *Chem Rev*, **110**, pp. 1857–1959.
[139] Caminade A. (2016). Inorganic dendrimers: recent advances for catalysis, nanomaterials, and nanomedicine. *Chem Soc Rev*, **45**(19), pp. 5174–5186.
[140] Mignani S, Rodrigues J, Tomas H, Zablocka M, Shi X, Caminade A, Majoral J. (2018). Dendrimers in combination with natural products and analogues as anti-cancer agents. *Chem Soc Rev*, **47**(2), pp. 514–532.
[141] Ren L, Lv J, Wang H, Cheng Y. (2020). A coordinative dendrimer achieves excellent efficiency in cytosolic protein and peptide delivery. *Angew Chem Int Ed*, **59**(12), pp. 4711–4719.
[142] Andreas Archut GCA, Balzani V, De Cola L, Vo1gtle F. (1998). Toward photoswitchable dendritic hosts. Interaction between azobenzene-functionalized dendrimers and eosin. *J Am Chem Soc*, **120**, pp. 12187–12191.
[143] Junge DM, McGrath DV. (1999). Photoresponsive azobenzene-containing dendrimers with multiple discrete states. *J Am Chem Soc*, **121**, pp. 4912–4913.

[144] Xu X, Zhang P, Wu B, Xing Y, Shi K, Fang W, Yu H, Wang G. (2020). Photochromic dendrimers for photoswitched solid-to-liquid transitions and solar thermal fuels. *ACS Appl Mater Interfaces*, **12**(44), pp. 50135–50142.

[145] Xu X, Wu B, Zhang P, Xing Y, Shi K, Fang W, Yu H, Wang G. (2021). Arylazopyrazole-based dendrimer solar thermal fuels: stable visible light storage and controllable heat release. *ACS Appl Mater Interfaces*, **13**(19), pp. 22655–22663.

[146] Han GD, Park SS, Liu Y, Zhitomirsky D, Cho E, Dincă M, Grossman JC. (2016). Photon energy storage materials with high energy densities based on diacetylene–azobenzene derivatives. *J Mater Chem A*, **4**(41), pp. 16157–16165.

[147] Colburn TW, Delmastro AC, Figueroa M, Lopez F, Cooper CB. (2022). Embedding azobenzene-functionalized carbon nanotubes into a polymer matrix for stretchable, composite solar thermal devices. *J Phys Chem C*, **126**(37), pp. 15565–15572.

[148] Zhao R, Zhang Y, Mu J, Yu C, Pan Y, Zhao J, Wang Q, Zhang S, Yang B, Liu F. (2020). Preparation of photoresponsive film via electrodeposition approach for ready-to-use solar thermal fuel device. *Adv Mater Interfaces*, **7**(23), pp. 2001079.

[149] Goulet-Hanssens A, Rietze C, Titov E, Abdullahu L, Grubert L, Saalfrank P, Hecht S. (2018). Hole catalysis as a general mechanism for efficient and wavelength-independent Z→E azobenzene isomerization. *Chem*, **4**(7), pp. 1740–1755.

[150] Surampudi SK, Patel HR, Nagarjuna G, Venkataraman D. (2013). Mechano-isomerization of azobenzene. *Chem Commun*, **49**(68), pp. 7519–7521.

[151] Wang ZH, Wu ZH, Hu ZY, Orrego-Hernandez J, Mu EZ, Zhang ZY, Jevric M, Liu Y, Fu XC, Wang FD, Li T, Moth-Poulsen K. (2022). Chip-scale solar thermal electrical power generation. *Cell Rep Phys Sci*, **3**(3), pp. 100789.

[152] Franz E, Kunz A, Oberhof N, Heindl AH, Bertram M, Fusek L, Taccardi N, Wasserscheid P, Dreuw A, Wegner HA, Brummel O, Libuda J. (2022). Electrochemically triggered energy release from an azothiophene-based molecular solar thermal system. *ChemSusChem*, **15**(18), pp. 202200958.

[153] Yogitha SN, Kumar B, Raghavendra, Imranpasha, Gupta SK. (2021). Effect of local electric field on trans to *cis* photo-isomerization of azobenzene containing polymer. *Mater Sci Eng B*, **267**, pp. 115094.

[154] Waidhas F, Jevric M, Bosch M, Yang T, Franz E, Liu Z, Bachmann J, Moth-Poulsen K, Brummel O, Libuda J. (2020). Electrochemically

controlled energy release from a norbornadiene-based solar thermal fuel: increasing the reversibility to 99.8% using hopg as the electrode material. *J Mater Chem A*, **8**(31), pp. 15658–15664.
[155] Singleton TA, Ramsay KS, Barsan MM, Butler IS, Barrett CJ. (2012). Azobenzene photoisomerization under high external pressures: testing the strength of a light-activated molecular muscle. *J Phys Chem B*, **116**(32), pp. 9860–9865.
[156] Dong Z, Seemann NM, Lu N, Song Y. (2011). Effects of high pressure on azobenzene and hydrazobenzene probed by raman spectroscopy. *J Phys Chem B*, **115**(50), pp. 14912–14918.
[157] Tanaka S, Kim H K, Sudo A, Nishida H, Endo T. (2008). Anisotropic photomechanical response of stretched blend film made of polycaprolactone-polyvinyl ether with azobenzene group as side chain. *Macromol Chem Phys*, **209**(20), pp. 2071–2077.
[158] Berson J, Moosmann M, Walheim S, Schimmel T. (2019). Mechanically induced switching of molecular layers. *Nano Lett*, **19**(2), pp. 816–822.
[159] Hagen S, Kate P, Peters MV, Hecht S, Wolf M, Tegeder P. (2008). Kinetic analysis of the photochemically and thermally induced isomerization of an azobenzene derivative on Au(111) probed by two-photon photoemission. *Appl Phys A*, **93**(2), pp. 253–260.
[160] Jung U, Schutt C, Filinova O, Kubitschke J, Herges R, Magnussen O. (2012). Photoswitching of azobenzene-functionalized molecular platforms on au surfaces. *J Phys Chem C*, **116**(49), pp. 25943–25948.
[161] Raimondo C, Kenens B, Reinders F, Mayor M, Uji IH, Samori P. (2015). Au nanoparticle scaffolds modulating intermolecular interactions among the conjugated azobenzenes chemisorbed on curved surfaces: tuning the kinetics of *cis-trans* isomerisation. *Nanoscale*, **7**(33), pp. 13836–13839.
[162] Titov E, Lysyakova L, Lomadze N, Kabashin AV, Saalfrank P, Santer S. (2015). Thermal *cis*-to-*trans* isomerization of azobenzene-containing molecules enhanced by gold nanoparticles: an experimental and theoretical study. *J Phys Chem C*, **119**(30), pp. 17369–17377.
[163] Angelini G, Scotti L, Aceto A, Gasbarri C. (2019). Silver nanoparticles as interactive media for the azobenzenes isomerization in aqueous solution: from linear to stretched kinetics. *J Mol Liq*, **284**, pp. 592–598.
[164] Kanyak JL, Pointer CA, Achey DC. (2018). Control of intramolecular isomerization reactivity of an azobenzene derivative anchored to zro2 nanoparticle thin films. *J Phys Chem C*, **122**(3), pp. 1589–1594.
[165] Yano A, Sato Y, Dachimba K, Yano R. (2020). Catalysis of thermal isomerization of methyl yellow by salts. *ACS Omega*, **5**(14), pp. 7956–7961.

[166] Wang H, Feng YY, Yu HT, Dong LQ, Zhai F, Tang JW, Ge J, Feng W. (2021). Utilisation of photo-thermal energy and bond enthalpy based on optically triggered formation and dissociation of coordination bonds. *Nano Energy*, **89**, pp. 106401.

[167] Hallett-Tapley GL, D'alfonso C, Pacioni NL, Mctiernan CD, Gonzalez-Bejar M, Lanzalunga O, Alarcon EI, Scaiano JC. (2013). Gold nanoparticle catalysis of the *cis-trans* isomerization of azobenzene. *Chem Commun*, **49**(86), pp. 10073–10075.

[168] Ludwanowski S, Ari M, Parison K, Kalthoum S, Straub P, Pompe N, Weber S, Walter M, Walther A. (2020). Ph tuning of water-soluble arylazopyrazole photoswitches. *Chem — Eur J*, **26**(58), pp. 13203–13212.

Chapter 4

Templated Solar Thermal Fuel Composites

Abstract

Solar thermal fuel (STF) is considered an ideal choice for closed-cycle and long-term utilization of solar energy because of its advantages of variable heat storage temperature, controlled heat release, and stable constitution. Azobenzene, the most widely investigated and attractive molecular photoswitches, exhibits several obvious advantages that make it very attractive for STF applications, such as the facile synthesis and good reactivity for a large number of chemical functionalizations, low cost compared to other expensive STFs, adjustable excitation wavelength for different types of sunlight, outstanding cycling stability in repeating photoisomerization, and controllable reverse isomerization between the stale state and metastable state. However, several great challenges exist, including the low energy density, short storage life, limited exothermicities, and high costs, which still need to be addressed for future large-scale STF applications. To obtain high-performance STF composites, a variety of novel templates have been exploited in the STF system, such as the metal templates, carbon nanotemplates, polymer templates, and organic phase-change templates. Herein, we emphasize that the different templated Azo-STFs, especially the carbon nanotemplated azo-STF composites, can improve not only the energy density and storage lifetime but also the photoelectric performance and long-term cycling structure stability, as well as the triggered conditions (such as the appropriate visible light), when the stored energy is released. Furthermore, although we focus on the Azo-STF composites using various templates mentioned in research studies, the same strategy can be applied to designing high-powered and long-term stable STF composites by other kinds of photoswitches using suited templates.

STFs have garnered significant interest as promising options for efficiently utilizing solar energy over extended periods. This is attributed to their characteristics of variable storage temperatures, controlled heat release, and structural stability. Functioning as latent heat storage mediums, STFs can capture solar energy within their molecular structures and subsequently release it as heat energy under specific conditions. This characteristic ensures a consistent and dependable solar energy output.

Optimizing the utilization of solar energy necessitates alignment between the excited states of STFs and the solar spectrum. Additionally, emphasis should be on wavelengths within sunlight that fulfill optimal excitation criteria, particularly by redshifting the excitation range toward visible light. In particular, STFs need to incorporate chemical bonds with high light chemical activity, such as conjugated C=C and N=N bonds.

Key factors for realizing effective solar energy application include power density/energy density (PD/ED), energy storage duration, and conversion efficiency. Essentially, an effective STF usually satisfies a set of evaluation criteria, including the following: (a) absorption of low-energy states should overlap with the most intense regions of solar emission (solar spectrum matching), (b) the difference of energy level between the stale state and metastable state relative to molecular weight/volume should be as large as possible (high energy storage ability), (c) the photoinduced conversion to the metastable state should be carried out efficiently (high quantum yield), (d) there is no photon cover of two isomers because the photoisomerization and reverse conversion processes of most photoactive molecules can be activated by different types of light (no photon cover), (e) the post-conversion barrier of the metastable state back to ground state should be large enough for long-term storage with strong and stable cyclic property (long storage ability), and (f) the cost of the raw materials, synthesis, and combination should have the potential for large-scale applications (low cost).

Considering these requirements, "azobenzene" molecular photoswitches have gained attention due to their considerable advantages. Initially, azobenzene can absorb sunlight (365 nm, UV), converting it into latent chemical energy through *trans–cis* isomerization and releasing the energy as heat through *cis–trans* isomerization. Despite limited energy storage capacity, low quantum yield, and short storage half-life, azobenzene possesses attributes such as ease of synthesis, reactivity for various chemical functionalizations, low cost, adjustable excitation wavelength, cycling stability, and controllable reverse isomerization. Meanwhile, the initial azobenzene shows a low energy storage capacity of 49 kJ mol^{-1}, a low quantum yield of 0.49, and a short storage half-life time of 4.2 days for a high-energy state (metastable state). For different azobenzene and azobenzene derivatives, the *trans* to *cis* isomerization mechanisms are complex and usually summed up as the following four kinds: (1) The rotation mechanism involves a 180° rotation around the N=N double bond. (2) The inversion mechanism proceeds through a dipolar transition state. (3) The concerted inversion involves the increase of the two N=N–C bond angles to 180° with a linear transition state. (4) The inversion-assisted rotation leads to large changes in the C–N=N–C dihedral angle and smaller but significant changes in the N=N–C angles simultaneously. It is important to note that the transition state in the concerted inversion lacks a net dipole moment, unlike the other three pathways, which have polar transition states. The relaxation of all four transition states can yield either *cis* or *trans* isomers. Therefore, the aforementioned mechanisms can be used to predict light homeostasis consisting of these two isomers.

Although the initial azobenzene shows a low energy storage capacity and quantum yield with a short storage half-life time, previous studies have proved that the scientific molecular engineering of azobenzene with adjustable structures can effectively regulate the absorptive character (UV to visible range) and improve the molecular energy storage performance (ED, half-life time, controlled release, and long cycling property) for high-performance STF molecules by different substitutions, especially four ortho positions (methoxy groups, fluorine, bromine, and so on), which exhibit great potential in particular applications. Nonetheless, several great challenges, including the low ED, short storage life, limited exothermicities, and high costs, still need to be addressed for future large-scale STF applications [1].

Recent studies have explored intermolecular interactions by employing templates to design STF composites. Such close-packed ordered structures result in enhanced intermolecular interactions, leading to improved ED, storage duration, and other physical performance parameters. Among various templates, carbon nanotemplated Azo-STF composites have exhibited promising results, addressing challenges related to ED, storage lifetime, photoelectric performance, and cycling stability.

As opposed to focusing solely on increasing the ED of azobenzene derivatives through individual molecular design, researchers have also invested considerable effort into studying intermolecular interactions within close-packed ordered structures. This strategy can result in the enhanced intermolecular interactions due to the relatively ordered azobenzene molecules in comparison to the weak intermolecular interactions via the disordered arrangement of azobenzene molecules. Moreover, the relatively ordered azobenzene molecules supported by the templated structures can further form the sterically hindered close-packed ordered structures compared to random disordered azobenzene derivatives, exhibiting a significant increase in both the ED and half-life time of azobenzene. In particular, this simultaneously improves the other important physical performances of azobenzene, such as thermal stability, cycling stability, and processability. We stress that the membrane energy storage device based on the templated STF composites shows great physical performance and broad potential for further practical applications compared with the azobenzene derivatives.

To obtain high-performance STF composites, a variety of novel templates have been exploited in STF systems, such as the metal templates, carbon nanotemplates, polymer templates, and organic phase-change templates. Herein, we emphasize that the different templated Azo-STFs, especially the carbon nanotemplated Azo-STF composites, can improve not only the ED and storage lifetime but also the photoelectric performance and long-term cycling structure stability, as well as the triggered conditions (such as the appropriate visible light), when the stored energy is released.

4.1 Metal-templated Azo-STF composites

Azo-STF composites have been widely studied and utilized in the different energy storage fields because of their outstanding performance and great potential for practical STF applications. Over the past 10 years, numerous studies have reported the design and construction of advanced Azo-STF composites that have greatly improved the storage ability using different special templates. In this section, we firstly discuss metal-templated Azo-STF composites that can be classified as metal-doped Azo-STF composites and metal-coordinated Azo-STF composites according the combination method. The former generally introduces the photothermal conversion materials (especially metal nanoparticles) to harvest and convert sunlight irradiation at a special wavelength into heat, while the latter often offers coordination bonds into the traditional STF storage system to increase the energy storage ability and structure stability. Thus, metal-templated Azo-STF composites show great advantage and potential for further large-scale applications.

4.1.1 *Metal-doped Azo-STF composites*

Although Azo-STF composites have attracted significant attention for energy storage and conversion in the past few years, the current STF applications still have several issues based on the constraint of azobenzene, especially the energy storage and release process. In the charging procedure, the Azo-STF composites often transfer from the low-energy and stable (*trans*) structure to the high-energy and metastable (*cis*) structure via light-introduced isomerization; meanwhile, the solar energy is chemically stored in the N=N bond by absorbing the photons from the sunlight (mainly the UV light). It is worth noting that the cis-rich Azo-STFs have the capacity to retain a high-energy state for varying periods known as energy storage times, ranging from seconds to years. This duration is determined by the thermal barrier that must be overcome during the discharging process in order to return to the trans-rich Azo-STFs.

However, this recovery from *trans*-rich to *cis*-rich Azo-STFs often needs external stimulus (such as the heat, light, or catalysts) if the thermal barrier is quite high. Actually, the heating temperature of to revert *cis*-rich Azo-STFs back to *trans*-rich Azo-STFs is more than sixty centigrade, which is a temperature difference before and after energy release of 2–15°C, which is uneconomic and a huge loss of energy, greatly limiting large-scale STF applications. To accelerate the recovery of *cis*-rich Azo-STFs back to *trans*-rich Azo-STFs for fast and controlled energy release, some catalysts (for example, the Cu(I)-containing heterogeneous catalyst) were applied, but the reaction rate is really low and not suitable for practical STFs. Compared to external heat resources and catalysts, appropriate light stimulus is identified as a promising way for controlled heat release of Azo-STF composites because this method is convenient, low energy, steerable, and clean. However, the actual research results demonstrate that it is difficult to realize the complete conversion of *cis*-rich Azo-STFs to *trans*-rich Azo-STFs for most Azo-STF composites by only light stimulation. Therefore, it is very important to expand the heat release method with complete conversion of *cis*-rich Azo-STFs to further establish practical STF applications.

Fortunately, photothermal conversion materials (PTMs) are promising in the field of harvesting and converting light irradiation into heat energy at a specific wavelength [2], which is also appropriate for releasing the stored energy of charged *cis*-rich Azo-STFs. We stress that a specific wavelength of light could synchronously induce the *cis*-to-*trans* transformation of Azo-STFs and the photothermal effect of the PTM. Thus, the main point is to find the appropriate Azo-STFs and PTM to simultaneously stimulate the conversion of *cis*-rich Azo-STFs by light and heat energy by PTM. It is worth noting that the heat release of cis-rich Azo-STFs must fulfill the requirements of being speedy, complete, and at a suitable excitation temperature, and this can only be achieved under the condition of exposure to light.

For example, metal nanoparticles (MNPs) among the well-known photothermal materials have gained wide attention for their remarkable photothermal capabilities and the plasmon resonance effect they exhibit. These unique properties make MNPs excellent candidates for efficient solar energy storage and conversion. Feng's group [3] designed and prepared a metal-doped Azo-STF composite for rapid, optically triggered heat release at room temperature. The controlled and complete heat storage and release process can be observed by only light stimulation at room temperature, which is astonishing. This special metal-doped Azo-STF composite (PVA/Azo@AuNP STF) is composed of a tri-Azo molecule as the STF material, gold nanoparticles (AuNPs) as the PCM material, and a transparent polyvinyl alcohol (PVA) polymer as the flexible template. Here, the tri-Azo molecule exhibited great long-term stability of the charged *cis* state owing to the steric hindrance and difficult

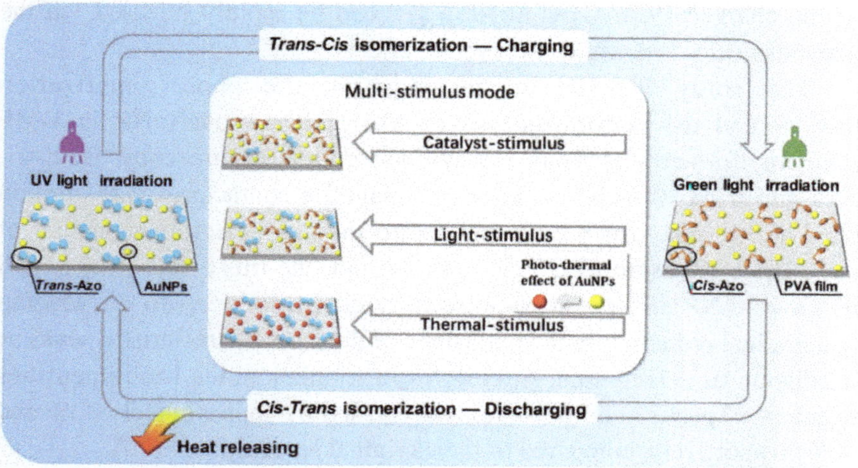

Figure 4.1. Schematic illustration of the charging–discharging procedure of PVA/Azo@AuNP STFs. Reprinted with permission from Ref. [3]. Copyright 2020, Elsevier.

recovery process of *cis*-to-*trans* transformation. However, achieving rapid energy release for this molecule is still a major challenge. Thus, AuNPs were further introduced into the above tri-Azo STF material with the excellent photothermal conversion effect, as they are appropriate catalysts of light–chemical reaction and light-triggered activity at the same wavelength. In particular, Figure 4.1 shows the energy storage mechanism of this metal-doped Azo-STF composite for rapid, optically triggered heat release at room temperature. Currently, UV light (about 365 nm) is an effective excitation means for the energy storage of a vast majority of Azo-STF composites and is also suitable for PVA/Azo@AuNP STFs. Moreover, visible light at about 520 nm is applied for the release of heat energy based on the structure characteristics of the tri-Azo molecule. Therefore, the great heat output performance of the PVA/Azo@AuNP STFs can be controlled by effective light excitation based on the following three cooperative effects: (1) irradiated trigger of *cis*-to-*trans* transformation of tri-Azo by light irradiation at 520 nm; (2) additional release of heat based on the plasmonic photothermal effect of the AuNPs by light irradiation at 520 nm, which can further stimulate the *cis*-to-*trans* transformation; and (3) the catalytic effect of the AuNPs that can accelerate the *cis*-to-*trans* transformation process. In brief, the heat energy of the charged PVA/Azo@AuNP STFs can be rapidly released via the abovementioned multiple cooperative effects.

The study focused on investigating the photoisomerization process and the performance of tri-Azo and triAzo@AuNP in DMF (Dimethylformamide) using UV–Vis absorption spectroscopy. Both tri-Azo and triAzo@AuNP solutions displayed a characteristic absorption band at 338 nm, which corresponded to the π–π* electron transition of Azo. Upon exposure to UV light at 365 nm, the *trans*-state tri-Azo and triAzo@AuNP underwent photoisomerization from *trans* to *cis*, leading to a gradual reduction in the intensities of the π–π* characteristic peaks as irradiation time increased. After 4–5 s of irradiation, the band intensities of the π–π* peaks in both solutions decreased by approximately 30% and 40%, respectively, compared to their original levels.

Further irradiation did not result in decreased intensities of the characteristic peaks associated with π–π* transitions, indicating that the samples had reached a *cis*-rich optical steady state. Both solutions also exhibited a broad absorption band at 428 nm, attributed to the n–π* transition of Azo. AuNPs displayed a localized surface plasmon resonance peak at 525 nm. The release of heat from solid-state *cis*-Azo was explored using differential scanning calorimetry (DSC), measuring heat-induced *cis*-to-*trans* isomerization. Charged *cis*-rich Azo powder exhibited significant exothermic heat flows during initial heating stages at different heating rates, confirming energy release through *cis*-to-*trans* transformation. Energy densities of *cis*-rich Azo were calculated as 31, 39, and 32 W h kg^{-1} at heating rates of 1°C min^{-1}, 5°C min^{-1}, and 10°C min^{-1}, respectively. However, these energy densities were not as high as reported for the latest Azo-based energy storage materials, largely due to the mixture of (E, E, Z), (E, Z, Z), and (Z, Z, Z) isomers in *cis*-Azo.

Moreover, solid tri-Azo@AuNP-based STF films were prepared using flexible PVA as the substrate. These films were studied under various stimulation conditions. The energy release mechanism of AuNPs was analyzed through temperature changes in PVA/Azo@AuNP films. Slight temperature changes were observed due to the catalytic action of AuNPs, facilitating *cis*-to-*trans* isomerization and moderate thermal energy release.

Notably, *cis*-to-*trans* isomerization of Azo occurred at visible light at 420 nm with minimal thermal energy release in PVA/Azo@AuNP films. In contrast, irradiating the STF film at 520 nm for 5 min led to complete energy release, aligned with thermal stimulation results. These findings indicated that all stored energy in PVA/Azo@AuNP could be released using 520 nm light at room temperature, although 520 nm light required a longer release time compared to 80°C heat treatment.

This design principle represents a customizable and cost-effective approach for high-performance STFs with rapid heat release. The strategy can also be applied to designing effective STFs for various applications at room temperature, using different Azo molecules and photothermal materials.

4.1.2 *Metal-coordinated Azo-STF composites*

In terms of metal-coordinated Azo-STF composites, optimization of energy levels and isomerization kinetics is crucial to achieve high-capacity energy storage. The standard free energy of isomerization (ΔE) and the degree of reversible isomerization (DI) play key roles in ED at the molecular level. Reversible molecular interactions, such as coordination bonds, offer promise in tuning energy levels and isomerization kinetics. These bonds, with relatively high binding energy and controlled formation/dissociation processes, allow for efficient energy storage and release through photoisomerization. Coordination bonds contribute to adjusting the energy gaps between isomers through binding energy and reversible interactions. The interplay between isomerization-induced bond formation and dissociation determines cyclic utilization of photothermal energy and bond enthalpy.

It is important to mention that high-density cross-linking of hydrogen bonds typically introduces a significant thermal barrier to isomerization, resulting in a limited degree of reversible isomerization. Moreover, due to their low selectivity, hydrogen bonds tend to reform before complete recovery. Consequently, the bonding enthalpy is released spontaneously during the transition, and the lack of control over the formation/dissociation process hampers further advancements in heat storage capabilities.

In contrast, coordination bonds, also known as dynamic metal–ligand interactions, exhibit great potential in expanding the energy gap due to their relatively high binding energy and controllable formation/dissociation process. This optically triggered reversible formation/dissociation is essential for the efficient storage and release of heat caused by isomerization. Additionally, coordination bonds with relatively suitable bond strengths facilitate reversible interactions with organic ligands and enable structural transformations, thanks to their low steric hindrance. Therefore, the rational design and construction of coordination bonds offer a promising approach to achieve Azo-STF composites with an enlarged energy gap. These bonds contribute to tuning the molecular energy level and the kinetics of isomerization [4,5]. It is worth mentioning that a substantial number of coordination bonds can provide high bond enthalpy through coordination and dissociation during isomerization. This implies that further investigation into coordination bonds could provide valuable insights to address potential challenges in the field of closed-loop utilization of photothermal energy.

In their study, Feng's group [6] developed a reversible dynamic-bond system using coordination bonds formed between magnesium (Mg) metal ions and sulfonate groups. They first designed and synthesized a sulfonate (SO_3H)-substituted azobenzene molecule (Azo), which was covalently grafted onto reduced graphene oxide (G-Azo). This G-Azo hybrid material had the ability to coordinate with metal ions. Upon coordination with Mg metal ions, the G-Azo hybrid formed a loose layered self-assembly structure, referred to as G-Azo-Mg.

The coordination bonds between the sulfonate groups and Mg ions, as well as any electrostatic interactions, contributed to the formation of this cross-linked structure. The researchers then investigated the optical triggering of the formation and dissociation of coordination bonds in G-Azo-Mg. By inducing structural transformations, such as *cis–trans* isomerization of the azobenzene moiety, they could control the formation and dissociation of coordination bonds. This controlled structural transformation allowed for tuning of the molecular energy level and the kinetics of isomerization. Overall, Feng's group demonstrated the use of reversible coordination bonds between Mg metal ions and sulfonate groups in a graphene-templated azobenzene hybrid. The ability to control the formation and dissociation of these coordination bonds through

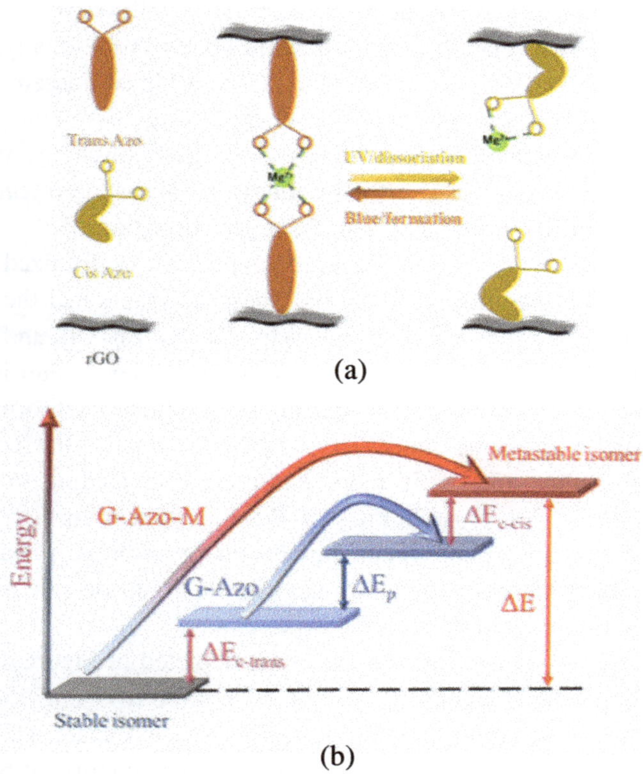

Figure 4.2. (a) Schematic illustration of reversible photoisomerization of G-Azo-Mg film in a solid state with formation and dissociation of coordination bonds. (b) Schematic energy diagram of G-Azo tuned by coordination bonds. Reprinted with permission from Ref. [6]. Copyright 2021, Elsevier.

optical triggering opens up interesting possibilities for tuning of energy levels and isomerization kinetics in hybrid materials.

As shown in Figure 4.2, the stability of *trans* isomers is enhanced through the formation of coordination bonds, resulting in a low energy level while introducing a thermal barrier. Upon photoisomerization, the dissociation of coordination bonds accelerates the reversion of the unstable *cis* isomers. This unique property allows the solid-state G-Azo-Mg self-assembly to effectively store both photothermal energy and bond enthalpy. Furthermore, this self-assembly system demonstrates the ability to release these stored energies at a high rate (720.1 kJ kg^{-1}) through the reversible formation and dissociation of coordination bonds. Additionally, we have successfully showcased the ability of the G-Azo-Mg self-assembly to collect, utilize, and transmit photothermal energy using phase-change materials at different temperature ranges. These findings highlight the remarkable potential of the G-Azo-Mg self-assembly as a solid-state integrable heat source for multifunctional distributed energy utilization.

In an earlier study [7], researchers designed and synthesized polymers based on sulfonic-grafted azobenzene. These polymers had the ability to interact with various metal ions (M = Mg, Ca, Ni, Zn, Cu, and Fe). This interaction allowed for the optimization of energy levels and isomerization kinetics of the polymers. By utilizing the photoinduced formation and dissociation of dynamic metal–oxygen (M–O) bonds, the energy gap between the *trans* and *cis* isomers was increased, leading to enhanced energy storage. The cross-linking of these polymers further promoted rapid isomerization, thus enhancing the energy storage capacity. This design enabled increased heat storage capability in the polymers while enabling low-temperature heat release.

The process involved in this system is depicted in Figure 4.2(a). The PAzo-M molecule absorbed UV light (365 nm, 80 mW cm^{-2}) to undergo E–Z isomerization, causing the dynamic bonds to dissociate. This transformation converted photoenergy into latent chemical energy. Subsequently, absorption of blue light (450 nm, 40 mW cm^{-2}) induced Z–E isomerization, and the dynamic bonds reformed through thermal relaxation.

The photographs in Figure 4.3(b) illustrate the interaction between metal ions and the -SO_3H group on the PAzo-M molecule. The theoretical energy gap (ΔE) between *E*- and Z-PAzo-M, controlled by isomerization-induced formation and dissociation of dynamic bonds, was calculated (Figure 4.3(c)). Dynamic bonds introduced into the system significantly increased ΔE by altering energy levels of isomers. The equation defining ΔE (Equation (1)) considers both isomerization energy (ΔEI) and dynamic-bond enthalpy (ΔEC). Notably, PAzo-Mg exhibited the highest ED at 113.5 W h kg^{-1}.

To demonstrate practical application, a wearable self-heated textile coated with PAzo-Mg was fabricated for temperature control in low-temperature environments. This textile selectively released heat,

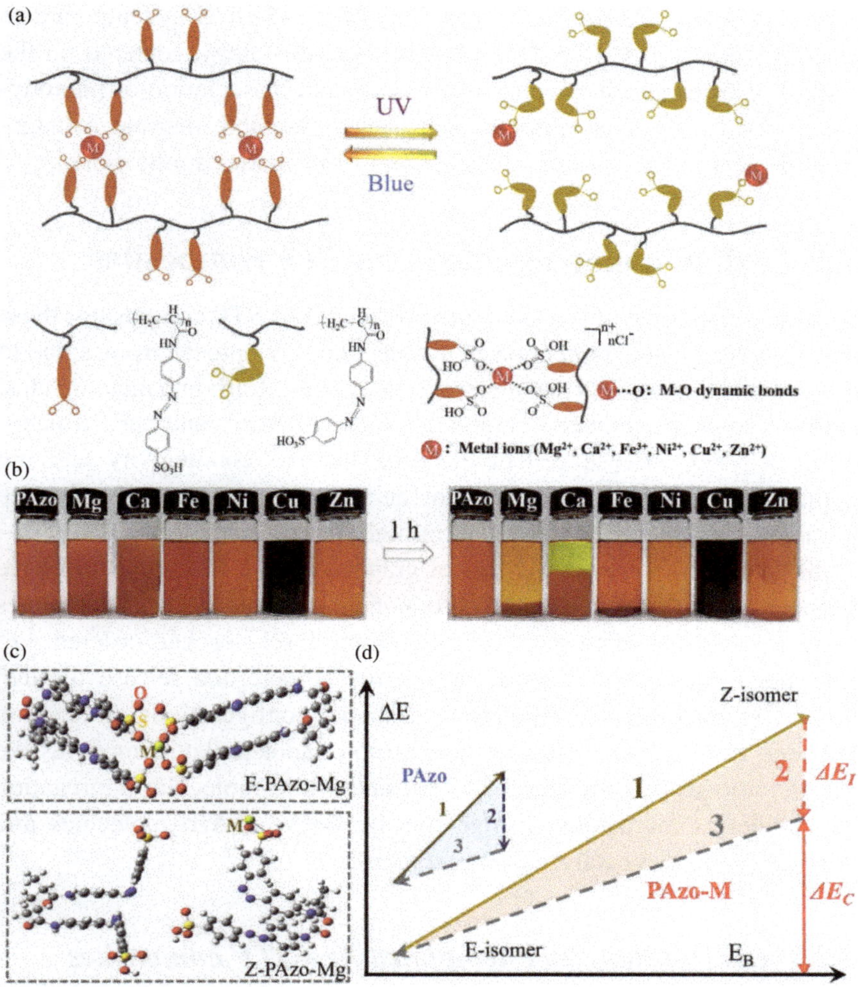

Figure 4.3. (a) Chemical structure and schematic illustration of PAzo-M with isomerization-induced formation and dissociation of M–O dynamic bonds formed by different metallic ions, with *n* being 2 or 3. (b) Photographs of PAzo solution after adding the same molar amount of MCln ($MgCl_2$, $CaCl_2$, $FeCl_3$, $NiCl_2$, $CuCl_2$, and $ZnCl_2$) for 1.0 h. (c) Optimized geometric structure of PAzo-Mg with the E- and Z-isomers. (d) Schematic energy diagram of PAzo and PAzo-M during the energy utilization process of (1)–(3). Reprinted with permission from Ref. [7]. Copyright 2022, Wiley-VCH.

achieving a temperature difference (ΔT) of 7.7–12.0°C at temperatures ranging from −5.0°C to 5.0°C. This innovative strategy, centered on the control of isomerization-induced formation and dissociation of photodynamic bonds, holds promise for self-heated wearable thermal management devices, smart on-skin clothing, and soft robotics applications.

4.2 Carbon nanotemplated Azo-STF composites

Shifting to the topic of carbon nanotemplated Azo-STF composites, these materials have gained attention in nanotechnology optoelectronics due to their outstanding optical and electronic properties. Carbon nanotemplates, such as graphene, reduced graphene oxide, fullerene, and carbon nanotubes (CNTs), offer excellent electrical/thermal conductivity and are applicable in various fields due to their conductive characteristics. In particular, graphene possesses ultra-high thermal conductivity (4800–5300 W $(m\ K)^{-1}$), surpassing that of graphite. Azo-STF composites have been explored using different carbon nanotemplates. These templates serve as effective platforms for assembling closely packed ordered Azo molecules, allowing for controlled energy storage and release through precise interactions and regulation of spatial configurations. Interaction between azobenzene molecules and carbon nanotemplates improves the energy storage performance of STFs, and while molecular aggregation can hinder isomerization, charge transfer between Azo molecules and nanotemplates can catalyze isomerization.

4.2.1 *Carbon nanotube-templated Azo-STF composites*

CNTs are structurally characterized by a closed cage of delocalized electron clouds along the tube axis. Due to their unique electrical and photoconductive properties, they are studied as ideal one-dimensional materials for the structure of optical nanodevices. Due to their importance in polarization-sensitive optics, recent studies have focused on the optical polarization of the outstanding CNT cylindrical nanostructures.

In 2007, the concept of photoresponsive Azo-multiwalled carbon nanotube composites (Azo-MWCNTs) was first proposed by Feng's group. As shown in Figure 4.4, Azo-MWCNTs were synthesized by

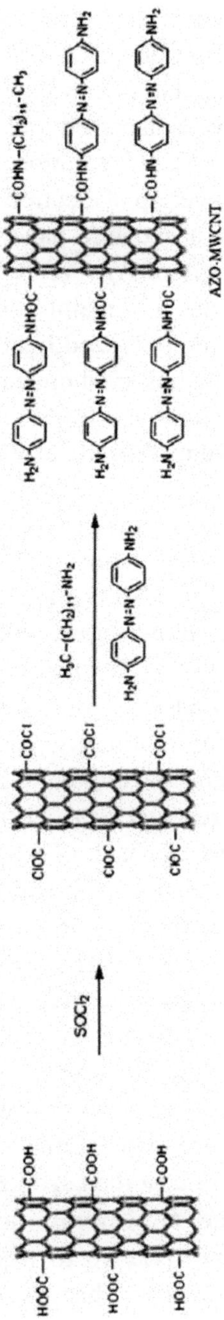

Figure 4.4. Synthesis route to Azo-MWCNT. Reprinted with permission from Ref. [9]. Copyright 2007, AIP Publishing.

covalently bonding Azo to the side wall of the MWCNTs [8,9]. They investigated the optical polarization of Azo-MWCNT composites prepared by functionalizing Azo onto the side walls of MWCNTs via covalent bonds. TEM images show well-dispersed individual Azo MWCNTs. The redshift band of the Azo-MWCNT solution in the UV–Vis absorption spectrum is attributed to the overlap of electron conjugations. An enhanced luminescent emission with a redshift band was observed in the PL spectrum. The photoanisotropy of Azo-MWCNT thin films in the cross-polarization absorption spectrum is mainly determined by the enhanced dipole transition parallel to the polarized light and the reduced transition perpendicular to the light. The observed optical anisotropy of polarization provides insight into polarization.

In 2011, Grossman's group theoretically simulated the azobenzenes and single-walled carbon nanotubes (SCNTs) to covalently bond with each other by Density Functional Theory (DFT) [10]. The potential advantage of Azo/CNT hybrid nanostructures as STFs stems from the tightly packed, highly ordered array of adsorbed photomolecules imposed by the CNT substrate. The existence of this crystal-like state, which, importantly, persists when the nanostructure is in solution, has two key effects. First, the number of photoactive molecules per volume increases significantly relative to the solution of free light molecules, resulting in an increase in volumetric ED of 57 orders of magnitude (depending on the solubility of a given substituting azobenzene species). Second, and more importantly, the proximity and ordered arrangement of adsorbed molecules are able to systematically manipulate intermolecular and intramolecular interactions, providing a set of efficient regulatory parameters to maximize the energy storage capacity and storage life of STFs. Additional regulatory parameters include the molecular packing density, the chemical properties of the linking groups, the diameter of the CNT, and the orientation of the adsorbed molecule and any functional groups on the adsorbed molecule with respect to the CNT. In addition to providing a template for ordered interactions, the CNT substrate also exerts a significant influence on ΔH by disrupting the symmetry of adsorbed azobenzene molecules. When the molecule binds to CNT, there are several orientations depending on the position of the linking group relative to the nitrogen double bond in azobenzene. The large geometric changes of the molecule and the corresponding changes in ΔH when three different Azo derivatives are bonded to CNT substrates by ortho-site, meta-site, and para-site connections are described. In most of the Azo/CNT systems studied, the ortho-connections are energetically superior to other possible connections in both the *cis* and *trans* states. This stability results from strong adjacent-induced interactions between the linker and azobenzene, and between azobenzene and CNT, resulting in a relatively small increase in ΔH of unsubstituted azobenzene.

Conversely, systems in which the linker is located in the para-position tend to have the most unstable *cis* configuration, resulting in larger values for ΔH. As shown in Figure 4.5, Grossman *et al.* [10] investigated the hydroxyl substitution of one or more hydrogens on the benzene ring with the aim of changing the relative stability of the two isomers by adding hydrogen bonds. In general, structures with more hydrogen bonds and/or stronger (shorter) hydrogen bonds are more stable; thus, maximizing the number of hydrogen bonds in the *trans* state while minimizing the H-bonds in the *cis* state will provide greater ΔH. In addition, by OH... N is usually found to be stronger than those with OH... O. Because of the symmetry breakdown imposed by CNT, ΔH can be optimized by selecting the

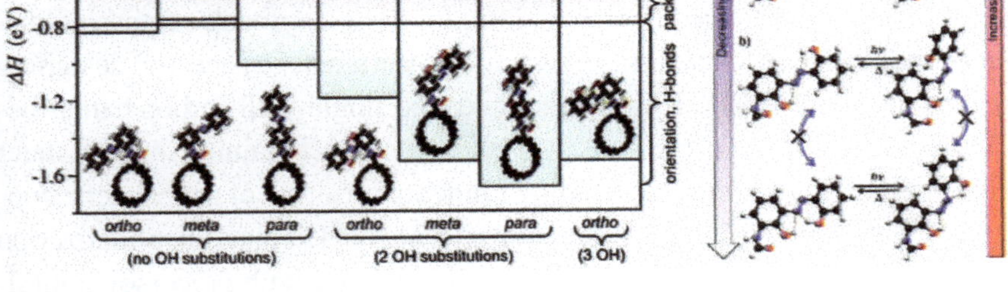

Figure 4.5. (a) DFT calculation values for typical SCNT-templated azobenzene composites. (b) Hydrogen bonding effects. Reprinted with permission from Ref. [10] Copyright 2011, American Chemical Society.

location of hydroxyl (or other functional groups). For example, Figure 4.5 shows several possible modes of the two OH groups on azobenzene, where the intermediate site has a linker, illustrating the dependence of ΔH on the relative positions of the two hydroxyl groups. These locations determine the number and strength of hydrogen bonds in both *cis* and *trans* configurations. As shown in Figure 4.5, these interactions provide a very efficient means of regulating the characteristics of STFs. Considering only two hydroxyl substituents, the energy stored per molecule increases by up to 3 times compared to the gas-phase azobenzene. Additional hydroxyl and/or other functional groups can further increase ΔH. The calculation results show that enhanced intermolecular interactions (especially H-bonds), via the highly ordered and closely packed arrays at a high grafting density, may greatly improve the ED and half-life time.

Based on DFT calculations, molecular interactions such as hydrogen bonds and adjacent-induced interactions were found to significantly increase ΔH and $t_{1/2}$ per Azo molecule. Through proper intermolecular separation, the interaction between the π electrons of the benzene ring adjacent to the *trans*-Azo molecule results in a net increase of 0.2 eV in ΔH per molecule. However, these calculations are mainly based on nanocarbon templates with high functional densities. This type of template can be used to expand solar thermal storage and get high ED, by combining the increased number of intermolecular interactions per unit volume with the increased weight per unit storage capacity. The lower storage capacity of these nanomaterials compared to the theoretical model (1/4–1/8) is mainly due to the lower ΔH and lower functional density of a single Azo molecule, i.e., 1/40–1/80 (each Azo molecule consists of 40–80 carbon atom carriers).

In 2014, Grossman's team [11] introduced an approach surpassing discrete molecular mechanisms, elevating the parameter space for designing solar thermal materials. They synthesized Azo single-wall carbon nanotubes (SWCNTs) using two methods: (1) adding ester-containing linkers, later converted to Azobenzamide; (2) direct functionalization of azobenzene connected with amide using organic groups. Route (1) led to a functionalization density of 1/19. However, route (2) faced challenges with precursor solubility, limiting high-density functionalization. Nonetheless, repeated free-radical-initiated processes increased functionalization density, exemplified by Azo-SWCNT 1, 2, and 3. Once functionalized, Azo and ester SWCNTs suspended well in polar organic solvents.

This innovative approach decoupled molecular properties to adjust thermodynamic and storage life parameters while preserving photochemical parameters and cyclic durability. It marked the first instance of enhancing ΔH and the thermal stability of photochromic isomers solely through templating. High-density Azo-SWCNT composites exhibited superior ED (56 W h kg^{-1}) compared to single Azo molecules (44 W h kg^{-1}) and significantly prolonged storage lifetimes without material degradation during cycling. Although these composites lacked certain hydrogen bonds from previous theoretical models, interactions between adjacent templates seemed more crucial. This work expanded potential chromophore–template combinations and underscored the importance of reinforcing stereostrains through templates.

However, the ED of Azo-SCNTs remains insufficient for practical STF applications due to the absence of H-bonds in this composite. More recently, Luo and Feng *et al.* [12] introduced H-bonding-regulated Azo-SCNT and Azo-MWCNT composites with high grafting density and suitable half-lifetime. These developments allow short-term solar thermal storage for peak power shifting within a single day. The elevated functional density on the surfaces of SCNTs and MCNTs leads to heightened intermolecular interactions and bundling effects in solid Azo-SCNT and Azo-MCNT composites. This enhancement significantly boosts storage capacity to approximately 77.1 W h kg^{-1} and 80.7 W h kg^{-1}, with a short-term lifetime of about 14 h and 16 h, respectively, while maintaining excellent cycling stability over 30 cycles.

These breakthroughs establish a promising category of solar thermal storage materials with high ED and suitable short lifetimes for daily utilization.

These results demonstrate the practical potential of molecular CNT-templated Azo-STF composites to mitigate daily sunlight periodicity by storing solar energy during the day and releasing heat at night.

4.2.2 *Graphene-templated Azo-STF composites*

While CNTs offer a feasible nanotemplate for supporting Azo molecules and optimizing molecular structures to enhance intermolecular interactions, achieving high-energy Azo-STF composites with long-term cycling stability remains challenging. This is mainly due to the difficulty in adjusting H-bonds in specific Azo-CNT configurations with significant steric hindrance. Consequently, a more suitable carbon nanotemplate is necessary to achieve high energy and long-term stability in Azo-STFs.

Two-dimensional graphene, comprising a single sheet of sp^2-bonded carbon atoms, presents promising potential for Azo-STFs [13]. Its rigid and orderly structure serves as an ideal nano-substrate to support Azo molecules without complex separation, purification, or organization requirements. Researchers turned their focus to graphene oxide (GO) to better leverage intermolecular forces. GO, as a graphene derivative, maintains a two-dimensional sp^2 hybrid network with delocalized π electrons. Distinctively, GO features carboxylic, hydroxyl, and carbonyl groups on its edges, along with epoxy and hydroxyl groups on its bottom surface. Given the abundance of –OH groups on GO's edges and surface, employing GO as a coupling agent to produce Azo-GO hybrids seems reasonable. Furthermore, the p–electron interaction facilitates efficient charge transfer and transport through optical switching or modulation when linked by light-responsive Azo molecules. Graft density and chemical structure play pivotal roles in determining molecular interactions between adjacent Azo molecules. Theoretical calculations confirm that intermolecular hydrogen bonds between substituted Azo (E and Z isomers) on carbon nanotemplates are regulated by controlling the spatial configuration of E and/or Z isomers and appropriate molecular distances based on high grafting.

In 2013, Feng *et al.* [14] reported the use of covalently functionalized azobenzene-reduced graphene oxide (Azo-RGO) for long-term solar heat storage by optimizing molecular hydrogen bonding at high functional densities. Their findings highlighted the impact of intermolecular distance, electron interaction, steric hindrance, and hydrogen bonding between adjacent Azo molecules on graphene, leading to increased ED and half-life. Subsequently, Wen *et al.* [15,16] designed and synthesized several Azo-RGO composites with different functional groups at various positions (ortho-site, meta-site, and para-site) to enhance ED and half-life by increasing molecular interactions. Specifically, the RGO nanotemplate for covalently attaching photographic meta- and ortho bis- substituted AZO molecule was presented as molecular solar heat storage material, which was prepared by a triple iterating functionalization. Azo-RGO hybrids achieved graft densities of up to 1/16–17, confirmed by TEM observations showing large rough surfaces covered with additives. Multiple intermolecular hydrogen bonds were validated through FTIR and DFT calculations. Fourier Transform infrared spectroscopy (FTIR) and DFT calculations also verified multiple intermolecular hydrogen bonds. The stacking interaction between adjacent graphene nanosheets, confirmed by the topological microstructure of the SEM, and the reduced layer spacing calculated from X-ray diffraction (XRD) data, further improved storage capacity and lifetime in solid forms. These nanotemplates achieved a remarkable ED of 112–138 W h kg^{-1} and a half-life exceeding one month (33–52 days), confirmed by density functional theory calculations. These materials maintained excellent cyclic stability under 50 cycles of visible light conditioning. These breakthroughs unveil a promising class of high-energy-density STFs that surpass previous nanomaterials and rival commercial flexible-package Li-ion batteries.

The incorporation of push–pull groups has been documented to heighten the difference in dipole moment between *trans*-isomerized and *cis*-isomerized azobenzene. Consequently, the assembly of multibranched azobenzene fosters the establishment of intramolecular/intermolecular interactions, magnifying the energy difference (ΔH) between *trans* and *cis* isomers. Moreover, this component also wields control over the *trans–cis* isomerization of distinct Azo units (either unrestricted or limited) by modulating the molecular structure.

Feng's group [17] concentrated on crafting specially structured Azo-graphene composites by optimizing spatial configuration, elevating grafting density, and linking azobenzene with varying branch counts. Their [18] proposal centered on densely packed 2-chloro-4,6-bis (4-(phenyldiazene)phenoxy)-1,3,5-triazine (bisazobenzene chromophore) that covalently bonds to reduced graphene oxide (RGO bisAzo). Thermogravimetric Analysis (TGA) and X-ray Photoelectron Spectroscopy (XPS) spectra revealed substantial graft density, with approximately one Azo unit covalently bound for every 25 carbon atoms. The spatial configuration and energy of bisazobenzene's *trans* and *cis* isomers were influenced by intramolecular and/or intermolecular steric hindrance amid two neighboring Azo units on identical and/or adjacent nanosheets. This phenomenon, attributed to high graft density and solid-state bundling effects, led to a noteworthy surge in enthalpy and activation energy of isomerization. The RGO double Azo compound exhibited a remarkable ED of 80 W h kg^{-1}, a peak PD of 2230 W kg^{-1}, and an adaptable heat release span of 2 min to 5520 h. These outcomes lay the groundwork for devising chromophore/hybrid templates for STFs through meticulous optimization of molecular interactions.

Subsequently, they [19] proposed a closely knit template assembly of bisazobenzene covalently grafted onto reduced graphene oxide (RGO-bisAzo) with a unit graft density of 1/23. They devised two distinct branched-chain structures of Azo assemblers (RGO-bisAzo-1 and RGO-diAzo-2) and probed the impact of molecular interaction on photothermal attributes. The intensive grafting of bisAzo molecules ushered in intermolecular hydrogen bonding and steric hindrance. The RGO-bisAzo-2 assembly showcased a high ED of 131 W h kg^{-1}, a substantial PD of 2517 W kg^{-1}, and an extended half-life of 37 days, bolstered by robust cycling performance over 50 cycles due to inter- and intramolecular H-bonding and steric hindrance. In solid RGO biAzo film, the restricted isomerization (98 W h kg^{-1}) yielded a 25% lower ED compared to the powder sample, stemming from steric hindrance. The complete cycle of UV irradiation, storage, and heat release of photothermal fuels (PTF) films was scrutinized. RGO-bisAzo films were competent in both releasing and accumulating heat, achieving a maximum temperature difference (DT) of 15°C. When the ambient temperature divergence exceeded 100°C, the film managed DTs of over 10°C for 30 min. Molecular engineering and optimization of microstructure are crucial methods for achieving high energy and long-term storage as well as high degree of isomerization in high-performance polymer thermal storage materials. The capacity to govern heat release within solid-state components lays a robust foundation for successful PTF device development. As depicted in Figure 4.6, the aforementioned studies underscore that optimizing the molecular structure and steric configuration of azobenzene on the graphene nanotemplate is of paramount significance to facilitate isomerization and intra/intermolecular interactions, leading to substantial enhancements in both ED and long-term stability of Azo-graphene STF.

In recent developments, Feng *et al.* achieved a breakthrough in templating components with triazobenzene (tri-Azo1:68/RGO), demonstrating impressive metrics including a maximum ED of 150.3 W h kg^{-1}, a

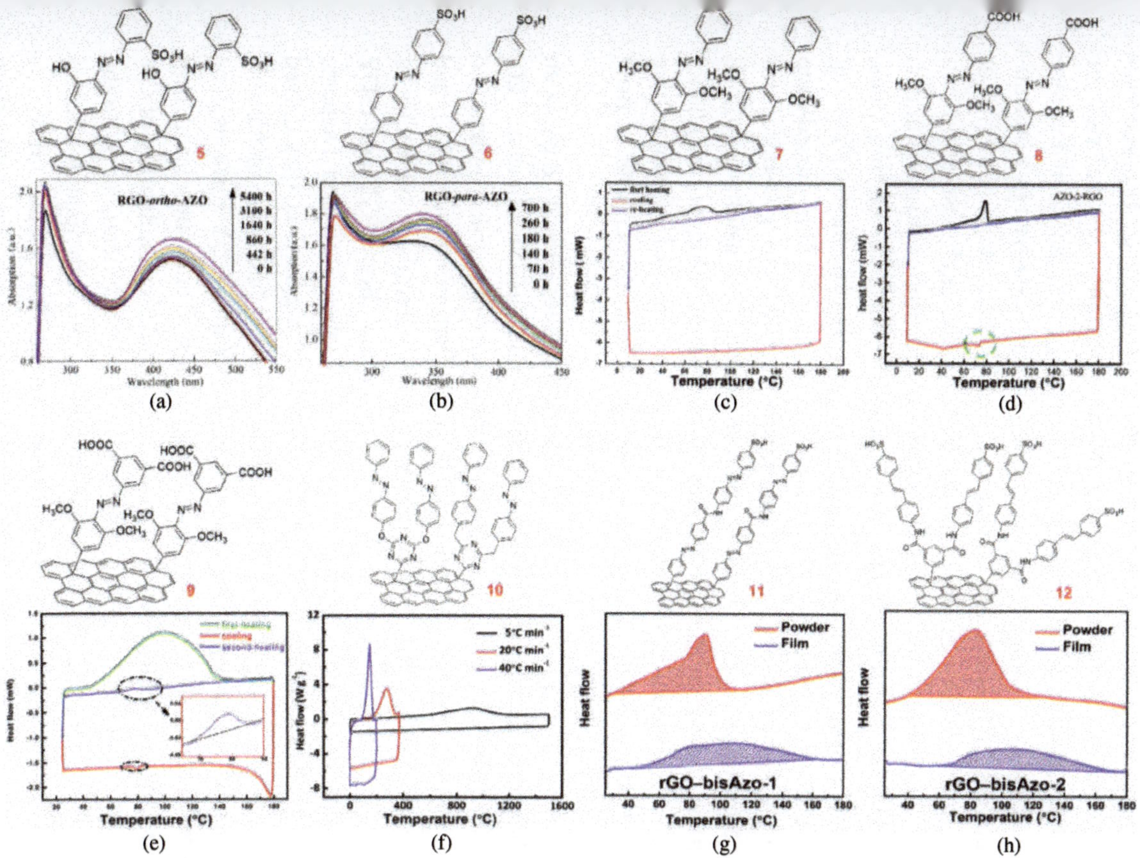

Figure 4.6. (a, b) Chemical structures and time-evolved absorption spectra of RGO-ortho-Azo and RGO-para-Azo. (c–f) Chemical structures and DSC curves of RGO-ortho-Azo, RGO-para-Azo, and RGO-bis-Azo. (g, h) Chemical structures and DSC curves of the first heating stage of RGO-bis-Azo-1 and RGO-bis-Azo-2. Reprinted with permission from Ref. [17]. Copyright 2016, Springer.

lengthy half-life (1250 h), and a high-power density of 3036.9 W kg^{-1}. The films crafted from tri-Azo1:68/RGO were capable of releasing 23.6–69.7% of stored heat, inducing a temperature rise of 2–7°C. This successful optimization of solid-state rapid thermal output paves the way for closed-loop utilization of photothermal energy, particularly in thermochromic displays. Notably, the thermal output was employed in a thermochromic display, effectively triggering color changes in patterns at temperatures lower than previously achieved. The high resolution and broad applicability demonstrated by these thermochromic displays underscore the film's ability to uniformly discharge heat at varying rates. Remarkably, even after 5 days of storage at a chilly 0°C, the tri-Azo1:68/RGO membrane retained a utilization rate of 10.2%. This closed-loop efficient utilization serves as a promising thermal energy source for management applications, thereby providing valuable insights into STF membrane development. However, despite substantial progress in enhancing ED and extending storage life, the performance of photosensitive chemical heat storage materials composed of raw azobenzene is from meeting practical application requirements.

Beyond CNT and graphene, other carbon nanomaterials have been investigated as nanotemplates in Azo-STF composites. Swathi *et al.* [20] established through theoretical analysis that Azo molecules adhered to graphene-like structure nanotemplates, including graphyne (GY) and graphdiyne (GDY), and hold potential as STF materials in the realm of data storage devices. Their study explored the viability of GY and GDY as substrates for two prototype molecular switches, Azo and Disperse Orange 3 (DO3). Employing electronic structure calculations, they delved into the structure, energetics, kinetics, and reaction pathways of thermal *cis–trans* isomerization of Azo and DO3 on GY and GDY surfaces. The calculated minimal alteration in activation barrier and substantial heat storage capacity upon adsorption indicated the practical suitability of GY and GDY substrates alongside graphene in device applications. With the recent fabrication of GDY sheets and nanotubes, they anticipate forthcoming experimental realization of molecular switches for GY and GDY.

Huang *et al.* [21] introduced an innovative approach, fabricating RGO/b-CD/C60 through targeted interaction between RGO/b-CD and Azo-C60. The hybridization of RGO/b-CD and Azo-C60 impeded

Azo-C60 aggregation and facilitated cellular uptake, yielding collaborative protection against NO-induced cytotoxicity by neutralizing RNS and ROS. Notably, utilizing ultraviolet light to switch Azo-C60 on and off on RGO/b-CD nanocarriers amplified cellular protection. The host–guest strategy seamlessly amalgamated the outstanding attributes of each component within RGO/b-CD/C60. This multifunctional graphene carrier, boasting an expansive surface area, thwarted Azo-C60 aggregation while promoting cellular uptake owing to hydrophilic b-CD loading. The substantial loading of C60 on the graphene surface, orchestrated by b-CD as a linker between RGO and Azo-C60, facilitated controlled Azo-C60 release under ultraviolet light. Despite small aggregates formed by the released Azo-C60 in aqueous solutions, the heightened flexibility and activity of Azo-C60 culminated in enhanced safeguarding against NO-induced cytotoxicity. This work holds tremendous potential in the realm of developing light-responsive multifunctional graphene-based materials.

4.3 Polymer-templated azobenzene

The application of polymer templates to azobenzene is also explored in the context of STFs. Polymer templates enable the formation of uniform thin films with controllable thickness, rendering them suitable for large-scale exothermic applications. Such homogeneous Azo polymer films capture photons efficiently and release heat by governing the *cis–trans* isomerization of solids. This offers opportunities for integrating STF materials into existing heating devices. While notable advancements have been made in modifying molecular transitions and controlling the liquid crystal order or packaging in individual or polymerized azobenzene derivatives, the combination of these effects remains unexplored for enhancing solar energy storage across a broad range of organic molecules and polymers.

Solar energy, an abundant yet underutilized resource, presents potential for both large-scale and small-scale applications. While significant strides have been made in large-scale solar energy collection and storage solutions, such as expansive solar cells, there remains substantial scope for smaller devices catering to industrial and consumer needs.

Closed-cycle systems, such as those employing recyclable energy storage within organic chromophore bonds, possess a unique advantage of collecting and storing energy within the same material. Furthermore, in certain applications, thermal energy might be more advantageous than electrical energy, making these materials particularly promising. Practical applications of STFs encompass the absorption of photons that induce electronic transitions, subsequently driving isomerization or reversible chemical processes. The high-energy state achieved after isomerization by actual STFs stores a significant portion of photon energy, countering substantial thermal reduction during storage. While the norbornadiene/tetracyclic system's high ED ($\approx$280 W h kg^{-1}) has been explored for STFs in solution, its limited recyclability rendered it unsuitable for extended use. Efforts to develop new materials, such as those with moderate energy storage and enhanced recyclability like ruthenium-based ones, are underway but there are still challenges in terms of cost-effectiveness for widespread adoption.

Harnessing the capabilities of advanced computational tools, a novel approach emerges, capitalizing on the spatial interaction between carbon nanostructures and solid-state STF_S to elevate ED. This method capitalizes on the utilization of azobenzene derivatives with highly recyclable attributes and moderate energy densities (60–70 W h kg^{-1}). Enhancing the ED of individual molecules through stenciling entails intricate multi-step reactions with low yields, resulting in materials with limited solubility in most organic solvents. Recently, room-temperature liquid azobenzene fuels have been developed by integrating substantial ligands onto the molecules, while computational investigations indicate the potential for heightening their ED via benzene ring functionalization. This platform signifies a promising avenue for future STF advancements. Despite the rapid progress in STF material development, solid-state platforms and their related applications remain largely unexplored, with recent explorations limited to semi-solid photoliquefied ionic crystals achieving an ED of 35 W h kg^{-1}. Fully transitioning to the solid state opens doors for the integration of STF materials into diverse established solid-state devices, such as deicing coatings, or innovative applications like solar blankets and consumer-oriented heating devices.

4.3.1 *Solid-state polymer-templated Azo-STF composites*

Precision engineering at the molecular level enables the controlled customization of STF materials in the solid state. However, the efficient realization of this has been lacking. Recent STF studies primarily revolve around carbon frameworks, introducing synthetic intricacies that hinder uniform film deposition, augment optical density without inducing photocharging, and introduce uncontrollable morphological influences that potentially restrict reversible switching and charging within solid states.

The ideal material class for constructing a solid STF coating should ideally possess (1) the ability to form smooth films of controllable thickness, (2) elasticity at elevated temperatures, (3) characteristics of retention of heat release akin to molecular STF counterparts, thereby maximizing ED, (4) the capability for charging, storage, and discharge within device-relevant time frames, and (5) viability for widespread large-area heat release applications. In this context, we opt to forge a solid-state polymer platform that, when thoughtfully designed, can be adjusted and scaled to meet these prerequisites. Significantly, the rich chemical composition within monomer and polymer frameworks permits the application of the same computational design principles employed for single-molecule STF_S, unveiling novel avenues for polymer–polymer chain interactions, cross-linking degrees, and conjugations within polymer scaffolds.

Grossman *et al.* [22] introduced a polymer-based STF storage platform for solid-state STF synthesis. Their design entails a polymer structure with an alkyl chain forming the main chain and azobenzene composing the side chain. Upon photon irradiation of appropriate energy, low-energy *trans*-azobenzene molecules ideally transition to the *cis* configuration, unaffected by adjacent side-chain interactions. An acrylate group attaches to the azobenzene derivative, enabling polymerization with common free radical initiators. Given the substantial impact that new functional groups can exert on energy and optical properties of azobenzene, the expected monomer was validated through density functional theory (DFT) simulations. These simulations indicated a moderate potential ED of 68 W h kg^{-1} for the monomer, implying that the additional substituents minimally affected the expected ED compared to azobenzene (76 W h kg^{-1}). Additionally, size exclusion chromatography

(SEC) revealed an estimated average of 45 monomer units per polymer chain (in reference to polystyrene), with scalability to gram-scale production, achieving a total reaction yield of ≈60%. Despite potential steric hindrance during synthesis, these polymers attain a sufficient size and solubility in organic solvents compatible with spin coatings to facilitate solid film preparation. The straightforward two-step chemical process for producing these materials promises scalability and cost-effectiveness. The crux of effective material design lies in leveraging the exceptional properties of specific small-molecule STFs and transitioning them into a solid state in the form of polymers. Consequently, comparing monomer and polymer components is imperative to ensure the retention of STF properties. Optical properties showcased consistent patterns between the two materials, with the polymer exhibiting slightly diminished absorption. Analogous to azobenzene, a dominant *trans* peak at 325 nm and a *cis*-dominant peak at 450 nm emerged. Excitation of these materials through $\pi > \pi^*$ transitions, as widely observed in parazobenzene, led to the attenuation of high-energy peaks and emergence of low-energy peaks. At room temperature (25°C), these materials exhibited half-lives (the time taken for 50% of the *cis* isomer to revert to *trans*) of 92 ± 1 and 55 ± 1 h, respectively, suitable for energy storage applications necessitating daily cycles. For prolonged utility, material recyclability was assessed, showcasing consistent optical and material properties across more than 100 cycles involving cycling between *trans* and *cis* states using a photon source.

To facilitate the transition of the material into a solid state, a solid-state STF platform was formulated, employing a straightforward solar thermal energy capacitor (STEC) construction. This entailed employing a transparent 1-inch quartz substrate and depositing STF material using rotational coating with toluene. This device design ensured optical property monitoring during charge and discharge cycles while allowing for the generation of highly reproducible samples. As depicted in Figure 4.7, a spatial infrared heat map was utilized to vividly illustrate macro energy release, showcasing temperature fluctuations of up to 10°C. These ground breaking findings lay the foundation for the advancement of efficient STFs with high ED within solid states. Leveraging the rich chemical composition of monomer and polymer scaffolds, future methodologies

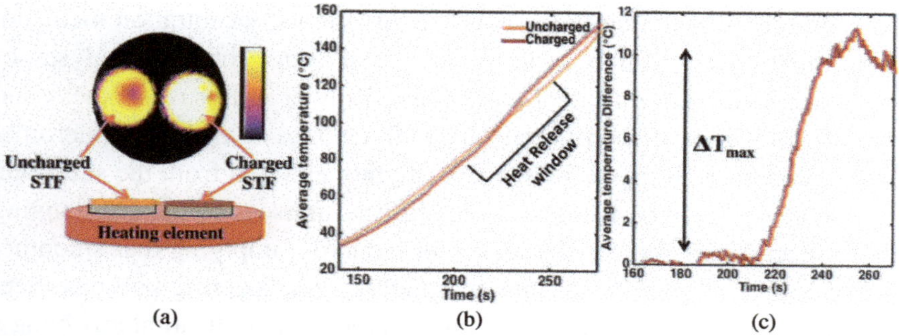

Figure 4.7. Macroscopic STF polymer heat release. (a) Top-view IR heat map of uncharged and charged solar thermal energy capacitors (STECs) placed on a heating element. (b) Average temperature recorded on the surface of each STEC plotted as a function of time. (c) Average temperature differences between the two STECs. Reprinted with permission from Ref. [24]. Copyright 2016, Wiley-VCH.

could capitalize on these additional degrees of freedom to amplify ED, enhance optical rechargeability and light equilibrium, and extend photon collection across a broader range of the solar spectrum. The remarkable flexibility of these materials, coupled with their potential for solid-state applications, promises an enticing trajectory for novel solid-state endeavors encompassing light–matter interaction science, as well as renewable energy storage and heat release.

Furthermore, an alternative approach was established employing electrodeposited copolymers comprising small functional molecules for solar energy collection and storage [23]. Electrodeposition on conductive substrates offers a versatile methodology, characterized by conformal form factors, capacity to employ low-concentration solutions, and optimal utilization. This streamlined process involves dissolving the target material in an electrolyte and applying a bias between the working and counter electrodes, inducing migration and deposition of dissolved material onto the oppositely charged electrode. By regulating deposition conditions (through voltage or current application), film thickness and morphology can be tailored. Importantly, this method is both scalable and cost-effective, devoid of the need for intricate infrastructure. Electrodeposition has found utility in various energy-related applications, including diverse electrode applications. In Grossman *et al.*'s study [23], they harnessed Azo

STF polymers, enabling their deposition in a low-concentration methanol solution, yielding a homogenous large-area film. Through analyses of cross-sectional stainless steel electrodes, films of uniform thickness and morphology were demonstrated. These films, measuring a few micrometers in thickness, form distinct layers, slightly elevated from the substrate. It is noteworthy that stainless steel surfaces possess roughness ranging from several hundred nanometers to one micron, implying that smoother substrates could further enhance morphology. Considering polymer electrodeposition and the resultant coatings, a potential application involving a form of STF can be envisaged. This entails coating a thin (~50 μm) metal wire with STF material, integrating it into fabrics. Given the STF polymer's inherently low thermal conductivity, amalgamating it with high thermal conductivity metals like copper (up to 400 W $m^{-1}K^{-1}$) can facilitate efficient heat transfer to other sections of the fabric or device. This method could be extended to diverse conductive substrates, offering controlled thin film growth through electrodeposition conditions. Relative to other solid-state processing techniques, this method exhibits an ED of approximately ~90 J g^{-1}, featuring solid-state rechargeability and excellent material utilization, thereby maintaining the stellar performance of STF. This innovative approach is apt for applications involving photon upconversion, photovoltaics, photosensitivity, light emission, and other systems where small functional molecules can be applied in solid-state applications.

Furthermore, Grossman [24] explored symmetrical diacetylene and polydiacetylene compounds, achieving high-energy-density materials with energy storage of up to 176.2 kJ mol^{-1} (or 200.2 kJ mol^{-1} when fully charged); this surpasses the original azobenzene's energy storage capability. In addition to the isomeric enthalpy of the azobenzene unit, extra energy storage in the material results from different ground state phases (crystalline solid) and metastable states (amorphous solid/liquid). Notably, the phase transition characteristics of organic materials can be considered as a parameter while designing high-energy-density photonic energy storage materials. This study delves into the potential to attain high-energy storage in photonic energy storage materials through diverse levels of intermolecular interaction between trans and cis isomers, proposing diacetylene containing azobenzene as an effective storage platform. Diacetylene exhibits facile functionalization, self-assembly, and crystal-like packing due to robust intermolecular interactions. Additionally, these compounds photopolymerize rapidly under ultraviolet light, generating polydiacetylene with azobenzene side chains on a rigid conjugated framework. This structure closely resembles a rigid template adorned with tightly packed photochromic units. Previous investigations of azobenzene-modified diacetylene and polydiacetylene primarily focused on asymmetric structures where azobenzene groups were functionalized on one side, while long alkyl chains on the other side self-assembled into vesicles, tubes, or bilayer films. These asymmetric structures found use in light-controlled molecular recognition and reversible optical switching. Leveraging insights gleaned from studies on structural changes in azobenzene-functionalized diacetylenes, symmetrical diacetylenes were devised, facilitating high loads and orderly arrangement of azobenzene groups. Compact systems of azobenzene derivatives lacking extensive groups hold the potential to maximize the weight and volumetric ED of photonic energy storage materials. The proposed high-energy storage material, rooted in azobenzene-functionalized diacetylene, exhibits a 113% increase over the original azobenzene, attributed to the loss of robust intermolecular interactions in the ordered *trans* state and prominent interactions in the *cis* conformation. To explore this novel material design space, a range of aspects were investigated, encompassing alkyl spacing lengths between the diacetylene core and terminal azophenyl groups, intermolecular hydrogen bonds, and diverse functional groups on azophenyl groups. X-ray diffraction analysis offered insights into material aggregation and intermolecular interactions, while ab initio calculations shed light on the relative energy storage mechanisms of this molecule series.

Venkataraman *et al.* [25] showcased a substantial ED of approximately 510 J/g in azo polymethacrylate polymer. This remarkable ED stems from several factors: (a) the strategic attachment of azobenzene to the polymer backbone, resulting in heightened functional group density along the polymer chain; (b) introduction of end groups onto the polymer, enabling self-assembly into cylindrical structures akin to carbon nanotubes; and (c) tailoring polymer–polymer interactions through the selection of diverse processing solvents. To explore this premise, polymethacrylate (PMA) was chosen as the foundational framework due to its pronounced synchronicity in radical polymerization. PMA exhibits a cocanonical triplet with an isoester group distance of ~5 Å, akin to the spacing between azobenzene units on a carbon nanotube. Employing hexaphenophenocolonene (HBC) as the end group facilitated self-assembly into cylindrical structures through π–π stacking interactions. Additionally, two reference solvents with low boiling points and high solubility, dichloromethane (DCM) and tetrahydrofuran (THF), were assessed to elucidate the role of polymer–solvent interactions and the optimal structure for high ED. The interplay between processing solvents and polymer–solvent interactions critically shapes the optimal morphological structure of high-density materials.

An assessment of the effect of the solid state on E–Z isomerization was conducted by comparing the isomerization free energy of azobenzene in the gas phase and the condensed state at room temperature. Condensed matter isomerization free energy was calculated by simulating two 125-molecule containers, each containing E- and Z-isomers, in equilibrium over 100 ps at 1 atm and 300 K to compare total energy. The results reveal that the free energy of solid-state isomerization (~14.8 kcal/mol, 340 J/g) is nearly doubled in comparison to gas-phase isomerization (~8.0 kcal/mol, 183 J/g). This trend underscores that E-isomers possess relatively potent stabilizing effects in intermolecular interactions when contrasted with Z-isomers. Moreover, this offers a direct theoretical foundation for manipulating solid-state effects to regulate the isomerization thermodynamics of azo-containing phenyl materials. Computational energy analysis underscored that the primary driver of high heat release is the stability of the E-isomer state. Employing THF as a solvent mitigates polymer aggregation in the solution state, facilitating coordinated isomerization in high-energy-density solids. Therefore, the hypothesis arose that T_g discrepancies existed between AzoPMA prepared using DCM and THF when the DSC was subjected to a second heating cycle (after discharge). To verify this, DSC measurements were set up similarly to prior experiments. In the initial heating curve, T_g was challenging to discern, as Z–E isomerization heat release commenced prior to the T_g of the original AzoPMA 3 (T_g = 62°C). However, in the DSC's second heating cycle, discernible T_g differences emerged between DCM and THF-prepared azopma (~130°C for DCM samples and ~90°C for THF samples). This implied dissimilarity in the polymer chain's surrounding environment.

This was congruent with WAXS data. Therefore, it is inferred that the heightened free volume not only expedites Z–E isomerization but also supplies ample room for E-isomers to attain robust stability through π–π stacking. Analysis indicated that Z-AzoPMA 3 films from THF exhibited stacking after isomerization, allowing E-isomers to partake in π–π stacking, fostering stability conducive to synergistic isomerization, as previously observed in polymers containing projecting azobenzene units. Conversely, in DCM-prepared films, inadequate room exists for E-isomer π–π stacking. These findings underscore that the origin of these packing differences is tied to solvent–polymer skeleton interactions. The outcomes underscore that morphological shifts of solid polymers resulting from distinct solvent treatments hold a pivotal role in governing Z–E isomerization activation energy and DSC exothermic heat profiles. Hence, this research underscores the fundamental role of processing solvents and thin film structures in achieving heightened energy densities.

4.3.2 *Crystal-to-liquid polymer-templated Azo-STF composites*

The increasing interest in ionic liquids stems from their distinctive physical and chemical attributes, customizable based on their chemical composition. Overcoming the challenge of controlling physical properties and functions through the phase transition between ionic crystal (IC) and ionic liquid (IL) phases necessitates exploration [26]. In the pursuit of exhibiting the photomechanical effect, photoresponsive polymers must incorporate photochromic groups like coumarin, spiropyrane, styrene, and azobenzene. Azobenzene exists in two configurations, a stable *trans* structure akin to a rod and a metastable curved *cis* isomer. The interplay of ultraviolet and visible light irradiation induces *trans* and *cis–trans* isomerization.

Distinguished by possessing both the elasticity of polymer networks and the ordered interelements characteristic of liquid crystals (LC), cross-linked liquid crystal polymers (CLCPs) offer a distinct advantage. Inclusion of azobenzene chromophores in CLCPs prompts substantial photodeformation due to a domino effect arising from microscopic changes in azobenzene's structure. This cascade effect emerges as the LC mesogen and polymer segment synergistically mobilize, leading to large-scale photodeformation of CLCP. The LC system undergoes a decrease in arrangement order due to *trans*-photoisomerization of azobenzene, culminating in an LC-I phase transition. This shift is attributed to the stabilizing influence of the rod-like *trans*-azobenzene segment on LC alignment, while the curved *cis*-form diminishes LC order parameters. After visible light exposure or heating, the shrunk CLCP film reverts to its initial size, facilitated by *cis*-inverse isomerization of azobenzene.

Yu *et al.* [26] employed a thermal polymerization process to craft a single-domain cross-linked liquid crystal polymer (CLCP) film, merging mono-acrylate and diacrylate within three cross-linking densities. This matrix, which integrated azo-toluene chromophores, underwent rigorous characterization. Differential scanning calorimetry assessed the thermodynamic properties of monomers, cross-linkers, their mixtures, and CLCP films. Polarizing optical microscopy and polarization UV–Vis absorption spectrometry delved into subcrystalline attributes. The extended conjugated structure of the side chain's tolane segment prompted photobending of the independent CLCP film upon exposure to 436 nm short-wavelength visible light. Wavelength shifts to 577 nm reverted the curved film to its original state. Visible light at 530 nm also induced bending, albeit with diminished bending velocity as the actinic light's wavelength increased. Notably, bending rate correlated with order parameters, light intensity, and temperature. At 436 nm visible light, film mechanical force was augmented with greater cross-linking density and heightened light intensity, yet diminished with escalating temperatures.

TMA analysis determined the mechanical force of azolidine CLCP film after light irradiation. The film's ends were clamped, an external force of around 200 kPa was applied to maintain film length, followed by exposure under 436 nm light of varied intensity and temperatures. Consistent film length coupled with heightened force signified force generation under light irradiation. At a light intensity of 25 mW cm^{-2}, the maximum force shift reached approximately 300 kPa, 200 kPa, and 100 kPa at temperatures of 30°C, 50°C, and 70°C, respectively. Notably, the film's maximum force generated under light irradiation declined with temperature elevation, attributed to a reduction in the elastic modulus, a pivotal mechanical property indicator. In essence, as film stiffness waned, mechanical force engendered by light irradiation diminished. Further findings revealed that the time necessary for the film to attain maximum force reduced with rising temperatures, as heightened temperatures augmented CLCP membrane fragments' flexibility, making it easier to achieve the microscopic changes of azobenzene molecules and their macroscopic deformation in the entire polymer network. This increased the overall polymer network's ease of deformation. Simultaneously, maximum force generation rose with heightened light intensity. For instance, at 25 mW cm^{-2} and 15 mW cm^{-2} light intensities, F8020's maximum force shift at 30°C was approximately 300 kPa and 200 kPa, respectively. Increased light intensity accelerated the *trans*-isomerization rate and azolium alkyl group's degree, prompting pronounced membrane surface region contraction. Temperature and light intensity exerted similar effects on photomechanical forces. Additionally, heightened film cross-linking density corresponded to increased force generation, as more cross-linking points elevated film rigidity and elastic modulus. Elastic moduli were about 98 MPa, 96 MPa, and 84 MPa, respectively. A heightened elastic modulus signifies stronger interaction between the medium and the framework, channeling greater force to the skeleton during CLCP film phase transitions. Consequently, higher elastic modulus CLCP films exhibit amplified photomechanical efficacy.

Kimizuka *et al.* [27] introduced the pioneering concept of photoresponsive integrated circuits (ICs) that can reversibly photoliquefy into ionic liquids (ILs). Figure 4.8 illustrates a typical instance of a 1(6,4)-Br XRD pattern coupled with polarization optical microscope (POM) imaging under cross polarizers. Initially, the yellow 1(6,4)-Br sample demonstrated birefringence in POM images under cross polarizers, coherent with the crystal lamellar structure observed via XRD measurements (Figure 4.8(a)). Subsequent exposure to ultraviolet light from a high-pressure mercury lamp [band pass filter, $\lambda = (365 \pm 10)$ nm] caused the yellow crystal sample to melt within a few minutes, transforming into a red liquid phase (Figure 4.8(b)). POM images under cross polarizers displayed darkening in the irradiated sample portion, paralleled by amplified *n*–p* absorption intensity at $\lambda = 450$ nm. These changes coincided with the obliteration of the XRD pattern (Figure 4.8(b)). These findings unmistakably confirm that *trans*-1(6,4)-Br's photoisomerization prompts its transition to an isotropic IL phase. The *cis* isomer molar ratio, as assessed

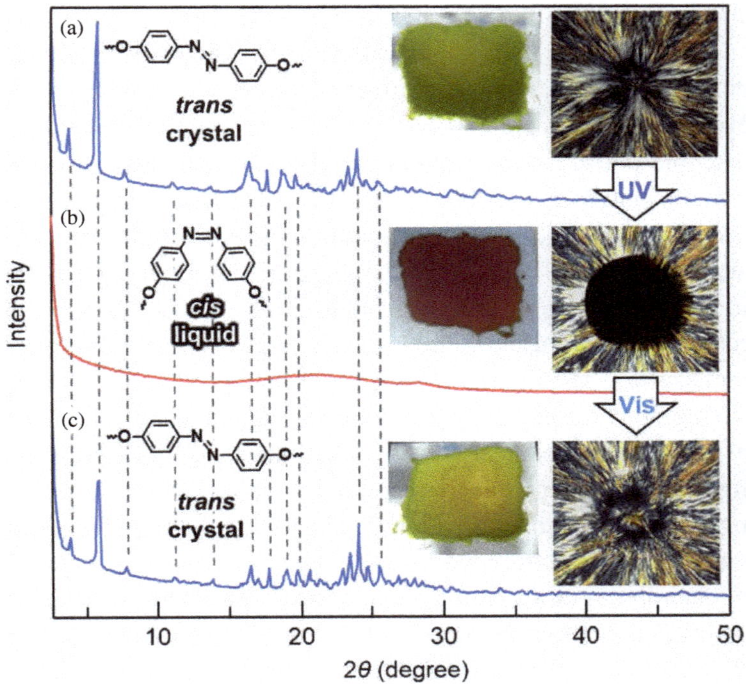

Figure 4.8. X-ray diffraction patterns and optical microscopy images under crossed polarizers. (a) As prepared film, (b) after UV irradiation, and (c) after Vis irradiation. Reprinted with permission from Ref. [27]. Copyright 2015, Wiley-VCH.

through 1h NMR spectroscopy (in $CDCl_3$), is approximately 80%. Furthermore, reversible crystallization occurred within 1 min of photoliquefying ILs under visible light irradiation [λ = (470 ± 20) nm], accompanied by diminished *n–p** absorption strength of the *cis* isomer at λ = 450 nm. This faithfully reproduced the unique XRD pattern of the *trans*-azobenzene compound (Figure 4.8(c)). At room temperature, 1(6,4)-Cl, 1(4,6)-Br, and 1(8,2)-Br exhibited reversible photoliquefaction and photocrystallization, showcasing fully reversible photoliquefaction and photocrystallization in integrated circuits. This superior trait surpasses preceding solid–liquid phototransition systems, considerably enhancing energy storage capability. Moreover, the photoinduced IC-IL phase transition induces substantial alterations in ionic conductivity, affording the azo phase transition system diverse functionalities. The amalgamation of azobenzene photoisomerization and ICs-IL phase transition not only transcends existing azobenzene STFs limitations but also introduces a novel perspective for phase crossover materials design.

Furthermore, Wu's group [28] introduced the synthesis and photoresponsive attributes of innovative Azo polymers, varying the intervals between azophenyl groups on both the polymeric main chain and side chain. Employing free radical polymerization, P-n-Azo polymers, designated as polyacrylates, were derived with distinct spacing lengths between the polymeric main chain and azobenzene units (ranging from no spacing to 2, 6, 12, or 20 methylene units; Option 1). The average molecular weights of P-N-azo were 7.6, 8.9, 9.9, 8.6, and 10.6×10^3 g mol^{-1}, respectively. In this study, all P-N-Azo polymers exhibited a reversible inverse oblique photoisomerization in the solid state, verified by ultraviolet/visible absorption spectra of P-N-Azo films before and after light irradiation. Prior to irradiation, ultraviolet bands in the *p–p** absorption range and weak *n–p** absorption bands in the visible spectrum characterized all P-N-Azo polymers. Following ultraviolet light exposure, the p–p* band decreased, and the *n–p** band increased, signifying *trans*-to-*cis* isomerization of P-N-Azo, consistent with the Azo chromophore's known behavior. Subsequent irradiation with visible light absorbed by the *cis*-N–P* range could induce *cis–trans* isomerization, effectively reversing the isomerization process. This *trans*-to-*cis* isomerization could be reversibly induced after several light exposure cycles. The absorbance of P-6-Azo and P-12-Azo films before irradiation was lower than that after ultraviolet and visible light irradiation due to chromophore stack and orientation alterations. Initially, the Azo groups were preferentially oriented perpendicular to the substrate. After ultraviolet and visible light irradiation, this vertical orientation diminished, resulting in distinct absorption values and shapes for P-12-Azo in both THF and thin films. The absorption band's blue shift indicated Azo group aggregation.

P-6-Azo and P-12-Azo demonstrated photoswitchable T_g (glass transition temperature) and photoinduced reversible solid–liquid transitions. For Azo polymers with shorter or longer intervals between the azophenyl group and the polymeric backbone chain, no light-induced solid–liquid transition occurred at room temperature. P-0-Azo did not manifest a photoinduced solid–liquid transition due to insignificant T_g reduction after irradiation. While P-20-Azo did not undergo a photoinduced solid–liquid transformation at room temperature, it exhibited such a transition at 508°C, significantly lower than its phase transition temperature of approximately 408°C. This was attributed to its lengthy side chains, promoting solid-state crystallization. Nevertheless, cisP-20-Azo's phase transition occurred above room temperature, rendering observation of photoinduced solid–liquid transition unfeasible. Heating *cis*-P-20-Azo to 508°C enabled photoinduced solid–liquid transition. P-12-Azo mirrored P-6-Azo's behavior closely. Remarkably, P-12-Azo was leveraged for real-time photoinduced healing, a groundbreaking achievement. This photoresponsive polymer holds promise as a smart material for printing and microstructure fabrication. It can undergo controlled flow or processing via UV irradiation, followed by curing through visible light irradiation to obtain a robust material applicable in diverse contexts. This study underscores the pivotal role of spacers in achieving photoinduced reversible solid–liquid transitions in azo polymers, offering insights into designing Azo polymers for photoinduced healing and processing.

4.3.3 *Full-spectrum polymer-templated azobenzene-based STF*

In a world of escalating energy demand, the quest for alternative energy sources grows more pressing. Among the possibilities, the Sun stands out as a promising reservoir, given that current energy consumption constitutes a mere 0.1% of the Earth's surface annual solar output. In harnessing the Sun's potential, Azo-STF emerges as an energy storage contender. Here, photon absorption from sunlight transforms low-energy isomers into their high-energy counterparts, effecting chemical solar energy storage. The release of this stored solar energy, in the form of heat, can be triggered via heating or catalysis, yielding low-energy isomers for recharging. Notably, the reversible *cis*-to-*trans* isomerization of azobenzene derivatives has captured interest in solar energy storage field due to its stability, straightforward synthesis, and tunable absorption band.

However, a key challenge with Azo-STFs arises from the fact that *trans*-to-*cis* isomerization typically necessitates ultraviolet light, which constitutes a mere fraction (4.5%) of total solar irradiance, hampering the efficiency of Azo-STF. Furthermore, visible light within the solar spectrum hampers ultraviolet energy storage within conventional Azo-STFs, as it triggers the reconversion of charged *cis*-isomers to their neutral *trans* form. This visible-induced reverse isomerization impedes broad-spectrum sunlight from yielding substantial *cis*-isomer content (approximately 25%) in azobenzene, hence diminishing storage potential. This limitation results in the currently reported Azo STFs solely harnessing monochromatic ultraviolet light as a proof of concept.

Recent strides have seen the synthesis of novel azobenzene derivatives with visible-induced reverse *cis*-isomerization, offering the potential to store visible light. However, the challenge persists, as blue light still induces *cis*-to-*trans* isomerization, undermining full-spectrum sunlight storage (mAzo). To surmount this hurdle, Wu *et al.* [29] devised an integrated Azo-solar thermal cell device adept at efficiently storing both ultraviolet and visible light across the full spectrum of solar irradiation. The device configuration, depicted in Figure 4.9, encompasses coumarin 314,

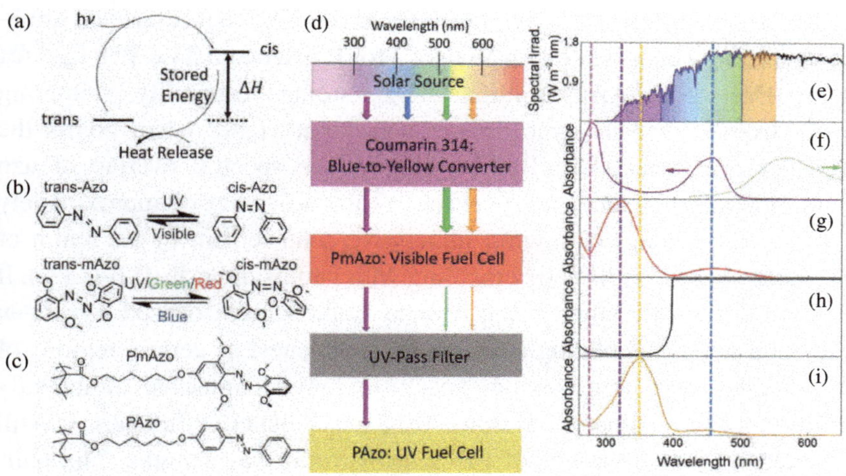

Figure 4.9. (a) Cyclic process of energy storage by photoinduced isomerization of azobenzene. (b) Schematic representation of Azo and mAzo. (c) Structures of PAzo under UV and PmAzo under Vis. (d) Schematic of 4-layer UV–vis solar thermal cell. (e) Solar spectrum (AM 1.5), UV–vis spectra of (f) Coumarin 314 film (reduced conc.), (g) PmAzo in THF (2×10^{-4} m), (h) UV-pass (UG-11) filter, and (i) PAzo in THF (6×10^{-5} m). Reprinted with permission from Ref. [29]. Copyright 2017, Wiley-VCH.

PmAzo, a UV-pass filter, and PAzo layers. At the top layer, a triple layer of coumarin 314 drip film (≈1–2 μm) within a poly (4-vinyl) phenol (PvPh) matrix absorbs near 460 nm (blue) light, while transmitting ultraviolet and green wavelengths. This layer also features fluorescent properties, actively converting absorbed blue light into 560 nm fluorescence that is captured by the subsequent PmAzo layer. This conversion aligns with the objective of countering *cis*-to-*trans* isomerization. The second layer, PmAzo, efficiently stores visible light through *trans*-to-*cis* isomerization, stimulated by wavelengths corresponding to *trans* $\pi^*\leftarrow\pi$ (λ_{max} = 323 nm) and $\pi^*\leftarrow n$ (λ_{max} = 460 nm) bands. This layer capitalizes on the spatial separation of *trans* and *cis*-$\pi^*\leftarrow n$ transitions to facilitate energy storage from longer visible wavelengths. A UV-pass filter precedes the final PAzo layer to prevent long-wavelength light absorption by PmAzo, which could cause undesirable *cis*-reverse isomerization in PAzo.

Noteworthy results emerge, with a *cis–trans* enthalpy difference of 54.8 kJ mol^{-1} for Pazo and 12.2 kJ mol^{-1} for PmAzo. While slightly lower than calculated enthalpy differences, these values align closely with theoretical computations. Applying these results to STF metrics yields gravitational energy densities of 4.4 W h kg^{-1} for PmAzo and 34.8 W h kg^{-1} for PAzo. Despite PmAzo's limited energy storage capability, stemming partly from its sensitivity to longer wavelength light, PAzo boasts the highest gravitational ED. By factoring in the spectral overlap of azo chromophores with the solar spectrum, a solar efficiency of approximately 0.4% is calculated, marking an impressive achievement in the realm of Azo-solar thermal cells and competing with photosynthesis (0.1–0.3%). It is important to account for the reverse isomerization caused by $\pi^*\leftarrow n$ transition when calculating solar energy efficiency, as certain regions of the standard azo chromophore spectrum remain unusable, potentially inflating prior estimates. Despite this, recalculated efficiency is still bounded by the feasibility of blocking harmful wavelengths. The half-lives of PmAzo (16(2) h solution) and PAzo (12.45(8) h solution) closely align with the thermal relaxation of *cis*-to-*trans* isomerization. This property enables both PmAzo and PAzo to offer storage capabilities lasting approximately half a day. In practical terms, this means the entire device can help counteract the diurnal variation of solar irradiance, charging throughout the day and discharging during the night.

The results show that the *cis–trans* enthalpy difference is 54.8 kJ mol^{-1} for PAzo and 12.2 kJ mol^{-1} for PmAzo. These enthalpy values are slightly smaller than the calculated (DFT/6-311G(d,p)/B3LYP) enthalpy differences of 69.9 and 29.5 kJ mol^{-1}. An oft-cited STF metric converts this enthalpy difference to gravitational ED: 4.4 W h kg^{-1} for PmAzo and 34.8 W h kg^{-1} for PAzo. In addition to the greater mass of PmAzo relative to PAzo, the small gravitational ED of PmAzo comes in part from the increased spatial interactions present in the transformation. Theoretical calculations show that this spatial effect reduces the enthalpy difference during isomerization. While PmAzo cannot store as much energy/mass, in part due to its sensitivity to longer wavelength light, PAzo exhibits the highest gravitational ED. In addition, combining the gravimetric energy density with the spectral overlap of the azo chromophores and the solar spectrum, the solar efficiency was calculated of about 0.4%, the highest to date for Azo-solar thermal cells, and is competitive with photosynthesis (0.1%–0.3%). When calculating solar energy efficiency, attention must be paid to the inverse isomerization caused by $\pi^*\leftarrow n$ transition. As a result, much of the spectrum of the standard azo chromophore is unusable, inflating previous solar efficiency estimates. Taking these regions out of the equation, the efficiency of recalculation is still limited by the conditions under which harmful wavelengths can actually be blocked. The half-lives of PmAzo (16(2) h solution) and PAzo (12.45(8) h solution) approximate the thermal relaxation of *cis*-to-*trans* isomerization. Both PmAzo and PAzo achieve half a day's storage time. Thus, the entire device can mitigate the daily periodicity of solar irradiance by charging during the day and discharging at night.

To summarize, Wu *et al.* present an innovative approach employing an Azo-solar thermal cells capable of effectively harnessing both ultraviolet and visible wavelengths from solar simulators, optimizing light collection efficiency. By judiciously blocking wavelengths responsible for *cis*-to-*trans* isomerization, they achieve a compelling *cis* conversion rate (60–70%) that competes with monochromatic radiation. This success is attributed to a well-conceived device design, along with enhanced spectral overlap, leading to a remarkable boost in solar efficiency of 0.4%, marking the highest efficiency reported for Azo-solar thermal cells thus far. Remarkably, PAzo's chemical structure showcases the potential to elevate ED per weight (40.9 W h kg^{-1}). An estimated solid-state conversion rate of up to 72.7% signals promising progress in solid-state solar thermal cells. However, solid-state influences, encompassing polymerization and polymer crystallinity, emerge as influential factors, underscoring the significance of fine-tuning the "solid-state" nature.

4.4 Phase-change templated Azo-STF composites

Polymers constitute integral components of everyday life. The glass transition temperature (T_g) characterizes the temperature range at which polymers shift from rigid, glassy states to pliable, rubbery states. While polymers with T_g below room temperature are amenable to room-temperature processing and healing, their soft nature impedes their use in applications demanding sturdier elements. Polymers boasting higher T_g values are more suitable for such applications. Traditional approaches for processing or mending high T_g polymers entail heating them above T_g or exposing them to plasticizing solvents, which can have environmental repercussions. A solution to this quandary lies in developing polymers capable of reversible softening and hardening at room temperature. While the concept of polymers with switchable T_g holds promise, it remains an underexplored territory, despite its potential to address challenges like crack healing in high T_g polymers and the processing of rigid polymers at room temperature without resorting to plasticizing solvents [26,29].

Azo compounds, featuring reversible *cis* isomerism, serve as promising candidates. Photoresponsive materials containing azobenzene find applications in this context. The T_g exhibited by Azo polymers, relative to small molecules containing azobenzene, determines whether the polymer remains solid or turns liquid. Intriguingly, evidence supports light-induced cis-to-trans cycling that induces polarization-dependent mass transfer (termed directional photofluidization) in Azo polymers. This phenomenon finds applications, including creating surface relief gratings, effecting photodynamic healing of wearable electrical conductors, and smoothing surface roughness. The mechanism behind directed photofluidization in Azo polymers remains a topic of debate, with suggestions that changes in the local environment during Azo photoisomerization correspond to temperature increases. An alternative notion posits that photoisomerization modifies the T_g values of Azo polymers. However, experimental proof of optical T_g switching in Azo polymers remains absent.

It is noteworthy that combining azobenzene with a phase transition mechanism is a popular strategy to enhance energy storage efficiency. Phase-change materials (PCMs), classified into inorganic and organic categories, offer stable temperature spans for latent heat release storage. While inorganic PCMs face issues like supercooling and phase separation, organic PCMs like paraffin and sugar alcohols mitigate these problems. The integration of a phase transition mechanism imparts azobenzene materials with the capacity for high energy storage density and broad melting temperature ranges. Recent advancements in the theory of room temperature photochemical crystal–liquid transition (PCLT) have improved understanding in this domain. The process involves Azo molecules capturing photons for *trans*-→ *cis*-photoisomerization, overcoming the molecular stack interaction that characterizes crystal → liquid-phase transition. This transition taps into ambient heat absorption as the *cis* phase→ anticrystalline transition releases stored energy in the form of chemical enthalpy and crystallization enthalpy (ΔH_{isom} and ΔH_{cryst}), ultimately generating high-temperature heat as needed.

The synergistic interplay between these two energy forms allows the collection of ambient heat in the low-melting-point metastable isomer liquid during photochemical melting, which is then released as high-temperature heat via recrystallization of the high-melting point-parent isomer. This approach yields higher energy capacity than conventional solar/thermal energy storage systems based solely on molecular photoisomerization or phase transition.

Wu's team [30] showcases the ability of light to modulate the T_g of azopolymers, inducing reversible solid–liquid-phase changes in the polymer matrix. Notably, the azophenyl group displays reversible *cis*-isomerism within the polymer framework. As depicted in Figure 4.10, the Azo polymer undergoes a captivating light-induced solid–liquid transition. Alternating irradiation with 365 nm ultraviolet and 530 nm visible light prompts transformation from irregularly shaped powder to droplets, with adjacent droplets coalescing. Both scenarios highlight UV irradiation's ability to induce polymer flow. Interestingly, visible light irradiation does not cause any alteration in powder morphology. Furthermore, the Azo polymer's reversible solid–liquid transition due to photoswitching T_g is demonstrated. This concept finds application in mending damage on azo polymers, reducing surface roughness in patterned azo polymers, and facilitating

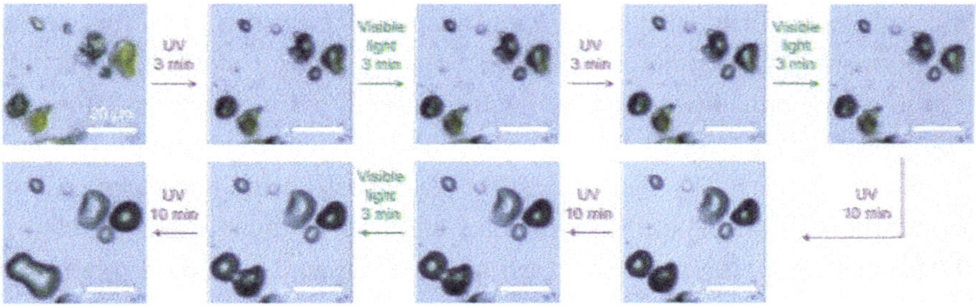

Figure 4.10. Optical microscopy images of photoinduced solid-to-liquid phased change of Azopolymer powders under alternating ultraviolet and visible light irradiation. Reprinted with permission from Ref. [30]. Copyright 2016, Springer Nature.

transfer printing. Similarly, P1 powder subject to alternating 365 nm ultraviolet and 530 nm visible light exhibits analogous morphological shifts, solidifying UV light's efficacy in liquefying higher molecular weight azo polymer (Mn = 2.7×10^4 g mol^{-1}).

Furthermore, three supplementary experiments were undertaken to investigate the light-induced solid-to-liquid transition. First, infrared thermometer measurements revealed that the surface temperature of P1 under ultraviolet irradiation (67 mW cm^{-2}) was 35°C, lower than P1's T_g (48°C). This indicates that the photothermal effect is inadequate for liquefying P1. Second, annealing P1 at 38°C for an hour did not result in liquefaction. Third, employing lower-intensity UV light (30 mW cm^{-2}) managed to liquefy P1 by elevating its surface temperature to 28.5°C. These systematic experiments conclusively establish that the photoinduced solid–liquid transition is not a photothermal process, but rather a photoisomerization process. Notably, the *trans*-azophenyl group within the P1 film exhibits a liquid crystal order. Ultraviolet irradiation transforms P1 from an anisotropic state to an isotropic *cis* state, constituting a photochemical phase transition. However, this solid–liquid transition in P1 is not attributed to the loss of liquid crystal ordering. Intriguingly, even amorphous P1, prepared through quenching above the isotropic temperature, displays the photoinduced solid–liquid transition.

DSC is the standard method for the direct measurement of T_g. However, due to the strong absorption of *trans* P1 at 365 nm, only the top layer of the powder is converted to *Cis*-P1, so ultraviolet irradiation does not produce enough *Cis*-P1 for DSC measurements. Therefore, Wu's team [31] prepared enough *cis* P1 for DSC measurements using this route. First, *trans*-P1 dissolves in CH_2Cl_2. The polymer solution was then irradiated with ultraviolet light under stirring and CH_2Cl_2 was removed under vacuum (25 mbar, 25 min). 1H NMR analysis of the photoisomerization of CD_2Cl_2 under the same irradiation conditions showed that 95% of the *trans* isomers were converted to *cis* isomers. CH_2Cl_2 has a boiling point as low as 40°C. TGA showed that CH_2Cl_2 was completely removed during sample preparation. *Cis*-p1 is a liquid. Two control samples were prepared to further demonstrate that liquefaction was not due to possible residual solvents in the polymer. Control-1 is prepared by adding 2.5 wt% CH_2Cl_2 to *trans* P1, and Control-2 is prepared by removing CH_2Cl_2 from *trans* P1 solution at 25 mbar. Both Control-1 and Control-2 are solids. Subsequently, *trans* P1, *cis* P1, and two control samples were studied by DSC. The T_g of *Trans* P1 on the first and second heating curves is 48°C, and two phase transitions occur at 94°C and 108°C. In contrast to *trans* P1, *cis*-P1 has T_g at −10°C and a wide tropical zone at ~94°C, due to the *cis–trans* isomerization that occurs in the first heating curve. The low T_g of *cis*-P1 is confirmed by Dynamic Mechanical Analysis (DMA) measurements, indicating that the T_g of *cis*-P1 is −6°C. The low T_g of *cis* P1 is consistent with the observation that *cis*-P1 is a liquid at room temperature. The DSC curve of the second heating of the sample was the same as that of *trans* P1, indicating that *cis*-P1 was thermally converted to *trans* P1 during the first heating process. The DSC curves of the two control samples were similar to *trans* P1. These data confirm that P1's photoinduced solid– liquid transition is due to its photoswitchable T_g.

In summary, this optical modulation of T_g presents a pioneering approach for designing high T_g polymer with self-healing and hard coatings processable at room temperature sans plasticizing solvents. An advantage lies in light's precise spatiotemporal resolution, enabling localized healing and polymer processing. The authors anticipate the possibility of other stimulus-responsive polymers displaying switchable T_g, akin to azo polymers. This study not only opens a path to control T_g in polymers, benefiting fundamental T_g comprehension and practical applications, but also reveals potential avenues for reprogrammable encryption technologies, dual-security anti-counterfeiting materials, and flexible applications.

Moreover, Grossman *et al.* [31] further delve into optically regulated phase-change azo-STF polymers, capable of serving as rewritable dual-mode patterns. Illustrated in Figure 4.11, they introduce a unique orthogonal optical patterning method for creating interference-free two-mode

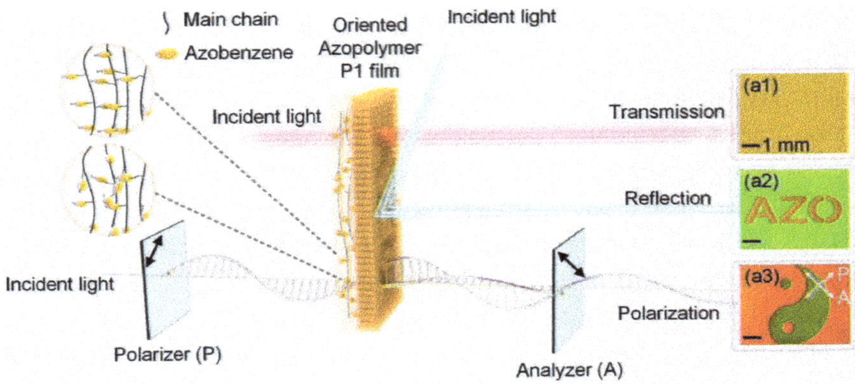

Figure 4.11. Schematic and photographs of a patterned film of the Azopolymer P1 observed in (a1) transmission, (a2) reflection, and (a3) polarization modes. Reprinted with permission from Ref. [31]. Copyright 2022, Wiley-VCH.

patterns on novel azo polymer P1. This Azo polymer's photoresponsiveness stems from the reversible *cis*-reflective isomerization of the Azo phenyl group. The selection of Azo polymers for photopatterning stems from their intriguing photoresponse properties suitable for intricate pattern generation. Moreover, certain Azo polymers have been found to exhibit a photoinduced reversible solid–liquid transition, a phenomenon leveraged in imprinting photonic structures. To prepare two-mode patterns involving photonic structure and color polarization on Azo polymers, conditions necessitate good tensile properties and a photoinduced reversible solid–liquid transition, properties rarely combined in Azo polymers so far. Success also hinges on ensuring non-interference between the two modes.

To fulfill these requirements, a specialized Azo polymer P1 was conceived, boasting exceptional stretchability and manifesting a light-induced reversible solid–liquid transition. They devise an orthogonal optical patterning technique for fabricating two-mode patterns on P1 films. Controlled pattern processing and diverse pattern designs are utilized to eliminate interference. In effect, P1 films demonstrate an "Azo" photon structure in reflection mode and a "Tai chi" color polarization mode. Additionally, both modes remain hidden when viewing the film in transmission mode. The dual-mode pattern's rewritability and reprocessability are successfully demonstrated, facilitating reuse, recycling, and material reprogramming. The accomplishments in crafting interference-free rewritable two-mode patterns stem from P1's distinctive attributes and novel patterning methods, marking uncharted territory.

In conclusion, this study unveils the utilization of P1 for producing bimodal patterns through orthogonal optical patterning. The tailored design of P1 and the innovative orthogonal light graph technique permit the creation of two-mode patterns on a single film while circumventing interference issues. Furthermore, the demonstrated rewritability and reprocessability in a dual-mode setup offer reusable and reprogrammable material applications. Unlike traditional optical storage, which records a sole information type, the dual-mode setup elevates information storage density. Combining these features with cryptography for future reprogrammable encryption technologies is anticipated, securing flexible dual-mode patterns as double security anti-counterfeiting materials and enabling versatile applications. This research serves as a gateway to optical molding material advancement and multi-molding technology development, carrying relevance for both fundamental research and practical application.

It has been observed that reversible solid–liquid transitions can be induced by light when the melting point (T_m) and crystallization temperature (T_c) of both *trans*-azobenzene and *cis*-azobenzene compounds are above and below room temperature, respectively. This phenomenon not only increases heat storage capacity but also presents a theoretical basis for achieving low-temperature heat release. This suggests that addressing low-temperature heat release requires the careful design of Azo compounds with appropriate T_c or T_m before and after photoisomerization. Building on this premise, Feng *et al.* [32] explored the solar energy utilization potential of long-chain azo compounds (T-Azo, F-Azo) that integrate photoinduced solid–liquid conversion and low-temperature photoisomerism. By optimizing the alkyl chain length, it becomes feasible to adjust the T_c and T_m of enriched *cis*-azobenzene within a suitable temperature range. This lays a robust foundation for T_c and T_m control via molecular engineering. DSC characterization was employed to assess T_m and T_c of long-chain Azo compounds before and after isomerization. In their initial *trans* states, T-Azo and F-Azo exhibited comparable T_m and T_c due to their similar molecular structures, both above room temperature. Following UV charging (365 nm), T_c and T_m exhibited significant reduction. Charged T-Azo plummeted to 6.3°C, while charged F-Azo stood at 8.2°C. UV charging at room temperature led

T-Azo and F-Azo to shift from yellow solids (*trans* state) to red liquids (*cis* state), returning to a solid state under 420 nm blue light. These experiments solidified the photoreversible solid–liquid transition characteristics of T-Azo and F-Azo. Notably, the alkyl chain length influenced the temperature change, with longer chains leading to relatively smaller decreases in T_c and T_m due to stronger intermolecular interactions. Depending on the application temperature, the molecular design can be tailored to meet specific requirements. Another significant factor in assessing heat storage performance is ED. For charged T-Azo and F-Azo, the stored energy can be categorized into the enthalpy of phase transition (ΔH_p) and the enthalpy of isomerization (ΔH_{iso}). These specific values can be determined by integrating their respective exothermic peaks. The total ED (ΔH_{total}) for T-Azo and F-Azo was found to be 170.5 J g^{-1} and 198.4 J g^{-1}, respectively, a substantial improvement compared to pure Azo compounds. Additionally, simultaneous storage of ΔH_{iso} and ΔH_p contributes to an enhanced ED, particularly valuable for achieving low-temperature heat release. Simultaneously storing phase-change enthalpy and light heat energy effectively boosts the ED of STFs, yielding prominent heat release effects. Upon blue light irradiation, the long-chain Azo compound showcased a temperature difference of up to 2.1°C at room temperature, effectively accomplishing photocontrolled heat release at lower temperatures. This study thus contributes novel insights into efficiently exploiting Azo compounds with photoinduced solid–liquid phase variations and photoisomerism to surmount the high exothermic temperature challenge posed by Azo STFs.

4.5 MOF-templated Azo-STF composites

An intriguing avenue surfaced upon a comprehensive literature review — the potential of metal–organic frameworks (MOFs). MOFs offer extensive prospects due to their design flexibility. MOFs' structural attributes, porosity, and surface functional groups can be tailored to suit specific applications. Their applications in molecular storage and separation are extensively explored. The functional group in azobenzene can undergo conformational changes triggered by light or heat, notably involving alterations in carbon atom pair distances (from 9 A in *trans*- isomers to 5.5 A in *cis* isomers). This isomerization can be harnessed in the organic linker of MOFs for molecular-level photochemical tasks [33–35].

Zhou [36] crafted a light-switching MOF named PSN-123, utilizing the ligand 2-(phenyldiazyl) terephthalate (shown in Figure 4.12). Crystals of PSN-123 exhibit quasi-high symmetry. The study revealed that UV or heat treatment led PPCN-123 to absorb varying amounts of CO_2,

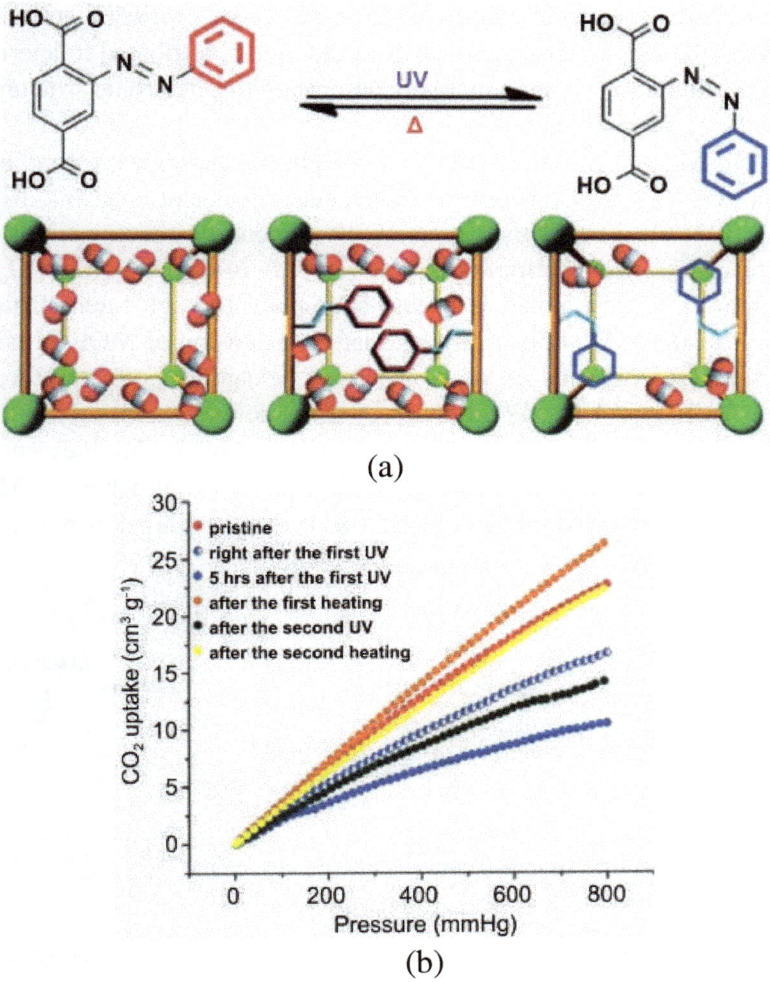

Figure 4.12. (a) (Top) *Trans*-to-*cis* isomerization of the ligand of PCN-123 induced by UV irradiation and the *cis*-to-*trans* isomerization induced by heat treatment. (Bottom) Schematic illustration showing the suggested CO_2 uptake in MOF-5, PCN-123 *trans*, and PCN-123 *cis*. (b) CO_2 adsorption isotherms (at 295 K) of PCN-123 showing reversible conformational change. Reprinted with permission from Ref. [36]. Copyright 2012, Wiley-VCH.

underscoring azo-MOFs' optical switching potential. Recent years saw the calculation of *cis* and *trans* optical absorption spectra of PPCN-123 via periodic and non-periodic models using the Bethe–Salpeter equation and GW approximation. *Cis* isomer exhibited maximum selective absorption around 3.2 eV, while *trans* isomer showed 4.2 eV. Notably, oscillation intensity ratio between *cis–trans* isomers of S1 varied significantly between the molecular model and periodic model, attributed to aperiodic models' limitations in quantitatively describing the electronic structure of periodic systems at the DFT level.

Azo MOFs exhibit remarkable adsorption capabilities, extending to gases, dyes, alkanes, and various molecules. Demonstrating this, Brown *et al.* [37] showcased the potential of azobenzene-functionalized isolattice metal–organic frameworks (Azo-IRMOF-74-III) [$Mg_2(C_{26}H_{16}O_6N_2)$] in controlling guest molecule release from pores through external stimuli [37]. Azo-IRMOF-74-III is an equal-lattice extension of MOF-74 with a B-lattice topology and a one-dimensional hexagonal pore structure (as shown in Figure 4.13). When the propyl iodide dye was loaded into the MOF, spectral studies showed that no emission of the luminescent dye was observed under environmental conditions. However, when the MOFs were irradiated at 408 nm, the rapid oscillating motion inherent in the

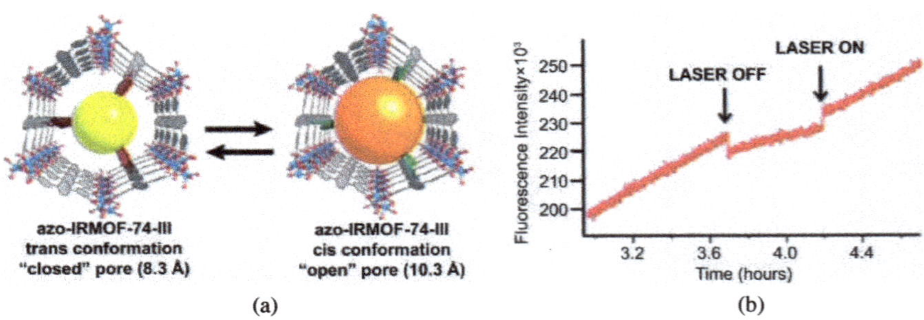

Figure 4.13. (a) View of the idealized Azo-IRMOF-74-III down the *c*-axis displays one-dimensional pores and azobenzene functional groups projecting into the pores. The yellow and orange balls represent pore aperture in idealized Azo-IRMOF-74-III, when the azobenzene functional groups are in *trans* and *cis* conformation, respectively. (b) An on/off trial, where the excitation beam was temporarily removed mid-release. Reprinted with permission from Ref. [37]. Copyright 2013, Wiley-VCH.

repeated isomerization of the azobenzene functional groups triggered the release of the dye from the hole. The light-induced release of the cargo can be modulated between the on-and-off states using the appropriate wavelength of light. The optical switch has different effects on the adsorption of different guest molecules. Wang [38], meanwhile, investigated the impact of photoelectric switching of azobenzene within MOFs on adsorption properties, revealing varying results depending on the guest molecule. This research paves the way for adaptable smart materials tailored to advanced applications.

The poor cycle performance of Azo energy storage system may be caused by its frequent isomerization during the cycle. The template method is to insert Azo units into carbon nanotemplates by grafting, which is one of the most active research directions at present. However, frequent isomerization can cause irreversible damage to the chemical bond or structure of the Azo unit, and there is no effective means to replace it on the carbon nanotemplate. Azo-MOF materials provide important insights for combining azobenzene derivatives and template methods. By inserting azobenzene into the Azo MOF template as a guest, the intermolecular forces such as coordination bonds can be fully utilized and the substitutability of azobenzene molecules can be realized at the same time. This new strategy is interesting and possible. In addition, Azo energy storage systems require good film-forming performance to ensure full use of solar energy during charging. Since MOFs are usually grown in the form of monocrystalline or polycrystalline powders, the preparation of photoswitched integrated MOF films has become a research hotspot in recent years. In 2017, Meller [39] first proposed Azo-MOF, whose key properties can be reversibly modulated only by visible light irradiation. The fluoroazobenzene side group in the MOF structure can be reversibly switched between the *trans* and *cis* states using green and violet light (no need to use ultraviolet light). It is interesting that azo MOFs have certain film-forming properties. Wang [40] combined surface-mounted MOF (surMOF) with quartz crystal microbalance to prepare five different photoluminescent columnar layer MOF structures containing azobenzene. The study also revealed a trend that the azobenzene portion density of each MOF cell battery is high and can achieve a large switching effect. In addition, Khayyami *et al.* [41] used atomic/molecular layer deposition technology to prepare various azophenyl metal–organic crystal films in situ. These are very interesting materials because azobenzene is one of the prototype organic molecules that has undergone photoisomerization. In 2020, Rice's review [42] emphasized photophysical studies and their dependence on MOF types or photoactive components, highlighting Azo-MOF's emergence as a smart material. The photoswitching mechanism suggests the potential to integrate carbon nanotemplates into Azo-STFs for controlled azobenzene derivatives or catalyst dispersion between carbon nanotemplate layers, enabling flexible, versatile Azo-STF and composite controlled energy release.

In conclusion, azobenzene-based STF has been shown to be suitable and efficient for solar thermal conversion and storage. In the current urgent need for clean energy, azobenzene has been paid more and more attention. At present, the derivative and template methods of molecular engineering and microstructure optimization are used to solve the problems of low ED, low quantum yield, short half-life, narrow overlap with solar spectrum, and irreversible degradation of Azo STFs.

In fact, azobenzene is still a long way from commercialization, but the biggest obstacle is not the energy storage density and half-life issues. In fact, some research results show that the energy storage efficiency of Azo-STF is already comparable to that of lithium batteries. We believe that there are three main issues that stand in the way of commercialization of Azo-STF. First, an improvement in performance in one area of Azo-STFs usually results in a sacrifice in other areas. For example, the realization of full-spectrum Azo-STFs requires the large redshift of the wavelength absorbed by the azobenzene optical switch, which inevitably leads to poor ED and thermal stability. Prolonging the half-life of Azo-STFs by substitution leads to a decrease in energy storage density. Therefore, the different parameters of Azo-STFs (such as ED, lifetime, absorption efficiency) need to be balanced to adapt to different use environments. This adjustment results in the actual energy storage of the prepared Azo-STF being lower than theoretically expected. Second, the control performance of Azo-STF is poor. At present, most studies on Azo-STFs focus on improving energy storage performance and prolonging half-life, and the reported energy storage performance values are only theoretical. Since the photoisomerization of azobenzene is an equilibrium reaction, Azo-STFs exhibit incomplete isomerization even under the irradiation of artificial short wavelength laser. Under the sun, which varies with the climate, the results will be even milder. In fact, the photoisomerization of azobenzene has been regulated by many methods such as metal nanoparticle catalysis. Unfortunately, there are not many studies that combine energy storage research with regulatory photoisomerization research, and this area deserves further research. Finally, the poor performance of the Azo-STF cycle is also its biggest drawback. The Azo STF with a carbon nanotemplate has the stability of hundreds of cycles, while the Azo derivative has the stability of only a dozen cycles. In addition, based on the energy storage mechanism of Azo-STFs, these materials are often used in the form of thin films. In the process of sunlight irradiation and exothermic cycle, the film will be irreversibly damaged in a short time. The poor stability of the material also increases its use cost, and there is no effective solution at present. Studies have shown that due to the difference in T_g of *cis*-azobenzene derivatives and *trans*-azobenzene derivatives, certain azobenzene materials can be used to repair damaged or cracked materials, and thus can be used to extend the service life of Azo STFs films. The authors believe that the poor cycle performance of Azo energy storage system may be caused by its frequent isomerization during the cycle. Although the theoretical recycling rate of Azo units is 100%, frequent isomerization can cause irreversible damage to the chemical bond or structure of Azo units. The template method is to insert Azo units into carbon nanotemplates by grafting, which is one of the most active research directions at present. The extremely high intermolecular forces in this material can result in a high energy storage capacity, comparable to commercial lithium batteries. However, if the Azo unit is damaged, there is no effective means to replace it on the carbon nanotemplate. Therefore, a strategy of alternating heat release through MOF materials is proposed. It is true that replacing two-dimensional structures with three-dimensional structures may compromise the energy storage capacity of Azo energy storage systems to a certain extent, but the method may also be worth further development as a potential way to effectively reduce its cost.

In recent years, in addition to storing light and heat energy, azobenzene materials have also been used in thermodynamic cycles. Azobenzene molecules undergo isomerization under light irradiation, resulting in a small change in molecular size (molecular axis from 9 A to 5.5 A), resulting in macroscopic visible deformation of the material. At present, it is already known that tension can promote the release of heat energy, which means that there is a link between azobenzene photothermal energy storage materials and azobenzene drive materials, which may become a future area of research within azobenzene materials.

While the Azo-STF field is still in its infancy, Azo-STFs offers a new approach to solar energy collection and storage. Future research will focus on achieving high ED, long storage life, controlled heat release, and excellent cycle stability. Azobenzene-related materials have great potential and are expected to be widely used in the field of solar energy collection and storage such as deicing, seawater desalination, and heat-driven generators in the era of high sustainable energy.

References

[1] Kolpak AM, Grossman JC. (2013). Hybrid chromophore/template nanostructures: a customizable platform material for solar energy storage and conversion. *J Chem Phys*, **138**(3), pp. 1103–1128.

[2] Hu Z, Zhang DY, Yu L, Huang, YD. (2018). Light-triggered C60 release from a graphene/cyclodextrin nanoplatform for the protection of cytotoxicity induced by nitric oxide. *J Mater Chem B*, **6**(3), pp. 518–526.

[3] Dong LQ, Zhai F, Tang L, Gao WC, Tang JW, Feng YY, Feng W. (2020). Azobenzene-based solar thermal energy storage enhanced by gold nanoparticles for rapid, optically-triggered heat release at room temperature. *J Mater Chem A*, **8**(36), pp. 18668–18676.

[4] Luo YN, Jiang HY, Yu LY, Liu ZC, Yu XY. (2017). Synthesis, crystal structure, and photoelectrical property of a new (3, 6) connected 3D Zn (II) coordination polymer. *Inorg Chem Commun*, **86**, pp. 304–307.

[5] Luo, YN, Jiang HY, Liu ZC, Yu LY, Yu XY. (2018). A new zinc coordination polymer constructed from 4-[(8-hydroxy-5-quinolinyl)Azo]-benzenesulfonic acid: synthesis, structure, and photoelectrical property. *Struct Chem*, **29**, pp. 977–982.

[6] Wang H, Feng YY, Yu HT, Dong LQ, Zhai F, Tang JW, Ge J, Feng W. (2021). Utilisation of photo-thermal energy and bond enthalpy based on optically triggered formation and dissociation of coordination bonds. *Nano Energy*, **89**, pp. 106401.

[7] Wang H, Feng YY, Gao J, Fang WY, Ge J, Yang XY, Zhai F, Yu YF, Feng W. (2022). Metallic-ion controlled dynamic bonds to co-harvest

isomerization energy and bond enthalpy for high-energy output of flexible self-heated textile. *Adv Sci*, **9**(20), pp. 2201657.

[8] Feng YY, Feng W, Noda H, Sekino T, Fujii A, Ozaki M, Yoshino K. (2007). Synthesis of photoresponsive Azobenzene chromophore-modified multi-walled carbon nanotubes. *Carbon*, **45**(12), pp. 2445–2448.

[9] Feng YY, Feng W, Noda H, Fujii A, Ozaki M, Yoshino K, Yoshino K. (2007). Photoinduced anisotropic response of Azobenzene chromophore functionalized multiwalled carbon nanotubes. *J Appl Phys*, **102**(5), pp. 053102.

[10] Kolpak AM, Grossman JC. (2011). Azobenzene-functionalized carbon nanotubes as high-energy density solar thermal fuels. *Nano Lett*, **11**(8), pp. 3156–3162.

[11] Kucharski TJ, Ferralis N, Kolpak AM, Zheng JO, Nocera DG, Grossman JC. (2014). Templated assembly of photoswitches significantly increases the energy-storage capacity of solar thermal fuels. *Nat Chem*, **6**(5), pp. 441–447.

[12] Huang J, Jiang Y, Wang J, Li CH, Luo W. (2017). A high energy, reusable and daily-utilization molecular solar thermal conversion and storage material based on Azobenzene/multi-walled carbon nanotubes hybrid. *Thermochim Acta*, **657**, pp. 63–169.

[13] Jiang Y, Huang J, Feng W, Zhao X, Wang T, Li CH, Luo W. (2019). Molecular regulation of nano-structured solid-state Azo-SWCNTs assembly film for the high-energy and short-term solar thermal storage. *Sol Energy Mater Sol Cells*, **193**, pp. 198–205.

[14] Feng YY, Liu HP, Luo W, Liu EZ, Zhao NQ, Yoshino K, Feng W. (2013). Covalent functionalization of graphene by Azobenzene with molecular hydrogen bonds for long-term solar thermal storage. *Sci Rep*, **3**, pp. 3260.

[15] Luo W, Feng YY, Cao C, Li M, Liu EZ, Li SP, Qin CQ, Hu WP, Feng W. (2015). A high energy density Azobenzene/graphene hybrid: a nano-templated platform for solar thermal storage. *J Mater Chem A*, **3**(22), pp. 11787–11795.

[16] Luo W, Feng YY, Qin CQ, Li M, Li SP, Cao C, Long P, Liu EZ, Hu WP, Yoshino K, Feng W. (2015). High-energy, stable and recycled molecular solar thermal storage materials using Azo/graphene hybrids by optimizing hydrogen bonds. *Nanoscale*, **7**(39), pp. 16214–16221.

[17] Li M, Feng YY, Liu EZ, Qin CQ, Feng, W. (2016). Azobenzene/graphene hybrid for high-density solar thermal storage by optimizing molecular structure. *Sci China Technol Sci*, **59**(9), pp. 1383–1390.

[18] Feng W, Li SP, Li M, Qin CQ, Feng YY. (2016). An energy-dense and thermal-stable bis-Azobenzene/hybrid templated assembly for solar thermal fuel. *J Mater Chem A*, **4**(21), pp. 8020–8028.

[19] Zhao XZ, Feng YY, Qin CQ, Yang WX, Si QY, Feng W. (2017). Controlling heat release from a close-packed bisazobenzene-reduced-graphene-oxide assembly film for high-energy solid-state photothermal fuels. *ChemSusChem*, **10**(7), pp. 1395–1404.

[20] Shekar SC, Swathi RS. (2018). Molecular switching on graphyne and graphdiyne: realizing functional carbon networks in synergy with graphene. *Carbon*, **126**, pp. 489–499.

[21] Hu Z, Zhang DY, Yu L, Huang YD. (2018). Light-triggered C60 release from a graphene/cyclodextrin nanoplatform for the protection of cytotoxicity induced by nitric oxide. *J Mater Chem B*, **6**(3), pp. 518–526.

[22] Zhitomirsky D, Cho E, Grossman JC. (2016). Solid-state solar thermal fuels for heat release applications. *Adv Energy Mater*, **6**(6), pp. 1502006.

[23] Zhitomirsky D, Grossman JC. (2016). Conformal electroplating of Azobenzene-based solar thermal fuels onto large-area and fiber geometries. *ACS Appl Mater Interfaces*, **8**(39), pp. 26319–26325.

[24] Han GD, Park SS, Liu Y, Zhitomirsky D, Cho E, Dinca M, Grossman JC. (2016). Photon energy storage materials with high energy densities based on diacetylene–Azobenzene derivatives. *J Mater Chem A*, **4**(41), pp. 16157–16165.

[25] Jeong SP, Renna LA, Boyle CJ, Kwak HS, Harder E, Damm W, Venkataraman D. (2017). High energy density in Azobenzene-based materials for photo-thermal batteries via controlled polymer architecture and polymer-solvent interactions. *Sci Rep*, **7**, pp. 17773.

[26] Cheng F, Zhang Y, Yin R, Yu Y. (2010). Visible light induced bending and unbending behavior of crosslinked liquid-crystalline polymer films containing Azotolane moieties. *J Mater Chem*, **20**(23), pp. 4888–4896.

[27] Ishiba K, Morikawa M, Chikara C, Yamada T, Iwase K, Kawakita M, Kimizuka N. (2015). Photoliquefiable ionic crystals: a phase crossover approach for photon energy storage materials with functional multiplicity. *Angew Chem Int Ed*, **54**(5), pp. 1532–1536.

[28] Weis P, Hess A, Kircher G, Huang S, Auernhammer GK, Koynov K, Butt HJ, Wu S. (2019). Effects of spacers on photoinduced reversible solid-to-liquid transitions of Azobenzene-containing polymers. *Chem Eur J*, **25**(46), pp. 10946–10953.

[29] Saydjari AK, Weis P, Wu S. (2017). Spanning the solar spectrum: Azopolymer solar thermal fuels for simultaneous UV and visible light storage. *Adv Energy Mater*, **7**(3), pp. 1601622.

[30] Zhou HW, Xue CG, Weis P, Suzuki Y, Huang SL, Koynov K, Auernhammer GK, Berger R, Butt HJ, Wu S. (2017). Photoswitching of glass transition

temperatures of Azobenzene-containing polymers induces reversible solid-to-liquid transitions. *Nat Chem*, **9**(2), pp. 145–151.
[31] Xu, WC, Liu CW, Liang SF, Zhang DC, Liu YZ, Wu S. (2022). Designing rewritable dual-mode patterns using a stretchable photoresponsive polymer via orthogonal photopatterning. *Adv Mater*, **34**(31), pp. 2202150.
[32] Liu H, Feng Y, Feng W. (2020). Alkyl-grafted Azobenzene molecules for photo-induced heat storage and release via integration function of phase change and photoisomerization. *Compos Commun*, **21**, pp. 100402.
[33] Liu N, Chen Z, Dunphy DR, Jiang YB, Assink RA, Brinker CJ. (2003). Photoresponsive nanocomposite formed by self-assembly of an Azobenzene-modified silane. *Angew Chem Int Ed*, **42**(15), pp. 1731–1734.
[34] Muraoka T, Kinbara K, Aida T. (2006). Mechanical twisting of a guest by a photoresponsive host. *Nature*, **440**(7083), pp. 512–515.
[35] Koshima H, Ojima N, Uchimoto H. (2009). Mechanical motion of Azobenzene crystals upon photoirradiation. *J Am Chem Soc*, **131**(20), pp. 6890–6891.
[36] Park J, Yuan D, Pham KT, Li JR, Yakovenko A, Zhou H. (2012). Reversible alteration of CO_2 adsorption upon photochemical or thermal treatment in a metal-organic framework. *J Am Chem Soc*, **134**(1), pp. 99–102.
[37] Brown JW, Henderson BL, Kiesz MD, Whalley AC, Morri W, Grunder S, Deng H, Furukawa H, Zink J, Stoddart J. (2013). Photophysical pore control in an Azobenzene containing metal-organic framework. *Chem Sci*, **4**(7), pp. 2858–2864.
[38] Wang Z, Grosjean S, Bräse S, Heinke L. (2015). Photoswitchable adsorption in metal-organic frameworks based on polar guest-host interactions. *ChemPhysChem*, **16**(18), pp. 3779–3783.
[39] Müller K, Knebel A, Zhao F, Bléger D, Caro J, Heinke L. (2017). Switching thin films of Azobenzene-containing metal-organic frameworks with visible light. *Chem A Eur J*, **23**(23), pp. 5434–5438.
[40] Müller ZK, Valášek M, Grosjean S, Bräse S, Wöll C, Heinke L. (2018). Series of photoswitchable Azobenzene-containing metal–organic frameworks with variable adsorption switching effect. *J Phys Chem C*, **122**(33), pp. 19044–19050.
[41] Khayyami A, Philip A, Multia J, Karppinen M. (2020). Composition-tuned metal-organic thin-film structures based on photoswitchable Azobenzene by ALD/MLD. *Dalton Transact*, **49**(32), pp. 11310–11316.
[42] Rice AM, Martin CR, Galitskiy VA, Berseneva AA, Leith GA, Shustova NB. (2017). Photophysics modulation in photoswitchable metal-organic frameworks. *Chem Rev*, **120**(16), pp. 8790–8813.

Chapter 5

Solar Thermal Fuel Devices

Abstract

After years of research, the development of photothermal materials has become increasingly mature. The high-energy storage density and better cycling stability make azobenzene a material with extremely high practical application value. For this reason, azobenzene materials have been used in thin-film energy storage devices, fabric-type devices, encapsulated devices, and hybrid solar energy conversion devices. As a typical application of solid-state materials, thin-film energy storage devices can be used as de-icing coatings, thermal blankets, and other large-area exothermic devices. Thin films of solar thermal fuels (STFs) based on azobenzene can efficiently trap photons and perform heat release in the solid state. Unfortunately, the solvent-assisted charging process is unavoidable due to the tightly packed structure of the azobenzene derivatives and the high molar extinction coefficient that limits the depth of incident light. In addition, the use of organic solvents undoubtedly aggravates environmental problems and increases costs. Currently, there is a lack of research on solid-state STFs based on azobenzene derivatives. The reported monomer polymer films are prepared by layer-by-layer cross-linking, which requires a complex process. To solve this problem, liquid azobenzene derivatives have been combined with porous flexible templates to give additional functionality to the designed materials while solving the complexity of film preparation. Flexible wearable solar thermal fuel devices (STFDs) have the potential to become heating or heat-generating devices in everyday life in the future. During the research process, it was found that the bonding strength between the azobenzene derivatives and the attached fabric is small and prone to leakage during practical use, which greatly affects the further development of photothermal fabric-type devices. For this reason, the microfluidic chip solar collector draws people's attention. It can be made into a transparent flat plate with meandering channels inside, and the photothermal material can flow freely inside the channels. Recently, the total energy storage efficiency of the liquid-based triple microfluidic chip system, inspired by multi-junction solar cells, can reach 0.02%. The overall energy storage efficiency of the whole operating device is higher than the efficiency of any single-layer device considering that renewable energy sources are mostly intermittent and unevenly distributed geographically. Hybrid solar energy conversion storage devices were created, which can store solar energy and heat water at the same time. With a multi-junction design, an ideal hybrid solar energy conversion storage system can have a maximum storage efficiency of 21% (simulation results). Therefore, investigating how the remaining part of the solar spectrum can be used is worth noting. One approach is to combine it with other technologies to form a hybrid device to improve the utilization of solar energy. Related reports have also demonstrated the feasibility of this type of design.

5.1 Thin-film energy storage device

The progress made in solar thermal fuel (STF) materials has been remarkable. However, it is astonishing that solid-state platforms and their associated applications have received limited exploration thus far. Although recent studies have achieved energy densities of 35 Wh kg^{-1} with semi-solid photolivatable ionic crystals, the full transition to the solid state remains largely unexplored. This transition holds immense potential for integrating STF materials into a wide range of existing solid-state devices. For instance, they can be used as coatings for de-icing or in innovative applications like solar thermal blankets and consumer-oriented heating devices. Thin film energy storage devices exemplify the practicality of solid-state materials in enabling large-scale heat release applications. Homogeneous STF films play a crucial role in efficiently capturing photons, and the subsequent heat release can take place in the solid state or be controlled through *cis*-to-*trans* isomerization. This presents a unique opportunity to seamlessly integrate STF materials into a wide array of existing heating devices.

5.1.1 *Polymer template solar thermal fuel film*

STF materials have the potential to be precisely tailored in their solid state through molecular-level design. However, achieving this goal has proven to be a challenge thus far. Existing methods for creating STFs often involve carbon scaffolds, which not only complicate the synthesis process but also fail to form uniform films. Moreover, these carbon scaffolds increase optical density without achieving optical charging, leading to uncontrollable morphological effects that limit the charging and reversible switching capabilities in the solid state. Similarly, single-molecule films are unable to form uniform layers and tend to crystallize and melt at low temperatures (e.g., azobenzene at approximately 70°C), rendering them less suitable for solid-state applications. Nevertheless, extensive research on azobenzene materials in the solid state for microswitches, microactuators, and sensors provides a valuable foundation. Our hypothesis is that an ideal class of materials for solid-state STF coatings must fulfill the following criteria: (1) enable the formation of smooth films with controlled thickness, (2) exhibit high-temperature flexibility, (3) maintain the heat release properties of molecular STFs to maximize energy density, (4) enable charging, storage, and discharging on relevant timescales, and (5) demonstrate feasibility for large-area heat release applications. To address these requirements, we have chosen to develop a polymer-based solid-state platform that can be finely tuned and scaled to meet the desired objectives. Crucially, the vast array of chemistry available for both the monomer and polymer backbone allows us to adopt the same computational design principles utilized for single-molecule STFs, while also exploring new possibilities in terms of polymer–polymer chain interactions, cross-linking degree, and conjugation within the polymer backbone.

Grossman *et al.* [1] synthesized a polymeric template STF by utilizing free radical polymerization. The main chain of the polymer consisted of a carbon chain, while azobenzene was employed as the side group (depicted in Figure 5.1). The resulting material was transformed into a solid state by

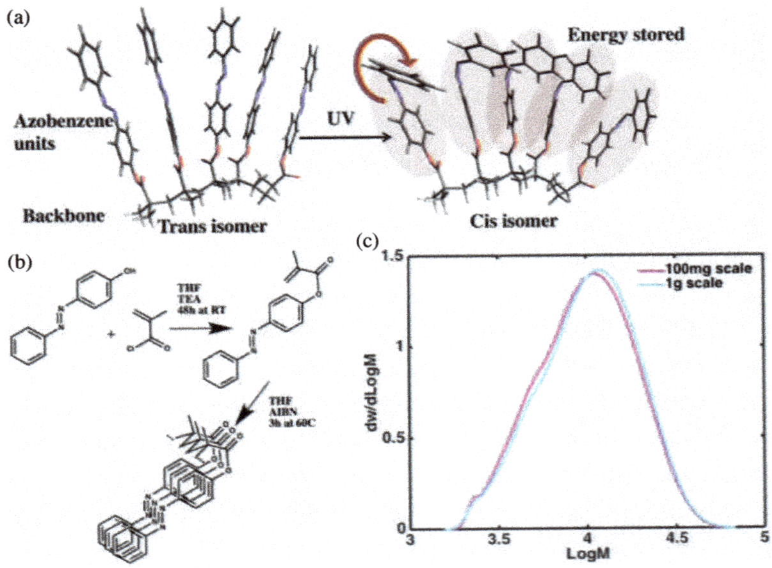

Figure 5.1. Solid-state solar thermal fuel polymer concept. (a) A schematic of an azobenzene polymer consisting of four monomers in the *trans*-state being converted to the *cis*-state upon UV illumination. The result is the apparent rotation of that azobenzenes about the N=N double bond. (b) Chemical synthesis scheme for generating the homopolymer by employing a radical polymerization of an azobenzene monomer. (c) Size exclusion chromatography was performed to analyze polymer samples and deduce the distribution, as well as reproducibility with a scaled-up synthesis. Reprinted with permission from Ref. [1]. Copyright 2013, American Institute of Physics.

dissolving it in toluene and spin-coating it onto a film with a flat surface, no cracks, and uniform thickness at a rotation speed of 700 rpm. This method facilitated the preparation of a STF film. Furthermore, the development of an STF capacitor platform utilizing thin polymer films allowed for charging, discharging, and thermal release in the solid state through optically rechargeable molecules. By employing UV-activated crosslinking, polymeric STF materials achieved uniform morphologies ranging from 100 nm to tens of microns in thickness, providing increased tunability. Studying the charging and discharging properties of the solar thermal energy capacitor (STEC) and the stored thermal energy within it enabled the construction of a macroscopic device capable of generating temperature differences of up to 10°C. This demonstration confirmed the feasibility of employing these devices for solid-state applications. Future approaches could explore the vast potential of the monomer and polymer backbone chemistry, allowing for increased energy density, improved optical chargeability and photostability, as well as the utilization of a wider range of solar spectrum. The exceptional material flexibility and feasibility for solid-state applications make STF materials an attractive option for groundbreaking research in light–matter interactions and the storage and release of renewable energy.

However, it is important to note that azo polymer films present certain limitations compared to other templated STFs. These limitations include relatively low-energy storage density and limited heat release at room temperature. These factors can be attributed to the small number of Azo groups (<30 wt%), the slow rate and low degree of isomerization, and weak intermolecular interactions.

Creating high-energy Azo polymer films while preserving molecular alignment poses a challenge. Enhancing the Azo content to achieve this goal is hindered by the strong interactions among Azo molecules, resulting in diminished film flexibility and formability. As a consequence, these films exhibit inhomogeneity, slow heat release, reduced overall energy, and limited deformability. Although self-assembled Azo polymers have been documented, they suffer from inadequate energy density and subpar film-forming properties. Consequently, the deposition of these polymers onto substrates such as fiber or metal is necessary to form films. Another exploration involves blending Azo polymers with other flexible polymers in the preparation of STF films. Nevertheless, this approach further diminishes energy and power density, and it could potentially influence heat release due to the possibility of phase separation.

The isomerization rates and degrees of Azo polymer films at room temperature are usually very low. To enhance the rate and degree of isomerization for Azo chromophores, optimizing the microstructure in terms of molecular interactions (orientation) and steric hindrance (free volume) has proven to be highly effective. In the realm of STF films, their chargeability (photo-induced *trans*-to-*cis* isomerization) and dischargeability (*cis*-to-*trans* reduction) under various deformations (such as shape, strain, and static/dynamic changes) play a crucial role in the recycling of

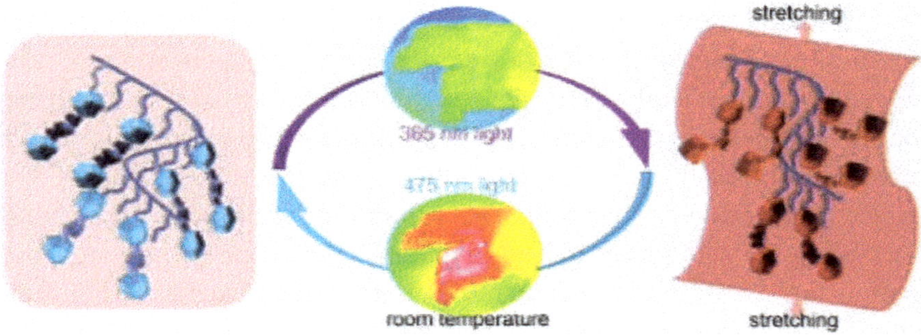

Figure 5.2. Photothermal energy storage and release of polymeric template azobenzene obtained by ring-opening translocation polymerization. Reprinted with permission from Ref. [2]. Copyright 2019, ACS Publications.

solar heat. These flexible STF films can be integrated into various heating systems, accommodating different shapes and deformations. However, research on deformation-controlled solar heat storage and release is limited. In light of this, Feng *et al.* [2] undertook a study as depicted in Figure 5.2. In their work, two azobenzene (PNB-Azo) STFs were developed using poly(norbornene) as the template. The researchers covalently grafted two different Azo compounds onto poly-2-(6-hydroxyhexyl)-norbornane-2,3-dicarboxylic anhydride (PNB) templates, utilizing flexible alkyl spacers. Subsequently, they obtained two homogeneous layers of flexible PNB-Azo films through solution treatment. To investigate stretch-enhanced solar thermal utilization, PNB-Azo-1 and PNB-Azo-2 films were prepared and characterized. PNB was synthesized via open-loop bias synthesis polymerization. Azo-1 and Azo-2 were then covalently grafted onto the PNB template through esterification, yielding long flexible spacers and peak Mn values of 66000 and 68000, respectively. PNB-Azo-1 had a graft density of 48%, exhibiting a relatively high degree of *trans–cis* isomerization (72%) attributed to favorable intermolecular interactions and steric hindrance. This film demonstrated remarkable storage stability. In contrast, PNB-Azo-2 displayed a low degree of D_I (19.2%) and was rapidly reduced due to pull-push electronic interactions.

Moreover, stretching the films further amplified the degree and rate of isomerization by increasing the free volume and facilitating structural transformation. The PNB-Azo films with 20% strain exhibited a combination of high photo-charging efficiency (85%), energy density (49.0 Wh kg^{-1}), power density (161.0 W kg^{-1}), and heat release rate, all induced by blue light (475 nm) irradiation at room temperature. Furthermore, dynamic stretching led to a rapid temperature rise compared to static deformation, resulting in a higher ΔT.

As shown in Figure 5.3, the reversible electrically driven dynamic stretching causes the PNB-Azo-1 finger to release heat at a high rate, resulting in a temperature rise of 1.5°C. This indicates that PNB-Azo-1 STF films can be used to control the temperature change at room temperature in both static and dynamic stretching states. By controlling the deformation, flexible stretchable PNB-Azo films can be developed for high-power dynamic solar heat sources.

5.1.2 *Solar thermal fuel film with nanocarbon template*

In addition to the previously mentioned polymer templates utilized for azobenzene STFs, nanocarbon-based materials have also been employed in the creation of STFs containing azobenzene compounds. The intermolecular interaction and steric hindrance improve the energy density and stability of these materials. Concerning Azo polymers, the photothermal properties typically depend on the spatial configuration homogeneity and microstructure of the film. In recent years, the synthesis of homogeneous Azo polymer films has been achieved by introducing a spacer layer between the Azo group and the polymeric backbone chain, effectively reducing the rigidity of the Azo polymer. While the energy density of azo polymer films is relatively low, the heat stored within them can be released by controlling the isomerization process. Comparatively, one-dimensional (1D) or two-dimensional (2D) carbon nanotemplates are regarded as favorable platforms for assembling densely packed ordered Azo molecules. By controlling the spatial configuration and functional groups, high molecular density grafting enhances intermolecular interactions and steric hindrance. This interaction not only decreases the energy of the *trans*-isomer but also stabilizes the *cis*-isomer, leading to significant improvements in the molecule's ΔH and t1/2. Furthermore, the charge transfer between Azo and graphene can catalyze isomerization. Thus, templated Azo graphene components exhibit substantial potential in the

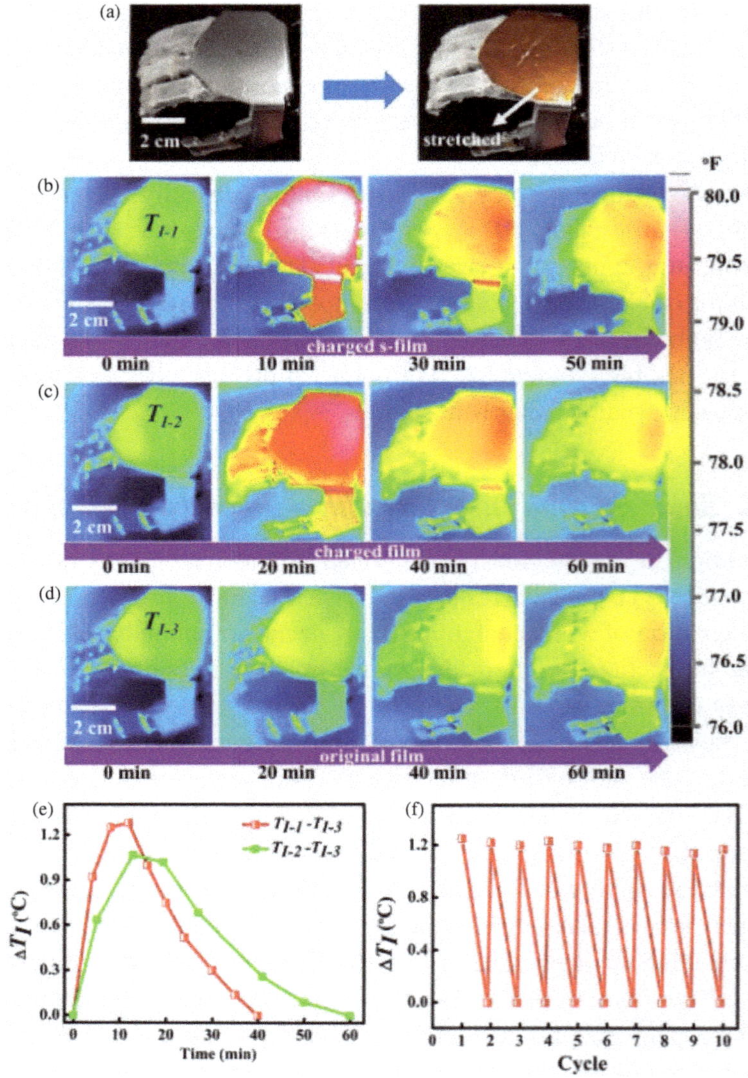

Figure 5.3. Heat-release performance of PNB-Azo film in static state under blue-light irradiation at room temperature. (a) Optical image of PNB-Azo-1 film on the back of the robot; time-evolved IR thermal imaging of (b) the charged PNB-Azo-1 s-film, I-1, (c) the charged PNB-Azo-1 film without stretching, I-2, and (d) uncharged PNB-Azo-1 film without stretching, I-3. (e) ΔTI at different times and (f) its change in 10 cycles. The temperature is obtained by averaging the data four times. Reprinted with permission from Ref. [2]. Copyright 2019, ACS Publications.

development of high-energy stable STFs through optimized interactions between molecules. Previous studies have demonstrated that the aggregation of Azo molecules can impede isomerization, resulting in prolonged photon absorption. Hence, optimizing the molecular structure and spatial configuration of Azo on carbon templates holds great significance in promoting isomerization and enhancing both intramolecular and intermolecular interactions, consequently improving the photothermal properties of the material. It remains a challenge to control heat release and accumulation in the solid state, hindering the utilization of STF applications unless materials are designed at the molecular level. Achieving closed-cycle STF systems based on Azo graphene films for light collection, storage, and heat release has been scarcely reported to date. The formation of homogeneous films and the improvement of isomerization in the solid state present ongoing obstacles in research. These challenges, combined with limited knowledge in the field, impede the further advancement of STF films. Therefore, proper molecular engineering for high-energy storage and well-designed integrated films for heat release are crucial prerequisites to effectively utilize high-energy solid-state STFs.

Zhao *et al.* [3] present a templated assembly method involving densely packed bisazobenzene covalently grafted onto reduced graphene oxide (rGO-bisAzo). Two types of Azo complexes, namely rGO-bisAzo-1 and rGO-bisAzo-2, were prepared to examine the impact of intermolecular interactions on their photothermal properties. The researchers successfully fabricated homogeneous STF films with a high graft density of 1/23 using compactly arranged rGO-bisAzo components. The reduced rGO-bisAzo-2, synthesized through intermolecular and intramolecular hydrogen bonding as well as steric hindrance, exhibited remarkable energy density and power density, namely 131 Wh d kg^{-1} and 2517 W kg^{-1}, respectively. This material also displayed a prolonged half-life of 37 days and 50 cycles. The solid rGO-bisAzo film, on the other hand, demonstrated lower isomerization (98 Wh kg^{-1}) due to steric hindrance, resulting in a 25% reduction in energy density compared to the powdered sample. The researchers conducted investigations into the closed cycles of UV irradiation, storage, and heat release in the STF films. Figure 5.4 demonstrates that the rGO-bisAzo films efficiently release and accumulate heat, with a maximum temperature difference (ΔT) of 15°C. Furthermore, when

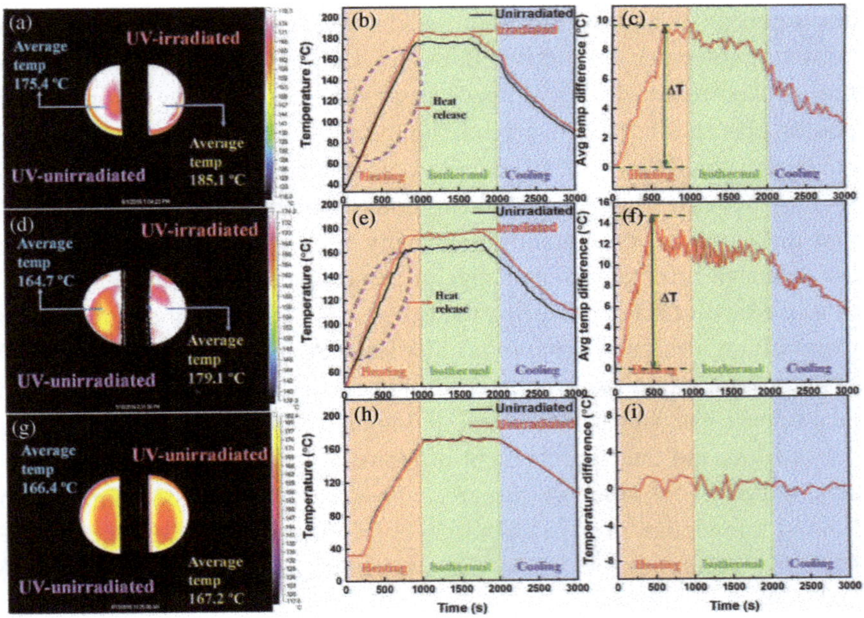

Figure 5.4. Macroscopic solid-state rGO-bisAzo heat release. First column: top-view IR heat map of films of (a) rGO-bisAzo-1, (d) rGO-bisAzo-2 with and without UV irradiation; (g) both semicircular rGO-bisAzo-2 films absorbed no UV light. The heat maps show the maximum rising temperature between areas; the color bar indicates the relative magnitude of the heat released. Second column: average temperature of (b) rGO-bisAzo-1, (e) rGO-bisAzo-2, and (h) unirradiated rGO-bisAzo-2 at different stages. Third column: average ΔT of (c) rGO–bisAzo-1, (f) rGO–bisAzo-2, and (i) unirradiated rGO-bisAzo-2. Reprinted with permission from Ref. [3]. Copyright 2022, Elsevier.

the ambient temperature difference exceeds 100°C, the ΔT of the film remains above 10°C for a duration of 30 min. To fabricate high-performance STFs, molecular engineering with high energy and long-term storage properties, alongside optimized microstructure that promotes significant isomerization, are essential considerations. The ability to control heat release in solid-state components serves as a solid foundation for the successful development of STFDs.

In addition to the challenges of molecular engineering for high energy density, achieving high rates of solid-state heat release and reversible temperature control is of paramount importance for the advancement of STF applications. The limited isomerization rate and hindered mobility, primarily due to steric hindrance, constitute crucial barriers impeding the rapid release of solid heat. The conventional approach entails the gradual release of a small amount of heat over a short period of time, known as high-temperature induction, or an extremely slow release at a constant rate, known as visible light induction. However, this approach results in a significant heat loss to the surrounding atmosphere, thus hindering the capture of heat energy required for achieving substantial temperature increases. Consequently, the temperature rise of STFs films at high temperatures, such as 100°C, is merely 2°C due to their low power density. Moreover, the low rate of heat release at temperatures ranging from 25°C to 60°C further diminishes the efficiency of STF utilization. Hence, exploring exothermic phenomena at lower temperatures holds practical significance for the development of thermal management applications. A feasible and effective approach to modulating reversible temperature changes through heat release represents a crucial milestone toward advancing STFs applications. Despite extensive research on the relationship between performance (energy density and power density) and isomerization (degree and rate), there is a limited number of research reports delving into the charge–discharge process of STFs. Furthermore, the control of heat release at different rates for varying temperature increases remains an unexplored area. Current literature lacks reports on the utilization of exothermic STFs for inducing reversible thermochromic displays through multiple photothermal energy cycles. Harnessing solar energy as a heat source represents a significant method for storing and providing clean thermal energy. However, the challenge lies in effectively combining high energy storage and high rate heat release in STFs, which impedes their application in temperature-sensitive or temperature-responsive systems. In a recent study by Yang *et al.* [4], the authors pioneered the utilization of photothermal energy to achieve closed-loop utilization of thermochromic displays by optimizing the solid-state high-rate thermal output of STF films. Through careful control of molecular interactions, they were able to prepare films with a maximum energy density of 150.3 Wh kg^{-1}, a long half-life of 1250 h, and a power density of 3036.9 W kg^{-1}. By releasing heat and increasing the temperature by 6–7°C, these STF films demonstrated reversible color changes in complex thermochromic pattern displays. The variable heat release was achieved through precise control of the heating rate and temperature, effectively utilizing photothermal energy. Additionally, the incorporation of the tri-Azo module enabled efficient recycling of photothermal energy, thus enabling its application in thermal management through photothermal power supply.

The researchers studied a closed cycle of light collection ("charge"), storage, and then heat release ("discharge") of STF films to demonstrate their ability to use photothermal energy to raise temperatures. Homogeneous STF films (30 mm × 30 mm × 25 mm, 365 nm transmittance of 3.17%) were prepared. The right half (30 mm × 15 mm) is "charged" (12 h) with UV light, while the left half is covered with a mask. The researchers conducted an investigation on the closed cycle of STF films, encompassing light collection (charging), energy storage, and subsequent heat release (discharging), to showcase the utilization of photothermal energy for temperature elevation. Homogeneous STF films (30 mm × 30 mm × 25 mm, with 3.17% transmittance at 365 nm) were fabricated. The right half of the film (30 mm × 15 mm) was subjected to UV light exposure for 12 h, representing the charging process, while the left half was masked. Following 24 h of darkness, Film-I was gradually heated on a hot plate, with temperatures ranging from 40°C to 63°C at a rate of 1°C per min, and subsequently maintained at 63°C for 23 min (condition I). Film-II underwent a similar heating process, from 40°C to 77°C, and then maintained at 77°C for 37 min (condition II). To monitor the temperature difference (ΔT) between the left and right halves, a high-resolution (±0.03°C) infrared thermal imaging camera was employed. The time evolution ΔT under both conditions is presented in Figure 5.5. Notably, the exothermic process of tri-Azo/reduced GO STF films commenced at temperatures ranging from 50°C to 55°C, consistent with the results of differential scanning calorimetry (DSC). Upon reaching the first stage temperatures of 63°C and 77°C, the maximum ΔT of the film (Figures 5.5(a) and 5.5(b)) was observed to be 2–3°C under condition I and 6–7°C under condition II. This signifies that the heat output of the STF film contributed to temperature elevation. Previous investigations into STFs have reported a rapid exothermic increase of 2°C ΔT at high temperatures (>100°C). However, by employing high-energy storage and controlled isomerization, the researchers achieved a high-rate heat output, resulting in a ΔT of 2–3°C at relatively lower temperatures (55–63°C), with an isomerization degree of only 24.0% below 63°C. Subsequently, during the second stage, when the

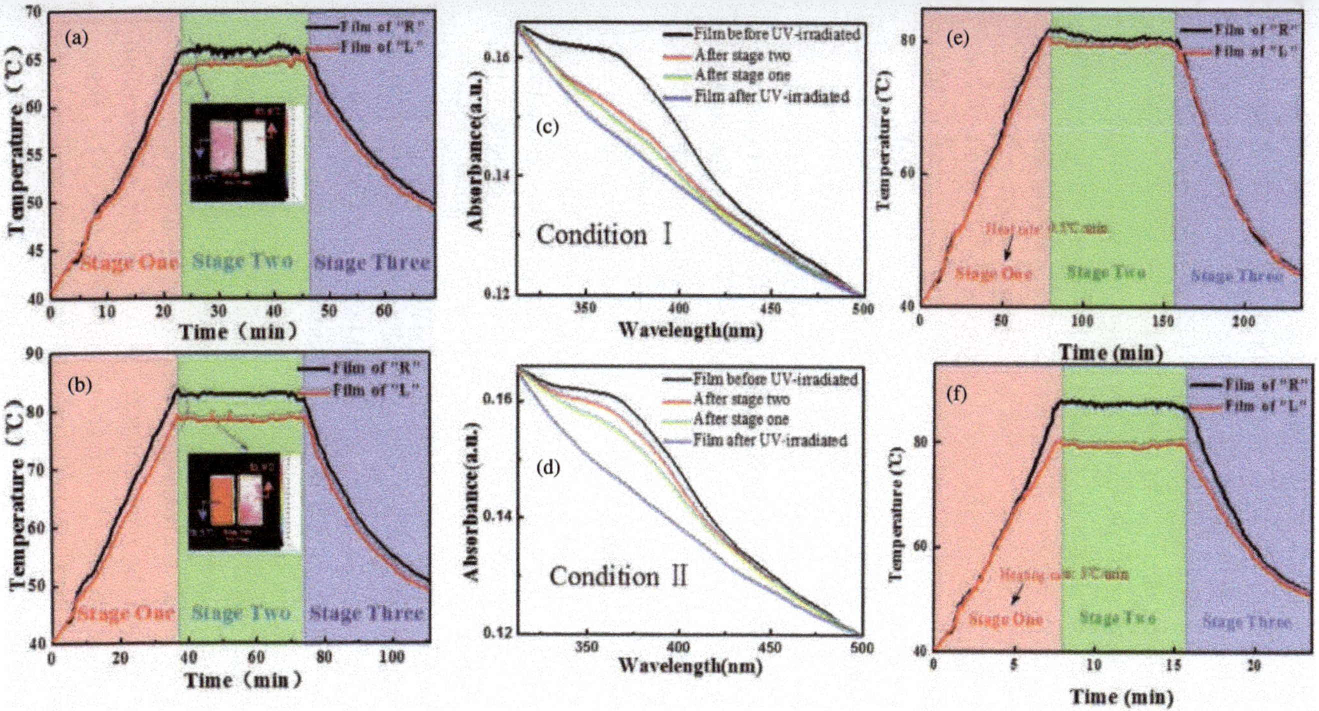

Figure 5.5. Solid-state controllable heat release of the tri-Azo1:68/rGO film under (a and c) Condition I and (b and d) Condition II, inducing temperature changes tracked by IR. The IR image of the tri-Azo1:68/rGO film after Stage one. The color bar indicates the relative magnitude of the temperature. The difference in the average temperature between the left and right sides is 2.5°C (Condition I) and 6.4°C (Condition II), respectively. Temperature changes in the tri-Azo1:68/rGO film are induced by the solid-state heat release. The film exhibits a maximum DT of 2–3°C and 6–7°C in Stage two (in which the temperature is held constant at 23 min and 37 min, respectively). Time-evolved absorption spectra of tri-Azo1:68/rGO films after different stages. And the average temperature of (e) the film with a heating rate of 0.5°C min^{-1} and (f) the film with a heating rate of 5°C min^{-1}. Reprinted with permission from Ref. [4]. Copyright 2019, The Royal Society of Chemistry.

hot plate temperatures were set at 63°C and 77°C, the film maintained a ΔT of 2–3°C and 6–7°C for 23 and 37 min, respectively. The heat release during the second stage was confirmed by time-evolved absorption spectra (Figures 5.5(c) and 5.5(d)). The utilization of solid-state heat output from low-temperature-induced composite films, rather than powders, holds pivotal importance for practical applications. Additionally, the researchers examined temperature changes at different heating rates (0.5°C/min and 5°C/min, Figures 5.5(e) and 5.5(f)), tracking the corresponding temperatures using high-resolution infrared images. Notably, higher heating rates expedited the heat release rate of the tri-Azo/reduced GO film. Specifically, for stage 1 and stage 2 (Figures 5.5(e) and 5.5(f)), the 0.5°C/min heating rate required 80 min and 8 min, respectively.

The unique structure of carbon nanotubes (CNTs) results in a closed cage with a delocalized π-electron cloud along their axis. This distinctive property, combined with their exceptional electrical and photoconductivity characteristics, makes CNTs ideal candidates for one-dimensional materials in optical nanodevices. Recent research on CNTs has focused on their optical polarization rather than their cylindrical nanostructure, particularly in relation to their significance in polarization-sensitive optical devices. In the interband energy region below 3.3 eV, single-walled carbon nanotubes (SWCNTs) exhibit weak optical absorption of cross-polarized light in both parallel and perpendicular orientations to the tube axis. However, the mechanism behind this effect and the baseline absorption from parallel polarized light have not yet been fully clarified. Previous studies wrongly assumed that polarized light perpendicular to the CNT axis is not absorbed, but this is not the case for interband and baseline absorption. Therefore, it is crucial to comprehend the fundamental anisotropic properties of polarized light in CNTs to advance their application in advanced polarization-sensitive optical nanodevices.

The use of azobenzene chromophores in optical data storage and liquid crystal applications has been extensively studied. However, limited experimental evidence exists regarding the correlation between the polarization of CNTs and the nanostructure, as well as the polarization absorption of photoresponsive chromophores attached to CNTs. Feng *et al.* [5] conducted an investigation into the photophysical processes of Azo attachment to polymers and organic molecules. In a previous study, they reported the synthesis of azobenzene chromophores covalently functionalized onto multiwalled carbon nanotubes (Azo-MWCNTs) through an amidation reaction. The optical properties of the azobenzene–CNT composites were characterized using UV–Vis absorption and photoluminescence (PL) spectroscopy. The polarization absorption of the Azo-multiwalled carbon nanotube (MWCNT) composite was examined through the phototropism of azobenzene under polarized UV light.

The optical polarization characteristics of composites consisting of Azo-functionalized MWCNTs were explored in this study. By covalently bonding Azo molecules to the side walls of the CNTs, well-dispersed Azo-MWCNTs were obtained, as confirmed by transmission electron microscopy (TEM) images. The redshift observed in the UV–Vis absorption spectrum of Azo CNT solutions can be attributed to the overlapping π-electron conjugation. Additionally, an enhancement of redshifted emission was observed in the PL spectra. The photoanisotropy of the cross-polarization absorption spectrum of the Azo-functionalized CNTs primarily arises from the increase in the dipole transition parallel to the polarization optical axis and the decrease in the transition perpendicular to the optical axis. By studying the anisotropy of polarized light, valuable insights into the polarized light properties of CNTs were obtained. In the investigation of solid STF films, it was discovered that azobenzene compounds not only enable photocontrolled heat storage and release but also exhibit photoconductive effects for light modulation. The photochromic properties of azobenzene molecules have been extensively studied in various light-driven applications such as optical switches, micropatterns, and reversible optical storage. Under UV light, azobenzene units can undergo *trans–cis*-photoisomerization, resulting in significant structural rearrangement. Importantly, recent studies have demonstrated that the electronic properties of CNTs can be modulated by conformational changes in the sidewall azophenyl groups. Therefore, the photoisomerization of azophenyl groups on graphene surfaces plays a vital role in potential applications of graphene in optical and photonic devices. However, the optical manipulation of electric properties in such hybrid materials remains largely unexplored.

Zhang *et al.* [6] synthesized a GO-Azo hybrid material through an amide linkage. TEM and XRD analysis showed that the equilibrium of

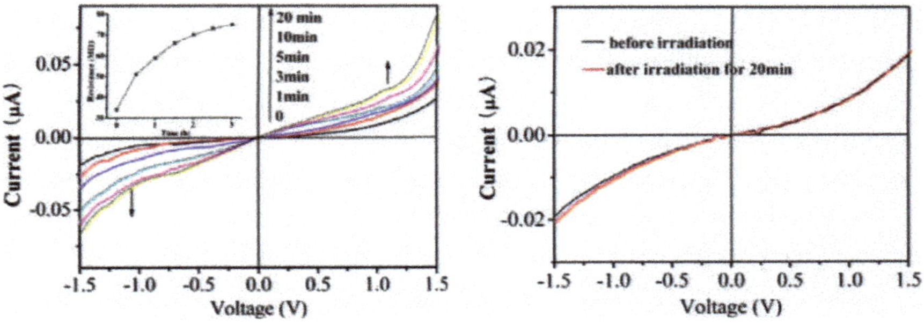

(a) (b)

Figure 5.6. *I–V* characteristics of the devices with the structures of (a) ITO/GO–Azo/ITO and (b) ITO/GO/ITO upon UV irradiation at 365 nm. Arrows indicate changes upon irradiation. Inset figure shows the resistance change of the GO-Azo film with different recovery time. Reprinted with permission from Ref. [6]. Copyright 2010, Elsevier.

electrostatic repulsion between the lattice of GO was destroyed, and a graphite-like structure was formed. GO-Azo hybridization has reversible photoisomerization due to covalent bonding of Azo hybrid groups. Due to the presence of Azo groups, optical modulation of conductance is achieved by irradiation of GO-Azo films for different times. It is believed that the isomerization of Azo group is the main cause of current change.

This study focuses on the conductance modulation of the GO-Azo complex through photoisomerization-induced conformational changes. A simple ITO/GO-Azo/ITO device was fabricated by sandwiching a thin film of GO-Azo between two ITO contacts. When exposed to UV light, the GO-Azo device exhibited distinct *I–V* characteristics, as shown in Figure 5.6(a). The current increased significantly under UV irradiation, accompanied by a conformational change of the Azo unit from *trans*-to-*cis*. After 20 min of UV irradiation, the current under both positive and negative bias doubled. Furthermore, a gradual recovery of the film resistance was observed, as depicted in the illustration, indicating a steady increase in resistance (approximately 45% after 3 h) at a positive bias of 1V. As a control experiment, GO without Azo groups was subjected to identical conditions, but no significant change in current occurred after 20 min of UV irradiation (Figure 5.6(b)). It has been established that the *cis*-Azo isomers exhibit better electrical conductivity than the *trans*-Azo isomers. In the case of GO-Azo, UV light triggers the conversion of *trans*-Azo to the more conductive *cis*-form, leading to a reduced length of the Azo molecules attached to the GO surface. Simultaneously, the conformational rearrangement of the Azo group on GO facilitates π-stacking within the GO lattice. The shortened tunneling barrier length of the Azo molecules, combined with structural rearrangement, creates a favorable environment for electron transfer and consequently increases the current. In the absence of UV light, the *cis*-Azo units revert to the *trans* form, resulting in an increase in film resistance. The reversible changes in membrane conductance offer a promising avenue for modulating the optical and electronic properties of graphene.

Azobenzene derivatives are widely recognized as photosensitive materials due to their ability to undergo photoactive motion upon excitation. This motion results in the anisotropy of the dimeric state and polarization of the chromophore portion, which can be directly applied to various "smart" material applications. These materials also find use in detecting and quantifying anisotropic external disturbances in "sensing" polymer systems. The photocontrollability of azophenyl polymers is a crucial characteristic, as it enables the generation of optical or electrical anisotropy based on molecular orientation. Polymers incorporating azobenzene molecules take advantage of directional interactions and exhibit anisotropic conductive, mechanical, or topographic responses. The exceptional photoconductivity and photosensitivity of azobenzene polymers are attributed to the electrical bistability of the azobenzene chromophore, primarily resulting from photoinduced *cis*-isomerization and subsequent charge capture. Previous research has proposed a photonic device driven by the photoisomerization of azobenzene, demonstrating dual-frequency switching properties and redirection of dielectric anisotropy under unpolarized UV light. Notably, the vertical permittivity of the material remains stable across different frequencies, whereas the parallel permittivity decreases with increasing frequency. The regulation of the permutation in azophenyl polymer systems has garnered significant attention in recent years. Hence, the structural design aimed at enhancing the optical switching conductivity of azobenzyl materials holds great potential for realizing ideal anisotropic membrane materials with optical switching conductivity.

The optical switching characteristics of anisotropic polymer azobenzene materials are influenced by several factors, including the properties of the external light stimulus (such as wavelength, intensity, and polarization), the molecular structure of azobenzene, and the morphology of the material (including orientation and thickness). Linearly polarized UV light or directional templates are commonly employed to orient the Azo phenyl polymer chains in a preferred direction. The use of alignment templates can overcome the limitations of light-induced redirection achieved by linearly polarized UV light alone. This makes aligned substrates an excellent solution for achieving strong, fast, reversible photoresponsive, and anisotropic materials. In the context of azobenzene materials, composites containing highly aligned CNTs exhibit anisotropic conductivity. To obtain stable anisotropic properties, it is crucial to improve the anisotropic photoresponse switchability of azophenyl materials. By doing so, the desired optical switching and anisotropic film characteristics can be achieved. These materials are expected to find applications in switchable molecular sensors, where their unique properties can be harnessed.

Controlling the optical switching conductance of anisotropic materials is a significant challenge in the development of advanced

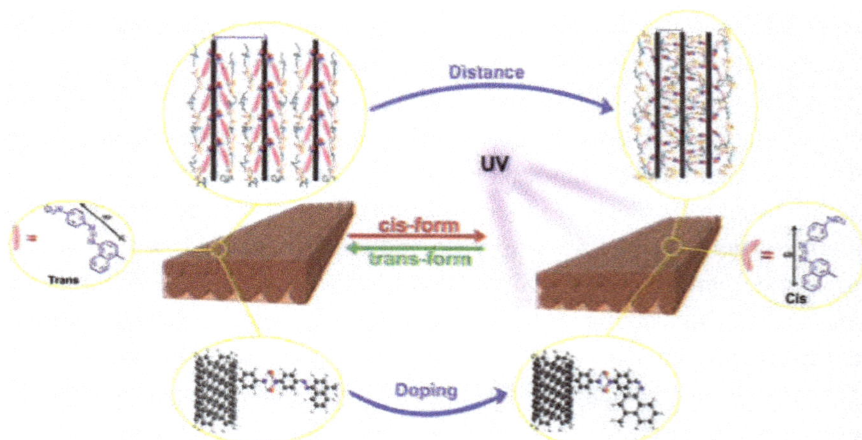

Figure 5.7. Conductive mechanism of the anisotropic Azo-PMMA/HACNT composite film. Reprinted with permission from Ref. [7]. Copyright 2017, The Royal Society of Chemistry.

photodetectors, photogated transistors, and energy storage devices. In a study conducted by Li *et al.* [7], the lateral walls of horizontally aligned carbon nanotubes (HACNTs) were densely filled with a transparent and flexible azobenzene–polymethyl methacrylate (Azo-PMMA) photosensitive film. This configuration resulted in an anisotropic conductivity (σ). Under UV irradiation, the composite film exhibited a steady increase in conductivity (σ) along both the longitudinal and transverse directions. However, after irradiation, the conductivity decreased continuously. The photoswitch conductance demonstrated excellent cycling performance for more than 25 cycles, which correlated with the isomerization of azophenyl groups in Azo-PMMA. Photoswitching conductance was observed in two directions, arising from the photoinduced switching to a larger proportion of conducting *cis*-Azo isomers, as well as the photodoping and shrinkage of the film during *trans* to *cis*-isomerization. Additionally, the anisotropic Azo-PMMA/HACNT composite films exhibited reversible photoswitching conductivity and stable switching amplitude in both the longitudinal and transverse directions. This behavior can be attributed to the combined effects of aligned nanotubes, increased conductivity of *cis*-Azo units, photodoping, and film shrinkage resulting from the photoisomerization of Azo units. Furthermore, the photoswitching conductance of the anisotropic Azo-PMMA/HACNT composite films is primarily influenced by differences in the longitudinal and transverse directions. The variation in photoinduced conductivity (σII) is primarily caused by *cis*-isomerization of azobenzene and doping modulation during film shrinkage. In the transverse direction, the change in conductivity (σI) is mainly influenced by film shrinkage and doping modulation. These findings offer promising prospects for the development of photo-switched aligned composite films. The photoresponsive anisotropic Azo-PMMA/HACNT composite films exhibit desirable photoswitching conductance and hold potential for applications in advanced organic photovoltaic devices and photoswitching nanoelectronic devices.

5.2 Fabric-type device

In the pursuit of increasing the energy density of nanohybrid STFs, the use of organic solvents for the charging process becomes inevitable due to the compact structure of azobenzene derivatives and their high molar extinction coefficient, which limits the depth of incident light. However, the reliance on organic solvents raises environmental concerns and escalates costs. Therefore, the exploration of optical charge and discharge in solid-state STFs holds greater promise and practicality, and further advancements are necessary. Unfortunately, the research on solid-state STFs based on azobenzene derivatives remains limited. Currently reported STF monomer polymer films are prepared using a complex layer-by-layer cross-linking process. Additionally, the creation of homogenous amorphous films of STFs using large volumes of AZ derivatives through spin-coating necessitates the use of inflexible quartz substrates. In this regard, attention should be directed toward photoliquefied AZ (PLAZ) derivatives, which can convert from a solid to a liquid state at room temperature and offer inherent advantages in solvo-free *trans* conversion. Photoinduced *cis*-isomers help lower the melting point of the mixture, while liquid PLAZ derivatives provide more free volume for further photoisomerization, creating a positive feedback loop. Furthermore, the phase transition characteristics of PLAZ derivatives improve the energy storage density (ESD) of the system, akin to the energy storage mechanism of phase change materials (PCMs). However, the fluidity of liquid PLAZ derivatives may pose challenges to the morphology of STFDs. To overcome this issue, combining PLAZ derivatives with porous flexible templates can address the concern of detrimental effects on device morphology while adding additional functionality to the designed material. The development of flexible wearable STFDs, serving as heating or cooling devices in daily life, is an area that lacks comprehensive research and exploration.

Yu *et al.* [8] successfully developed and fabricated a solid-state shape transition fluid device by leveraging a flexible fabric template in combination with a PLAZ derivative. This innovative design eliminates the need for organic solvents in the charging process, as the PLAZ derivative undergoes a photoinduced phase transition. To create the flexible shape transition fluids, the researchers utilized a cloth template and hot-melted the PLAZ derivatives at 60°C, without the requirement of any organic solvents (see Figure 5.8). In this study, the 4-methoxy-azobenzene derivative was selected as the energy storage component due to its *trans*- and *cis*-isomers having melting points of 53°C and 25°C, respectively.

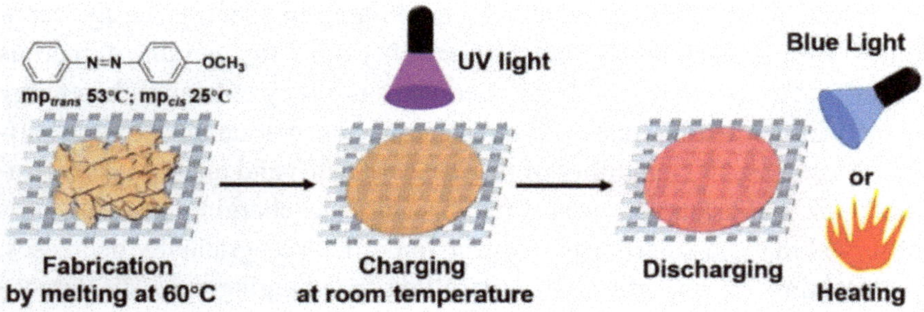

Figure 5.8. Schematic of the flexible STFs' fabrication, charging, and discharging processes. The PLAZ derivative (4-methoxyazobenzene) powder was melted at 60°C and absorbed by the fabric. Then, the room temperature charging under UV irradiation and discharging stimulated by blue light or elevating temperature were demonstrated. Reprinted with permission from Ref. [8]. Copyright 2019, Wiley.

Notably, Cammenga *et al.* discovered pseudobinary fusion graphs within the *cis*/*trans* AZ derivative system, allowing for the possibility of AZ mixtures having lower melting points compared to the individual *trans*- and *cis*-isomers. Consequently, while the *cis*-4-methoxy-azobenzene derivative's melting point is slightly above room temperature, the PLAZ derivatives can still be effectively liquified through photostimulation. The resulting composites, generated by incorporating the PLAZ derivatives into fabrics, exhibit remarkable flexibility and can be produced on a large scale using readily available fabric templates. Importantly, the PLAZ derivatives are evenly dispersed within the fabric texture's gaps. To accurately measure the energy storage capacity of the flexible shape transition fluids, a standardized set of DSC samples was prepared. Each sample consisted of a circular fabric template (approximately 3 mg) punched to fit an aluminum crucible, with the addition of PLAZ derivative powder (about 2 mg) applied onto the fabric and subsequently melted. By varying experimental variables such as UV irradiation time and intensity, DSC scanning rate, and dark time under identical charging conditions, the samples were further divided into three subgroups. This division allows for the exploration of the relationships among incident energy, energy density, solar energy conversion efficiency, and shape transition fluid lifespan without further external stimulation.

Flexible STFDs were created by combining the energy storage capability of PLAZ derivatives with the flexibility of fabric, yielding numerous benefits. The utilization of fabric templates not only enhances the energy density of the flexible materials but also extends their energy storage half-life. Moreover, the reliability of the measurements and analysis specifications for STF samples is ensured. During the charging process, it is observed that the solar energy conversion efficiency gradually decreases, highlighting the importance of controlling the charging time. To demonstrate the heat release of flexible STFDs, infrared cameras are employed under various stimuli. The heat release rate can be controlled by adjusting the intensity of blue light irradiation or the excitation temperature. Importantly, the entire charging and discharging process of STFs is accomplished without the need for solvents. Furthermore, the phase transition characteristics of PLAZ derivatives contribute to the enhancement of the energy storage capacity of AZ-based STFs, while the fabric template effectively addresses the issue of liquid-PLAZ derivatives' fluidity after charging. The energy storage capability of flexible STFs is durable, exhibiting promising potential for various applications in daily life. In future developments, the transparent or selective light absorption coating can be applied to the surface of flexible STFs. This coating would simultaneously address the problem of *cis–trans* isomerization within the solar spectrum and enhance solar energy conversion efficiency. Additionally, the introduction of alternative flexible templates and PLAZ derivatives with higher optical heterogeneous quantum yields and increased energy densities can further improve the energy storage capacity of the device. Ultimately, this flexible STFD holds significant promise for widespread implementation in areas that benefit from solar energy.

Flexible self-heating textiles play a crucial role in intelligent flexible equipment, skin clothing, and soft robots. In low-temperature environments, efficient energy collection and utilization are essential for achieving continuous and controllable heat output. Intelligent self-heating textiles possess a unique advantage in recycling light energy by means of light collection, low-temperature storage, and controlled heat release, rather than relying on the photothermal effect associated with electrically heated conductors and insulating materials.

Although the polymer template azobenzene, with its reversible *trans* (*E*) to *cis* (*Z*) photoisomerization and adaptable molecular interface interaction between the polymer and textiles, is an ideal photoresponsive material for uniformly coating flexible self-heating textiles, its limited molecular levels, low-energy capacity (<200.0 J g^{-1}), and inadequate low-temperature heat release ability (low isomerization rate) hinder its widespread application in self-heating textiles focused on thermal management in low-temperature settings. To address these limitations, Wang *et al.* [9] propose a novel strategy that involves controlling photodynamic bonds to optimize the energy levels of sulfonic acid-azobenzene grafted onto polyacrylic acid (PAzo) ions. These ions interact with various metal ions (Mg^{2+}, Ca^{2+}, Ni^{2+}, Zn^{2+}, Cu^{2+}, and Fe^{3+}), and the dynamic bonds formed include both dynamic ionic bonds and metal–ligand bonds. Alkali metal ions tend to form ionic bonds, while transition metal ions lean towards metal–ligand bonds. By leveraging the ability to photocontrol the reversible formation and deionization of dynamic bonds, the energy gap between *E*- and *Z*-isomers is widened, leading to enhanced energy storage capabilities. The impact of dynamic bonds with different binding energies on the kinetics of photoisomerization and bond enthalpy is thoroughly investigated. Additionally, the dynamic interaction reduces the thermal barrier of *Z–E* isomerization and enables reversible dynamic bond crosslinking of polymer chains. The cyclic utilization of isomerism energy and dynamic bond enthalpy results in PAzo-M demonstrating a high rate of structural transformation at low temperatures and the release of high-powered heat under light stimulus. Finally, the controlled self-heating characteristics of PAzo-M coating on flexible textiles at temperatures ranging from −5.0°C to 5.0°C are demonstrated, showcasing the potential of flexible self-heating textiles in wearable intelligent devices.

The research focused on investigating the self-heating properties of nylon fabric (NF) by coating it with PAzo-Mg to form NF@PAzoMg. Achieving a uniform coating on textiles is a crucial prerequisite for long-term thermal management in wearable devices (Figure 5.9(a)). Unlike mAzo and PAzo powders, PAzo-Mg has the ability to form a self-supporting film with a smooth surface. The SEM image in Figure 5.9(d) demonstrates that a PAzo-Mg layer, approximately 2 μm thick, was uniformly assembled on the surface of individual nylon fibers using a 5–10 mg ml^{-1} PAzo-Mg solution. The intermolecular hydrogen bonding between PAzo-Mg and polyamide facilitates strong adhesion of the PAzo-Mg coating to the nylon fiber, preventing its aggregation on the fiber or dispersal

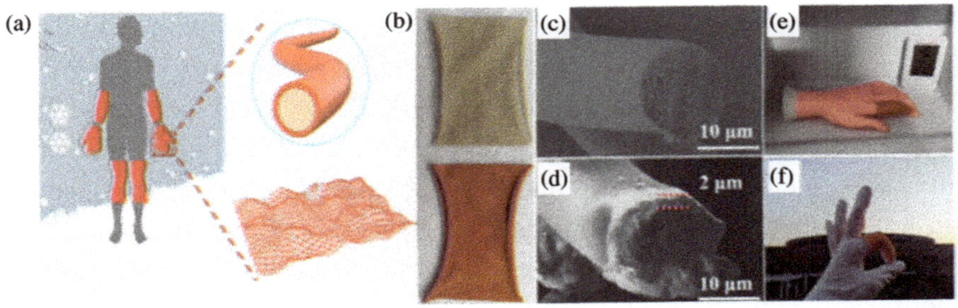

Figure 5.9. (a) Schematic and (b) optical image of the NF@PAzo-Mg textile. (c, d) SEM images of the cross-section of NF and NF@PAzo-Mg with a coating of 2 μm thickness. Image of NF@PAzo-Mg (e) on model hands (indoors) at 0.0–4.0°C and (f) real hands (outdoors) at –5.0 to –3.0°C. Reprinted with permission from Ref. [9]. Copyright 2022, Wiley.

within the fabric gaps. The concentration of the PAzo-Mg solution is carefully regulated to maintain coating uniformity. By achieving an appropriate thickness, PAzo-Mg adheres evenly to the fiber surface, promoting efficient light collection and facilitating a high degree of isomerization. The flexible NF@PAzo-Mg exhibits a tensile strength of 0.42 MPa at a high strain of 90% and demonstrates elastic deformation even after 100 cycles. This elasticity arises from the stretching properties of the woven structure. Therefore, the flexible NF@PAzo-Mg is well-suited for integration into smart wearable devices, ensuring a snug fit and compatibility.

Previous research has predominantly focused on the exothermic properties of thin films, specifically those based on photothermal conversion at relatively high temperatures (>25.0°C). However, the challenges associated with assembly and the limitations of high-temperature exotherm have hindered the potential applications of these films in smart wearable devices. A key obstacle faced by flexible self-heating textiles is the ability to regulate low-temperature exotherm without the need for an external power supply. To address this issue, a novel flexible NF@PAzo-Mg material was developed for indoor and outdoor use in self-heating applications, capable of operating within the temperature range of –5.0–5.0°C (Figure 5.10). The researchers conducted a three-step investigation into the low-temperature heat release of NF@PAzo-Mg. First, they exposed the material to UV light (80 mW cm^{-2}) to enable light absorption over a period of 90 min, achieving a light collection efficiency of 86.3%.

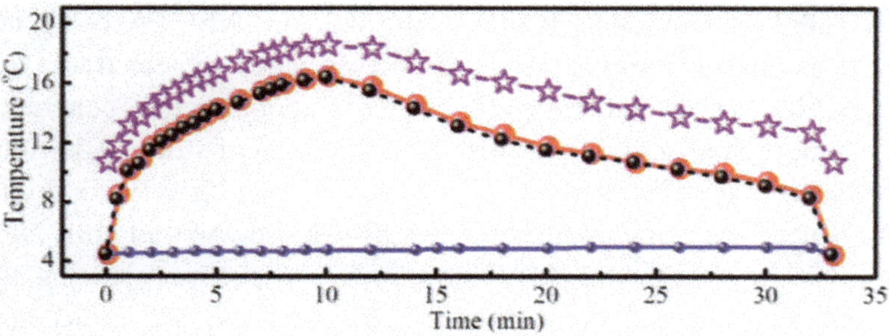

Figure 5.10. Time-evolved temperature of NF (blue, ●), NF@PAzo-Mg (red, ●), and NF@PAzo-Mg (black, ●) at a strain of 50% on model hands (indoors) and NF@ PAzo-Mg (purple, ☆) on real hands (outdoors) under blue-light irradiation for 32 min. The temperature — time curves are drawn by the temperature of the irradiation center. Reprinted with permission from Ref. [8]. Copyright 2019, Wiley.

Subsequently, the material was stored at temperatures ranging from −5.0°C to 5.0°C for 1 h, resulting in an energy loss of approximately 0.1%. Finally, heat release was induced by blue light irradiation at a power density of 40 mW cm^{-2} for a duration of 32 min. Throughout the third step, the researchers utilized a high-resolution infrared camera to monitor the temperature variations in the NF@PAzoMg fabrics, which were positioned on different hand models (indoors and outdoors). The hand models, measuring 30 cm × 10 cm × 8 cm in size, were placed in a refrigerator set at temperatures ranging from 0°C to 4.5°C. Upon exposure to blue light, the textile (initially at 4.5°C) initiated heat release through isomerization energy and bonding enthalpy, resulting in a continuous increase in temperature. Within 10 min, the temperature reached a peak value of 16.5°C, with a temperature difference (ΔT) of 12.0°C relative to the ambient temperature. It is important to note that a balance between self-heating and heat exchange with the environment was maintained. Subsequently, the temperature gradually declined to 8.5°C over the following 22 min due to a lower heat output rate compared to the rate of heat loss.

Upon termination of the blue light irradiation, the textile rapidly returned to its original temperature of 4.5°C. Remarkably, the self-heated textile exhibited excellent cycling performance, demonstrating consistent fluctuations in ΔT ranging from 11.5°C to 12.0°C over a span of

10 cycles. Moreover, it is worth noting that the NF@PAzo-Mg textile exhibits a consistent temperature profile even when subjected to a tensile strain of 50%. This suggests that the textile can continuously generate heat throughout human activities at this particular level of strain. Additionally, the temperature difference (ΔT) observed in the NF@PAzo textile is 6.3°C. Under the same conditions, the NF@PAzo-Mg and unlit NF textiles exhibit ΔT values of 4.0°C and 0.5°C, respectively, owing to the effect of photothermal conversion. Furthermore, we have successfully demonstrated the self-heating capabilities of the flexible NF@PAzo-Mg textile outdoors. These experiments involved wearing the textile as self-heating gloves under specific environmental conditions, including temperatures ranging from −5.0°C to −3.0°C, humidity levels of 10.0%, and wind speeds of 5.5–7.9 m/s. Remarkably, despite the low ambient temperature, the textile exhibited an onset temperature of 10.8°C on one finger due to the body heat of the wearer. Upon exposure to blue light, the temperature of the NF@PAzo-Mg textile increased from 10.8°C to 18.5°C within a 10-min duration, resulting in a ΔT of 7.7°C. Subsequently, the temperature gradually decreased in the following 22 min. Notably, the decrease in ΔT observed in the textiles on the hands was primarily attributed to significant heat loss caused by air cooling. Moreover, the temperature distribution of the NF@PAzo-Mg textiles on both the model hands and the actual hands exhibited a distinct gradient. As one moved farther away from the irradiation center, the temperature decreased due to the formation of a limited number of Mg–O dynamic bonds, resulting from a lower degree of isomerization. This remarkable property highlights the potential of self-heating through the precise control of blue light irradiation range and power density. It also enables the NF@PAzo-Mg textiles to selectively raise the temperature in specific regions based on the ambient temperature, thereby offering a versatile solution for localized warmth.

Zhao *et al.* [10] successfully fabricated a flexible STFD by combining IPDAZ and commercially available fabric cloth through the photo-liquation or thermal-liquation method. This method offers convenience, speed, and environmental friendliness, as it is solvent-free. The composite film obtained from the process exhibits a yellow color when rich in *trans*-isomers, while it turns dark red when rich in *cis*-isomers, as depicted in Figure 5.11(a). This color change is a result of photochromism during

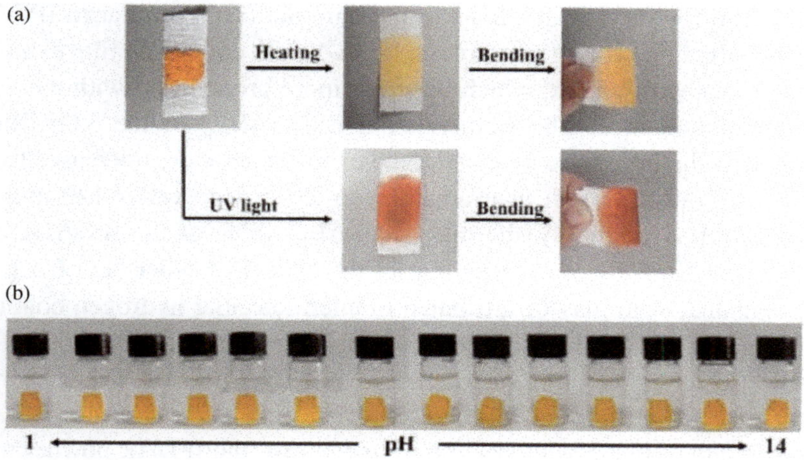

Figure 5.11. (a) The composite STF film fabricated by photo-induced and thermal-induced fusion method with IPDAZ powder and commercialized cloth, (b) the digital photos of IPDAZ STF film immersed in water solution with pH = 1–14 for 24 h. Reprinted with permission from Ref. [10]. Copyright 2017, Wiley.

the photoisomerization process. Furthermore, the composite membrane demonstrates excellent toughness and flexibility. Surface morphological analysis of the composite film using scanning electron microscopy (SEM) reveals a commendable surface morphology. The cloth-like surface of the *cis*-type composite film appears thick, yet smooth, without any notable aggregation phenomena. After undergoing five cycles of charge and discharge experiments, SEM observations demonstrate that the composite film maintains a smooth surface throughout. This indicates the stable uniformity of the composite film during cyclic use. Additionally, the composite film's pH stability is examined by immersing it in aqueous solutions with pH values ranging from 1 to 14 for 24 h. Remarkably, throughout the immersion period, the solutions remain colorless and transparent without any signs of dissolution or precipitation. These results indicate that the IPDAZ/cloth composite membrane possesses excellent pH stability. The energy storage capacity of IPDAZ/cloth STF films is investigated using DSC. Specifically, IPDAZ/cloth solar thermal film samples, measuring 5.4 mm in diameter and containing 2.0 mg of IPDAZ, are placed in an aluminum crucible and subjected to UV light irradiation at 355 nm for varying durations to measure the ESD. Similar to IPDAZ powder, the first heating curve of the IPDAZ/cloth composite film exhibits a wide exothermic peak, which becomes more pronounced with increasing irradiation time. The isomerization ESD (ΔH_{iso}) of the STF films, defined as the integrated area under the exothermic peaks, increases from 42.5 J g^{-1} (after 20 min of irradiation) to 210.5 J g^{-1} (after 120 min of irradiation). Accordingly, the maximum ΔH_{iso} value of the IPDAZ/cloth STF film (210.5 J g^{-1}) exceeds that of the IPDAZ powder (181.6 J g^{-1}). This enhancement may be attributed to intermolecular hydrogen bonding between IPDAZ and the substrate. Notably, the photocharge efficiency of the IPDAZ/cloth STF film surpasses that of IPDAZ powder. This disparity may arise from the poor transmission of UV light through the IPDAZ powder due to its uneven packaging. In contrast, the IPDAZ powder uniformly adheres to the cloth, allowing UV light to pass through the smooth surface of the IPDAZ/cloth STF film.

The infrared thermography technique was employed to investigate the instantaneous heat release process of IPDAZ STF films. The corresponding temperature rise data are illustrated in Figure 5.12. The STF films were exposed to 355 nm UV light for a duration of 2 h. A sample of STF film, unexposed to any irradiation, was utilized as a blank sample for comparison purposes. Both samples were placed on a planar thermostat set at 55.0°C to expedite the *cis–trans* isomerization process. Visual depiction of the samples can be observed in Figure 5.12(a) through digital photographs. In order to capture the temperature variations at different time intervals, an IR thermograph was positioned at a vertical distance of 25 cm above the sample. Figure 5.12(b) presents an infrared thermograph that displays the surface temperature dissimilarities between the charged and uncharged STF film samples. Evidently, the temperature of the charged STF film is considerably higher than that of the uncharged sample. This discrepancy reflects the rapid heat release generated by the stimulation of the *cis–trans* isomerization of IPDAZ. The temperature change versus time curves for both samples are depicted in Figure 5.12(c). Following a 3-min duration on the heater, the temperature of the charged STF sample exhibits a rapid escalation from 33.0°C to 54.5°C. Subsequently, the temperature continues to rise steadily, ultimately reaching 57.7°C after 19 min. The red curve in Figure 5.12(c) represents this

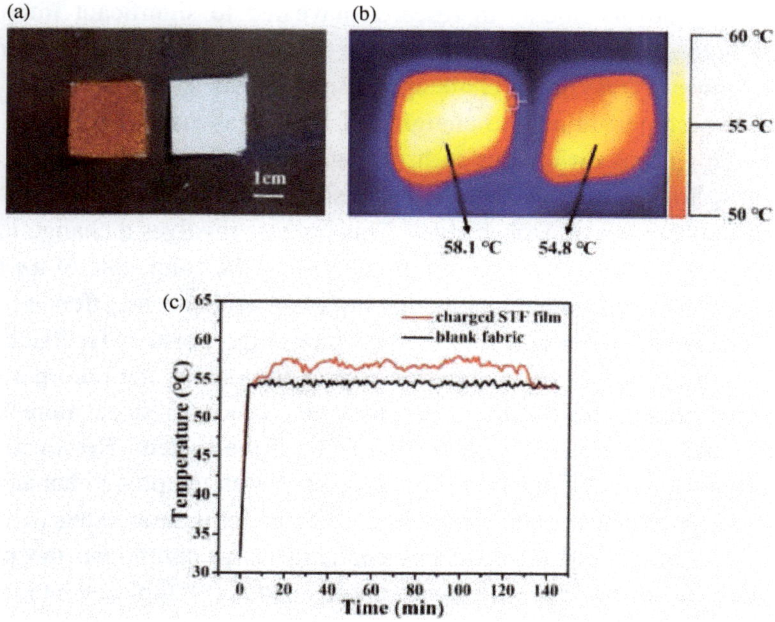

Figure 5.12. (a) The digital photo of charged IPDAZ STF film and blank fabric cloth, (b) the infrared thermal imager photo of the two samples on 55°C heater, (c) the detailed temperature — time curves of the two samples. Reprinted with permission from Ref. [10]. Copyright 2017, Wiley.

behavior, highlighting the nearly constant temperature over a span of approximately 110 min due to the continuous energy release arising from the *cis–trans* isomerization. Eventually, a gradual decline to 54.4°C occurs after a subsequent 5-min period. In contrast, the blank sample also undergoes a rapid temperature rise, reaching 54.5°C, but maintains a relatively stable temperature without persistent overheating. This behavior is denoted by the black curve in Figure 5.12(c). The charged STF film is able to sustain a higher temperature compared to the blank sample for approximately 130 min. This prolonged duration is attributed to the stimulated *cis–trans* isomerization of IPDAZ, which exhibits an enduring heat release capability. Consequently, the Azo/cloth composite film demonstrates an exceptional capacity for stimulated rapid energy release, thus rendering it highly suitable for high-performance STFs.

Recent advancements in research have led to significant improvements in the energy density and half-life of azobenzenes and their derivatives. However, one major challenge arises from the steric hindrance that limits the isomerization and charging of solid azobenzenes solely in solvents. This poses difficulties in handling these substances using organic solvents. To address this issue, a promising solution lies in the photoinduced liquefaction of azobenzenes. This process involves a change in the melting point after isomerization, transforming the solid state of azobenzenes into a liquid state. Notably, the liquefied azobenzene offers a larger free volume for isomerization, thereby promoting positive feedback and enhancing the overall isomerization rate. Additionally, the incorporation of phase change heat storage provides an avenue to store more solar energy and consequently improve the ESD of the system. However, it is important to note that the mobility associated with the phase change process can potentially compromise the system's structural integrity. One effective approach to mitigate this challenge is to compound the phase change azobenzene molecules with porous and soft templates. This technique enables the preservation of the system's structural integrity while harnessing the benefits of the phase change properties of azobenzenes.

5.3 Encapsulated device

The field of optical energy storage is still in its early research stage, and the main challenge currently lies in developing high energy density and high stability energy storage materials. The further challenge lies in obtaining high-performance encapsulated device, which puts forward more requirements for energy storage materials and devices. For example, it is necessary to increase the molecular materials properties without damaging the ESD, photoisomerization efficiency, and isomer stability of the molecules. The device needs to ensure heat release, but it must also ensure the efficiency of light energy storage.

Encapsulated solar collectors with increased light-concentrating capacities can be designed for charging STFDs systems. The sun is an abundant energy source on earth, providing approximately 3.4×10^6 Exajoules of solar energy to the Earth's surface annually through the absorption and scattering of photons. This amount is 7000–8000 times higher than the world's annual primary energy consumption. Therefore, solar energy technologies have the potential to meet global energy demands while minimizing harmful emissions. However, the utilization of solar power is still relatively low compared to other energy resources, and in 2019, solar power accounted for less than 3.6% of the world's total energy production. One of the challenges of solar energy is the intermittent nature of its production and storage, which necessitates the development of scalable energy storage technologies for solar power. Various methods for solar energy storage, including carbon dioxide conversion, supercapacitor, hydrogen energy storage, and batteries, are being implemented. A less explored method is the utilization of photo-switchable molecules known as STFs or molecular solar photothermal energy storage systems. At present, the "photothermal fuel" based on photoactive materials has recently attracted more and more attention. Its characteristic is that under the irradiation of sunlight, the photoactive materials can undergo photo isomerization, transition from low-energy state to high-energy state, and store the radiant solar energy in the form of chemical energy in the chemical bond of the materials. On necessary, the stored energy is then released in the form of thermal energy through other stimulation methods such as heat, light, and catalysts. The design and development of STFDs have been the subject of several recent reviews. The primary objective of this chapter is to provide an overview of the current status and challenges pertaining to STFD-based devices in terms of energy capture, storage, and release utilizing different molecular systems [11,12].

Firstly, a flat panel photoconversion device with meander-shaped channels inside can be utilized. In 2019, Moth-Poulsen *et al.* [13] designed a similar microfluidic chip solar collector with 100 mm deep channels to investigate various conversion conditions of azobenzene derivatives, particularly Azo1 (refer to Figure 5.13(a)). Azo1 exhibited higher photon absorption at 514 nm compared to the absorption of the diruthenium complex. Additionally, it demonstrated a greater photoisomerization quantum yield of 21% and a $\Delta H_{storage}$ of 167.5 J g^{-1}. Nevertheless, Azo1 functions as a positive photochrome and its photoisomer absorption spectrum is red-shifted in comparison to its parent state, making it susceptible to back-conversion by visible light (refer to Figure 5.13(b)). Consequently, achieving 100% conversion becomes challenging. To address this issue,

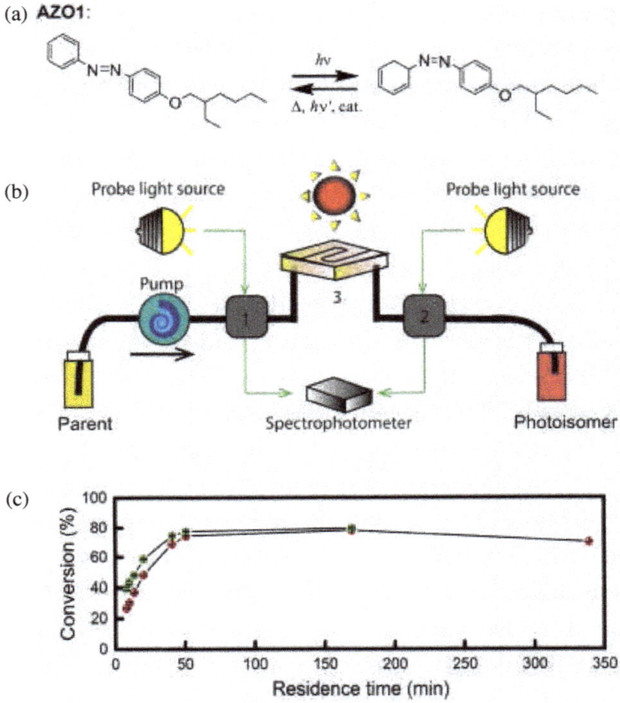

Figure 5.13. (a) Illustration of the structural changes of the Azo1 derivative. (b) The setup of the MOST charging device, which includes two flow cells (gray boxes 1 and 2) connected to a UV–Vis spectrophotometer, a quartz chip with an optical path length of 100 mm. And the total volume is 33.9 mm^3. (c) The experimental data for two different concentrations of Azo1: 0.2 mM in green and 0.5 mM in red. Reprinted with permission from Ref. [13]. Copyright 2019, The Royal Society of Chemistry.

the utilization of a bandpass filter was proposed to prevent light-induced back-conversion from *cis* to *trans* states. Based on these findings, negative photochromism, where the photoisomer exhibits a more blue-shifted absorption onset than the parent state, is preferable for STFD applications due to the reduced inner filter effect.

Experimental results indicated that even with low concentrations of 0.2 mM and 0.5 mM in toluene, the maximum conversion rate reached approximately 80% in both cases, with an experimental maximal energy storage efficiency of 0.01% (refer to Figure 5.13(c)). Recently, a liquid-based multijunction STFD, inspired by multijunction solar cells, was experimentally demonstrated using a triple microfluidic-chip system. This system exhibited a total energy storage efficiency of 0.02%, surpassing the efficiency of any single-layer device. These results suggest that this approach may enhance efficiency in future STFD systems.

The microfluidic chip employed for Azo1 indoor conversion experiments, which was made from quartz, with channels etched isotopically 100 μm deep wet into the substrate, resulting in a lateral bias of 100 μm in the actual optical path length. The chip had a total volume of 33.9 mm^3. For spectroscopic characterization measurements in solution, a Cary 100 UV–Vis spectrophotometer equipped with a 1-cm path length cuvette was used. Kinetic studies involved a temperature control system using the Luma 40TM/Univ-Short sample holder from Quantum Northwest. Azo1 conversion for absorptivity measurements and kinetic studies was performed using a 340 nm LED light from Thorlabs. Back-conversion of the Azo1-*cis* state was achieved using a 455 nm LED light from Thorlabs. A UG11 bandpass filter with a 1 mm thickness from SCHOTT was employed during Azo1 conversion tests to filter out all light above 400 nm, assuming 100% transmission below 400 nm. The microfluidic chip was customized with 100 μm deep channels. Absorption spectra for the microfluidic chip tests were recorded using two micro flow cells with calibrated path lengths of approximately 0.8 mm and 0.7 mm, respectively.

In terms of catalysts, several options have been explored for inducing the back conversion of azobenzene derivatives. These include mineral acids such as perchloric acid, Cu(II) salts like $CuCl_2$ and $Cu(OAc)_2$, gold nanoparticles that operate through a redox mechanism, and electrocatalytic methods. However, for the development of a closed cycle system suitable for device operation, it is necessary to develop a heterogeneous catalyst that can be immobilized at a reaction center. Thus, the need arises for the development of a heterogeneous or insoluble homogeneous catalyst. For the specific case of Azo1, the $[Cu(CH_3CN)_4]PF_6$ salt was chosen as incorporation into the STFDs systems for further testing.

Based on the aforementioned findings, a small-sized reaction center was constructed. In this setup, 5 mg of Cu(I) salt was introduced into a Teflon tube with a 1 mm inner diameter. A 5×10^{-4} M solution of 79% *cis*-Azo1, obtained from the microfluidic chip experiments, was then passed through the catalytic bed at a flow rate of 1 mL h^{-1}. As a result, 48% of the *cis*-Azo1 species were successfully converted back to *trans* state. These continuous fluidic chip experiments demonstrate the fulfillment of requirements for the comprehensive application of the STFDs concept, encompassing the processes of photon capture, storage, and energy release. In 2019, Grossman *et al.* [14] compounded the long-chain alkyl azobenzene derivatives with the PCM tridecylic acid and realized the reversible solid–liquid phase change under the excitation of UV and visible light. And by utilizing the difference in force between azobenzene and phase change molecules before and after *cis*/*trans*-isomerization, the crystallization temperature of the azobenzene phase change composite material before and after charging changes, resulting in a certain degree of undercooling. As a result, the PCM can stably store energy at a lower temperature than the initial crystallization temperature, while achieving controllable heat release of the PCM. This provides new ideas for the long-term energy storage and controllable release of PCMs. In addition, this method is expected to achieve energy storage and release of energy storage materials at low temperatures. The excellent thermal storage capacity of this Azo STF is achieved by doping metastable *cis*-Azo, which can disrupt the stacking of PCM molecules through spatial repulsion and dipole interactions. By adjusting the doping level and cooling rate of azobenzene molecules in the Azo STF composite system, the crystallization temperature of azobenzene can be reduced, thus locking the charged *cis*-Azo in the liquid phase of PCM, so that the stored latent heat can be stored at a lower temperature. In this study, the introduction of azobenzene as a dopant into conventional organic PCM not only changed the PCM's intermolecular dynamics, but also provided stability for the phase state of stored heat energy and enabled the composite to release energy under controlled light excitation. This technology can retain approximately 200 J g^{-1} of thermal energy when placed at a temperature below the original crystallization point for more than 10 h. This method provides many inspirations for the design of portable thermal energy storage systems.

Long aliphatic compounds like *n*-fatty acids or *n*-paraffin waxes can form lamellar structures through hydrocarbon side packing, driven by van der Waals forces. The interaction between the CH_3 end groups of these compounds can also influence the molecular arrangement, leading to polymorphism. The azobenzene dopants used in this study exhibit strong π–π^* interactions between adjacent aromatic cores and van der Waals interactions with alkyl chains. Additionally, the ester linkers in the molecules are expected to contribute to intermolecular interactions.

To demonstrate the concept on a larger scale, an experiment was designed where 3 g of a UV-charged PCM composite was optically triggered to release heat, which then transferred to 1 g of water, effectively raising its temperature (refer to Figure 5.14). The decrease in the critical temperature (Tc) of the PCM from 38°C to 29°C upon the addition of the azobenzene dopant and UV activation is clearly observed (Figure 5.14(a)). The optical irradiation also triggers the crystallization of the PCM above 29°C (Figure 5.14(b)). Comparing the heat release from the optically triggered composite to that of the UV-charged composite in the dark, it can be seen that the former exhibits a higher heat release due to the additional contributions from dopant crystallization and azobenzene isomerization energy. Despite the continuous and spontaneous heat loss to the cooler surroundings, the longer crystallization process and prolonged heating of the water demonstrate significant heat release from the composite and its transfer to the water. A control experiment involving pristine PCM and water confirms that the impact of blue light irradiation on the temperature of bulk materials, including the PCM and water, is negligible.

The LED lamp was positioned 20 cm away from the PCM flask. As illustrated in Figure 5.14(c), pristine PCM was able to heat water to a higher temperature in a shorter period due to its higher T_c (38–39°C) and faster heat loss to the surroundings at high temperatures. The crystallization process of pristine PCM was slower when not in contact with a water container (Figure 5.14(d)). Control experiments were conducted to highlight the influence of phase change and latent heat on water-heating dynamics. In these experiments, other substances were used instead of PCM, which only transferred sensible heat to water within the temperature range of the study. Octadecane, a paraffin with a thermal conductivity and heat capacity similar to tridecanoic acid but with a low T_c of 28°C, was chosen (Figure 5.14(e)). Water, a common heating medium with a high heat capacity, was also used (Figure 5.14(f)). In both cases, the heated water cooled down rapidly, in contrast to water in a solidifying PCM composite bath, demonstrating the

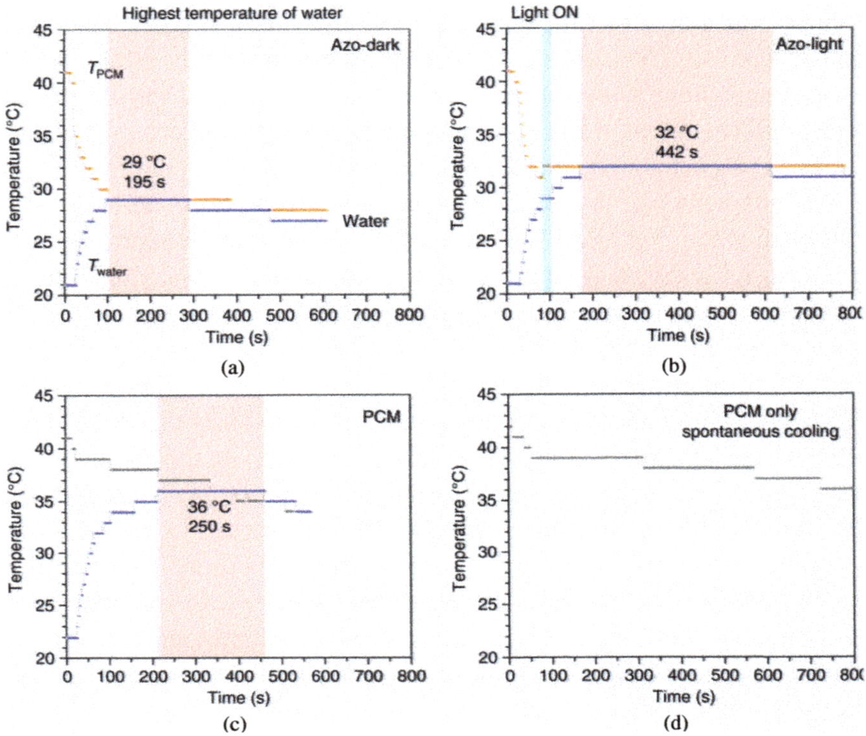

Figure 5.14. (a) The experiments conducted on water-heating solid-to-fluid PCMs. In these experiments, the temperature changes of 1 g of water in contact with various materials were measured. The blue curve represents the temperature changes of water in contact with a composite material consisting of 3 g of UV-charged phase-change materials (PCM) with 35 mol% of *cis*-azobenzene dopant (compound 1). The experiments were conducted in the dark. (b) To maximize the discharging of *cis*-azobenzene, the composite material was exposed to a blue-light-emitting diode lamp at 31°C for 500 s. For comparison, the temperature changes of water in contact with (c) pristine PCM (tridecanoic acid), (e) octadecane, and (f) a water bath were also measured. In the experiments, each thermocouple was submerged in a heating medium, and a vial containing water was immersed in the heating medium. The temperature of the water in the vial under ambient conditions was 21–22°C, and it increased quickly when the vial was immersed in the heating medium at 40°C. The temperatures of the heating medium (T_{PCM}) and water (T_{water}) were recorded simultaneously after T_{PCM} reached 41–42°C while the heating fluid was stirred. (d) In a separate experiment without the process of dipping a water container into the PCM heating bath, the temperature change of pristine PCM under ambient conditions was measured. Reprinted with permission from Ref. [14]. Copyright 2017, the authors.

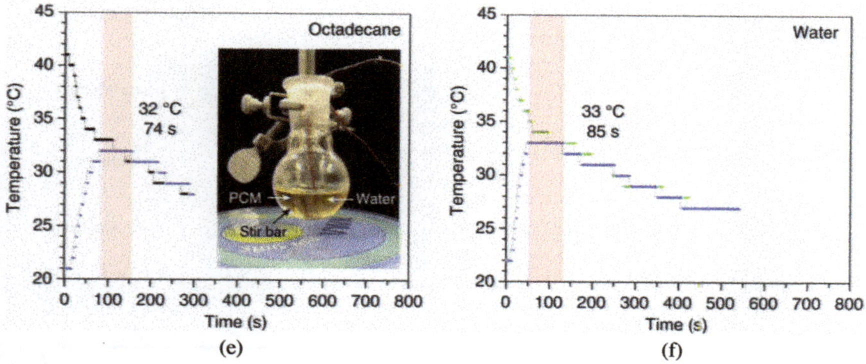

Figure 5.14. (*Continued*)

significant advantage of PCMs over heated liquid as a heating medium. The experiments showed that light-triggered phase change allowed for over five times longer heating of water compared to sensible heat transfer from other heated fluids, highlighting the significance and practicality of latent heat storage and release mechanisms.

In their study, Wang *et al.* [15] designed some azobenzene derivatives that exhibited reversible structural and chromatic changes. These derivatives were functionalized with different alkoxy chains (C_{14}Azo, C_8Azo, and C_4Azo) in order to achieve high miscibility with similar-structured PCMs like tetradecanol (C_{14}OH), dodecanol (C_{12}OH), and hexadecanol (C_{16}OH). The researchers aimed to investigate the energy release through optical regulation and thus designed solid-to-fluid phase change devices. In these devices, C_{14}OH@C_{14}Azo served as a solar energy storage shell, analogous to walls of a room or clothing, enclosing a container (e.g., a room or the human body) (Figure 5.15(a)). The temperature control range of 20–45°C Ccorresponds to the working environment for individuals, enabling regulation of body or surrounding temperatures. When the outdoor temperature is cooler, the charged C_{14}OH@C_{14}Azo provides over three times longer cooling of the inner water compared to pure water, demonstrating the practicality of both latent heat storage and protection of the inner temperature (Tin) from a rapid decline (Figure 5.15(b)). Due to its lower crystallization point (7:3) in comparison to pure water, the charged C_{14}OH@C_{14}Azo exhibits a slower cooling rate within the range of

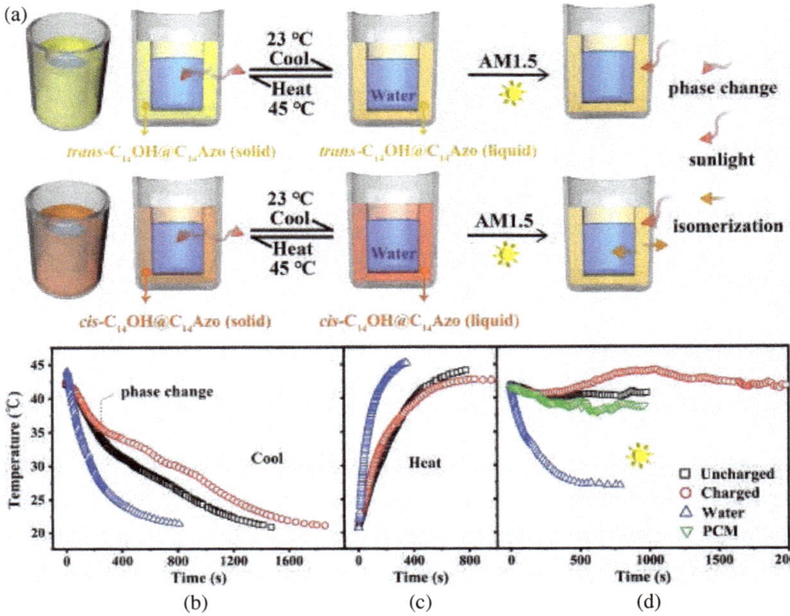

Figure 5.15. Illustration the energy release of the charged composite of $C_{14}OH@C_{14}Azo$ (7:3) through optically regulation. (a) The testing conditions. The temperature of the inner water (1.5 g) is monitored while contacted different materials, including 1.5 g water, 1.5 g the charged, or 1.5 g uncharged composite, and 1.5 g PCM. (b) The cooling process without light illumination, where the temperature decreases from approximately 43–20°C. (c) The heating process without light illumination with the temperature increasing from around 20–44°C. (d) The cooling process with simulated sunlight (AM1.5, 200 mW cm^{-2}) illumination. Reprinted with permission from Ref. [15]. Copyright 2020,Wiley.

–29–32°C. On the other hand, if the outdoor temperature increases, $C_{14}OH@C_{14}Azo$ also maintains a cooler state for an extended period in a hot environment (Figure 5.15(c)).

Significantly, when the environment temperature cools with the assistance of sunlight (AM1.5), both the phase change energy and isomerization energy occur within the charged composite, protecting the Tin from cooling and even resulting in a slight temperature increase due to the photothermal effect of the AM1.5 light source (Figure 5.15(d)). The Tin, protected by the charged composite, exhibits approximately 4°C higher temperature compared to the uncharged composite, which can be attributed to the additional isomerization energy and enhanced solar energy absorption within the charged composite.

To emphasize the impact of solar energy storage, the researchers designed a control group using pure PCM (white). In comparison, the uncharged $C_{14}OH@C_{14}Azo$ at 40°C exhibits a greater capacity for maintaining warmth compared to pure $C_{14}OH$ at 38°C. This confirms the composite's ability to warm itself and the surrounding environment. Consequently, the charged $C_{14}OH@C_{14}Azo$ efficiently stores and releases energy to regulate temperature when activated by light. Moreover, it facilitates improved solar energy capture by reducing reflection.

Microfluidic control device, referred to as microfluidics device, is a kind of commonly used powerful tool for transporting micro liquid in micro channels and conducting micro liquid experiments. Now we classify and analyze the components of common optical microfluidics devices: Microfluidics devices include solid components, a certain amount of driven liquid, and a certain amount of air that is closed or semi closed by the boundary of liquid and solid. Among these three types of objects, the influence of light energy on solid deformation and liquid surface tension is limited to the position of illumination, while light has almost no effect on air. However, as a fluid, the shape of a gas is determined by the solid liquid boundary that encloses it. Under the premise of constant pressure and mass, its volume is approximately proportional to temperature. This shows that in the microfluidics device that has closed the gas, as long as the volume of the gas is changed by changing the temperature, the original solid liquid boundary of the closed gas section can be changed, that is, the relative position of the liquid and the solid is changed to achieve the effect of driving the liquid. Although the photothermal effect of air is almost zero, a "photothermal compound" can be placed in the interval of closed air, which indirectly affects the temperature and volume of the air around the photothermal compound through the photothermal effect of the photothermal compound, so as to drive the liquid. On this basis, the light source only needs to aim at the photothermal polymer and adopt two modes of illumination or non-illumination to control the movement of the liquid without the need to move or rotate the light source.

Yu *et al.* [15] designed one photothermal pneumatic microfluidics device (PPMD). As shown in Figures 5.16(a) and 5.16(b), the feeding

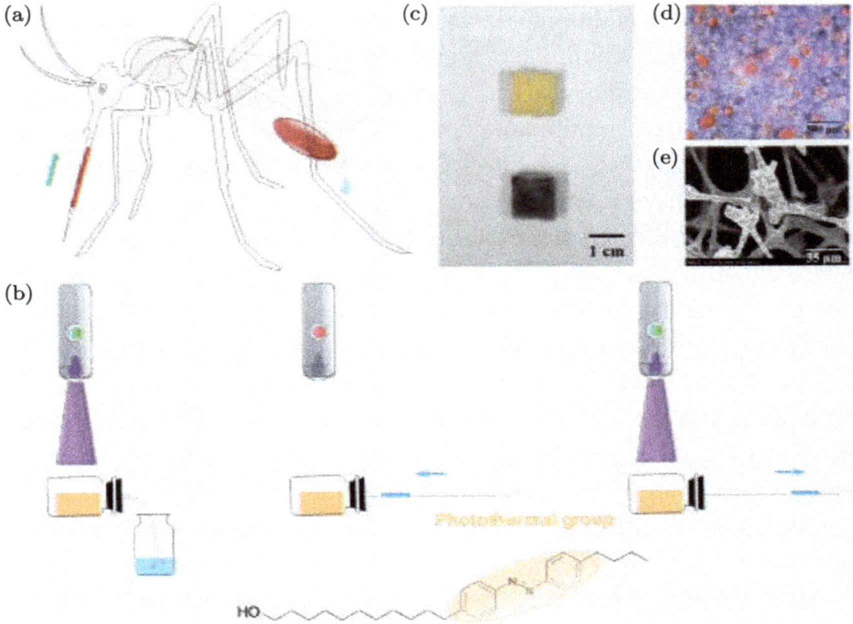

Figure 5.16. (a) Schematic showing the collection, transportation, and separation of fluids during mosquito feeding. (b) Structure and fabrication of PPMDs. Place the aforementioned photothermal composite sponge in a colorless and transparent glass sample bottle, drill a small hole at the plastic bottle cap, and connect a silica gel tube with a matching diameter to connect the inside of the glass bottle and the external atmosphere. Use epoxy resin glue to seal all connections, including between the bottle body and cap, the holes on the surface of the cap, and the outer wall of the silicone tubing. After the glue is cured, a photothermal pneumatic microfluidics device is obtained. A series of photothermal pneumatic microfluidics devices with different diameters and materials can be fabricated by a similar method. The inner diameter range of the silicone tubing used is 0.5 mm–5 mm–2.0 mm. A photothermal pneumatic microfluidics device containing a 0.5 ml capsule as an air chamber and a glass capillary with an internal diameter of 0.1 mm as a micro pipe. (c) Appearance of the photothermal composites. Immerse a sponge with a porosity of 98.6% into a tetrahydrofuran solution of $C_{11}AzC_4$. The solution preparation method is to dissolve 4.8 mg of $C_{11}AzC_4$ in 5 g of tetrahydrofuran. The sponge was taken from azobenzene solution and dried at 50°C to obtain an azobenzene doped photothermal composite sponge containing 0.8 mg $C_{11}AZC_4$ and 8.7 mg sponge. Azobenzene small molecule $C_{11}AzC_4$ attached to the node of sponge fiber: (d) optical microscope photos, (e) scanning electron microscope photograph. Reprinted with permission from Ref. [16]. Copyright 2019, Wiley.

process of female mosquitoes can be briefly divided into the following parts: inserting mouthparts into the capillaries of animals, injecting digestive enzyme into the capillaries, sucking blood into the abdomen, separating the high nutrient content red blood cells from the low nutrient content water, and finally expelling the low nutrient content water from the body. The entire process includes water collection, water transportation, and water separation. Correspondingly, the PPMDs designed in this study can be one-to-one corresponding to the above processes. Among them, the main step of the mosquito feeding process is to absorb blood by changing the volume and pressure of the abdominal cavity. It can be seen that the mosquito itself is also a pneumatic microfluidics device, which is the key to the combination of this research and female mosquito bionic technology. In the biomimetic design section of this study, the body structure of female mosquitoes can be simplified as an air chamber+pipeline structure. The pipeline mimics the mouthparts of female mosquitoes, and the air chamber mimics the abdominal cavity of female mosquitoes. A photothermal composite is placed inside the hard and transparent cabin, which regulates the temperature of the air inside the cabin and has a specific light absorption band. When the matched light source illuminates the photothermal complex inside the transparent gas chamber, the temperature inside the chamber increases and the gas volume increases. When the irradiation is stopped, the air inside the transparent chamber gradually cools down to room temperature under natural cooling, and the gas volume decreases. The photothermal composites used in this study include photothermal agents and substrates. The photothermal agent includes a MW CNT, which respectively represent a narrow absorption band photothermal agent and a wide absorption band photothermal agent (Figure 5.16(c–e)). The substrate is a polyurethane sponge with a porosity of 98.6%. The reason for using a sponge as a substrate is that it has a high surface area. After doping with a photothermal agent, it can increase the heat transfer area with air, forming a better heating effect when applying light, or an effect of being easily cooled by air when stopping light. The purpose of this study is to explore the operation law of photothermal pneumatic microfluidics devices, and to find a fast, accurate and reliable method to manipulate trace liquid using photothermal effect, as well as to realize trace chemical reaction using the manipulated liquid.

In the single chamber mode, the liquid column gradually moves away from the chamber during each cycle of repeated light switching. During the illumination phase, the liquid column is kept away from the gas tank; in the without illuminated phase, the liquid column retreats toward the gas compartment. However, at the end of the without light stage, the liquid column cannot return to the position before the start of light, resulting in a phenomenon where the liquid column moves further away from the gas tank after a reciprocating motion. During the entire process, the temperature rises during illumination and decreases after turning off illumination. However, the rate of natural cooling is slower than the rate of photothermal reaction heating. For example, if the temperature rises after 2 s of illumination, even if it cools naturally for 5 s, it cannot fall back to the temperature before illumination. This is because the heating stage takes 2 s to rise from room temperature A to temperature B, and the time required to reach the equilibrium temperature is far from enough. However, descending from temperature B to room temperature A is a complete Newtonian cooling process. As the temperature difference decreases, the cooling rate slows down, and theoretically, it takes an infinite amount of time to lower to room temperature. Therefore, even in cyclic testing, the without illumination stage cannot be brought back to room temperature for a longer period of time. At the end of each cycle, the temperature rises slightly compared to the previous cycle, causing the liquid to gradually move away from the gas chamber. Due to the limited length of the pipeline, the liquid may be completely pushed out of the pipeline mouth after multiple cyclic tests. To ensure that the liquid is not pushed out of the pipeline opening, it may be necessary to reduce the amount of testing liquid and increase its pipeline length, or reduce the number of cycles.

In the dual chamber mode, if two identical light sources alternately illuminate the two chambers at equal times, the equilibrium position of the reciprocating motion of the liquid will not change. When the light source of the left air compartment is turned on, the light source of the right air compartment is turned off, and the pressure on the left side rises, causing the liquid to move to the right. As the right side is also closed, the liquid compresses the gas on the right when moving to the right, and the gas on the right will have an opposite force; Similarly, when the left cabin light source is turned off and the right cabin light source is turned on, the liquid also receives pressure from both directions of gas and moves to the left. Due to the symmetrical energy input and environment of the two stages, as well as the symmetrical heating time and natural cooling time, the pressure changes on both sides of the gas tanks are also symmetrical and equal. Therefore, the gas pressure on the liquid during the two motion stages is also symmetrical and equal, and the motion state and distance are also symmetrical and equal. In summary, in the cyclic test of the dual chamber mode, under symmetrical lighting conditions, the liquid vibrates with the same amplitude around the same equilibrium position in the pipeline. The phenomenon of this mode can be applied to scenes where the movement of liquids needs to be limited to a specific range, such as under a fixed camera lens, in the field of view of a microscope, on a limited thermal stage, or in specific lighting with limited area, and so on.

In conclusion, a PPMD was prepared in this study, which realized the bionic feeding process of mosquitoes, including the collection, transportation, and separation of liquid; It achieves remote and precise control of liquid movement in pipelines, without the need for any coating or modification of the inner wall of the pipeline, without the need for special design of the chemical composition and geometric structure of the pipeline, and without the need to move the light source to follow the movement of the liquid, without bringing risks of temperature changes, chemical environment changes, and radiation damage to the driven liquid. We established a theoretical model of exponential function to study the relationship between the displacement of the driven liquid and the light time, and thus realized the fast, uniform, and accurate control of the liquid in the pipeline, that is, to reduce the interval between the light time and the non-light time, so that the exponential decay stops before amplification. The silicone tube with adsorption effect on organic solvent is used as a microfluidics pipe to realize the separation of oil–water two-phase liquid in motion and the photo controlled trigger of water-phase chemical reaction under the control of separate PPMD. Two modes were used for cyclic testing of PPMD, demonstrating the stability and repeatability of the device. In the integrated PPMD pipeline network, precise multi-directional liquid driving and precise triggering of chemical reactions in parallel pipelines have been achieved. Therefore, this bionic PPMD shows excellent efficiency and stability of driving liquid and has good application prospects in the field of microfluidics, gas driver, water–oil two-phase liquid separation, and micro reaction control.

In 2020, our research group developed an annular STFD for distributed energy utilization. In our study, we focused on achieving low-temperature heat release in PCMs through optically triggered mechanisms. The heat release process was controlled by the photoisomerization of azobenzene, specifically the *trans–cis* (*E–Z*) isomerization. To enable this heat release, we incorporated a photo-responsive alkyl-grafted azobenzene compound into tetradecane (Ted), resulting in the creation of a photosensitive energy barrier within the PCM. This photosensitive formulation allows for the regulation and control of the phase change process. To implement and demonstrate the concept of distributed energy utilization, we designed and fabricated an annular device. This device enables the delivery and release of latent energy through opticallytriggered heat output. The annular configuration of the device promotes efficient and uniform energy distribution, making it suitable for various applications in distributed energy systems.

Due to Azo/Ted-1 (Figure 5.17(a)) has the characteristic of photo-induced reversible solid–liquid transition similar to that of PCMs. Therefore, in order to prevent leakage during its use, it is necessary to choose appropriate materials for encapsulation. Compared to other materials, quartz materials have better light transmittance, especially for UV light, with a transmittance of over 92%, which is conducive to achieving

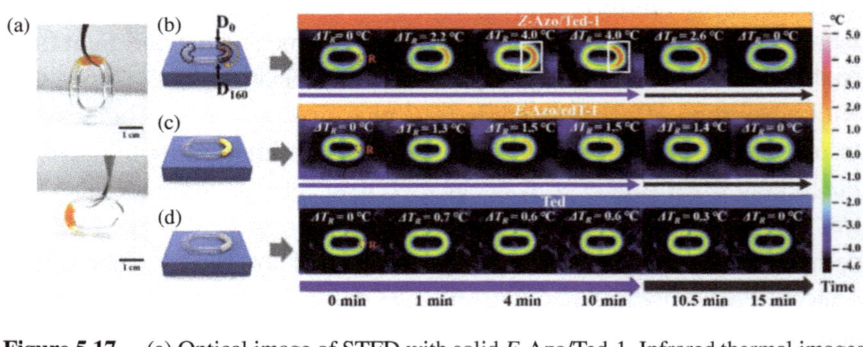

Figure 5.17. (a) Optical image of STFD with solid *E*-Azo/Ted-1. Infrared thermal images of temperature changes in the annular device (b) charged *Z*-Azo/Ted-1, (c) *E*-Azo/ Ted-1, and (d) the Ted. Reprinted with permission from Ref. [17]. Copyright 2020, Wiley.

efficient energy storage and release. Therefore, choosing a mold made of quartz material is beneficial for Azo/Ted-1 encapsulate. And in order to fully utilize the energy after charging Azo/Ted-1. The flowability of the customized U-shaped quartz tube is encapsulated. Take an appropriate amount Azo/Ted-1. The sample is filled into a U-shaped quartz tube, and then another empty U-shaped tube of the same size is taken to bond with the U-shaped tube filled with the sample, forming a circular circuit and achieving the encapsulation effect. As shown in Figure 5.17, after charging Azo/Ted-1 transformed into a red liquid, it can flow in a circular circuit and there will be no leakage at the junction of the two pipes. Through this device, this study designed a micro space thermal management system using alternating irradiation of UV and blue light, as well as changes in environmental temperature. The working steps of this system are shown in Figure 5.17(b).

Firstly, under room temperature conditions, the annular device is placed on a semiconductor cold bench, and 10% of the annular device is irradiated with an UV light source (365 nm) Azo/Ted-1 perform irradiation charging and continue irradiation for 10 min, Azo/Ted-1 transformed from an initial orange yellow solid to a flowing red liquid. Subsequently, the robotic arm gripped the annular device and controlled the flow of liquid in the tube to the target position by manipulating the movement of the robotic arm. At the same time, reduce the temperature of the cooling table to −5°C, ensuring that the temperature inside the tube is close to 0°C, thereby reaching 10% Azo/Ted-1. Within the operating temperature range of light-controlled heat release. After the liquid inside the tube flows to a specific position, the robotic arm releases it. At this point, use blue light (420 nm) to irradiate the liquid inside the tube to induce heat release. Under the illumination of blue light, the phase transition from a red liquid to an orange-yellow solid will occur, which also proves that Azo/Ted-1 from the rich cis structure to the rich *trans* structure, the phase transition potential and isomerization enthalpy are released during the phase transition and isomerization recovery processes. If recharged, restore to room temperature conditions and irradiate the device with UV light. Through the above process, a complete photo-controlled energy storage and induced release thermal management system has been constructed, which combines photo control, cycling, and positioning.

The temperature change during blue light-induced heat release was tracked by a high-resolution infrared thermal imager (Figure 5.17(b)). Before the blue light irradiation, the tube temperature is –0.3°C, and when the blue light is turned on, the temperature increases. When the blue light was continuously illuminated for 4 min, the temperature reached a maximum of 3.7°C, which was 4°C higher than before the irradiation. As a comparison, under the same conditions, using blue light irradiation filled in 10% of the uncharged energy Azo/Ted-1 (Figure 5.17(c)). The experimental results show that after blue light irradiation, the temperature of the uncharged annular device increases from –0.3°C to 1.4°C. This is the temperature change caused by the photothermal effect of the material. In order to eliminate the influence of blue light photothermal, this study used blue light to irradiate an empty annular device under the same conditions (Figure 5.17(d)). The results show that after blue light irradiation, the temperature change is very small compared to the annular device after charging, thus excluding the influence of blue light photothermal. Before the blue light is turned on, the annular device remains at –0.3°C. After the blue light is turned on, the temperature significantly increases. After 4 min of irradiation, Azo/Ted-1 reaching equilibrium state, the temperature tends to stabilize and can be maintained for more than 5 min. The response speed of empty devices and uncharged devices after blue light is turned on is faster, and they reach the equilibrium state faster than charged ring devices. When the blue light is turned off again, the temperature of the three sets of ring devices rapidly decreases and returns to the initial state.

The microthermal management system designed by this research institute also has important guiding significance for the effective storage and utilization of thermal energy in the future. In extreme environments with alternating cold and warm temperatures, such as deep space environments, such materials can serve as high-power thermal management systems for heating various types of equipment.

In 2021, Yu *et al.* [18] proposed a straightforward yet efficient strategy to generate a significant gravitational moment. This was achieved through the design of a photothermal pneumatic wheel (PPW) consisting of multiple photothermal pneumatic fluidic devices (PPFDs), as depicted in Figures 5.18(a) and 5.18(b). The driving principle of the PPW is initially, the center of mass of the runner is at the center. When the light source illuminates the photothermal complex in the No. 1 PPFD, the center of

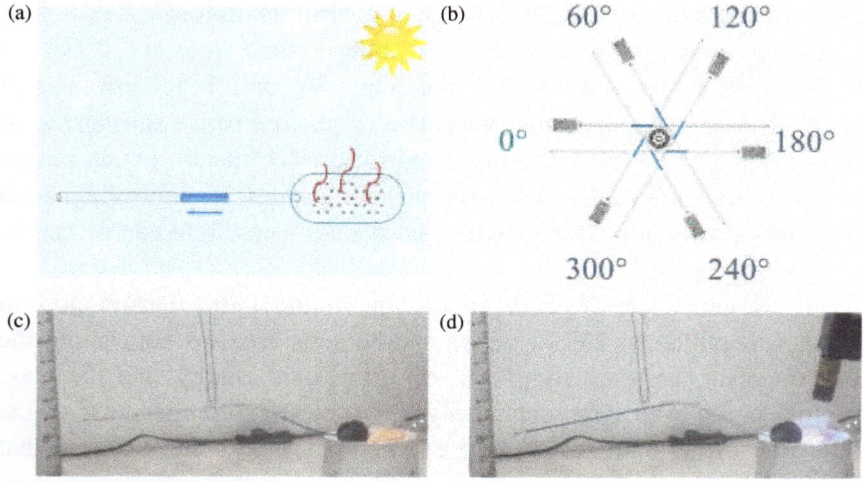

Figure 5.18. Schematic illustrations of (a) a PPFD under light irradiation, (b) a PPW with six PPFDs. (c and d) digital images of a light induced mass center shift of a PPFD. Reprinted with permission from Ref. [18]. Copyright 2021, Elsevier.

mass of the No. 1 PPFD moves to the left, and the center of mass of the runner also shifts to the left, causing the runner to rotate counterclockwise. The No. 1 PPFD air chamber gradually detached from the light source, and the liquid returned to the center of the No. 1 pole, but the runner still moved counterclockwise due to the deviation of the center of mass; When the 5th air PPFD rotates to the light source, the center of mass of the 5th PPFD also moves to the left, and so on. When each air chamber containing a photothermal composite rotates near the light source, it is equivalent to providing new rotational power for the runner. Every time the cabin is turned to the light source's illumination range, it is like the end of winter and the return of spring, when nature is once again full of vitality and the wheels return to running speed.

In the previous text, the use of light energy to control the movement of liquids in pipelines, the use of light energy to control the movement of photothermal pneumatic actuators in water, and the use of light energy to control the rotation of photothermal pneumatic wheels are actually the use of light fields to control the movement of liquids and solids. Due to the fact that light energy acts on photothermal composites, there is no need to dope photosensitive substances into liquids in research. Therefore, we have achieved the use of light energy to control the movement of liquids that do not contain photosensitive or photosensitive substances, but we have not yet achieved the use of light energy to control the movement of solids that do not contain photosensitive or photosensitive substances, as both PPFR and the photothermal pneumatic wheel itself contain photosensitive materials such as photothermal composites. So controlling solid substances without photoresponsive substances using light energy can be a problem to be solved.

In the chapter of PPFR, a glass sample bottle with a density close to water was pulled by PPFR, which can be considered as controlling the movement of the glass sample bottle using light energy, and the glass sample bottle does not contain any photosensitive components. In the "dynamic evolution trajectory of the technological system", we found that light energy drives the glass sample bottle through thin wires connecting the driver to the glass sample bottle, which also belong to a flexible body that provides tension through mutual contact. Therefore, although the use of light energy to drive solid substances without photoresponsive substances has been achieved, it is still in the stage of evolution. The direction of evolution is to use light energy to control a certain "field" to make solid substances without photoresponsive substances or neutral substances undergo controllable motion.

Common fields include gravity field, magnetic field, electric field, and light field. The magnetic field can only drive materials or conductors containing magnetism. Although light energy can change the magnetic field by providing thermal energy, this is not consistent with our goal of driving ordinary neutral substances; At present, the light field can only directly drive solid substances containing photoresponsive substances; The gravity field is determined by the relative position of the center of mass of the solid and the Earth, and the light field can only change the center of mass of the solid material with light response, which is not consistent with the target; An electric field can polarize small and light objects, thereby attracting them. The changing electric field can control the movement of small and light objects with the real-time distribution of the electric field. Small and light objects, such as small sponges, meet the definition of "solid matter without light responsive substances". Therefore, we have determined the target of manipulating light and small objects by controlling the changes in the electric field with light energy.

Through the bedding of PPMD, light energy can easily drive the liquid in the pipeline, and during the entire driving process, the liquid changes position. Conductive liquids, such as electrolyte aqueous solutions, can cause the redistribution of static charges in the surrounding environment. As the conductive liquid moves, the distribution of static charges also adjusts in real time, which is the reason why the liquid movement causes changes in the electric field. Therefore, it is possible to achieve the goal of controlling the movement of liquid with light energy, indirectly controlling the changes in electric field, and thus controlling the movement of small and light objects (Figures 5.18(c) and 5.18(d)).

To drive the photothermal pneumatic wheel (PPW) without the need for a moving light source, a self-shadowing mechanism was employed. A light barrier with a window was positioned between the light source and the PPW. When the photothermal pneumatic fluidic devices (PPFDs) were in the exposed area, the light from the source drove the PPW to rotate from 0° to 60°. As each PPFD entered the shadow area, the next PPFD would rotate into the exposed area, facilitating a continuous self-sustained rotation of the PPW. This self-sustained rotating PPW, when combined with a suitable photothermal agent, can be driven under a constant light source such as sunlight or other area light sources. For example, a self-sustained PPW utilizing graphene oxide (GO) as the photothermal agent can be driven by sunlight, making it ideal for outdoor applications. The PPW can be designed in various shapes and fabricated using different materials, providing a versatile platform for applications in robotics, biomimetics, and energy conversion.

Furthermore, STFDs have shown promise in harvesting and storing thermal energy. In the charging process, a photochromic molecule's stable isomer absorbs photon energy and undergoes a conversion to a high-energy metastable isomer, effectively storing solar energy within chemical bonds. Discharging the STFD system involves the metastable isomer transitioning back to the stable isomer through external stimuli, thereby releasing the stored energy as heat. While STFD systems share similarities with PCMs in terms of their energy storage capabilities, their mechanisms differ. STFD systems utilize photons and molecular thermodynamics to control energy storage and release, whereas PCMs rely on heat transfer. Recently, the combination of STFD and PCMs into a single component material (STFD–PCM) has been explored to enhance the storage capacity of STFD systems. The STFD–PCM system can charge not only through solar irradiation but also by extracting energy directly from the surrounding environment, resulting in a nearly 100% increase in energy density. Another advantage of the STFD–PCM combination is that the phase change is controlled by external stimuli rather than occurring spontaneously, eliminating the need for insulation to retain latent heat and significantly expanding the functionality of the system.

In 2022, Li *et al.* [19] introduced new arylazopyrazoles as solid-to-fluid phase change materials (STFD–PCMs). These STFD–PCMs are capable of being recharged below 0°C using visible light, as depicted in Figure 5.19. By simultaneously harnessing the energy from visible light and low-temperature ambient heat, an energy density of 0.25 MJ kg^{-1} was achieved. This represents a 67% increase compared to previous similar systems. Another notable characteristic of the arylazopyrazoles is the stable energy storage capacity of the *cis*-isomer, which has a half-life of 22 days at 0°C. This demonstrates the potential for long-term energy storage. The combination of high energy density, extended storage time, and the ability to charge the system at low temperatures presents an

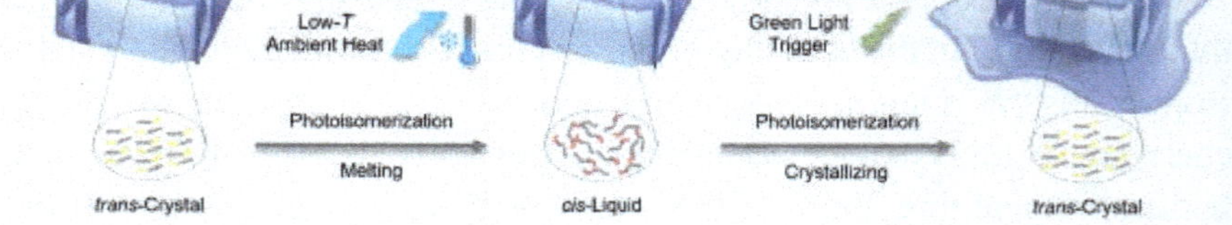

Figure 5.19. Schematic representation of the rechargeable molecular solar thermal system below 0°C illustrates its ability to store energy from both visible light energy and low-temperature ambient heat. This system captures energy from these sources and stores it within the molecular structure. At a later time, when high-temperature heat is desired, the stored energy can be discharged on demand. Reprinted with permission from Ref. [19]. Copyright 2022, the authors.

opportunity to explore the functionality of these materials in a new type of optically regulated STFD–PCM window.

To showcase the concept, the researchers prepared a miniature energy storage window by coating glass with arylazopyrazoles. This proof-of-concept study illustrates the potential function of the STFD–PCM system in an energy storage coating, allowing the glass to serve as an optically regulated window for energy management. Unfortunately, without access to Figures 5.19 and 5.20, I cannot provide specific visual

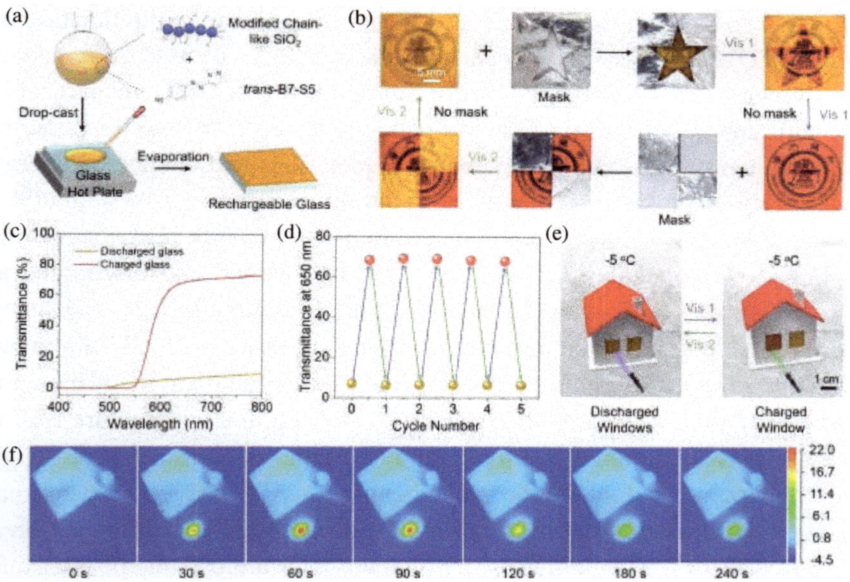

Figure 5.20. (a) Schematic illustration shows the fabrication process of rechargeable glass through drop-casting. (b) The transparency of the glass can be controlled or modulated by external stimuli such as light. (c) The changes in transmitted light through the glass before and after the charging/discharging process. (d) This graph or plot shows how the transmittance of light at the specific wavelength of 650 nm changes over multiple charging and discharging cycles. (e) The process of charging and discharging the glass windows below 0°C is shown. Two different wavelengths of light are used: Vis 1 (400 nm light with an intensity of 40 mW cm^{-2}) and Vis 2 (532 nm light with an intensity of 110 mW cm^{-2}). (f) These images capture the changes in infrared radiation emitted by the charged windows over time. It is mentioned that a maximum heat release temperature of up to 22.7°C was recorded. Reprinted with permission from Ref. [19]. Copyright 2022, the authors.

details regarding the experimental setup or results. New arylazopyrazoles report as STFD–PCMs, which are rechargeable below 0°C by visible light, as illustrated in Figure 5.19. In addition, by co-harvesting the visible light energy and low-temperature ambient heat, an energy density of 0.25 MJ kg^{-1} is achieved, which is an increase of 67% over previous comparable systems. Furthermore, the *cis*-isomer has a half-life of 22 days at 0°C, demonstrating its stable energy storage capacity. The combination of high energy density, storage time, and the fact that the system can be charged at low temperatures provides the opportunity to explore the function of the material in a new type of optically regulated STFD–PCM window as a proof-of-concept study, designed to illustrate the function of the STFD–PCM system in a coating. Glass coated with arylazopyrazoles is prepared as a miniature energy storage window. This rechargeable system enables the utilization of solar energy even in low-temperature environments, expanding its potential applications and enhancing energy efficiency. By combining energy from multiple sources, this system offers a versatile and sustainable approach to high-temperature heat generation.

The rechargeable glass windows can be charged and discharged at −1°C using specific wavelengths of light. Charging is achieved using 400 nm blue light, while discharging is accomplished with 532 nm green light. During the discharging process, the surface temperature of the window increases from −1°C up to 21.7°C, resulting in a temperature rise of 22.7°C. This significant temperature increase demonstrates the high thermal power output of the rechargeable glass windows. The thermal power output during the continuous 60 s discharging period is measured at 256.2 W m^{-2}. This indicates that the windows are capable of releasing a substantial amount of heat energy. The ability to charge and discharge the windows at low temperatures opens up potential applications in various areas. Functional clothing, advanced sunglasses, de-icing systems, and home heating under ice-cold conditions are mentioned as examples. By utilizing these windows, thermal comfort can be enhanced while reducing the energy consumption associated with conventional heating methods. Overall, the findings highlight the potential of the novel rechargeable glass windows in providing efficient energy management and thermal regulation in cold environments, offering benefits in terms of comfort and energy sustainability.

Trans-B_7-S_5 and diethoxydimethylsilane-modified chain-like silica were mixed in an ethanol and isopropanol solution. This mixture was drop-cast onto a glass substrate to create rechargeable coatings (Figure 5.20(a)). The modified silica formed a transparent porous network structure on the glass, effectively preventing the leakage of the *cis*-liquid. The rechargeable glass substrate allowed for the creation of various patterns through selective writing and erasing processes (Figure 5.20(b)). By placing a mask on the glass and irradiating it with 400 nm blue light, the exposed area transformed from opacity to semi-transparency. This resulted in a color change from orange to red, effectively writing the desired pattern on the glass. On the other hand, without a mask, exposure to 400 nm blue light led to an erasing process, making the substrate globally semi-transparent. Further patterning could be achieved by selectively irradiating the glass with 532 nm green light. Finally, the glass returned to its opaque state after exposure to 532 nm green light without a mask. The transmittance spectra of the rechargeable glass under different light irradiations were recorded (Figure 5.20(c)). In the discharged state, the glass had a low transmittance of less than 10% in the 500–800 nm wavelength range. However, after exposure to 400 nm blue light, the glass was charged, resulting in a transmittance of up to 70% in the 650–800 nm wavelength range. The rechargeable glass demonstrated good durability under alternating blue light and green light irradiations (Figure 5.20(d)). Based on the high transmittance of the charged state glass and the low-temperature phase transition of B7-S5, the rechargeable glass shows potential for use as a photochromic solar thermal energy storage window in daily life, particularly in cold winters. The ability to control transparency and store solar thermal energy makes it a promising technology for improving energy efficiency and thermal comfort in buildings or other applications.

The rechargeable glass sheets, with a size of 10 by 12 mm and a coating thickness of approximately 400 mm, were installed on a miniature house model as windows for the experiment. The house model was placed in –5°C surroundings to ensure that the surface temperature of the windows was around –1°C. Figure 5.20(e) illustrates that exposure to 400 nm blue light (40 mW cm^{-2}) caused the window to transform from opacity to semitransparency, storing visible light energy and ambient low-temperature heat. By triggering the window with 532 nm green light

(110 mW cm^{-2}), the stored energy was rapidly released as high-temperature heat. This transformation from semitransparency to opacity is shown in Figure 5.20(e). An infrared thermal imaging camera was used to track the temperature changes of the window during the 400 nm and 532 nm light irradiations (Figure 5.20(f)). During the 400 nm light irradiation (charging process), the temperature of the window showed a difference of approximately 3°C above the ambient temperature, indicating a weak photothermal effect. During the 532 nm light irradiation (discharging process), the charged window reached a temperature of 21.7°C at 60 s, which was 22.7°C higher than the cold surroundings. A control experiment, where the discharged window was irradiated with 532 nm light, showed a much lower temperature change of 9.3°C. This suggests that the temperature difference between the charged window and the environment was mainly due to *cis–trans* isomerization. Assuming that the *cis–trans* isomerization was fully completed at approximately 60 s, the corresponding thermal power output was estimated to be 256.2 W m^{-2}. This high-temperature heat release indicates that the B7-S5 molecules on the surface of the window can act as a photon-driven molecular heat pump, upgrading thermal energy from low to high temperatures.

The optically controlled heat release of the rechargeable glass enables temperature gradients that are about an order of magnitude higher than those achievable with traditional solar thermal façade window coating concepts. The solar thermal energy storage coatings demonstrated in this study offer exceptional performances, including visible-light trigger/storage, high energy density, and recyclable ice-cold charging/discharging. These findings hold great promise for future energy management systems. The study highlights the potential of rechargeable glass sheets as a novel and efficient approach to solar thermal energy storage, with wide-ranging applications in energy management and thermal regulation systems.

5.4 Hybrid solar energy conversion device

More than 70% of the total primary energy input worldwide is currently wasted as heat, with approximately 63% of this waste categorized as low-grade heat below 100°C. Despite the potential to convert such low-grade heat into high-grade electrical energy using pyroelectric technology, the existing economically viable means of energy conversion efficiency remain below 2%. Given the substantial demand for low-grade heat in various sectors such as building heating and industrial processes, it is crucial to explore efficient utilization methods for this significant amount of waste heat. Advanced thermal energy storage and upgrade technologies are required to address the temporal, spatial, and intensity mismatches between heat supply and demand.

Due to the intermittent nature and uneven geographical distribution of renewable energy resources, the development of new energy storage technologies has become imperative. STFs or molecular solar thermal energy storage (MOST) have emerged as a promising solution, involving molecules that undergo photoinduced isomerization reactions capable of absorbing light, storing it as chemical energy, and releasing it as thermal energy upon demand. Various molecular systems, including norbornadienes (NBDs), azobenzenes, stilbenes, ruthenium derivatives, anthracenes, and dihydroazulenes, have been investigated for their potential in STFD applications. Among these, the polycyclic strained molecule NBD has attracted significant attention due to its high-ESD and the ability to retain energy over extended periods. However, unsubstituted NBD poses certain limitations, including poor solar spectrum matching and low quantum yield. Despite these challenges, we have successfully demonstrated their functionality in laboratory-scale test devices for harnessing, storing, and releasing solar energy.

This section highlights the recent significant findings related to the engineering of key properties in the NBD/QC system and azobenzene STFDs. These properties include photochemistry, energy storage, heat release, stability, and synthesis. While it was previously known that the introduction of donor–acceptor groups could lead to a red-shifted absorption spectrum that better matches the solar spectrum, our research has achieved an unprecedented solar spectrum match and high energy density by incorporating low-molecular-weight donor and acceptor groups. By strategically introducing steric hindrance in certain systems, the storage time of the photoisomer QC has been dramatically increased. Additionally, the utilization of dimeric systems with independent energy barriers has resulted in improved solar spectrum matching, prolonged storage times, and higher energy densities. These discoveries offer a range of chemical modification strategies that can be employed to tailor the properties of NBD/QC and *trans*/*cis* azobenzene systems, making them more suitable for specific applications. This knowledge serves as a valuable toolbox for researchers interested in designing efficient STFD systems.

Several test devices have been developed to validate the performance of these systems. For instance, a hybrid STFD device has been constructed, capable of simultaneously storing sunlight energy and heating water. Furthermore, we have designed a device that enables the monitoring of catalyzed QC to NBD conversion or *cis*-azobenzene to *trans*-azobenzene, facilitating the quantification of significant macroscopic heat generation. Additionally, various formulations of polymeric composites have been tested, demonstrating their ability to absorb light during the day and release the stored energy as heat during the night, thereby showcasing their potential for future applications in window coating. These laboratory-scale realizations serve as important steps in driving the field closer to the practical implementation of STFD systems.

To accurately determine the solar power storage capacity of a STFD system, it is crucial to estimate the overall solar energy storage efficiency, which takes into account the optical absorption of both isomers and covers the entire solar spectrum. The maximum estimated energy storage efficiency for an ideal STFD system can exceed 15%, depending on the height of the energy barrier and the specific model being considered. However, achieving these numbers in practical applications is challenging, primarily due to the potential overlap in the optical absorption ranges of photoswitchable molecules.

In order to demonstrate the feasibility of STFD systems in a laboratory setting, various experiments have been conducted. DSC allows for the validation of heat release from STFD systems on a small scale. For larger-scale devices, one approach is to employ heterogeneous catalysts in flow reactors to trigger energy release in molecular isomers. This technique enables straightforward separation of the catalyst, as well as direct and precise control of the energy release rate. In the 1980s, it was already demonstrated that heat release from QC derivative solutions could potentially achieve temperatures above 50°C. In recent experiments, utilizing a fully charged push-pull QC compound dissolved in a low specific-heat capacity solvent like toluene, temperature changes of up to 63 K (equivalent to 85°C absolute temperature) were achieved in less than 1 min of reaction time.

STFD systems can also be integrated with other solar technologies to further enhance solar energy utilization. As mentioned earlier, with a multijunction design, an ideal STFD system may achieve a maximum energy storage efficiency of up to 21% (based on simulations). Therefore, exploring the utilization of the remaining portions of the solar spectrum becomes an interesting avenue to investigate. One approach is to combine the STFD concept with other technologies, creating hybrid devices that can enhance the overall solar energy utilization.

Theoretically, it is possible to initially store some solar energy as chemical energy within the molecules themselves. The residual light can then be absorbed by a thermochemical reactor, where methanol (CH_3OH) decomposition occurs, resulting in the production of hydrogen and carbon oxide gas. This hybrid system has an estimated maximum solar-to-chemical efficiency of 69%, as depicted in Figure 5.21(a). To further enhance this concept, a photovoltaic system was added, as shown in Figure 5.21(b). However, this addition did not lead to an improvement in the overall solar utilization efficiency. Under these conditions, the system achieved a solar utilization efficiency of only 67%. This means that 7% of the incoming solar energy was stored in the STFD, 45% was utilized in the thermal chemical process, and the remaining 15% of photons were transformed into electricity. Interestingly, when considering the Hebei area in China, the estimated daily average solar utilization of this device was calculated to be 58% in summer and 45% in winter. These findings suggest that this particular design holds promise as a renewable energy strategy.

One example of a hybrid solar thermochemical device is described by Ghasemi *et al.* [21] In this system, a STFD is coupled with PCMs to achieve full solar spectrum utilization. Specifically, the researchers combined NBD, a STFD molecule, with a localized PCM consisting of carbonized rayon (L-PCM) with a composition of 54.1% KNO_3, 20.0% $NaNO_3$, and 25.9% $LiNO_3$. This hybrid system holds the potential for 24/7 energy delivery, as depicted in Figure 5.22(a). To maintain an appropriate temperature difference, the STFD molecules and PCMs are isolated by a low thermal conductivity, highly transparent silica aerogel. A heat transfer fluid (HTF) is employed to harvest the heat. During the daytime,

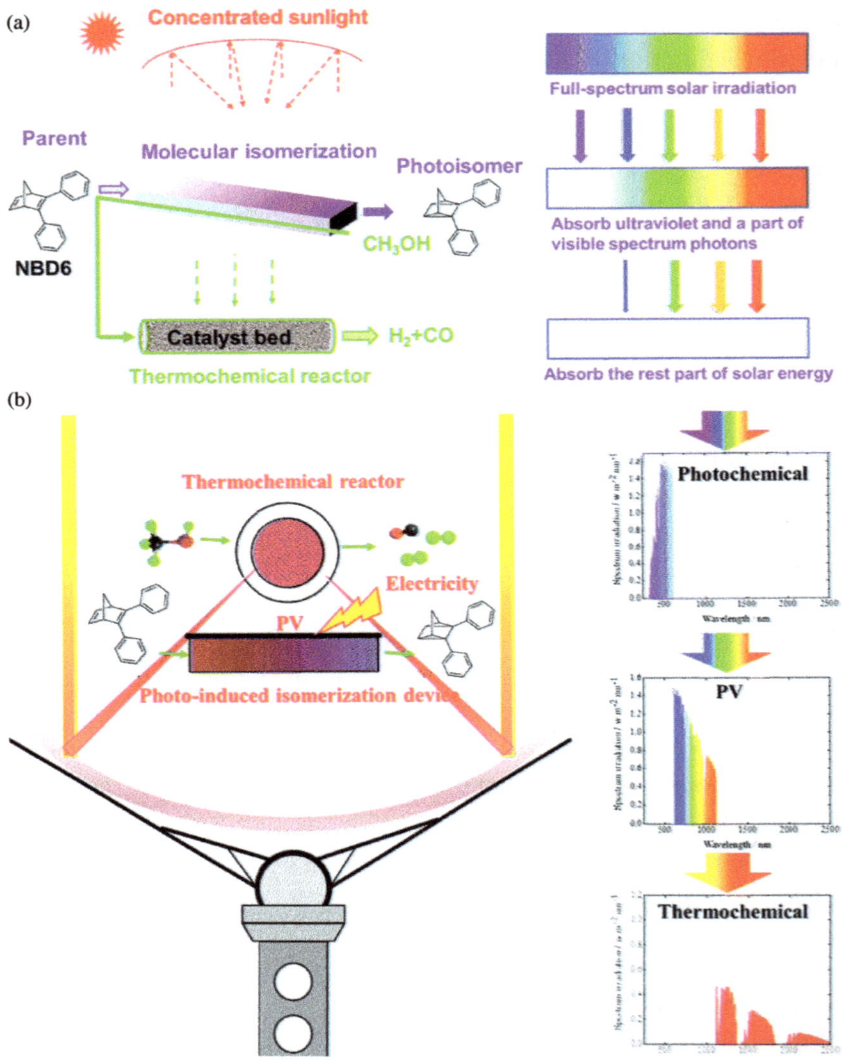

Figure 5.21. (a) Hybrid STFD device and the schematic utilization of sunlight. (b) Photovoltaic thermochemical device. Reprinted with permission from Ref. [20]. Copyright 2022, The Royal Society of Chemistry.

(a)

(b)

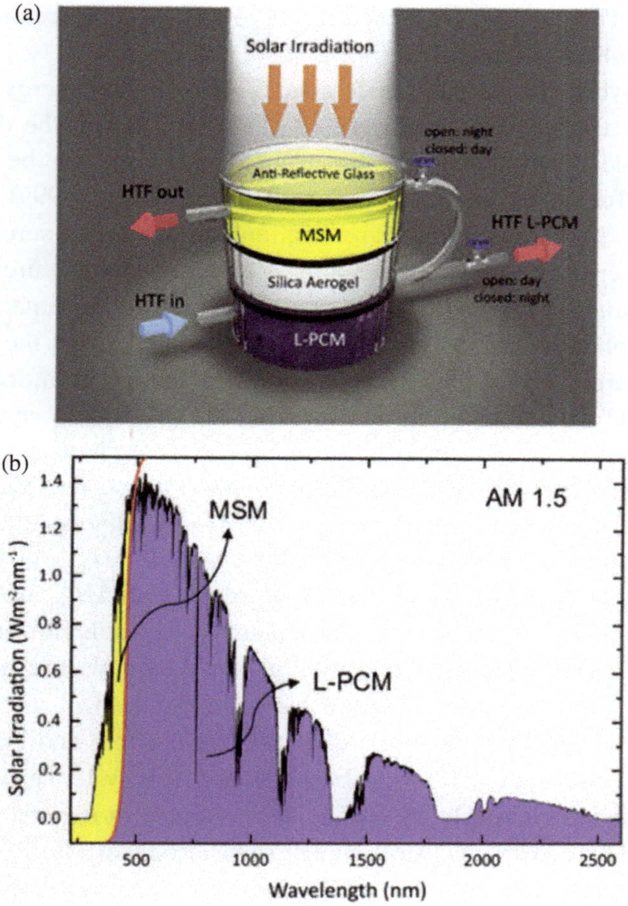

Figure 5.22. (a) STFD-PCM hybrid device. (b) The solar irradiation during the daytime. Reprinted with permission from Ref. [21]. Copyright 2019, Elsevier.

a valve connecting to the STFD layer is closed, allowing the STFD material to charge under solar irradiation. Simultaneously, the valve on the bottom PCM layer is opened, enabling the HTF to flow through the L-PCM, transporting the released heat for further heating, as shown in Figure 5.22(b). During the nighttime, the upper valve is opened, facilitating the release of the stored energy as heat from the charged STFD material. This release is initiated by the threshold temperature from the L-PCM, which is transferred by the HTF.

The hybrid device achieved a remarkable 73% solar energy harvesting efficiency from a solar irradiance of 2 kW m^{-2} during the daytime. In larger-scale devices, the two-dimensional side loss can be neglected, allowing for a maximum harvesting efficiency of up to 90%. The setup involving STFD and PCM hybridization utilized silica aerogel, which played a critical role in maintaining a constant temperature difference between the L-PCM and STFD material. During the daytime, up to 97% of solar photons can be absorbed by the L-PCM, with the remaining energy stored in the STFD material for release during nighttime use.

In addition to their applications in STFDs, MOST systems have been combined with material-based micro-electromechanical system thermoelectric generators (MEMS-TEG) to generate power. Moth-Poulsen *et al.* [22] conducted a study using an NBD derivative solution (0.78 M NBD4 in toluene) with specific enthalpy changes (ΔH_{isom}) of 93 kJ mol^{-1} and 0.37 MJ kg^{-1}, and an Azo derivative (Azo3) with ΔH_{isom} of 52 kJ mol^{-1} and 0.17 MJ kg^{-1}, together with a neat phase-changeable film and an ultrathin TEG. This setup resulted in the generation of electric power up to 1.3 W m^{-3}, as illustrated in Figures 5.23(a) and 5.23(b).

By utilizing both catalytic (CoPc@C fix-bedded) and optical (LED light irradiation) triggers for the back reaction, this system generated continuous power for nearly 30 min, as demonstrated in Figures 5.23(c) and 5.23(d). This remarkable achievement showcased that MOST technology can effectively convert solar energy into electricity, overcoming limitations imposed by time and geographical factors. Consequently, this technology can be considered a novel type of carbon-based, heavy metal-free solar battery. The physical properties of NBD and Azo couples led to the proposal of two heat-to-power generation device concepts, as depicted in Figure 5.23(b). In the NBD system, a saturated solution of NBD (QC) in toluene was prepared in Gothenburg, Sweden, using a photochemical flow reactor and transported to Shanghai, China. In one experiment, 5 mg of CoPc@C catalyst was loaded inside the catalytic reactor (in the central copper tube) with a thermally conductive pad filling the gap between copper plates. The reactor was placed above the MEMS-TEG chip, and the

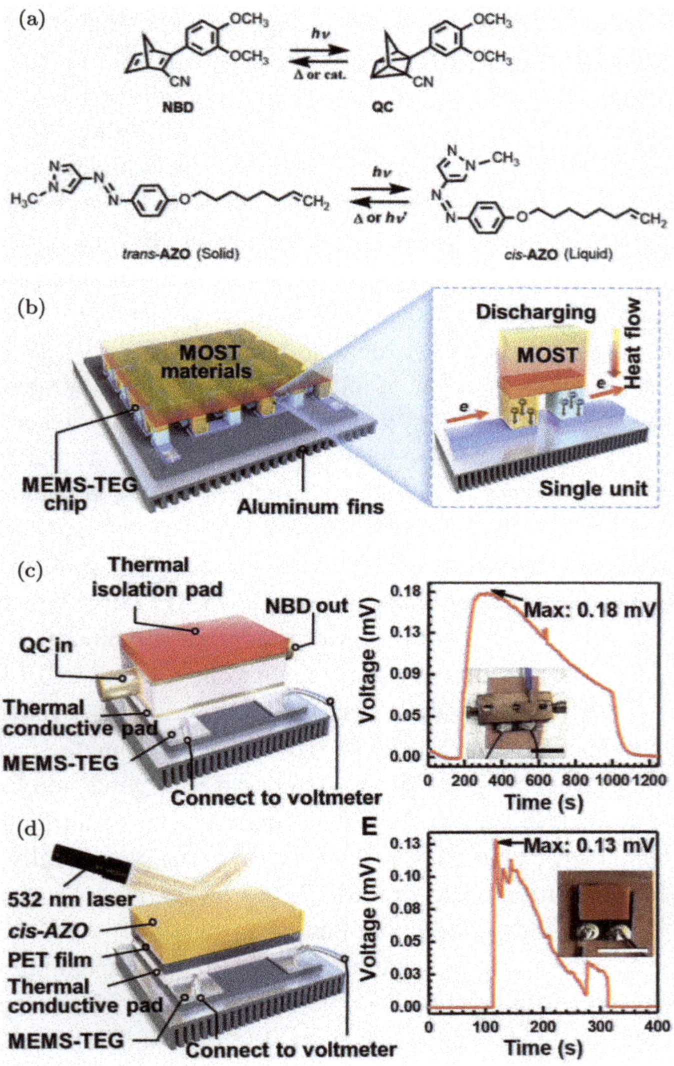

Figure 5.23. (a) molecular structures of NBD4 and Azo3. (b) MEMS-TEG device. (c) NBD4 couple-based MEMS-TEG experimental setup. (d) Schematic of Azo3 film-based MEMS-TEG experimental setup for power generation. Reprinted with permission from Ref. [22]. Copyright 2022, the authors.

complete setup was positioned on an aluminum brick heat sink, providing a stable ambient temperature.

To ensure efficient heat utilization, the hot side of the system was covered with thermally insulated plastic foam, reflecting the generated heat rather than dissipating it into the environment. With less than 1 ml of a 0.78 M NBD solution and a flow speed of 5 ml h^{-1}, a measured bias voltage of up to 0.18 mV was recorded. Based on the characterization of the MEMS-TEG chip, this voltage corresponds to a maximum power output of approximately 0.1 nW, resulting in a power output density of 1.3 W m^{-3}. Moreover, during a continuous operating test with a slightly lower concentration of 0.62 M in toluene and the same flow speed, a total electric energy of up to 0.44 W h m^{-3} was successfully generated over 24 min — a noteworthy achievement. In the case of Azo, a smaller amount of material was utilized. Specifically, 7 mg of *cis*-Azo, generated from a 365 nm LED light source, was loaded onto the thermally conductive pad. A polyethylene terephthalate (PET) film was carefully pressed onto the material, ensuring even coverage of the MEMS-TEG chip's surface. The back conversion was triggered by laser irradiation at 532 nm, as depicted in Figure 5.23(d), resulting in a bias output voltage of up to 0.13 mV.

An interesting observation was made regarding the generation of power from the NBD couple and the Azo couple. Despite the fact that the Azo couple reached a higher surface temperature, it generated less power compared to the NBD couple. This discrepancy can be attributed to the distinct heat release quantities of these two systems. Specifically, 1 mL of 0.78 M NBD solution could produce 72.5 J of energy, whereas 7 mg of *cis*-Azo could only produce 2.3 J of energy.

The researchers also noticed that the maximum temperatures attained by the NBD couple and the Azo couple were 37°C and 52°C, respectively, corresponding to generated voltages of 0.18 mV and 0.13 mV, respectively, as measured by the MEMS-TEG. However, when comparing the power density of the NBD and Azo couples' maximum power outputs to that of hot water at similar temperatures, a significant reduction in power density is observed. This disparity is likely due to the better thermal contact surface between the hot water rapidly flowing in the copper reactor and the pipe wall compared to the NBD solution and *cis*-Azo film being discharged. Consequently, even though the hot water possesses a temperature of 36.5°C, the power density generated by the MEMS-TEG is nearly five times higher than that of the NBD solution at a similar temperature (37°C). This highlights the importance of effective thermal management and the use of optimal thermoelectric (TE) materials, as they can potentially enhance the power output of future devices utilizing MOST materials.

In the experiments, an ultrathin-film-based MEMS-TEG device with high sensitivity was utilized to harness the released heat from the MOST system. This setup achieved a significant milestone by generating electricity for the first time. The power output reached up to 0.1 nW, resulting in a remarkable power output density of 1.3 W m^{-3}. It should be noted that this compact and localized solar energy storage-power generation system operates through a mechanism that is distinct from traditional combinations of photovoltaics (PV) and battery storage. Moreover, this device demonstrated the capability to continuously generate electricity for more than 25 min, representing a significant advancement in the field. It is important to acknowledge that this system is capable of charging and discharging across different geographic locations, as demonstrated in the study with the transport of the solution from Sweden to China over the course of a few months. This capability highlights the potential of solar energy storage for on-demand power generation. While there is room for improvement in terms of efficiency, this research paves the way for future applications of the MOST system in localized power generation. With further development and optimization, these systems hold promise for powering various local energy needs in the future.

The concept of STFDs involves collecting and storing sunlight through a photochemical reaction under flow conditions. When energy is needed, a solution containing the metastable molecule can be passed through a catalytic bed reactor to release heat energy. This heat can then be utilized for applications such as heating water or generating steam, as exemplified in a domestic implementation. The viability of STFD depends on the ability to store energy for an extended period, which requires a high energy barrier for the thermal conversion from the high-energy isomer to the parent molecule. To complete the energy storage and release cycle, an effective method of demonstrating heat release is crucial. From a molecular design perspective, solubility becomes a significant factor as the practical use of STFD depends not only on the energy density of the molecules but also the overall energy density of the system. Particularly in solutions, it is important to have a high weight percentage of molecules in the solvent.

For this purpose, the molecule 1-cyano-2-pmethoxyphenyl-norbornadiene 10 was designed to exhibit increased solubility in toluene. Experiments revealed that it could be dissolved at concentrations up to 1.5 M (334 g/l, Figure 5.24(a)). By testing several catalyst candidates, it was determined that CoPc could effectively trigger the release of energy from the molecule QC 10 to its NBD 10 form. To integrate the system into a small device, the catalyst was impregnated onto a solid activated carbon support (Figure 5.24(b)). The device was then placed in a high vacuum environment to minimize heat losses and exhibited temperature gradients

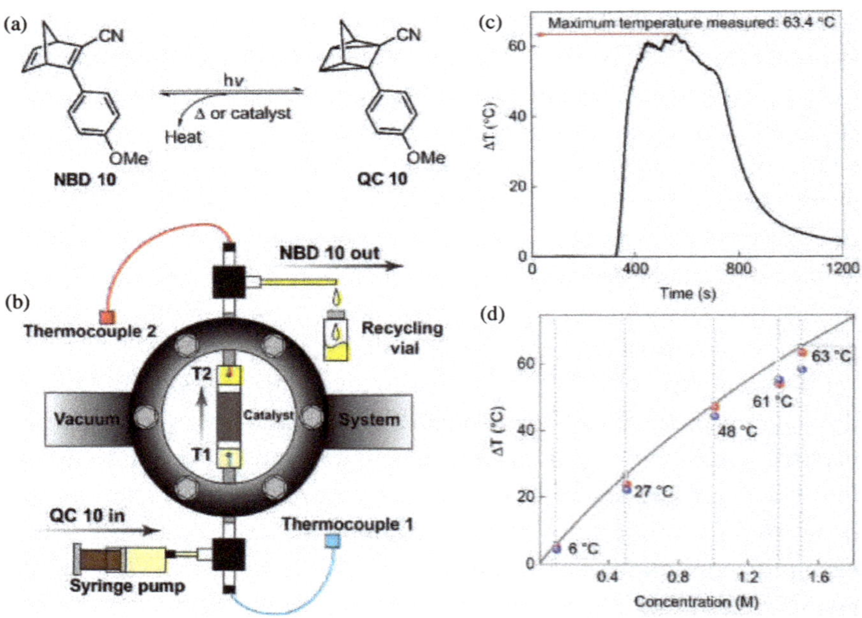

Figure 5.24. (a) Photoisomerization of NBD 10 to QC 10. (b) Illustration of the vacuum chamber. (c) Heat release thermogram of QC 10 with 63.4°C. (d) Graph of heat release vs concentration. Reprinted with permission from Ref. [22]. Copyright 2021, American Chemical Society.

of up to 63°C (85°C actual temperature, Figures 5.24(c) and 5.24(d)). Another advantageous feature of this system was its rapid ramp-up time of less than 1 min, which suggests potential use in load leveling alongside other heat sources for various applications.

The researchers in this study successfully synthesized and extensively characterized a scalable derivative of NBD/QC, which is crucial for the performance of a STFD system. The newly developed compound, NBD10, exhibited improved absorptivity with a maximum absorption wavelength at 326 nm and an absorptivity of 1.3×10^4 M^{-1} cm^{-1}. It also demonstrated a significant spectral difference from its corresponding QC10 counterpart. Other favorable attributes of NBD10 include a high photoisomerization quantum yield of 61%, a long half-life of 30 days at 25°C, and high solubility, with a maximum concentration (C_{max}) of 1.52 M for QC10 in toluene. Additionally, the researchers evaluated the robustness of NBD10 solution at a temperature of 85°C. The compound proved capable of enduring 43 storage and release cycles with minimal degradation, exhibiting a degradation rate of only 0.14% per cycle. These findings highlight the potential of NBD10 as a promising candidate for future STFD applications.

The ESD of the NBD10 to QC10 reaction was confirmed to be 0.4 MJ mol^{-1} through both DSC measurement and MP2 calculation. To demonstrate the feasibility of lab-to-site scale-up implementation, an outdoor test facility utilized a large ~900 cm^2 solar collector to drive the NBD10 to QC10 reaction. The successful achievement of quantitative photoconversion illustrates the potential for practical implementation. The efficiency of energy release was achieved by employing the CoPc@C heterogeneous catalyst with a small reaction center. This allowed for quick and efficient release of the stored energy. Notably, a macroscopic heat release of up to 63.4°C was measured using a 1.5 M solution of QC, surpassing previous measurements of STFD systems. Complementing the experimental work, a detailed computational study was conducted to investigate the catalytic reaction mechanism. The results indicated a low energy barrier in agreement with the observed fast conversion experimentally. In summary, a comprehensive proof-of-concept demonstration was achieved, showcasing both successful conversion under solar light and efficient heat release using a fixed-bed catalyst. Future molecular design efforts should focus on developing red-shifted derivatives of NBD to achieve better spectral overlap with the solar spectrum, while ensuring that other essential properties of STFD systems are maintained.

In a study conducted by Grossman *et al.* [24] in 2016, azobenzene STF polymers were utilized and functionalized to allow for deposition of uniform and large-area thin films from low-concentration solutions in methanol. This approach enabled the deposition to conform to various conducting substrates, whether they were flat or structured, depending on the intended application. The method employed in this study allowed for control over film growth by manipulating the conditions of electrodeposition, resulting in highly uniform films ranging from hundreds of nanometers to microns in thickness. The researchers demonstrated that this approach preserved the STF properties of the films, achieving energy densities of approximately 90 J/g. The films were also chargeable in the solid state and exhibited exceptional material utilization compared to other solid-state processing methods.

This novel approach holds relevance not only for STF systems but also for other applications that involve small functional molecules enabling solid-state functionalities. The development of a copolymer primarily comprised of an azobenzene monomer and sparsely incorporated 4-vinylbenzoic acid charging unit was a key aspect of this work (Figure 5.25(a)). It was important to minimize the content of the charging unit in the polymer while still enabling efficient electrodeposition (EPD). Once the polymer was dissolved, it could be electroplated onto the positive electrode from methanol with an applied bias ranging from 5 V to 10 V on various conductive electrodes. The process involved ionized carboxyl groups moving the polymer towards the anode, where an excess of cations induced a pH change, causing the polymer to become insoluble and form a solid-state coating (Figure 5.25(b)).

Based on the analysis of the electrodeposition (EPD) of the STF polymer and the resulting coatings, a potential application can be proposed in the form of STFDs. In this application, a thin metal wire, approximately 50 μm in diameter, is coated with the STF material and integrated into fabrics. The STF polymer used has a low thermal conductivity (<1 W $m^{-1}K^{-1}$). By combining it with high thermal conductivity metals such as copper (with conductivity up to 400 W $m^{-1}K^{-1}$), efficient heat

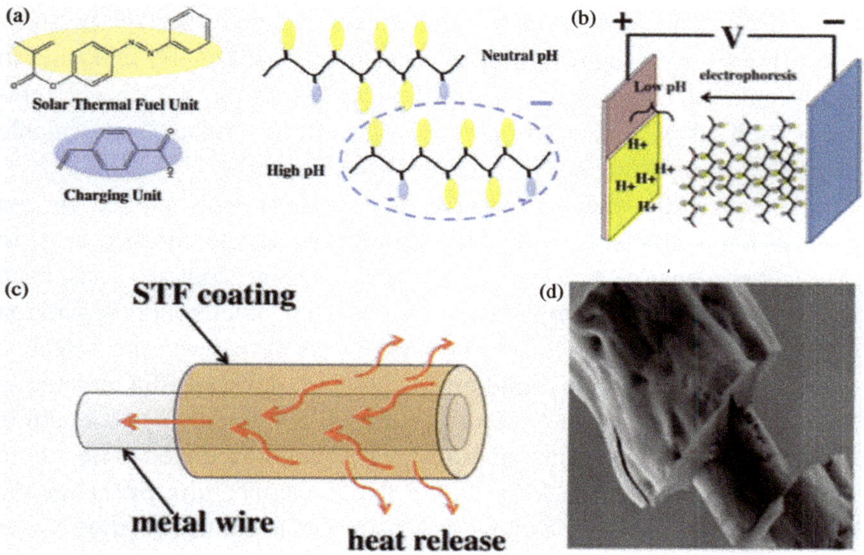

Figure 5.25. (a) Structures of the STF copolymer and charging units. (b) Diagrammatic sketch of the electrodeposition. (c) Structures of the STF fiber. (d) SEM morphology diagram of the fiber. Reprinted with permission from Ref. [24]. Copyright 2016, American Chemical Society.

transport to other parts of the fabric or device becomes possible. This is important since the area exposed to photon radiation (where the STF charges) may not necessarily overlap with the region that requires heat. The STF would charge with UV radiation and release heat into the wire upon triggering, enabling other portions of the fabric or device to be heated (Figure 5.25(c)). In Figure 5.25(d), the authors depict a wire with a STF coating as a proof-of-concept for such an STFD. The deposition conditions were optimized to achieve thicker coatings, comparable in thickness to the wire itself. It is noted that at higher film thicknesses, there may be greater film shrinkage during drying, leading to the introduction of cracks and deformities. Engineering higher molecular weight polymers or adding binders are suggested as potential remedies, representing promising future directions for optimizing such fiber materials.

The authors also discuss the implications of employing copolymers and their impact on STFD performance. The incorporation of charging units, while adding nonproductive mass from an energy storage perspective, allows for the introduction of novel functionalities. By choosing to incorporate the charging units instead of modifying the azobenzene molecule itself, they seek a compromise that offers a robust and scalable platform for imparting new properties to the materials.

Previously, the energy storage in STFDs using PCMs relied on π–π and van der Waals interactions, resulting in only finite enhancements to the energy density. However, a nanoconfinement approach has been reported to improve energy density by utilizing microphase separation (MPS) of a block copolymer. This approach allows for the selective dispersion of PCMs into nanodomains within a polymer film, creating a STF. In this approach, a liquid-crystalline block copolymer containing azobenzene as the continuous mesogenic phase is doped with photo inert organic PCMs. The MPS within the copolymer film maintains the periodic nanostructures, leading to a distinct nanoconfinement effect on the dopant PCM nanodomains. This results in a wider range of crystallization temperatures for the PCMs, from 24 to −38°C. The nanoscale MPS enables the simultaneous storage of natural sunlight and environmental heat, which can be released through mechanisms such as *cis*-to-*trans* isomerization and isotropic-to-liquid crystal phase transition of the azobenzene-containing polymer block at higher temperatures, as well as phase transition of the PCMs at lower temperatures, even below −30°C. The improved STFD performance achieved through this nanoconfinement approach has enabled the development of wearable warm fabrics and photothermal pneumatic actuators. These advancements pave the way for high-performance STFD systems utilizing MPS and nanoconfinement. This opens up possibilities for the application of STF materials in various fields, including wearable smart materials, warm fabrics, and actuators.

Utilizing azobenzene and its derivatives for photochemical storage and heat release in STFDs offers advantages such as higher energy density per unit volume, leading to smaller energy storage materials and reduced costs. The charging process of Azo molecules involves absorbing solar energy and converting it into thermal energy through photoisomerization (Figure 5.26(a)). Azobenzene moieties can be utilized as mesogens in liquid-crystalline polymers (LCPs), where they can undergo *trans*-to-*cis* isomerization upon photoirradiation, triggering an LC-to-isotropic

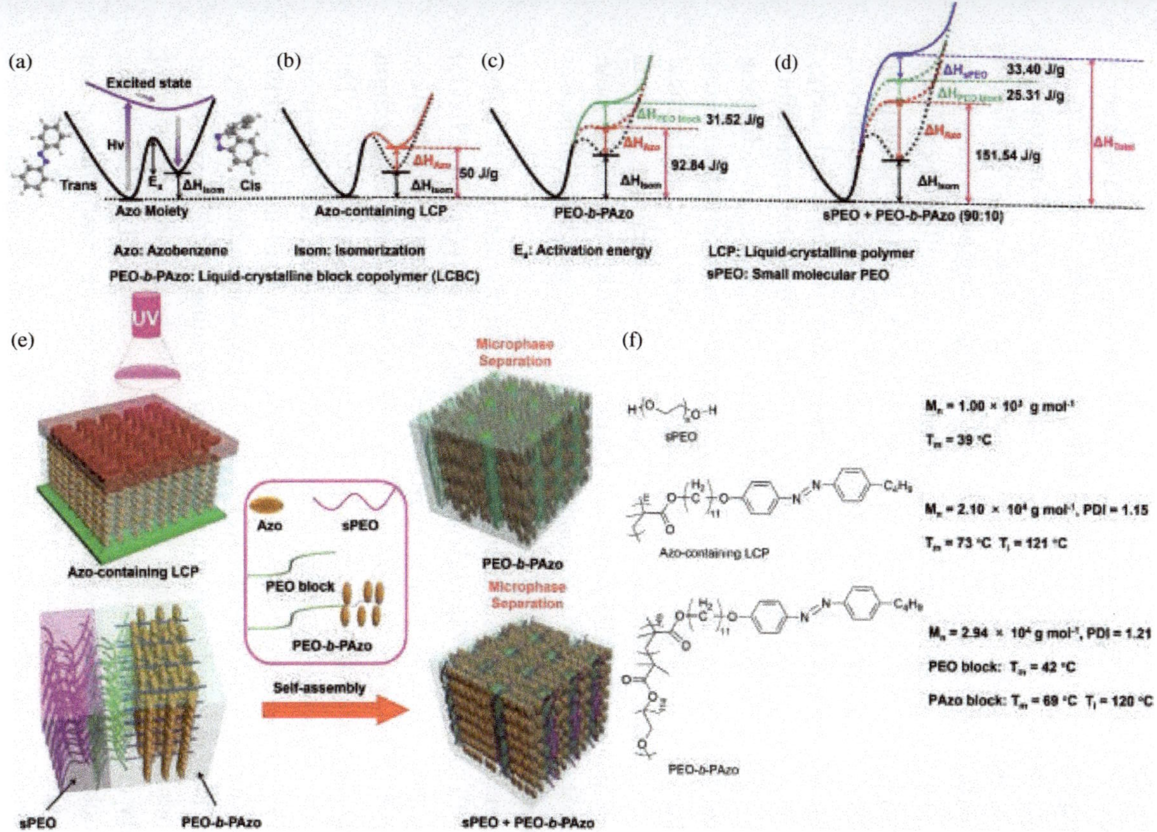

Figure 5.26. Solar thermal energy storage within (a) Azo moiety, (b) Azo-containing liquid-crystalline polymers (LCPs), (c) PEO-b-PAzo block copolymer (BC), and (d) composite of sPEO and PEO-b-PAzo. (e) Illustrations of Azo-containing LCP upon UV irradiation, MPS of PEO-b-PAzo, and the self-assembly of sPEO and PEO-b-PAzo composites. (f) Chemical structures of sPEO, Azo-containing LCP, and block copolymer PEO-b-PAzo. Reprinted with permission from Ref. [25]. Copyright 2021, American Chemical Society.

phase transition. This process allows for storing latent heat through the phase-change process, thus increasing the energy density of these LCPs (Figure 5.26(b)).

However, the low light transmittance of Azo-containing LCPs presents a challenge, hindering the effective utilization of light (Figure 5.26(e)) [25]. This inefficient charging process not only limits real-time energy harvesting and results in low energy density but also causes significant energy waste. To address these limitations, organic composites have been explored by incorporating photo inert PCMs such as paraffin waxes, fatty acids, poly(ethylene glycols), and sugar alcohols into Azo-containing molecules or polymers. This approach allows for the simultaneous absorption of ambient heat and light energy, leading to improved energy density. However, these composites often suffer from poor compatibility, resulting in macroscopic phase separation and difficulties in achieving film formation. Moreover, triggering heat release at low temperatures, particularly below 0°C, remains a challenge. Therefore, a chemical approach that effectively combines PCMs with Azo-containing LCPs is necessary. Nanoscale MPS presents a promising solution to address these issues, as it allows for precise control and organization of different phases at the nanoscale.

Figure 5.26(f) shows a designed LC block copolymer (LCBC), specifically PEO-b-PAzo, where the separated phase of poly(ethylene oxide) (PEO) serves as a light-inactive PCM, while the continuous phase of Azo-containing LCP serves to store solar energy through photoisomerization and latent heat via LC-to-isotropic phase transition (Figure 5.26(c)). The goal is to efficiently combine organic PCMs with the Azo-containing LCBC to improve energy density. To achieve this, a small-molecule PEO (sPEO) with a lower molecular weight (ca. 1 kDa) is introduced into PEO-b-PAzo. This is based on the principle that blending a block copolymer with a homopolymer identical to one of the blocks can undergo similar MPS. By comprehensively utilizing the Azo-containing LCP block, the PEO block, and the additive sPEO, both solar energy and ambient heat can be stored in the composites (Figure 5.26(d)).

Furthermore, it is known that melts can be supercooled below their melting points (T_m) prior to crystallization, and the degree of supercooling can be influenced by the properties of the solid in contact with the melt.

At the nanoscale, a suitable combination of solid and melt portions allows for capturing the melt and retaining it for an extended period. In this case, sPEO and the PEO block self-assemble to form a nanoscale microphase through MPS. This nanoscale confinement leads to the crystalline temperature of sPEO decreasing from 24°C to −38°C. Essentially, thermal energy derived from the crystallization of sPEO can be retained at low temperatures due to the nanoconfinement of MPS. Consequently, the composites are capable of storing natural sunlight and latent heat. This energy can then be released.

The phenomenon of photochemical crystal ↔ liquid transitions (PCLTs) combines reversible photochemical transformations with thermophysical phase transitions. One potential application of PCLTs is the development of photo-responsive smart materials capable of exerting reversible adhesion on specific surfaces at desired timing, which is not achievable with conventional adhesives. In most PCLT-based adhesives, UV light is used as the stimulus, which can cause material degradation and health concerns. However, Li *et al.* [26] have developed smart and robust adhesives that can be controlled by visible light. These adhesives utilize small molecule Azo photo switches capable of efficient *trans*-crystal → *cis*-liquid and *cis*-liquid → *trans*-crystal transitions under 405 nm and 532 nm light irradiation, respectively. In their *trans*-crystal state, these adhesives exhibit strong adhesion strengths on various substrates, such as 1.13 MPa on quartz/quartz and 1.58 MPa on wood/wood. The glued substrates can be easily separated within 1 second upon light-induced transition to the *cis*-liquid state, resulting in a loss of adhesion strength. Furthermore, these adhesives demonstrate robust switching of adhesion strength in multiple cycles and can also function effectively in underwater environments. By utilizing visible light-controlled reversible PCLTs, these adhesives offer a promising strategy for the development of STFDs and high-performance photo-responsive adhesives.

The researchers initially designed a visible-light-activated Azo switch called 4-dimethylamino-substituted azopyrazole (Figure 5.27). They investigated the application of three different molecules as photo-switchable and reversible adhesives based on efficient bidirectional PCLTs. In their experiments, they used two quartz plates as adherends with an adhesion area of 0.25 cm^2, containing 0.5 mg of

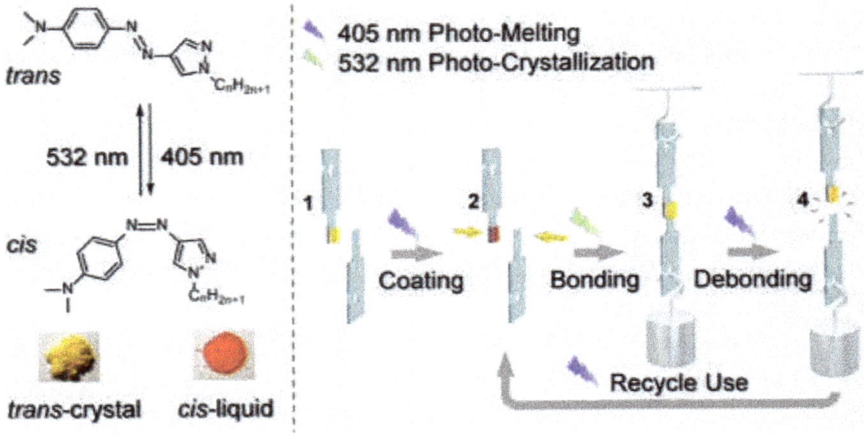

Figure 5.27. Structural of azopyrazole molecules and diagrammatic sketch of the azopyrazole adhesion procedure. Reprinted with permission from Ref. [26]. Copyright 2022, American Chemical Society.

4-dimethylamino-substituted azopyrazole. The adhesive was coated onto the substrates by converting the *trans*-crystal state to a *cis* liquid state through irradiation with 405 nm light (160 mW cm^{-2}, 4 min). Another quartz plate was then placed on top of the liquid sample, followed by applying pressure to bond them together. To initiate the photochemical crystallization process and strengthen the adhesive bond, 532 nm light (180 mW cm^{-2}, 2 min) was applied.

Debonding the adhered plates could be easily achieved by irradiating the adhesive with 405 nm light. This step only required photo melting near the interface between the 4-dimethylamino-substituted azopyrazole and the irradiated substrate. Additionally, complete photo melting allowed for the recycling and reuse of the adhesive material, enabling repeatable bonding and debonding behaviors. In summary, the researchers successfully demonstrated the photo-switchable and reversible adhesive properties of 4-dimethylamino-substituted azopyrazole. By employing specific light wavelengths, they could control the adhesion and debonding processes, thereby allowing for the recycling and repetitive use of the adhesive material.

The development of solid-state functional coatings is an emerging and promising direction for STFDs. Research is currently underway to design new solid STFD materials based on the principles discussed earlier. The goal is to integrate these materials into functional coatings for future window applications, enabling the storage of incoming sunlight. Future research directions could focus on developing methods to trigger the isomerization of molecules, such as the conversion from QC to NBD or the *cis* to *trans* isomerization of azobenzene, using external stimuli. Catalysts or heat have shown effectiveness in the past, and recent studies have demonstrated that modified *trans*/*cis* azobenzene and NBD/QC systems can be optically modulated, transforming them into complete photoswitchable systems. The vision for the future includes developing a heat-generating STFD battery that utilizes a carbon-based molecular material for solar energy capture, storage, and heat generation. Unlike traditional photovoltaics and battery combinations, this system would operate based on the principles of STFD, where light energy is converted and stored within the molecule itself without the need for an external energy storage unit. It is estimated that the maximum STFD energy storage efficiency (20.5%) surpasses that of photosynthesis (0.1–0.3%). However, further advancements are required to reach the conversion efficiency of recent solar cells, which can achieve up to 47.1% with a six-junction flat-plate terrestrial design under high concentrations of sunlight.

The STFD concept relies on the ability to convert and store light energy within the molecule itself, eliminating the need for an external energy storage unit [27]. This concept aims to create heavy metal-free and highly robust systems for future devices. The performance of an ideal STFD device is closely linked to the properties of the selected molecular candidate. Advancements in synthetic, physical, and organic chemistry hold the potential to reveal more red-shifted STFD molecules that have less spectral overlap with their photo isomers. To evaluate these molecules, future small- and large-scale cycling tests will be essential. Fortunately, as technology continues to improve, the cost of implementing large-scale industrial applications can be offset by the energy savings achieved. Additionally, the use of machine learning and artificial intelligence (AI) technology for molecular screening can expedite the discovery of new candidates. Furthermore, the automation of chemistry systems enhances the potential for scaling up production. While the primary focus for STFD systems is on large-scale applications, such as industrial energy storage, there are also unique opportunities for niche and small-scale applications. For instance, wearable STFD cloth or hybrid devices could be potential future areas where STFD systems find applications.

References

[1] Kolpak AM, Grossman JC. (2013). Hybrid chromophore/template nanostructures: a customizable platform material for solar energy storage and conversion. *J Chem Phys*, **138**(3), pp. 034303.

[2] Fu LX, Yang JX, Dong LQ, Yu HT, Yan QH, Zhao FL, Zhai F, Xu YH, Dang YF, Hu WP, Feng YY, Feng W. (2019). Solar thermal storage and room-temperature fast release using a uniform flexible azobenzene-grafted polynorborene film enhanced by stretching. *Macromolecules*, **52**(11), pp. 4222–4231.

[3] Zhao RY, Li YC, Bai JY, Bai JY, Mu JH, Chen LX, Zhang N, Han JS, Liu FS, Yan SK. (2022). Preparation of flexible photo-responsive film based on novel photo-liquefiable azobenzene derivative for solar thermal fuel application. *Dyes Pigm*, **202**, pp. 110277.

[4] Yang WX, Feng YY, Si QY, Yan QH, Long P, Dong LQ, Fu LX, Feng W. (2019). Efficient cycling utilization of solar-thermal energy for thermochromic displays with controllable heat output. *J Mate Chem A*, **7**(1), pp. 97–106.

[5] Feng WK, Feng YY, Wang SF, Feng W, Yi WH, Gong QH. (2010). Intramolecular photoinduce delectron-transfer in azobenzen *E*-perylene diimide. *Carbon*, **48**(11), pp. 3091–3096.

[6] Zhang XQ, Feng YY, Huang D, Li Y, Feng W. (2010). Investigation of optical modulated conductance effects based on a graphene oxide-azobenzene hybrid. *Carbon*, **48**(11), pp. 3236–3241.

[7] Li SW, Feng YY, Long P, Oin CQ, Feng W. (2017). The light-switching conductance of an anisotropic azobenzene-based polymer close-packed on horizontally aligned carbon nanotubes. *J Mater Chem C*, **5**(21), pp. 5068–5075.

[8] Hu J, Huang S, Yu MM, Yu HF. (2019). Flexible solar thermal fuel devices: composites of fabric and a photoliquefiable azobenzene derivative. *Adv Energy Mater*, **9**(37), pp. 1901363.

[9] Wang H, Feng YY, Gao J, Fang WY, Ge J, Yang XY, Zhai F, Yu YF, Feng W. (2022). Metallic-ion controlled dynamic bonds to co-harvest isomerization energy and bond enthalpy for high-energy output of flexible self-heated textile. *Adv Sci*, **9**(20), pp. e2201657.

[10] Zhao XZ, Feng YY, Qin CQ, Yang WX, Si QY, Feng W. (2017). Controlling heat release from a close-packed bisazobenzene-reduced-graphene-oxide assembly film for high-energy solid-state photothermal fuels. *ChemSusChem*, **10**(7), pp. 1395–1404.

[11] Orrego-Hernandez J, Dreos A, Moth-Poulsen K. (2020). Engineering of norbornadiene/quadricyclane photoswitches for molecular solar thermal energy storage applications. *Acc Chem Res*, **53**, pp. 1478–1487.

[12] Wu S, Li T, Zhang ZY, Li T. (2021). Photoswitchable phase change materials for unconventional thermal energy storage and upgrade. *Matter*, **4**, pp. 3385–3399.

[13] Wang ZH, Losantos R, Sampedro D, Morikawa M, Borjesson K, Kimizuka N, Moth-Poulsen K. (2019). Demonstration of an azobenzene derivative based solar thermal energy storage system. *J Mater Chem A*, **7**, pp. 15042.

[14] Han GD, Li HS, Grossman JC. (2017). Optically-controlled long-term storage and release of thermal energy in phase-change materials. *Nat Commun*, **8**, pp. 1446.

[15] Fei L, Yin YJ, Zhang JJ, Wang CX. (2020). A visible energy management by photochromic solar thermal fuel using a color display. *Sol RRL*, **4**, pp. 2000499.

[16] Lv XD, Wang WZ, Yu HF. (2019). A bioinspired photothermal pneumatic device enabling optical manipulation of microfluid toward precise control of microreactions. *Adv Eng Mater*, **21**, pp. 1900977.

[17] Liu H, Tang JW, Dong, LQ, Wang H, Xu TY, Gao WC, Zhai F, Feng YY, Feng W. (2021). Optically triggered synchronous heat release of phase-change enthalpy and photo-thermal energy in phase-change materials at low temperatures. *Adv Funct Mater*, **31**, pp. 2008496.

[18] Lv XD, Yu MM, Wang WZ, Yu HF. (2021). Photothermal pneumatic wheel with high loadbearing capacity. *Compos Commun*, **24**, pp. 100651.

[19] Shangguan Z, Sun WJ, Zhang ZY, Fang D, Wang ZH, Wu S, Deng C, Huang XH, He YX, Wang RZ, Li TX, Moth-Poulsen K, Li T. (2022). A rechargeable molecular solar thermal system below 0°C. *Chem Sci*, **13**, pp. 6950.

[20] Wang ZH, Holzel H, Moth-Poulsen K. (2022). Status and challenges for molecular solar thermal energy storage system-based devices. *Chem Soc Rev*, **51**(17), pp. 7313–7326.

[21] Kashyap V, Sakunkaewkasem S, Jafari P, Nazari M, Eslami B, Nazifi S, Irajizad P, Marquez MD, Lee TR, Ghasemi H. (2019). Full spectrum solar thermal energy harvesting and storage by a molecular and phase-change hybrid material. *Joule*, **3**, pp. 3100–3111.

[22] Wang ZH, Wu ZH, Hu ZY, Orrego-Hernandez J, Mu E, Zhang ZY, Jevric M, Fu XC, Wang FD, Li T, Moth-Poulsen K. (2022). Chip-scale solar thermal electrical power generation. *Cell Rep Phys Sci*, **3**, pp. 100789.

[23] Wang ZH, Roffey A, Losantos R, Lennartson A, Jevric, M, Petersen AU, Quant M, Dreos A, Wen X, Sampedro D, Borjesson K, Moth-Poulsen K. (2019). Macroscopic heat release in a molecular solar thermal energy storage system. *Energy Environ Sci*, **12**, pp. 187–193.

[24] Zhitomirsky D, Grossman J. (2016). Conformal electroplating of azobenzene-based solar thermal fuels onto large-area and fiber geometries. *ACS Appl Mater Interfaces*, **8**, pp. 26319–26325.

[25] Cai F, Song TF, Yang BW, Lv XD, Zhang LQ, Yu HF. (2021). Enhancement of solar thermal fuel by microphase separation and nanoconfinement of a block copolymer. *Chem Mater*, **33**, pp. 9750–9759.

[26] Huang XH, Shangguan ZC, Zhang ZY, Yu CY, He YX, Fang D, Sun WJ, Li YC, Yuan CR, Wu S, Li T. (2022). Visible-light-induced reversible photochemical crystal-liquid transitions of azo-switches for smart and robust adhesives. *Chem Mater*, **34**, pp. 2636–2644.

[27] Wang ZH, Erhart P, Li T, Zhang ZY, Sampedro D, Hu ZY, Wegner HA, Brummel O, Libuda J, Nielsen MB, Poulsen KM. (2021). Storing energy with molecular photoisomers. *Joule*, **5**, pp. 3116–3136.

Chapter 6

Applications of Solar Thermal Fuel

Abstract

Researchers have improved solar thermal fuel (STF) by changing the microstructure of the monomer or skeleton to address its shortcomings, such as low storage energy density, low quantum yield, and irreversible degradation. At present, STF is capable of being used in various fields, including developing practical deicing films/coatings for vehicles and buildings, preparing self-heating clothing, and solar thermal cells for storing and converting solar energy. Photothermal materials can also convert low-order solar energy into high-order heat and steam energy for seawater desalination, among other applications. Therefore, based on a comprehensive analysis of the advantages and disadvantages of STFs currently under research, this chapter introduces the application progress of STF in various fields and evaluates their comprehensive performance in these fields, which provides an important reference for their future development directions. By conducting research on these promising topics that have infinite possibilities, we can enhance our understanding of STF and promote the diversified application of solar energy collection and storage materials.

In recent years, in view of the shortcomings of STF, such as low storage energy density, low quantum yield, short half-life, narrow overlap with the solar spectrum, irreversible degradation and so on, scientific researchers have improved it by changing the microstructure of monomer or skeleton, so that STF can be used in some fields. As mentioned above, STF carries out sufficient charge and heat release cycle under low-temperature conditions, which can be used to develop practical deicing films/coatings for vehicles, buildings, and other deicing films or to prepare self-heating clothes, and can also be used in the field of heating, solar thermal cells can also be prepared to store and convert solar energy, solar evaporation of photothermal materials can also directly convert low-order solar energy into high-order heat and steam energy, such as seawater desalination and other applications.

However, it is still a challenging task to improve the storage energy density, energy barrier and absorption efficiency, inhibit degradation, and reduce costs at the same time. Therefore, there is still much room for improvement in the basic theory and practical application of STF. Through the research on these promising topics with infinite possibilities, we can not only enhance our understanding of STF, but also promote the diversified application of solar energy collection and storage materials.

6.1 Heating field

6.1.1 *NBD/QC*

After receiving sunlight, its photoswitchable parent molecule undergoes chemical isomerization to create a metastable high-energy species. By integrating the photoswitch into solid materials or employing liquid-based molecular optical switching technology (MOST) systems, this notion has been shown to be workable. When a system is constructed on a solution, a catalyst can be employed to release the energy that has been stored there "on demand" in the form of heat, thus reproducing the original molecule. Unlike conventional solar thermal concepts, this system operates in a distinct manner by converting photons into chemical energy, which can be stored at ambient temperature. As a consequence, practical devices don't require insulation materials. For a system to have the best possible photoswitchable molecular design, numerous elements must be taken into account. A robust absorption is ideal. The optimal scenario is a substantial ultraviolet–visible (UV–Vis) absorption by the original molecule with no absorption by the equivalent high-energy isomer. In order to increase the energy density, additional requirements encompass a high photoisomerization quantum yield, exceptional durability, and a low molecular weight.

For solutions, it is advantageous for the solvent to possess a moderate heat capacity while facilitating a substantial amount of photoswitching.

Under flow-related circumstances, sunlight is harnessed and reserved through a photochemical response. When energy is required, the metastable molecule solution can be directed through a catalytic bed reactor to release the stored energy as heat. In this specific scenario, the liberated heat can be utilized for heating water or generating steam.

The molecular solar thermal energy storage system, developed by Swedish Chalmers, is a promising way to store solar thermal energy. The system is an energy storage system based on a molecular light switch, where light-switchable molecules absorb sunlight and undergo chemical isomerization to a sub-stable high-energy state (Figure 6.1(a)).

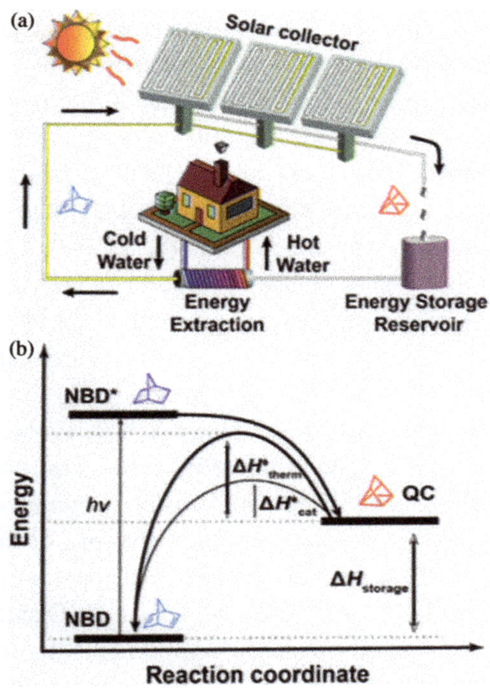

Figure 6.1. (a) Operating setup based on the system. (b) Illustration of molecules. $\Delta H^{\ddagger}_{therm}$ and $\Delta H^{\ddagger}_{cat}$ represent the activation enthalpy for quadricyclane-norbornadiene (QC-NBD) by heat and a catalyst, respectively. $\Delta H_{storage}$ corresponds to the energy storage enthalpy. Reproduced under the terms of the CC BY 4.0 from Ref. [1]. Copyright 2017, the author(s).

The liquid is a new "deglomerated diene derivative," a liquid composed of carbon, hydrogen, and nitrogen, which, when irradiated by sunlight, rearranges the bonds between atoms to form a high-energy isomer, where the energy from the sun is trapped between the strong chemical bonds of the isomer and the high-energy state is maintained despite the fact that the molecule cools to ambient temperature (Figure 6.1(b)). Isomer can be maintained despite the fact that the molecule cools to ambient temperature. When energy is needed, the catalyst fluid is simply pumped through the catalyst, which transforms the molecule back into its original state and releases heat as energy.

This renewable, emission-free energy device consists of a concave reflector with a central duct that tracks the sun like a satellite dish. The system works in a cyclic fashion, where fluid pumped through a transparent tube is heated by sunlight, converts molecular deglacial dienes into the isomer tetracyclic alkanes to store heat, and then stores the fluid at room temperature. When energy is needed, the fluid is filtered through a special catalyst, the fluid is heated 63°C, the fluid molecules are converted back to their original form for recycling, and the heat can be used in home heating systems to power water heaters, dishwashers, clothes dryers, etc. [1].

Researchers have demonstrated through 125 cycles that the liquid is reusable and that the process of heat absorption and release does not cause significant loss of molecules. The researchers claim the liquid can now hold 250 Wh kg^{-1} of energy, which is twice the energy capacity of a Tesla Powerwall battery.

This team then improved the MOST system for norbornadiene by substituting the norbornadiene and synthesizing a set of norbornadiene derivatives (Figure 6.2(a)). The NBD derivatives contain two substituents, producing a so-called "push–pull" system across one of the olefins, which causes a redshift in absorption and improves compatibility with the solar spectrum. Both approaches were utilized to make NBD and produced abundant amounts of the desired substances. Importantly, the majority of these compounds exhibit liquid behavior at normal temperature and possess essential qualities for energy storage applications. The molecular system can be pumped without the use of additional solvents because the neat deglomerated diene liquid has shear viscosity on the same order of magnitude as canola oil. This is a significant step toward a solvent-free system with high energy density. They have recorded energy storage densities of

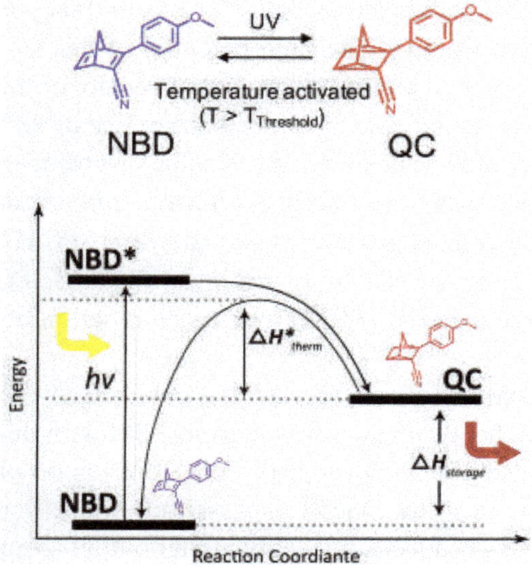

(a)

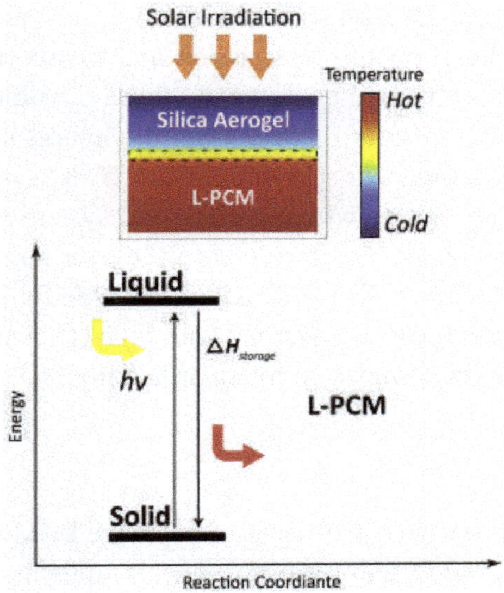

(b)

Figure 6.2. (a) The concept of molecular energy storage is illustrated. The NBD derivative undergoes photoisomerization to the corresponding QC derivative upon exposure to UV radiation and undergoes back-isomerization upon thermal initiation. (b) The concept of phase-change energy storage is shown. Energy is stored in the form of enthalpy. Reprinted with permission from Ref. [2]. Copyright 2018, Wiley.

high to 637 kJ kg^{-1} (estimated as 379 kJ kg^{-1}), and they can absorb up to 10% of the solar spectrum. The high potential of this substance for solar energy storage is demonstrated by the energy density of this system, which translates into a potential adiabatic temperature rise of 384°C [2].

Ghasemi's group has designed a full solar spectrum energy harvesting and storage device, which is mainly composed of molecular storage material (MSM) and localized phase-change material (L-PCM) (Figure 6.2(b)). MSM is chosen from the NBD-QC system, while the L-PCM is composed of a phase change material (PCM) with a combination of KNO_3, $NaNO_3$, and $LiNO_3$.

In order to start the phase change, the device receives solar radiation, transforms it into heat, and then supplies the PCM with this heat. Although the PCM exhibits an absorption peak in the UV region of the solar spectrum (λ = 300 nm), the MSM is essential for efficiently capturing UV energy. MSM produces noticeably higher output temperatures (T_{out}) at night while maintaining system efficiency by releasing absorbed energy during the back-isomerization process [3].

Using latent heat, photoisomerization and phase changes can store energy. capture solar thermal energy and offer a possible nonstop energy supply. Even after 35 cycles of operation, the hybrid system still effectively absorbs the entire solar spectrum. At 2 kW m^{-2} solar irradiation, we were able to harvest energy with a high hybrid system efficiency of 73% during the day, which can be enhanced to 90% for large-scale operation and 80% during the night. The new model available for installation opens the way for solar energy to be obtained with high efficiency and low operating costs, laying the foundation for solar heating [3].

6.1.2 *Azo*

One of the most promising approaches to manage PCM energy consumption throughout a wide temperature range is optically induced energy release. By incorporating photochromic molecules into organic PCMs, this technique facilitates photo-induced reversible structural changes in PCMs, governing their phase change at designated temperatures. The significance of photochromic molecules' phase change is pivotal in this process. Through this approach, both the enthalpy associated with phase change (PC) and the photo-thermal (PT) energy of PCMs can be concurrently released. The UV and visible light-triggered reversible isomerization of azobenzene derivatives, shifting from the metastable *cis* (*Z*)-isomer to the stable *trans* (*E*)-isomer, has established them as optimal candidates for orchestrating structural shifts on a molecular level. These derivatives have found widespread employment in harnessing light-induced thermal energy for storage and might hold promise in data storage applications. Consequently, the extent and pace of azobenzene isomerization profoundly influence the efficacy of heat release regulation and heat storage stability at different temperatures, particularly when integrating azobenzene molecules into organic PCMs.

The disparity in structure between azobenzene and organic PC molecules poses a hurdle to their utilization in achieving optically triggered latent heat release (both phase change and photo-thermal) across a wide temperature spectrum, despite the inherent benefits of azobenzene. Evident differences in PC temperatures between azobenzene and organic PC molecules, coupled with steric hindrance and molecular interactions at their juncture, impact the isomerization of azobenzene. Consequently, achieving simultaneous PC transition of azobenzene and PC molecules becomes a complex endeavor. Achieving consistent dispersion of azobenzene within PC molecules mandates a degree of resemblance in the molecular architectures of the two, contingent on the interaction between them. This approach entails modifying molecular alignment to introduce an energy barrier for PC crystallization. This observation underscores the potential for modulating the crystallization temperature (T_c) and PC enthalpy (ΔH) of PCMs by manipulating the quantity and structure (substituents) of azobenzene.

Organic PCMs face a noteworthy constraint when it comes to efficiently combining heat storage and release, particularly at extremely low temperatures, specifically below a critical threshold known as T_c. In the case of most organic PC molecules, heat (or enthalpy) is released through solid–liquid phase changes occurring above 0°C. However, as T_c is further decreased, it tends to diminish the enthalpy change (ΔH) associated with the PCM.

To address this limitation, an effective approach has been introduced: supercooling, which allows for the storage of PCM potential at temperatures below T_c. This is achieved through the integration of sub-stable photochromic molecules possessing remarkably low T_c values, thus significantly influencing the crystallization behavior of the PC molecules, ultimately resulting in the attainment of a supercooled state. Consequently, the extent of supercooling, indicated by the reduction in T_c, represents the minimum temperature threshold at which heat storage and release can occur.

Grossman's group reports photocontrolled composites composed of tridecanoic acid and azobenzene dopants that are functionalized through tridecanoic acid chains (Figure 6.3). These dopants have an activation energy barrier that switches between isomers, allowing the composites to

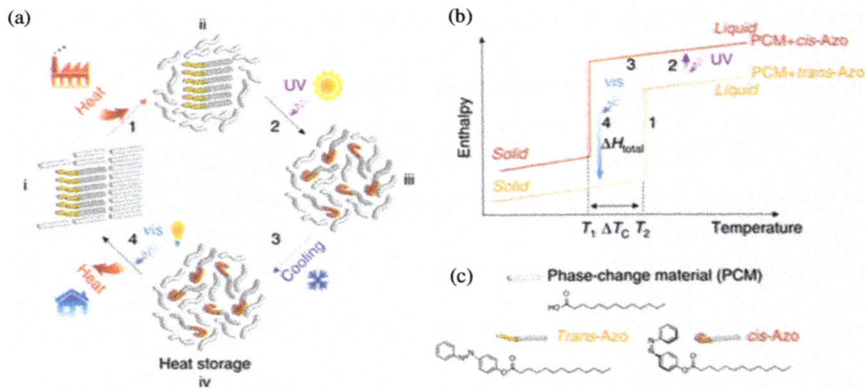

Figure 6.3. Optically controlled thermal energy storage and release cycle. (a) Schematic of (1) thermal energy absorption by phase-change materials (PCM) composite, (2) UV illumination for photoisomerization of azobenzene dopants, (3) cooling of liquefied PCM composite, and (4) visible-light (vis)-triggered reverse isomerization and heat release. The straight and curved rods exhibit different degrees of translational and rotational freedom, i.e., different phases. (b) Schematic energy diagram of PCM composites containing uncharged (*trans*-Azo) or charged (*cis*-Azo) dopants, showing enthalpy and temperature changes along (1)–(4) stepwise process for the thermal energy storage and release cycle. T_1 and T_2 are crystallization points of charged and uncharged composites, respectively, and ΔT_c is the difference between T_1 and T_2. ΔH_{total} represents the expected heat release from the optically discharged composite. (c) Chemical structures of PCM (tridecanoic acid) and the azobenzene dopant, functionalized with tridecanoic ester group, in each isomeric form.

release stored energy when optically triggered. UV irradiation raises the energy level of the molecular dopants, and even when the temperature cools below the original crystallization point, the energetic composites remain in the liquid phase and the stored heat can be released by visible light-induced crystallization. This subcooling energy storage confirms that the dopant can regulate the crystallization point of the composite by changing the strong intermolecular interaction forces. The energy storage density of the composites reached 0.20 MJ kg^{-1} and was maintained for more than 10 h. This foundational approach underpins the integration of functionality into conventional heat-storage materials, potentially extending to diverse PCMs, including inorganic variants. The strategy leverages appropriate photoswitches capable of effective interaction with PCMs. This proof of concept holds promise for enhancing thermal energy management across applications like waste heat recovery, solar thermal harvesting, and intelligent building temperature regulation systems [4].

Han's team synthesized a series of ortho-functionalized azobenzene derivatives capable of absorbing filtered solar radiation. This leads to the storage of latent heat and photon energy via concurrent phase transitions and photoisomerization (Figure 6.4).

This study marks the inaugural demonstration of spontaneous solar activation of azobenzene derivatives, yielding simultaneous phase changes from solid to liquid and isomerization from *E*-to-*Z* configurations. The n–π^* absorption of various ortho-functionalized azobenzene derivatives experiences significant redshifting, enabling efficient *E*–*Z* switching under filtered sunlight, obviating the need for external UV sources. The integration of tridecanoate and ethylhexanoate groups, along with ortho-substituents, crucially lowers compound melting points, facilitating solid-to-liquid phase shifts during photo-switching under filtered sunlight. *Z* isomer's liquid phase showcases remarkable stability within a wide temperature span (–40–110°C), effectively preserving latent heat and photon energy for nearly two months. Specific optical triggers induce prompt crystallization and heat release. This innovative class of optically regulated PCMs introduces fresh avenues for harnessing and storing solar energy using optically responsive organic compounds. It emerges as a complementary technology to organic photovoltaics and photocatalysts [5].

Wei Feng's team at Tianjin University introduces a novel strategy to achieve optically triggered low-temperature heat release in PCMs through

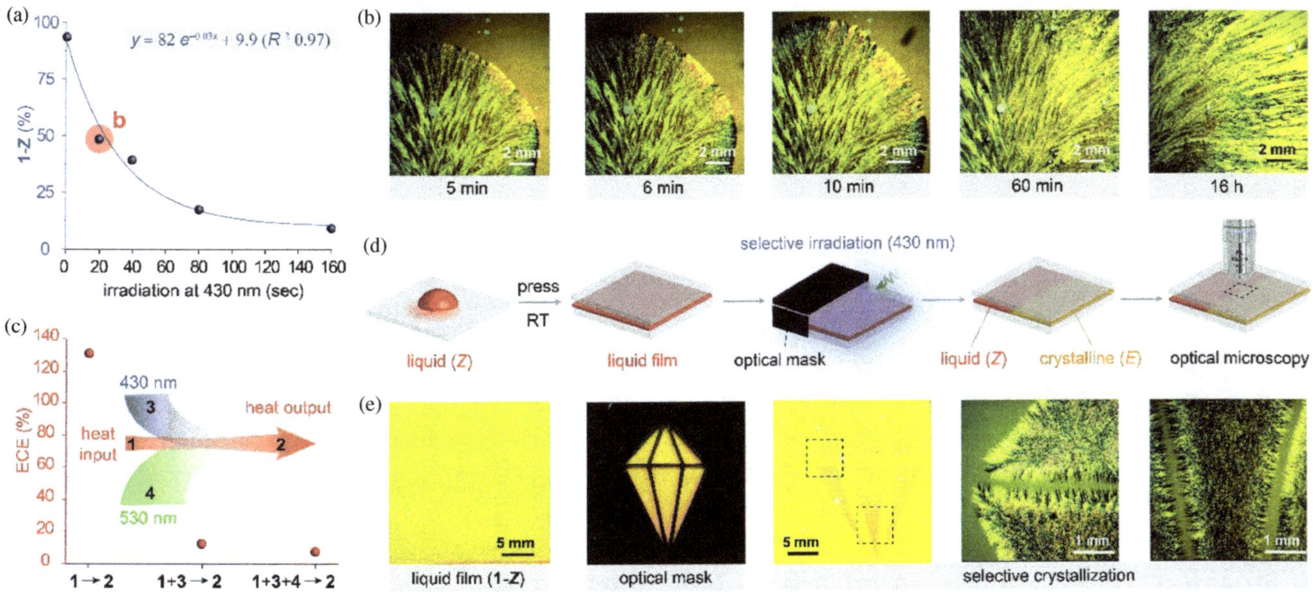

Figure 6.4. (a) Decreasing ratio of Z isomers (%) in the liquid films of compound 1 exposed to 430 nm LED at room temperature. Upon 20, 40, 80, and 160 s irradiation, each liquid film crystallized after various delays. The 80 s and 160 s irradiated films crystallized immediately. (b) A 20 s-irradiated film containing more than 50% *E* showed a relatively slower nucleation and gradual crystal propagation, which was captured by optical microscopy of the identical sample area. (c) Energy conversion efficiency (ECE, %) calculated for the latent heat storage and release, showing the large heat output than input, as well as substantially larger photon energy input required for photo-switching. (d) A schematic illustration of the selective photo-induced crystallization process of liquid sample. (e) Optical microscope images of a liquid film before and after selective crystallization via 430 nm irradiation, showing a clear distinction between the irradiated and covered areas of sample. The dotted boxes mark the areas that are further magnified under the microscope and shown on the right. Reprinted with permission from Ref. [5]. Copyright 2021, Royal Society of Chemistry.

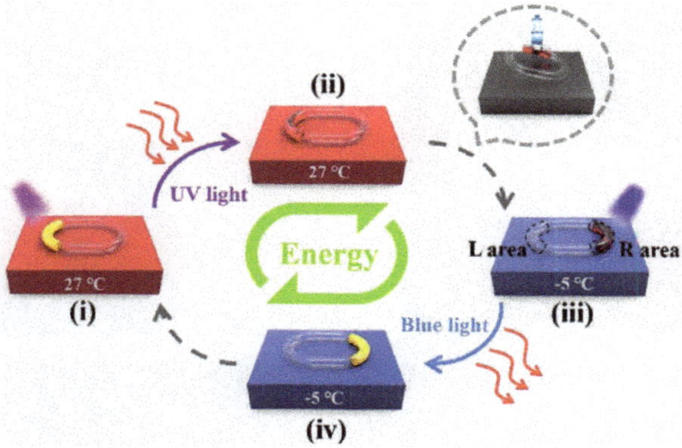

Figure 6.5. Schematic illustration of the distributed energy utilization system using an annular device. Reprinted with permission from Ref. [6]. Copyright 2021, Wiley.

azobenzene's *trans–cis* (*E–Z*) photo-isomerization (Figure 6.5). Their study incorporates a photo-responsive alkyl-grafted azobenzene (Azo) into tetradecane (Ted) to establish a photosensitive energy barrier for phase change control. The Azo/Ted composite demonstrates manageable supercooling in the 4.04–8.80°C range, enabling synchronized release of PC enthalpy and photo-thermal energy. Additionally, Azo molecules impede Ted crystallization by intercalating into its molecular alignment. Under light exposure, the Azo/Ted composite liberates a substantial heat quantity (207.5 J g^{-1}) at relatively low temperatures (−1.96–6.71°C). In a sub-zero environment (−5°C), an energy utilization device's temperature rises by 4°C. This research points towards an encouraging direction for developing advanced distributed energy systems capable of effectively managing energy storage and release in PCMs over a broad temperature spectrum for heating applications [6].

The research team led by Prof. Haifeng Yu at Peking University has achieved a significant milestone by developing solid-state flexible STF devices (Figure 6.6). These devices combine the energy storage capacity of Azo derivatives with the flexibility of fabric templates through thermal melting. The azobenzene derivatives used, known as photo-liquefied AZ (PLAZ) derivatives, can undergo a solid-to-liquid transition at room temperature. Furthermore, the inherent *trans→cis* conversion capability of

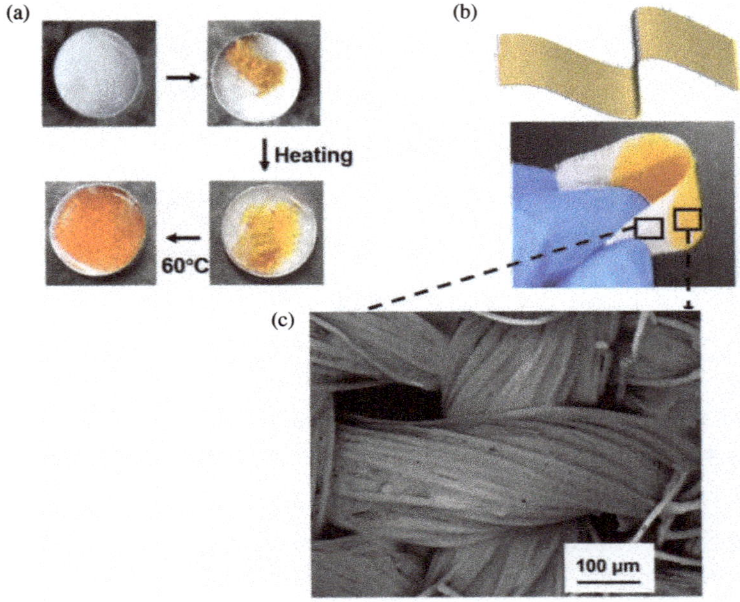

Figure 6.6. Fabrication and morphological characterization of flexible STFs. (a) Photos of preparing the flexible STF devices, specifically the DSC samples of STFs. (b) Schematic and real product of the large area flexible STFs. (c) SEM image for the surface morphology of the composite (2 mg PLAZ in 3 mg fabric, the filling rate was 0.07 mg mm^{-2} cloth fabric). Reprinted with permission from Ref. [7]. Copyright 2019, Wiley.

these derivatives without the need for solvents is advantageous since the *cis*-isomer helps to lower the melting point of the mixture.

The inclusion of liquid PLAZ derivatives increases the available free volume, facilitating further photoisomerization and creating a positive feedback loop. This enhances the energy storage capacity of the system, resembling the energy storage mechanism observed in PCMs. The dispersed azobenzene derivatives within the fabric template exhibit excellent phase transition properties, resulting in devices with a high energy storage capacity.

The presence of the fabric template not only increases the energy density of the flexible STF but also prolongs its energy storage half-life. The intermolecular interactions between the polyester fabric and azobenzene derivatives may contribute to inhibiting *cis*-to-*trans* isomerization. Moreover, the use of the fabric template addresses the issue of mobility of the liquid azobenzene derivative after charging. Importantly, the entire charging and exothermic process of the STF is completely solvent-free.

The temperature difference of 4°C achieved between the charged and exothermic samples under blue light stimulation demonstrates the ability of the STF to stabilize the external ambient temperature, highlighting its potential application in daily life. To further enhance the energy storage capacity, the study proposes the introduction of azobenzene derivatives with higher photoisomerization quantum yields and flexible templates with higher energy density.

In summary, the utilization of solid-state flexible STF devices incorporating PLAZ derivatives and fabric templates represents a significant step forward. These devices offer efficient energy storage, extended half-life, solvent-free operation, and the potential for applications such as warm-keeping or heating equipment in various everyday scenarios [7].

The group led by Prof. Haifeng Yu at Peking University also introduces a nanoconfinement approach to improve the energy density of STF by utilizing microphase separation of a block copolymer. This approach enables the selective dispersion of PCMs into nanodomains of a polymer film for use in STF applications.

By incorporating a liquid-crystalline block copolymer with azobenzene as the continuous mesogenic phase and doping it with a photoinert organic PCM (sPEO), the PCM selectively disperses into the nanodomains of the polymer film. The microphase separation of the block copolymer ensures the periodic nanostructures are maintained, thereby confining the PCM within nanodomains. This unique nanoconfinement effect achieved through microphase separation allows for a wide range of crystallization temperatures, spanning from 24°C to −38°C. The nanoscale microphase separation enables the simultaneous storage of light energy from natural sunlight and heat energy from the environment. The release of stored energy can occur through multiple mechanisms, including the *cis*-to-*trans* isomerization and isotropic-to-liquid crystal phase transition of the azobenzene-containing polymer block at higher temperatures and the phase transition of PCMs at lower temperatures, even below −30°C.

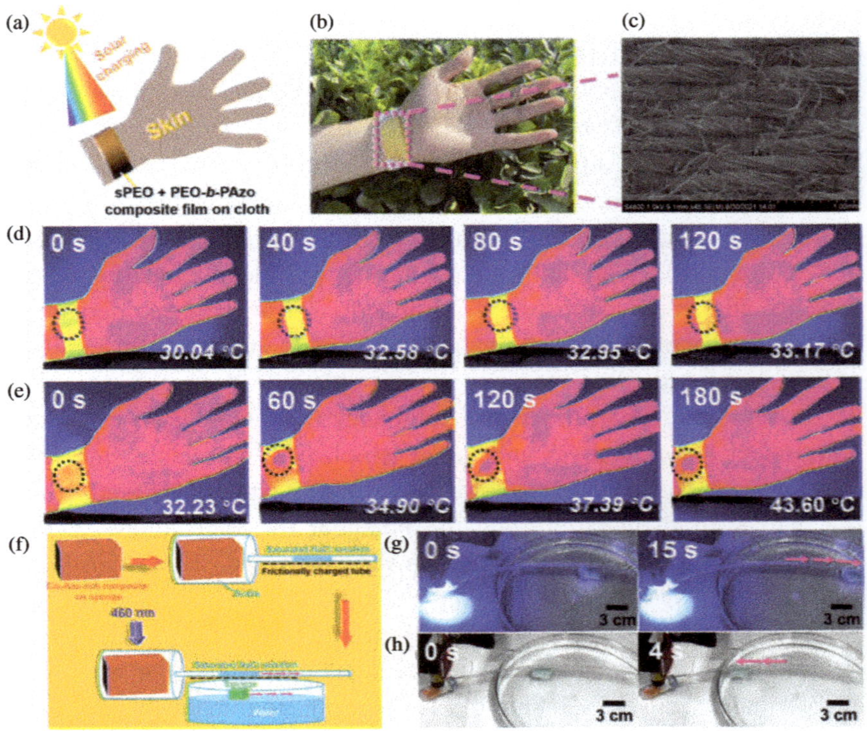

Figure 6.7. (a) Schematic illustration and (b) image of the sPEO and PEO-b-PAzo composites (90:10) within cloth charged by solar irradiation for human body thermal application. (c) SEM image for the surface morphology of the sPEO and PEO-b-PAzo composite (90:10) within cloth. (d) IR thermal images of pure cloth on the wrist during 460 nm irradiation for different times. (e) IR thermal images of the charged sPEO and PEO-b-PAzo composites (90:10) within cloth on the wrist during 460 nm irradiation for different times. (f) Schematic illustration for the preparation of actuator on the water surface under light energy conversion pneumatic electric field. (g) Movement of the sponge on the water surface by pneumatic electric field change with the 460 nm light illumination from 0 to 15 s. (h) Movement of the sponge on the water surface from 0 s to 4 s by pneumatic electric field change without the 460 nm light illumination. Reprinted with permission from Ref. [8]. Copyright 2021, American Chemical Society.

One practical application of this approach is the preparation of wearable thermal fabrics, as depicted in Figure 6.7(a) of the research paper. The cloth coated with the sPEO and PEO-b-PAzo composite absorbs sunlight and stores light energy, as shown in Figure 6.7(b). The research team also successfully fabricates flexible and rechargeable thin-film STF devices. This method can be extended to the development of wearable warm fabrics and actuators, where changes in pneumatic electric fields induce motion on the water surface.

Overall, this research paves the way for high-performance STF systems achieved through microphase separation and nanoconfinement. These advancements contribute to the utilization of STF materials in various fields, such as wearable smart materials, warm fabrics, and actuators, furthering their practical applications [8].

The research conducted by Si Wu's group at the University of Science and Technology of China focuses on the development of an Azo-based STF device capable of efficiently storing UV and visible light across the full solar spectrum. The device is constructed with four layers: Coumarin 314, PmAzo UV filter, and PAzo, arranged from top to bottom. Two of these layers function as filters, responsible for storing solar energy and modifying the effective incident light spectrum.

The two filter layers consist of different Azo polymers, namely PAzo and PmAzo, which store energy by facilitating the isomerization of the chromophores present in the side chains of the polymers. This isomerization process involves the conversion of low-energy *trans*-isomers to high-energy *cis*-isomers. Through careful device design and optimization of spectral overlap, the research team achieved a significant increase in solar efficiency, reaching 0.4%.

This work demonstrates the potential of Azo-based materials for efficient solar energy storage in STF devices, offering possibilities for enhanced performance and utilization of solar energy across the UV and visible light spectrum [9].

Guojie Wang's research group has developed a fascinating multifunctional photochromic dendrimer by incorporating azobenzene units onto dendrimers. This innovative dendrimer design offers a range of capabilities, including controlled adhesive switching, efficient repair of coating scratches, high-performance solar energy storage, and on-demand heat release. When exposed to UV irradiation, the dendrimers in their powdered form undergo a remarkable transformation, changing from a solid state to a red viscous liquid. This reversible phase transition allows for the controlled switching of adhesives when temporary bonding or adhesion is desired. As shown in Figure 6.8(b), the energy storage density of

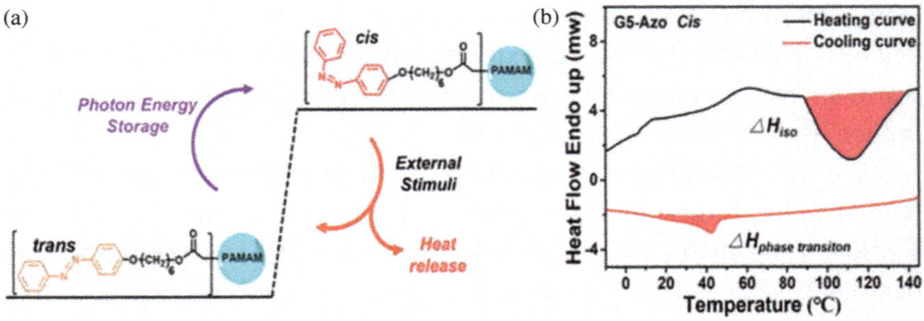

Figure 6.8. (a) Schematic diagram of photon energy storage and heat release of dendrimer phase-change azobenzene; (b) DSC plot of *cis*-dendrimer phase-change azobenzene. Reprinted with permission from Ref. [10]. Copyright 2020, American Chemical Society.

dendrimer phase change azobenzene is 0.21 MJ kg^{-1}. It should be noted that the energy density of dendrimers with photoinduced phase transition increases with the increase of the number of optical switches on the surface of macromolecules. The research conducted by Guojie Wang's group is indeed an exciting development in the field of phase-change azobenzene. The ability of *cis* liquid dendrimers to release stored energy as heat under visible light irradiation, resulting in a significant temperature difference between the *cis* and *trans* isomers, highlights their potential for practical applications [10].

Flexible self-heated textiles necessitate molecular light-harvesting capabilities and the generation of low-temperature heat. To accomplish this, incorporating photoresponsive molecular interactions, referred to as photodynamic bonds, emerges as an effective strategy. These interactions augment the energy storage capacity and optimize the photochromic kinetics at low temperatures. Introducing supramolecular interactions into azobenzene chromophores provides a promising approach to fine-tune their molecular energy levels. Polymer-templated azobenzene can capitalize on reversible isomerization-induced formation and dissociation of supramolecular bonds, enabling the storage and release of additional bond enthalpy. The potential utilization of bond enthalpy in azobenzene-based dynamic polymers is made possible by the dynamic-bond-controlled reversible crosslinking. Moreover, the steric configuration and structural transformation of azobenzene grafted onto flexible chains experience notable changes due to these interactions. Consequently, this effect plays a crucial role in the temperature-dependent isomerization kinetics, leading to heat output.

This research unveils the concurrent capture of isomerization energy and bond enthalpy from a series of photo-responsive PAzo-M species (Figure 6.9). This process involves light absorption, low-temperature storage, and eventual heat release. The incorporation of dynamic bonds into crosslinked PAzo-M species augments ΔE while diminishing the thermal barrier for Z–E isomerization. This enhancement is facilitated by photo-controlled reversible formation and dissociation. By judiciously selecting a suitable metallic ion for the M–O dynamic bond, an optimum binding energy is achieved. This, in turn, accelerates efficient photoisomerization at low temperatures, leading to the emission of notable photothermal energy and bond enthalpy. The PAzo-Mg variant showcases a peak energy density of 408.6 J g^{-1} (113.5 Wh kg^{-1}), accompanied by a relatively high ΔE (6.98 eV) and DI (86.3%). Moreover, the team fabricates a wearable self-heated textile coated with PAzo-Mg, enabling temperature control in cold environments. The uniform NF@PAzo-Mg textile selectively releases heat, generating a temperature difference (ΔT) of 7.7–12.0°C within the −5.0–5.0°C temperature range. This innovative strategy for modulating photothermal polymer heat release via control of isomerization-induced photodynamic bond formation and dissociation introduces avenues for self-heated wearable thermal management devices, smart on-skin clothing, and applications in soft robotics [11].

To enhance the energy storage potential of self-templating frameworks through molecular engineering of azobenzene photo-switches, certain challenges like short half-lives and limited exothermicities need addressing. A research group led by Grossman has recently developed a novel series of Azo-STFs, where robust carbon-based templates such as graphene, carbon nanotubes (CNT), and C60 are integrated with azobenzene photoswitches. Through first-principles calculations, they predict that these STFs can effectively and reversibly store solar energy, with energy densities comparable to Li-ion batteries. They anticipate potential external efficiencies of up to 35% and adjustable thermal stability spanning minutes to years. By employing carbon templates with varied

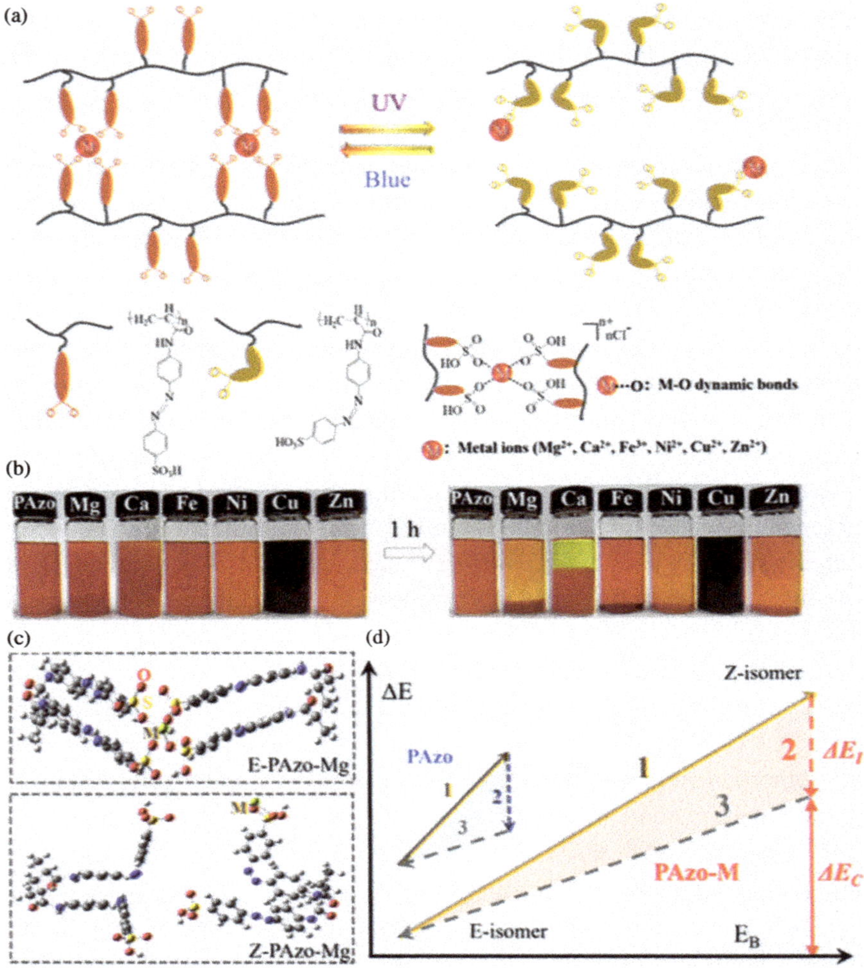

Figure 6.9. (a) Chemical structure and schematic illustration of PAzo-M with isomerization-induced formation and dissociation of M–O dynamic bonds formed by different metallic ions, while n is 2 or 3. (b) Photographs of PAzo solution after adding the same molar amount of MCln ($MgCl_2$, $CaCl_2$, $FeCl_3$, $NiCl_2$, $CuCl_2$, $ZnCl_2$) for 1.0 h. (c) Optimized geometric structure of PAzo-Mg with the *E*- and *Z*-isomers. (d) Schematic energy diagram of PAzo and PAzo-M during the energy utilization process of (1)–(3). *Note*: (1) *E*-to-*Z* isomerization, (2) *Z*-to-*E* isomerization, (3) bond enthalpy release by forming dynamic bonds or H-bonds. Reproduced under the terms of the CC BY 4.0 from [9]. Copyright 2022, the authors.

geometries and electronic properties, the fundamental characteristics of nanocarbon-templated Azo-STFs can be finely tuned, surpassing systems using unsubstituted azobenzene. The adoption of carbon-based materials as templates for azobenzenes is expected to enhance energy densities, storage durations, optical and electronic properties, thermal stability, chemical durability, and temperature control triggered.

In 2007, Wei Feng's research group [12] achieved a significant milestone by successfully synthesizing a photoactive multiwalled carbon nanotube (MWCNT) through covalent bonding of azobenzene onto the sidewalls of MWCNT. This innovative approach opened avenues for further exploration of the polarized optical properties exhibited by MWCNT-templated azobenzene composites. Through the application of UV–Vis and photoluminescence spectroscopies, the research group thoroughly investigated these optical properties. Although the specific investigation into solar energy storage was not conducted in their study, their findings provided valuable insights and guidance for the synthesis and characterization of azobenzene composites templated on CNTs, thereby contributing to the advancement of this field.

In 2011, Grossman and colleagues conducted a theoretical study that showcased how chemically linking azobenzenes with CNTs facilitated the creation of highly organized, densely packed arrays. This arrangement led to an augmentation in both storage half-life and energy density within the resulting self-templating frameworks, attributable to specific chemical interactions between adjacent azobenzene molecules. In contrast to standalone azobenzene molecules, the presence of CNT matrices induced an alteration in the molecular symmetry of azobenzene. This imposition of a closely packed crystalline-like state discouraged the isomerization reaction through phenyl ring rotation, thereby expanding the potential phase space for hydrogen bond (H-bond) formation.

As illustrated in Figure 6.10, multiple azobenzene derivatives featuring varying numbers of hydroxyl substituents on phenyl rings were incorporated onto the CNT surface to modulate the relative stability of the two isomers via the introduction of H-bonds. Computational outcomes demonstrated that distinct quantities of molecular H-bonds on *trans*- and *cis*-Azo at the CNT surface led to a notable increase in ΔH (enthalpy change) and $t_{1/2}$ (half-life). Typically, hybrid structures with a greater number of H-bonds

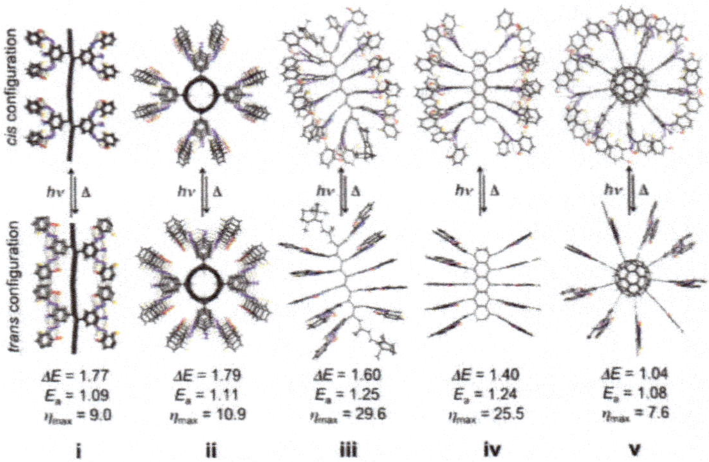

(a)

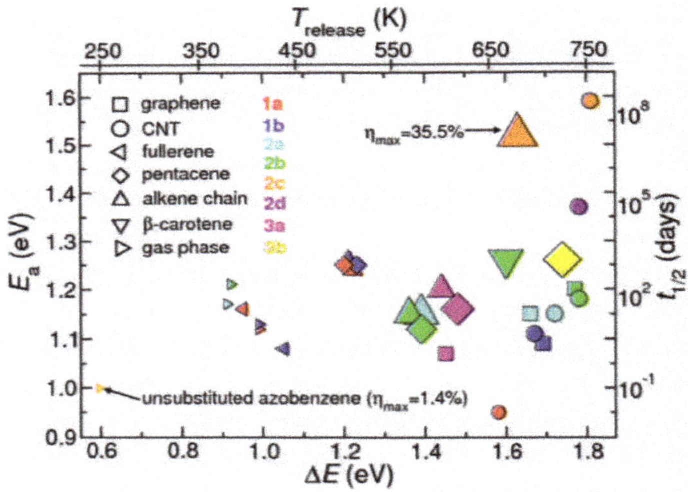

(b)

Figure 6.10. (a) Calculated *trans* and *cis* configurations and properties of nanocarbon-templated azobenzenes including graphene-templated azobenzene (i), CNT-templated azobenzene (ii), β-carotene-templated azobenzene (iii), pentacene-templated azobenzene (iv), and C60-templated azobenzene (v). Orange, red, blue, dark gray, and white balls/sticks represent F, O, N, C, and H atoms, respectively. (b) Calculated thermal activation barrier (E_a) vs. stored energy per molecule (ΔE) for nanocarbon-templated azobenzene and gas phase azobenzene derivatives. The alternate axes indicate half-life ($\tau_{1/2}$) and temperature of the released heat ($T_{release}$), estimated from E_a and ΔH, respectively. The size of the symbols illustrates the relative values of the maximum external efficiency (η_{max}). Reprinted with permission from Ref. [13]. Copyright 2011, American Chemical Society.

and/or more robust (shorter) H-bonds exhibited heightened stability. Consequently, maximizing the number of H-bonds in the *trans* state while minimizing them in the *cis* state allowed for achieving an elevated ΔH. The creation of H-bonds additionally elevated the stability of both *trans* and *cis* configurations, contributing to relatively prolonged half-lives.

Moreover, the study proposed various conceivable configurations for CNT-templated dihydroxy azobenzene derivatives. It was observed that the strength and quantity of H-bonds in the *trans* and *cis* configurations closely correlated with the relative positions of the two OH groups [13].

In 2014, Grossman's research group successfully realized the theoretical model of CNT-templated azobenzene hybrids, as depicted in Figure 6.11. The experimental results demonstrated an impressive energy density of 120 kJ mol^{-1} (~200 J g^{-1}, or ~56 Wh kg^{-1}) in the CNT-templated azobenzene system, surpassing the energy density of 58 kJ mol^{-1} (~160 J g^{-1}, or ~44 Wh kg^{-1}) observed in non-templated azobenzene. This signifies a significant enhancement in the energy stored per molecule through the application of CNT templates. Importantly, these CNT-templated

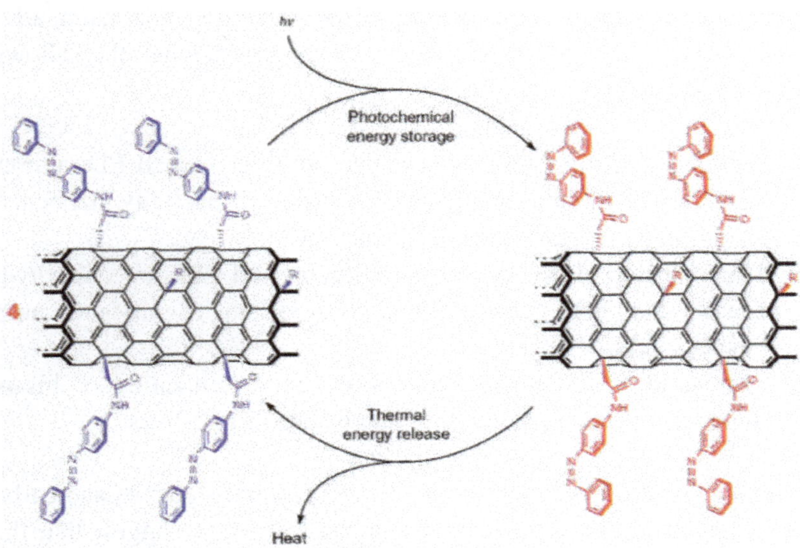

Figure 6.11. Reaction scheme for photo-chemical energy storage and thermal energy release in CNT-templated azobenzene systems. Reprinted with permission from Ref. [14]. Copyright 2014, Springer Nature.

azobenzene systems exhibited excellent cyclability and stability, further highlighting their potential as promising energy storage materials [14].

The efficient transfer of charge between azobenzene and graphene acts as a catalyst for driving the photoisomerization reaction. To refine the molecular structure of graphene-templated azobenzene hybrids, Grossman's research group engaged in extensive model construction. These models encompassed covalently binding photo-responsive azobenzene with distinct functional groups (ortho-, meta-, and para-) at various positions on graphene. Employing density functional theory calculations, the team established the suitability of the 2D graphene nanosheet as an optimal platform for supporting nanocarbon-templated Azo-STFs. Building on these calculations, the group successfully devised and synthesized a series of graphene-templated azobenzene materials. Their strategies were geared toward heightening energy density and extending half-lives through several effective methods, including augmenting molecular interactions, enhancing grafting density, optimizing steric arrangements, and attaching appropriate substituents at diverse azobenzene positions.

The research group presented an encompassing compilation of energy densities (EDs), half-lives, and grafting densities for nanocarbon-templated azobenzene STF materials. From their insights, the following conclusions emerge:

(i) Elevated grafting densities correlate with heightened ED, although limitations arise due to factors like reduced reactivity, azobenzene steric hindrance, and nanosheet stacking.
(ii) The inclusion of azobenzene substituents and intermolecular hydrogen bonds in both *trans*-Azo and *cis*-Azo states significantly amplifies ED and half-life.
(iii) Compared to individual azobenzene units, utilization of bis-Azo molecules yields superior EDs and extended half-lives.

While these studies focus on azobenzene photo-switches and carbon-based templates, the experimental outcomes strongly suggest that these design principles can be effectively extrapolated to create STFs using alternative molecular photo-switches and templated materials.

6.1.3 *Anthracenes*

The urgent need for alternatives to fossil fuels, which contribute to global warming, has driven researchers to explore zero-carbon intensity renewable energy sources. Among these options, solar energy stands out as a tangible and virtually infinite resource of carbon-neutral energy. In this context, MOST storage devices have emerged as a promising approach for storing solar energy.

MOST systems, which are potential candidates for future renewable energy applications, harness the phenomenon of photoisomerization in molecules. This process is triggered by UV–Vis radiation within the solar spectrum range of 300–700 nm. The key concept is the reversible conversion between pairs of photoisomers. When a low-energy isomer is exposed to UV–Vis radiation, it should be converted to a higher-energy metastable isomer with a significant quantum yield (Φ). This conversion allows for the storage of solar energy in the form of chemical bonds.

Crucially, the photoisomerization reaction must possess a large positive ground state enthalpy ($\Delta H_{\text{storage}}$) and an optimal kinetic barrier (E_a) to prevent thermal back reaction (TBR). This ensures that the stored energy can be held for a reasonable period of time. However, the kinetic barrier should not be too high, as it would hinder the release of energy through gentle heating or the use of a catalyst.

Therefore, an ideal MOST material should possess an optimally balanced combination of several unique attributes. It should be capable of repeated cycles of energy conversion and release, making it a highly efficient and sustainable solution for solar energy storage and utilization.

The notion of harnessing solar energy within molecular chemical bonds has been a subject of study for a considerable duration. Two pivotal photoisomerization reactions in this context are the $E \leftrightarrow Z$ isomerization of azobenzene (AB) and the $[2\pi + 2\pi]$ cycloaddition of NBD leading to QC formation. In addition, fulvene tetracarbonyl complexes incorporating transition metals ($FvRu_2(CO)_4$) have been identified as potential contenders for solar-thermal energy storage.

Nonetheless, existing systems exhibit certain limitations that diminish their viability as candidates for MOST applications. Common hurdles encompass UV–Vis-triggered back conversion, suboptimal quantum yield (Φ), absorption spectra outside the visible range, and modest energy storage density. Achieving absorption spectra within the visible range to activate photoisomerization has occasionally resulted in compromised energy storage density. In certain instances, the need for multiple photoisomerizations within a single MOST candidate may lead to a low quantum yield (Φ).

It is important to acknowledge that the number of known chemical motifs suitable for serving as ideal MOST systems is limited. Therefore, there is a need to identify and explore newer molecular motifs that can offer improved solar-thermal storage capabilities. By addressing these challenges and discovering innovative approaches, researchers aim to develop more efficient and effective MOST materials for optimal solar energy storage.

As early as the 1970s, researchers wanted to store solar energy as chemical energy in anthracene-based systems (Figure 6.12). Unfortunately, their study on n.n bis-anthracene cyclophanes ([n.n]BA, n = 2–6) using well-calibrated density functional methods has indeed revealed promising results contrary to the existing consensus view. In their research, they found that [n.n]BA cyclophane candidates with longer linker alkyl chains

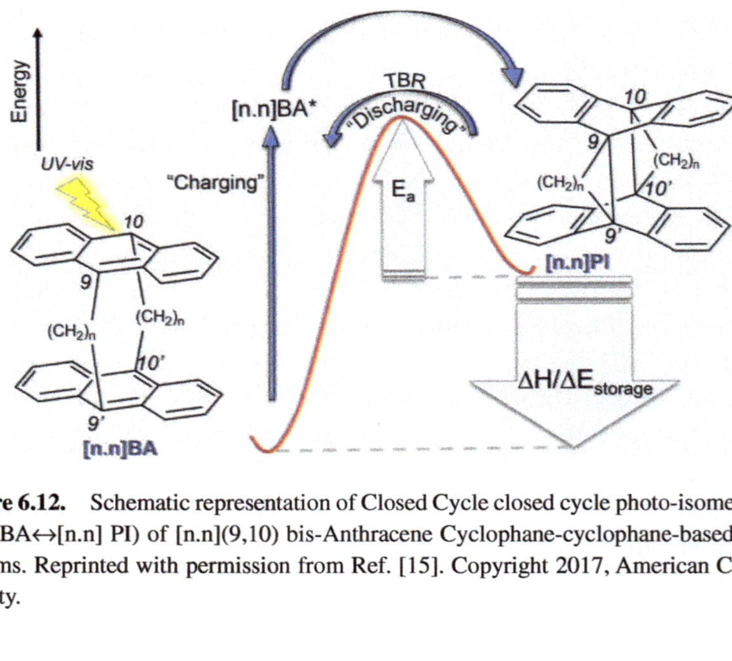

Figure 6.12. Schematic representation of Closed Cycle closed cycle photo-isomerization ([n.n]BA↔[n.n] PI) of [n.n](9,10) bis-Anthracene Cyclophane-cyclophane-based MOST systems. Reprinted with permission from Ref. [15]. Copyright 2017, American Chemical Society.

(higher values of "*n*") can be highly effective MOST systems, exhibiting the highest gravimetric energy storage density reported to date.

The presence of ring strain in the high-energy photoisomers of these [n.n]BA cyclophanes contributes to their exceptional energy storage capacity and weight energy storage density. This intriguing discovery has sparked renewed interest in anthracene-based systems as potential STFs, which were previously considered less promising.

These findings highlight the importance of continuous exploration and research in the field of solar energy storage, as new discoveries can challenge existing assumptions and open up new possibilities for more efficient and effective MOST systems. Further investigation and development of anthracene-based motifs could potentially lead to significant advancements in solar-thermal energy storage technology [15].

6.2 Deicing

The photochemical crystal-liquid transition (PCLT) exemplified by the reversible transition of Azo molecules between *trans*-crystal and *cis*-liquid states offers a new principle for energy storage and conversion. During the *trans*-crystal to *cis*-liquid transition, Azo molecules trap photons to undergo *trans* to *cis* photoisomerization, and the intermolecular stacking interactions are overcome, resulting in a crystal to liquid phase transition. In this process, heat is absorbed from the environment. This allows for the collection and storage of both solar energy and ambient heat in a single-component material, thereby achieving a high energy density by integrating both forms of energy.

On the other hand, during the *cis*-liquid to *trans*-crystal transition, which can be triggered externally by light irradiation or heating, the stored energy is released in the form of isomerization and enthalpy of crystallization (ΔH_{isom} and ΔH_{cryst}). The combination of these two heat energy streams enables the attainment of high power densities and allows for the release of heat at higher temperature levels.

The photochemical-thermo-physical coupling mechanism demonstrated by phase change azobenzene molecules presents a unique approach to energy storage that goes beyond traditional solar or thermal energy storage systems. In this hybrid system, the combination of both solar energy and ambient heat enables synergistic contributions to energy storage and conversion.

One advantage of using Azo switches is their ability to store heat at *cis*-T_m temperatures and release heat at *trans*-T_m temperatures. The *trans* T_m can be much higher than the *cis* T_m, resulting in a temperature difference (ΔT_m) of up to 60–70°C. This temperature difference allows for the release of heat at significantly higher temperatures compared to the storage temperature.

Additionally, the high subcooling behavior of *cis*-liquids means that they can retain a high energy capacity ($\Delta H_{\text{cryst}} + \Delta H_{\text{isom}}$) over extended periods of time at low temperatures without the need for insulation. This property facilitates on-demand photo-triggered heat release even in cold environments. It allows for the storage of thermal energy at one temperature and the selective release of that energy at a higher temperature when needed, enhancing energy efficiency and performance in energy storage and utilization systems.

By capitalizing on these unique characteristics of azobenzene-based systems, there is potential for achieving higher energy efficiency, temperature control, and targeted heat release, making them promising candidates for various applications in areas such as thermal management, energy storage, and temperature regulation.

The recent development by Han's research group involves the creation of a series of arylpyrazole azobenzene derivatives functionalized with a dodecanoate group. These derivatives have demonstrated the ability to release stored photon energy and undergo phase change at low temperatures, specifically at −30°C.

Arylazopyrazole derivatives featuring dodecanoate groups exhibit versatile activation mechanisms: thermal, optical, or a synergistic blend of both. Upon activation, these derivatives transform into stable liquid-phase *Z*-isomers across an extensive temperature spectrum. These liquid-phase compounds can subsequently be optically prompted to isomerize into crystalline structures, thereby releasing latent heat at notably low temperatures, including −30°C.

The arylazopyrazole core of these derivatives facilitates the enduring storage of latent heat in the liquid state. Simultaneously, the dodecanoate group enhances the crystallinity of the *E*-isomers and bolsters the stability of the polar *Z*-isomer's liquid phase. The pronounced polarity and augmented volume of the polar *Z*-isomers enable them to preserve their liquid state even at frigid temperatures. This phenomenon is achieved by effectively disrupting π–π interactions among aromatic groups and van der Waals interactions among alkyl chains.

This study marks the initial instance of phototriggered crystallization in molecular switches at temperatures substantially below 0°C. This discovery bears considerable significance for applications demanding heat release in exceedingly cold weather conditions. It offers potential advancements in areas where heat management and energy storage during extremely cold temperatures are crucial [16].

In a groundbreaking study, Prof. Tao Li's group at Shanghai Jiao Tong University developed a phase change heterocyclic azobenzene de-icer. This de-icer was subjected to de-icing experiments conducted at temperatures ranging from −10°C to −15°C (Figure 6.13).

During the experiments, light irradiation was applied to the de-icer for varying durations. After just 2 min of light irradiation, the region undergoing the *cis*-liquid → *trans*-crystal transition experienced significant melting of the ice flakes, while the ice remained intact in other areas. Increasing the irradiation time to 3 min resulted in a larger area of ice melting as the discharge expanded.

By moving the light spot to adjacent areas with thinner ice layers, the corresponding PCLT caused almost complete melting of the ice within the irradiated area. Additionally, the surrounding ice also exhibited melting due to the lateral diffusion of heat within the film.

The light discharge from the film was found to increase the surface temperature by more than 20°C, which proved to be sufficient to raise the temperature above 0°C. This significant temperature increase verified the de-icing capability of the phase change heterocyclic azobenzene deicer.

These findings have important implications, particularly in scenarios where traditional heat generation methods like combustion are limited or ineffective, such as in the defrosting of engine oils, mechanical parts, and personal heating under extreme cold conditions. The ability of the de-icer to effectively melt ice at low temperatures opens up possibilities for innovative applications in areas requiring efficient and controlled ice melting processes [17].

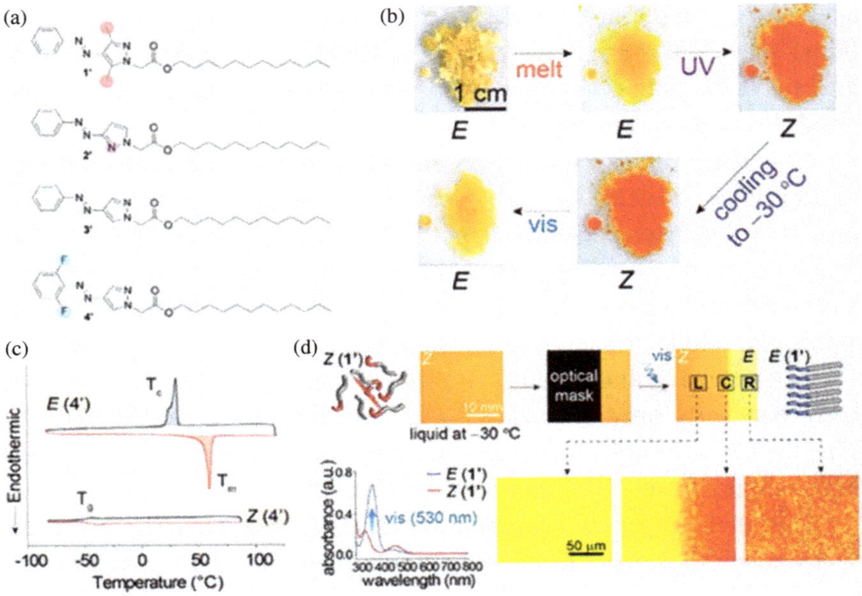

Figure 6.13. (a) Chemical structure of aryl pyrazoles as phase-change azobenzene. (b) Series of optical images: melting, UV irradiation, cooling and visible light Induced-induced crystallization. (c) DSC plot of the *trans*- and *cis*-isomers of phase-change azobenzene 4′4′. (d) Optical image of phase-change azobenzene 1′1′ liquid film under visible light irradiation. Reprinted with permission from Ref. [17]. Copyright 2020, American Chemical Society.

The current method of de-icing car windows is to install heated wires on the windows, which would block the driver's view, while 30% of an electric car's electricity would be used for heating and de-icing in winter, which would significantly reduce the vehicle's driving range. And Grossman team's objective to get a transparent Azo energy storage material can result in a transparent Azo compound film, which will be sandwiched between two layers of glass, with an appropriate adhesive to connect. This film can achieve de-icing function and a similar technology can also prevent the windshield shattering "splash." Currently, BMW Germany has sponsored the research and is very much looking forward to its application.

To overcome the challenges of photochemical conversion between isomers in solid-state systems, one approach is to incorporate molecular photoswitches into a porous structure. Metal–organic frameworks (MOFs) are a class of porous materials that are well-suited for hosting photoswitchable guest molecules. MOFs offer the necessary porosity to provide a defined void space where photoswitches can undergo isomerization.

Furthermore, certain MOFs exhibit solid-solid phase transitions, often referred to as "breathing" transitions, which involve changes in the pore volume. By incorporating photoswitches within these MOFs, it becomes possible to manipulate the porosity of the material through photoisomerization. This enables the regulation of the transition between solid phases and the control of energy storage and release.

By combining PCMs with molecules capable of photothermal conversion, such as azobenzene derivatives, it is possible to create a system that can store solar energy efficiently for extended periods at ambient temperatures and release it on-demand as heat. This concept has been previously demonstrated by using azobenzene derivatives to trigger solidification in liquid–solid PCMs through optical regulation.

Utilizing MOFs as the framework for hosting photoswitchable molecules allows for confinement while still providing sufficient steric freedom for isomerization. This integration of porous materials and photoswitchable molecules offers a promising strategy for developing solid-solid systems with optical regulation capabilities, opening up opportunities for enhanced energy storage and conversion applications [18].

The research led by John M. Griffin's team involved filling the porous structure of MOFs with azobenzene molecules. Through their experiments, they exposed the material to ultraviolet light, causing the azobenzene molecules to undergo a shape change within the pores of the MOFs, resulting in a strained structure. This process stores energy in a manner similar to the potential energy stored in a bent spring (Figure 6.14).

The crucial aspect of this research lies in the narrow pores of the MOFs, which trap the azobenzene molecules in their strained shape. This allows for the storage of potential energy for long durations at room temperature. The energy storage half-life of this material can extend to several years.

While the current energy density of the material is relatively low, the researchers recognize the potential for improvement through further work. Future endeavors will focus on enhancing the energy density of the material. Ultimately, this innovative material holds great promise.

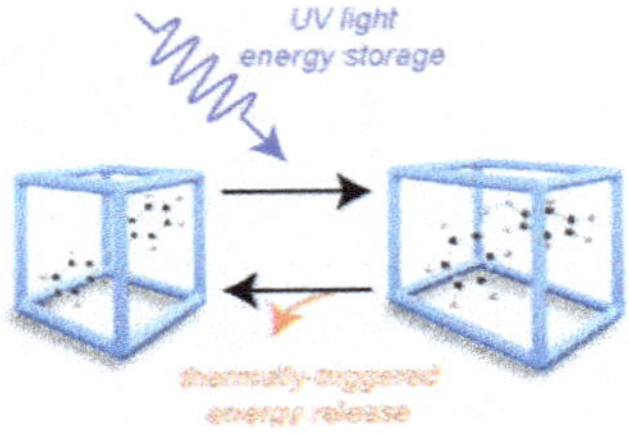

Figure 6.14. Hybrid STF combining molecular light switch and porous structure. Reprinted with permission from Ref. [18]. Copyright 2020, American Chemical Society.

The material's capabilities could include capturing energy from the sun during the day and releasing it as heat during the night. It could also be employed for storing summer energy for winter use. Additionally, the material could serve as a useful coating for heating buildings or rapidly de-icing windows and windshields on cold winter mornings.

This research offers an exciting avenue for advancing energy storage and utilization technology, with potential applications in diverse areas such as renewable energy, thermal management, and smart surfaces [18].

6.3 Self-heating clothes

Flexible self-heating fabrics with intelligent temperature control are key components of intelligent flexible devices, intelligent temperature-controlled garments, and color-changing garments. In low-temperature environments, efficient energy harvesting and utilization are particularly crucial for the continuous and controlled heat output of flexible self-heating fabrics.

Polymer template azobenzene is an ideal photothermal fuel, and its flexible chain segment template not only facilitates the enhancement of the isomerization transformation rate of azobenzene molecules, but also enhances the molecular interface interaction with the fabric substrate and adapts to a variety of morphological changes (stretching, bending, etc.), providing an important way to achieve human thermal management at low temperatures. At present, polymer template azobenzene has limited its application in human thermal management system due to low energy density of thermal storage and difficulty in low temperature heat release.

Recently, Wei Feng's research group at Tianjin University prepared a series of dynamically bonded modulated polymer template azobenzene photothermal fuels [19]. The introduction of dynamic bonding improved the thermal storage energy density of polymer template azobenzene and realized the storage of both isomerization energy and dynamic bonding energy. The human thermal management system was further constructed based on Azo-based flexible fabric. Based on molecular design and theoretical calculations, this work synthesized polyacrylamide-based Azo-benzene (PAzo-M) with dynamic bonding, which is partially formed by the interaction of sulfonic acid groups and different metal ions (Figure 6.15). Under light, dynamic bonding accompanied by isomerization and reversion of the azobenzene structure undergoes reversible breakage and formation. The binding energy of dynamic bonding determines the energy storage of isomerization and bonding energy, and it becomes critical to choose the suitable metal ions to maximize the photothermal energy. In this work, based on theoretical calculations and experimental results, the principle of choosing the appropriate metal ion is proposed.

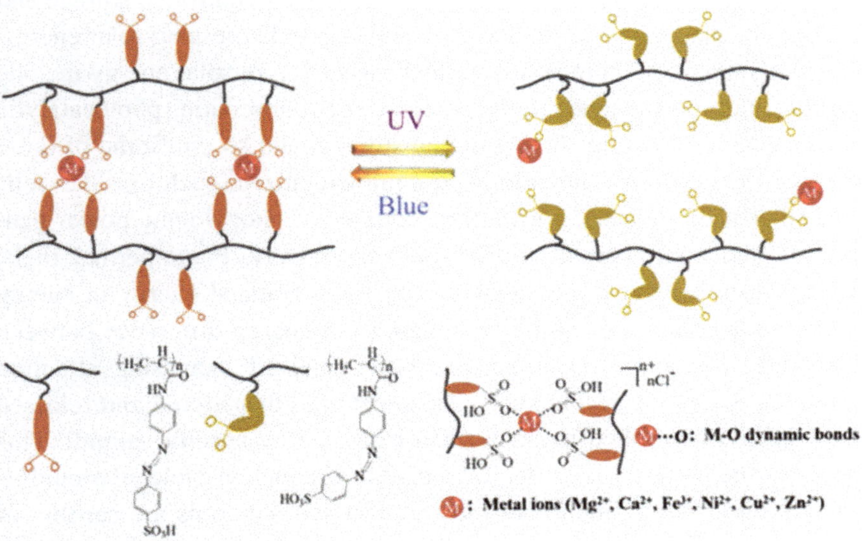

Figure 6.15. Chemical structure and mechanism of polyacrylamide-based azobenzene (PAzo-M M = Mg, Ca, Ni, Zn, Cu, Fe) based on dynamic bonding modulation. Reprinted with permission from Ref. [11]. Copyright 2022, the authors.

Small binding energy leads to small energy level difference of *cis–trans* isomer and low thermal energy storage density; large binding energy leads to large energy level difference of *cis–trans* isomer but low isomerization degree and low thermal energy storage density as well (Figure 6.16), therefore, choosing suitable binding energy can lead to large energy level difference and isomerization degree, and thus greatly improve its thermal energy storage density (Figure 6.16). The results show that magnesium metal polyacrylamide-azobenzene (PAzo-Mg) has the most excellent photothermal storage performance (113.5 Wh kg^{-1}, Figure 6.17).

Based on these excellent thermal properties, the Azo-based flexible fabric (NF@PAzo-Mg) human thermal management system was constructed in combination with nylon fabric (Figure 6.18). It can be used for human body heating in outdoor low-temperature environment (–5.0–5.0°C) with exothermic temperature up to 7.7–12.5°C (Figure 6.18). The light-controlled dynamic bonding-assisted Azo-based flexible fabric proposed in this study provides a new way to realize wearable thermal management devices at low temperatures.

Solar energy has received a lot of attention and research as a continuous renewable and clean energy source. Achieving controlled storage and release of solar energy is the most critical step in harnessing solar energy. Currently, the more commonly studied ways to store solar energy include photovoltaic power generation, photothermal conversion, photocatalytic decomposition of water for hydrogen storage, and so on. Some of these storage methods have already been commercialized, such as solar cells and solar water heaters, which are examples of photovoltaic power generation and photothermal conversion, respectively. Photothermal fuels, which work by converting solar energy into chemical energy in energy storage materials, especially by utilizing the energy difference between different molecular conformations, have received increasing attention from researchers in recent years. This energy is then stored and released as heat in a controlled manner when triggered by specific stimuli. This requires the material to undergo reversible changes in molecular conformation under the influence of light. Azo-based materials are capable of undergoing photochemically reversible *cis–trans* isomerization reactions. The *trans*-Azo isomer is a low-energy stable state, while the *cis*-Azo isomer is a high-energy metastable state. Under specific conditions,

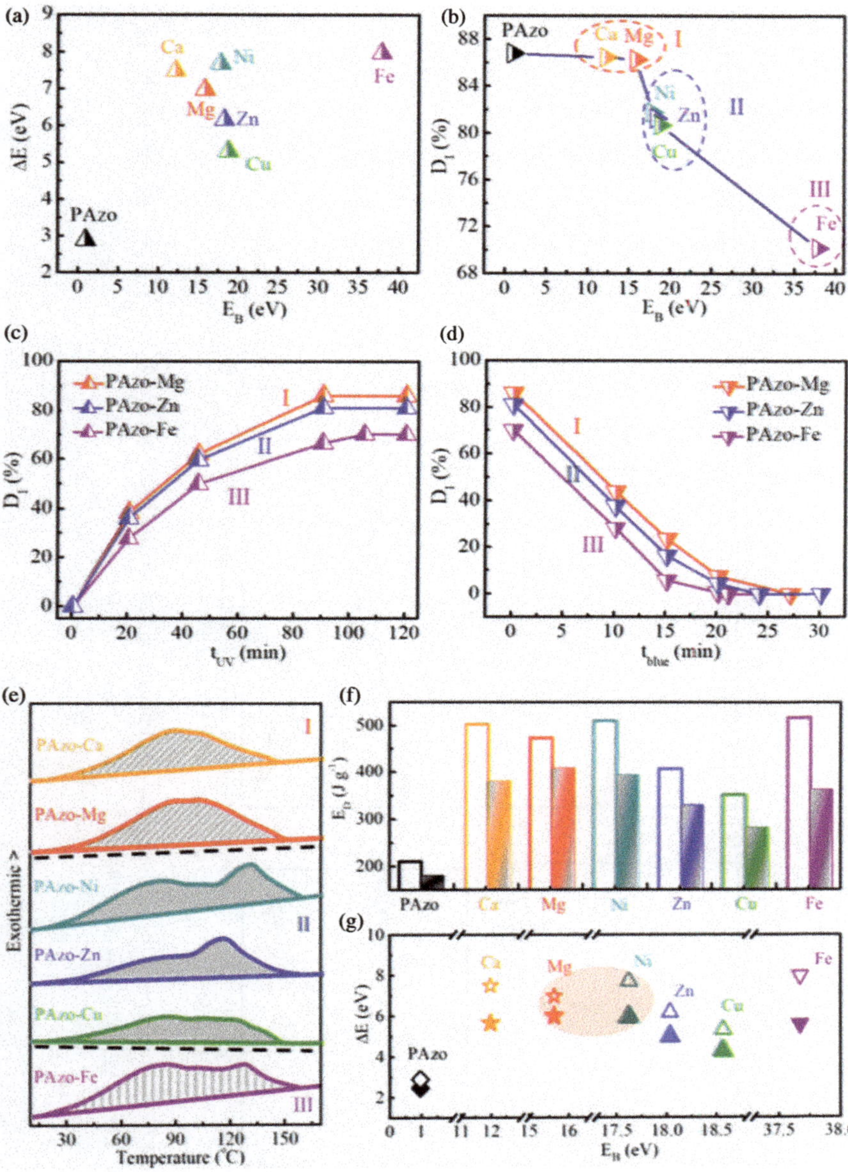

Figure 6.16. (a–d) Effect of different dynamic bonding species on the degree of photoisomerization and heat release rate of PAzo-M (M = Mg, Ca, Zn, Ni, Cu, Fe). (e–g) The effect of different dynamic bonding species on the photothermal performance of PAzo-M was investigated. Reprinted with permission from Ref. [19]. Copyright 2021, Elsevier Ltd.

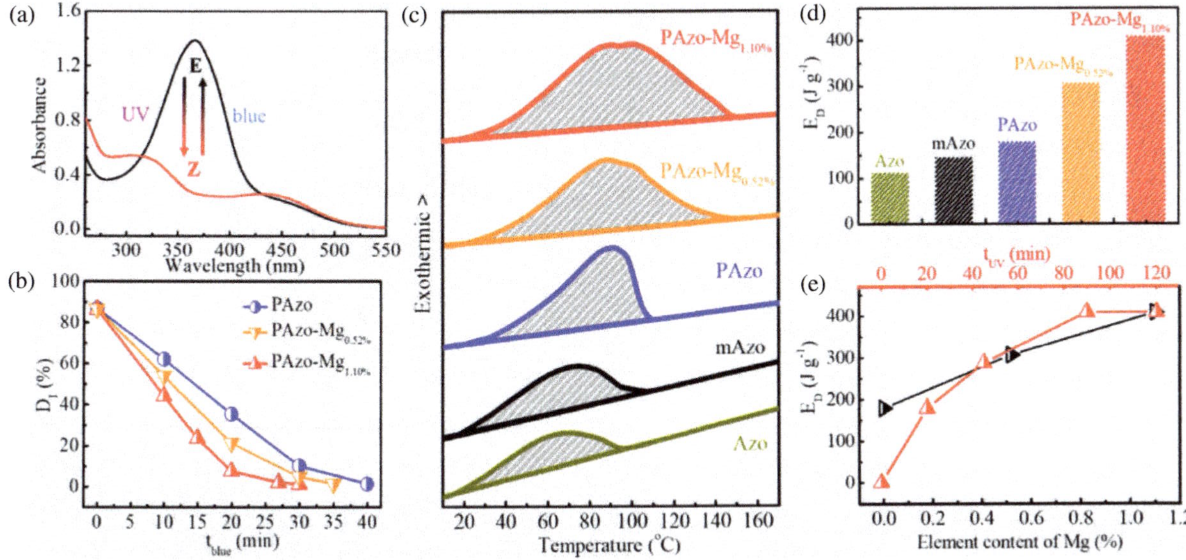

Figure 6.17. Comparison of the photothermal performance of Azo, mAzo, PAzo, and PAzo-Mg. Reprinted with permission from Ref. [11]. Copyright 2022, the authors.

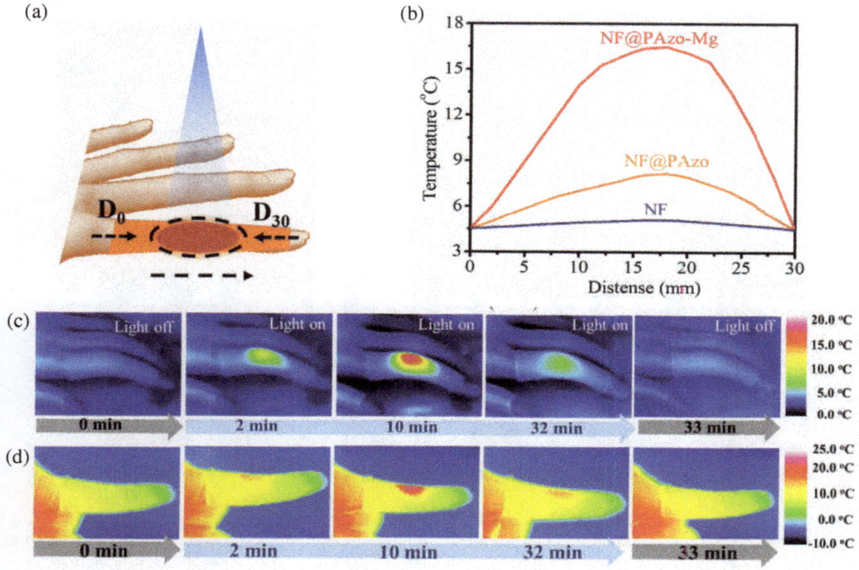

Figure 6.18. (a, b) Azo-based flexible fabric (NF@PAzo-Mg) human thermal management system. (c, d) Temperature regulation effect of Azo-based flexible fabric (NF@PAzo-Mg) human thermal management system at low temperature. Reprinted with permission from Ref. [11]. Copyright 2022, the authors.

these two isomers can interconvert, meeting the requirements of STFs. Additionally, compared to other systems such as the deglacial diene ↔ tetracycloheptane system or metal-ligated fulvalene system, STFs based on Azo materials offer significant advantages in terms of cost-effectiveness and reduced degradation.

Haifeng Yu's group at Peking University has developed a controllable and flexible STF cell based on polymer composite materials using a commercially available fabric as a template and compounding it with an azobenzene derivative molecule that can undergo a photogenerated solid–liquid phase transition, which enables reversible and controllable solar energy storage as well as thermal energy release [20]. Yu and his team have developed a flexible and wearable device that utilizes photo-switches with alkoxy and halogen groups for thermal energy storage. The device, depicted in Figure 6.19(a), consists of cloth fabric embedded with photo-switches, specifically 4-methoxyazobenzene. Under UV

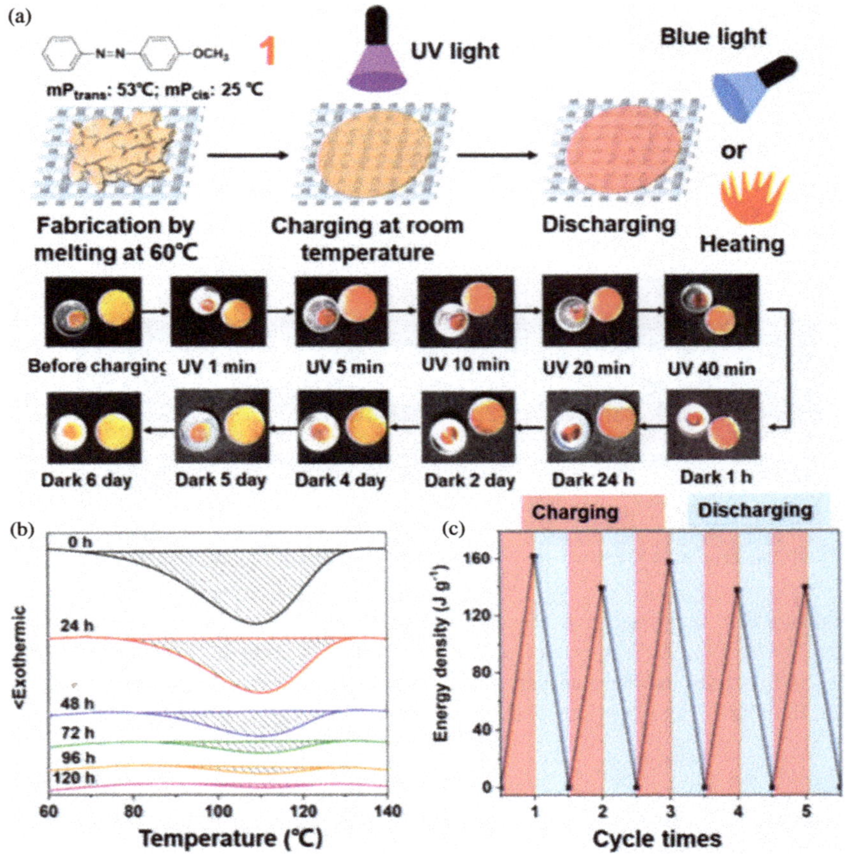

Figure 6.19. (a) Schematic diagram of energy storage and heat release processes of flexible device. Energy storage and heat release of the azobenzene photoswitch and flexible device. (b) DSC results of samples discharged in dark for different times. (c) The same charging and discharging cycles were repeated for five times. Reprinted with permission from Ref. [20]. Copyright 2019, Wiley.

light exposure, the photo-switches can transform from a solid to a liquid state. Subsequently, they release heat slowly in the absence of light or experience light-induced rapid heat release. This unique feature of the photo-switches allows for efficient thermal energy storage. The molecular photo-switches possess a characteristic known as PC, which significantly increases the energy density for storage to approximately 0.20 MJ kg^{-1}. This stored energy can last for 40 h due to the intermolecular interaction between the polyester fabric and the PC photo-switches, as illustrated in Figure 6.19(b). The device demonstrates excellent stability, as it maintained its performance after undergoing five cycles of energy storage and heat release, as shown in Figure 6.19(c). This implies that the photo-switches retain their functionality and efficiency over multiple usage cycles. The flexibility and wearable nature of this device make it suitable for everyday use as a warming or heating device. It presents a novel and practical solution for personal thermal comfort.

The unique characteristics of human skin, such as its softness, elasticity, and stretchability, have recently led to the development of soft elastomers that mimic human skin. These materials have found new applications in wearable electronic devices, flexible electronic packaging, and skin sensors. Soft elastomers are a type of cross-linked material with a much lower Young's modulus (approximately 0.1–1.0 MPa) compared to traditional elastomers (approximately 1.0–100 GPa). This mechanical property allows them to match the Young's modulus of human skin (in the kilopascal range) and seamlessly conform to the human skin for wearable applications. Recent studies have shown that light-controlled stimuli offer significant advantages in terms of the mechanical response of soft materials compared to other stimuli such as temperature, humidity, electric fields, and magnetic fields. Therefore, combining the elasticity of rubber with light-responsive properties presents new opportunities for obtaining soft elastomers that resemble human skin.

Industry-widely used SBS is a styrene-butadiene-styrene block copolymer, usually with a two-phase structure, namely, polybutadiene continuous phase and polystyrene dispersion phase. When heated, the material can flow and be easily processed and molded. When cooled to room temperature, the PS chain segments aggregate into glassy micro-regions, which act as physical cross-linking points for the PB chain segments, thus making the material exhibit rubbery properties, hence the term "thermoplastic elastomer". By mimicking the molecular and phase structures of SBS and combining the room temperature photo-responsive phase change properties of photo-responsive liquid crystal polymers, Haifeng Yu's group has achieved the first photo-controlled mechanical property interconversion of polymers [21]. The designed liquid crystal block copolymer has a triblock structure similar to SBS, except that the PB block is replaced by a photo-responsive liquid crystal polymer (PM11AZC4). The material is highly rigid at room temperature and shows a photo-controlled mechanical property transition to soft elasticity in the non-thermal state. Based on the photo-reversible switch between rigidity and elasticity at room temperature, the researchers demonstrated precise control of nano-patterning on non-planar substrates while being particularly suitable for the fabrication of encapsulated chalcogenide solar cells, thus enabling the use for mechanically adaptive optical display and electronics packaging applications through a simple, convenient, and controllable approach, as shown in Figure 6.20.

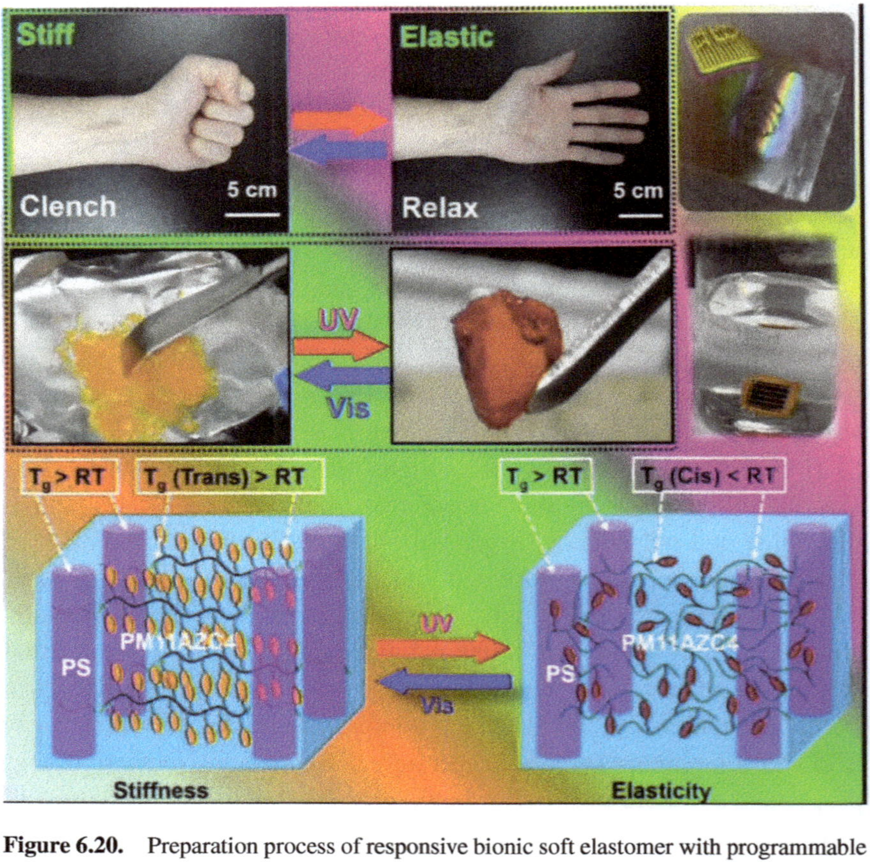

Figure 6.20. Preparation process of responsive bionic soft elastomer with programmable mechanics and its application. Reproduced under the terms of the CC BY 4.0 from [21]. Copyright 2022, the authors.

The photo-tunable mechanical behavior of the triblock copolymers was characterized using dynamic mechanical analysis (DMA). Both the photo-responsive block PM11AZC4 and the PS block containing azobenzene show glass transition temperatures above room temperature, so that the polymer behaves as a hard plastic at room temperature. Under UV irradiation, the glass transition temperature of PM11AZC4 drops below 0°C, allowing it to act as a soft segment that provides elasticity, while the photo-inert PS portion acts as a physical crosslinking point, giving the triblock copolymer a behavior similar to that of a thermoplastic elastomer, SBS, which researchers call a "photo elastomer". Moreover, this photo-induced mechanical interconversion occurs at room temperature, and the Young's modulus of the UV-irradiated triblock copolymer is 0.2 MPa, which is a typical soft elastomer that mimics human skin.

Using the light-controlled mechanical property transformation properties of this material, the researchers eventually achieved the preparation of micro-nano-patterns by pressure and light on complex geometry surfaces. Depending on the difference in the period and pattern of the two templates, different structural colors were produced, and the Chinese character "BeiDa" was finally printed on the forearm and the back of the hand (Figure 6.21(c)). Figures 6.21(d) and 6.21(e) show the scanning electron microscope images of the polymer films after the two-step imprinting process, respectively. It is clear that this method combines the advantages of micro-phase separation of photo-flexible elastomers and block copolymers and is very promising in the application cases of flexible displays in smartphones or computers, wearable technologies and even fabrics.

In addition, the researchers used liquid crystal block copolymer films to encapsulate the chalcogenide solar cells on a flexible PDMS substrate, and apparently the resulting devices can be used on different curved surfaces, such as arms. This encapsulated chalcogenide solar energy was studied for its water resistance by immersion in water. The unencapsulated device started to decompose when immersed in water for 10 s and completely decomposed at 180 s. When the encapsulated chalcogenide solar cell device was immersed in water for 1 h, no significant discoloration was observed. Subsequently, the stability of the unencapsulated and encapsulated devices in water was quantified by UV–Vis absorption spectroscopy. The absorption intensity of the unencapsulated devices started to decrease

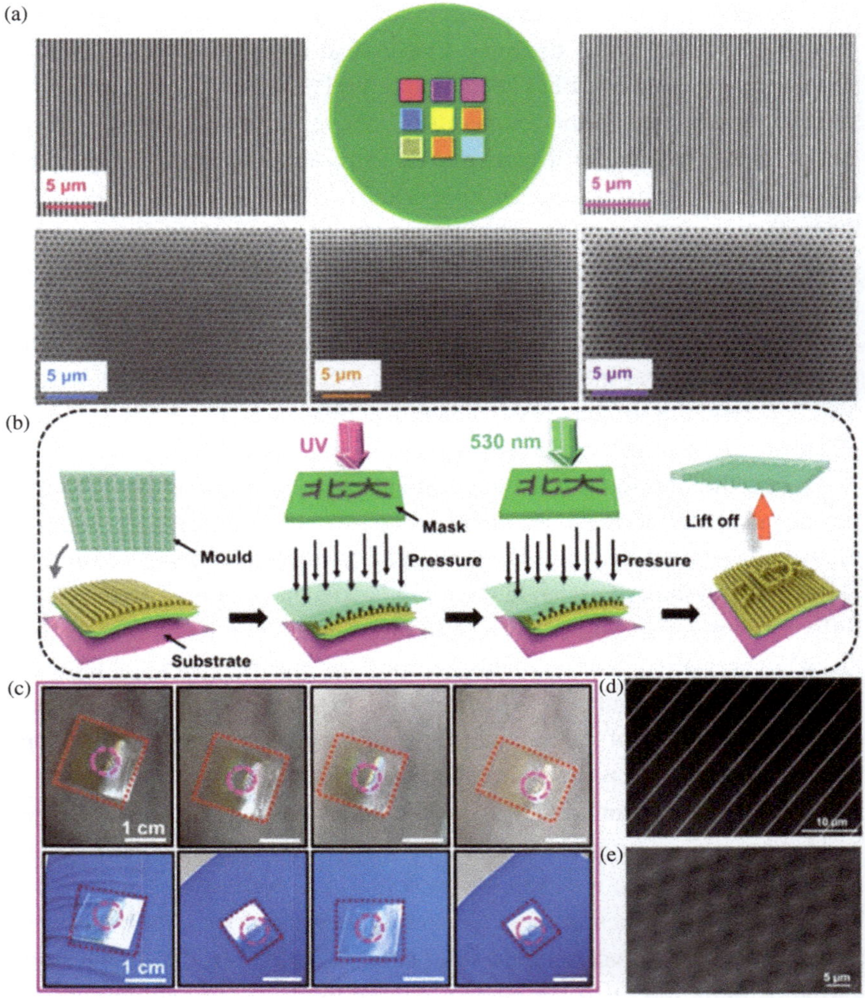

Figure 6.21. Precise programmable embossing process of triblock copolymer films illuminated by different wavelengths of light. Reproduced under the terms of the CC BY 4.0 from [21]. Copyright 2022, the authors.

significantly at 30 s and dropped to a minimum value at 600 s. Even after the encapsulated chalcogenide solar cells were immersed in water for 3600 s, the absorption intensity remained essentially unchanged, indicating that the block copolymer with light-controlled mechanical property interconversion can effectively isolate the cell devices from water and exhibit a good sealing effect.

MOST is able to store solar energy in molecular structures, and then turn it into thermal energy for release through external stimuli (light, heat, etc.) (Figure 6.22). Most of the azobenzene molecular solar thermal energy storage materials can only absorb light in the UV band, and this part of energy only accounts for about 3% of the total solar energy spectrum, so it is still a challenge to realize the full solar energy spectrum utilization to improve the utilization of solar energy.

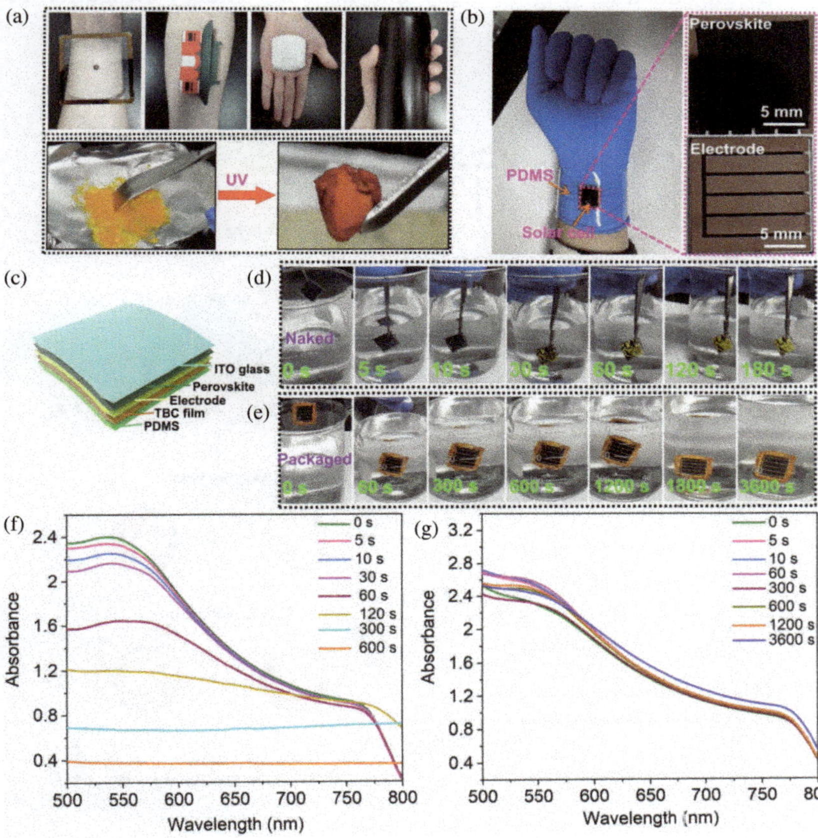

Figure 6.22. Preparation process and characterization of encapsulated PSC. Reproduced under the terms of the CC BY 4.0 from [21]. Copyright 2022, the authors.

Prof. Wang's team at Jiangnan University proposed a full solar spectrum absorbing fabric (VSSF) based on molecular solar energy storage, which can effectively increase the utilization of the solar spectrum [22]. By mixing tetradecyloxyazobenzene molecules and tetradecanol phase change molecules, the absorption of isomeric energy as well as phase change energy is achieved, and the two types of energy can be released in a synchronized light-controlled manner. The principle of phase change point of azobenzene photo-controlled PCM was further explained by molecular binding energy calculation. The team established a closed-loop full solar spectrum absorption, conversion, storage, and release system on fabric as the substrate (Figure 6.23), and the solar energy utilization rate reached 4.8%.

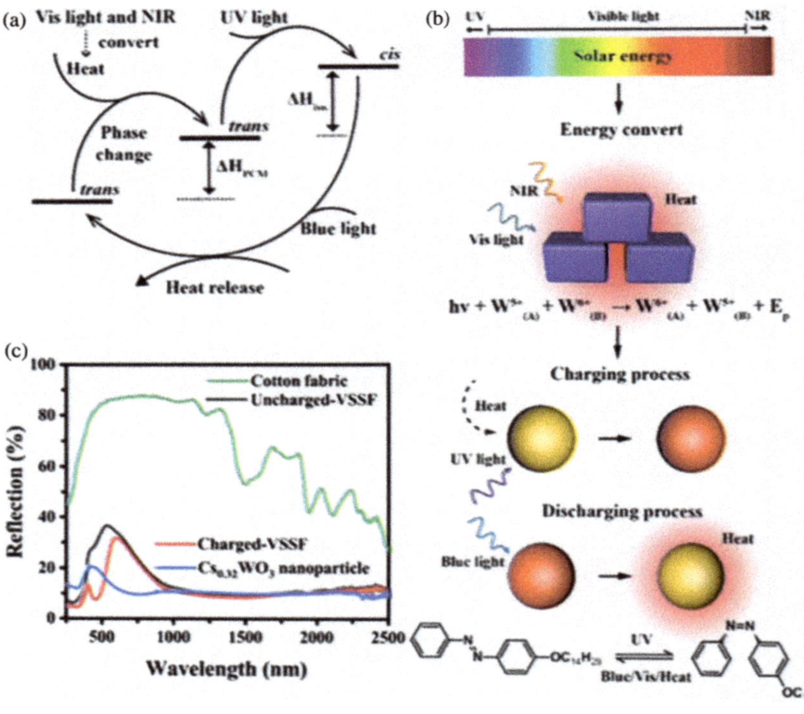

Figure 6.23. (a) UV–Vis–NIR solar spectrum absorption, storage and release system; (b) UV–Vis–NIR reflectance spectrum; (c) Solar energy absorption and conversion process. Reprinted with permission from Ref. [22]. Copyright 2021, Elsevier B.V.

This work applies VSSF to photoexcited self-heating wrist guards, which stimulate heat generation in response to simulated sunlight (AM1.5, 40 mW cm^{-2}) and blue light (460 nm, 40 mW cm^{-2}), respectively. This heat generation was attributed to the release of isomeric heat from the azobenzene molecule, the release of the latent heat from the PCMs, and the photothermal conversion properties of the nano-cesium tungsten bronze powder. The thermal insulation and heat retention properties of the PCM contribute to slowing down the heat dissipation during the cooling process by turning off the light source (Figure 6.24). Relative to the blank cotton fabric, the human body surface temperature was elevated by 4°C (to 37.2°C) within 400 s of light exposure.

Interestingly, by adjusting the intensity of the stimulus response light can be used to construct a photothermal lifter. By the PLA shape memory material having the property of thermally induced deformation (response temperature of 60°C), the VSSF is designed to effectively control the deformation of the PLA, and this deformation can be used as a force to lift the heavy object. Under the stimulation of simulated sunlight (AM1.5, 80 mW cm^{-2}), the heating temperatures of the VSSF in the charged and uncharged states are different, 65°C and 55°C, respectively. Therefore, the energized VSSF can excite the PLA to produce deformation recovery under solar stimulation, thus lifting an object three times its own weight and realizing photothermal drive (Figure 6.25).

The team analyzed the relationship between color properties (K/S value) and the degree of isomerization to visualize and monitor the energy storage and discharge process of VSSF. Under UV light stimulation, the color changes regularly due to the photoisomerization property of azobenzene, and the color "returns" during the isomerization recovery of azobenzene. The color change was attributed to the photochromic property of azobenzene, that is, the ratio of *cis–trans* isomerization of azobenzene affected the overall color change. At the same time, the ratio of azobenzene isomerization is proportional to the energy storage state, so by establishing the relationship between the color and the degree of isomerization, the energy storage state can be reflected quickly, visually, and in real time (Figure 6.26).

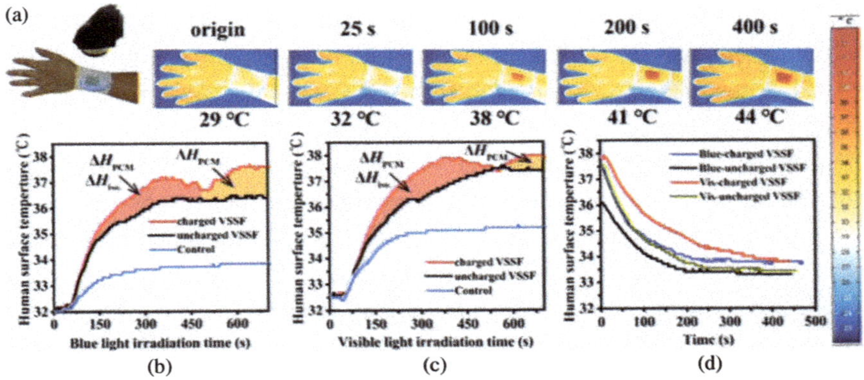

Figure 6.24. (a) Infrared thermographic image of the surface temperature of the light-excited self-heating wrist guard; human body surface temperature monitoring curve under the stimulation of blue light (b) and simulated sunlight (c); (d) human surface temperature cooling curve. Reprinted with permission from Ref. [22]. Copyright 2021, Elsevier B.V.

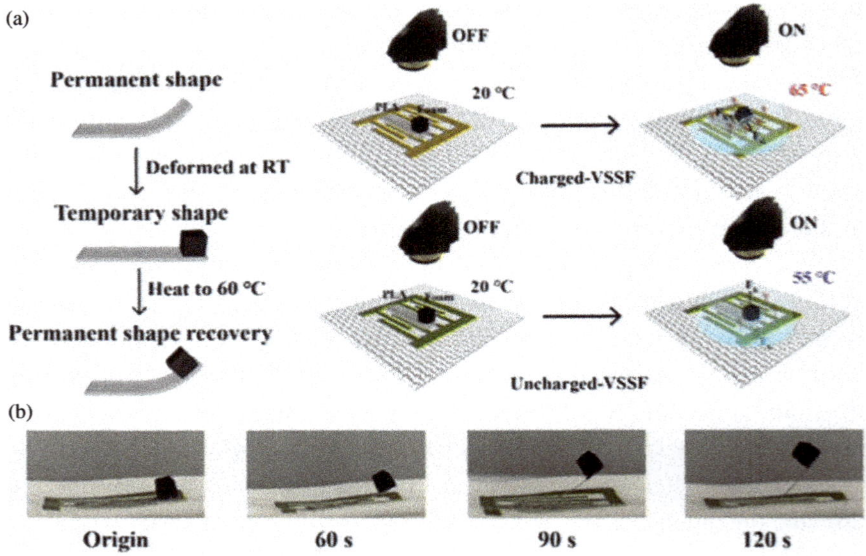

Figure 6.25. (a) Schematic diagram of the photothermal starter. (b) Optical photographs of the photothermal starter under different lighting times. Reprinted with permission from Ref. [22]. Copyright 2021, Elsevier B.V.

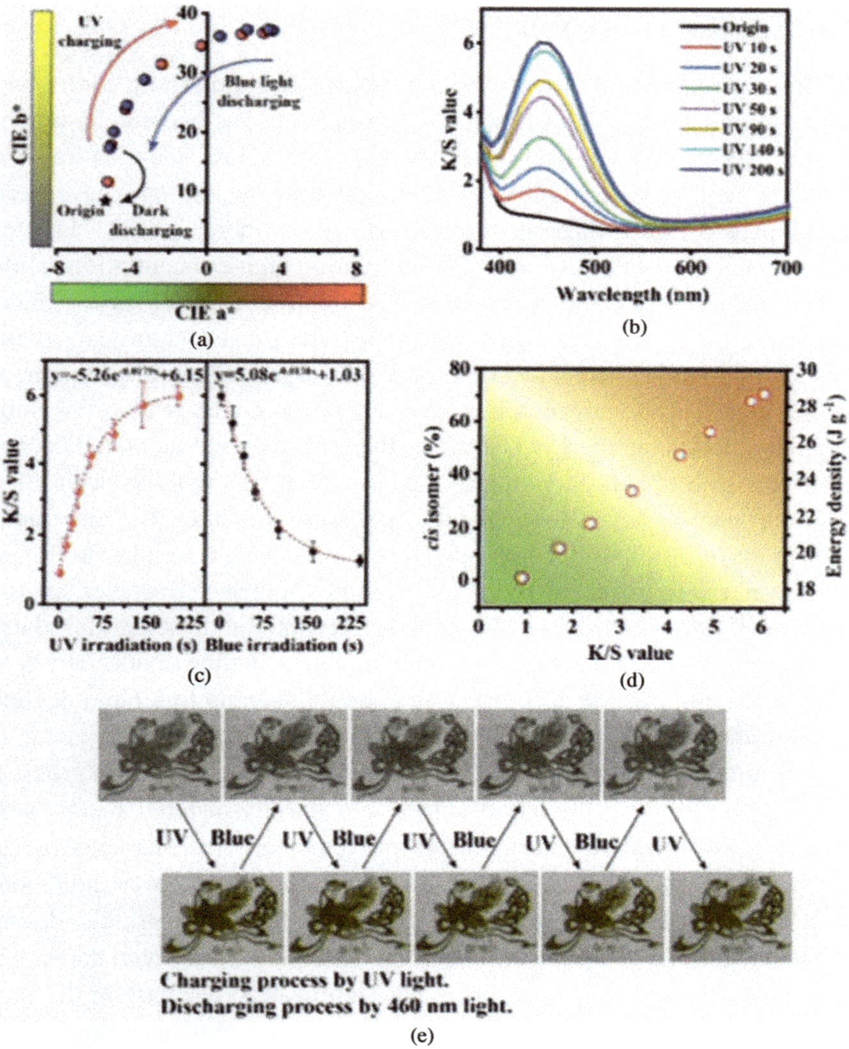

Figure 6.26. (a) Changes in color parameters (a* and b*) under light stimulation; (b) K/S spectral changes under UV light; (c) curves of K/S values at 450 nm with UV light and blue light; (d) curves of K/S values, degree of isomerism, and energy storage relationships; (e) color cycling performance. Reprinted with permission from Ref. [22]. Copyright 2021, Elsevier B.V.

6.4 Seawater desalination

With rapid population growth and increasing water pollution, addressing water scarcity and treating wastewater effectively has become an urgent global issue. Various water treatment technologies, such as reverse osmosis membranes and ultrafiltration membrane technology, have been developed to tackle these challenges. However, there is still a lack of efficient and cost-effective methods for treating high-concentration saline water (salinity >7%) or wastewater. The treatment processes for high-concentration saline water and wastewater often cause significant damage to membranes used in commonly employed membrane treatment technologies, while concentration/crystallization methods consume substantial amounts of energy. Therefore, there is an urgent need to discover efficient, stable, and economical methods to treat highly concentrated seawater and wastewater. In recent years, solar-driven evaporation techniques have offered new possibilities for seawater desalination and wastewater treatment for their low carbon footprint. However, in the process of treating high-concentration saline water and wastewater, these devices face severe fouling and scaling issues, which hinder sunlight absorption and impede adequate water supply, leading to a rapid decline in photothermal conversion efficiency and even device failure.

Inspired by the structure of lotus leaves, Prof. Zhu's research group at Nanjing University [23] has designed a water-lily-inspired hierarchical structure (WHS) photothermal–vapor conversion device that achieves an 80% photothermal conversion efficiency when treating 10 wt% high-concentration saline water and 30 wt% wastewater. Of particular note, during the treatment of high-concentration saline water and wastewater, the photothermal conversion efficiency does not significantly decrease until complete evaporation ("zero emissions") and the surface remains clean.

As seen in Figure 6.27(a) and 6.27(b), the natural water lily leaves have an interesting structure. Firstly, the surface of the leaf absorbs sunlight and has sufficient stomata for the escape of vapors. At the same time, its surface is hydrophobic and can effectively self-clean. Secondly, the inner part of the water lily blade consists of many hollow chambers, which allows the whole blade to float naturally above the water surface. Finally, the water lily leaf is connected to a slender rhizome, which provides the entire leaf with an adequate water supply.

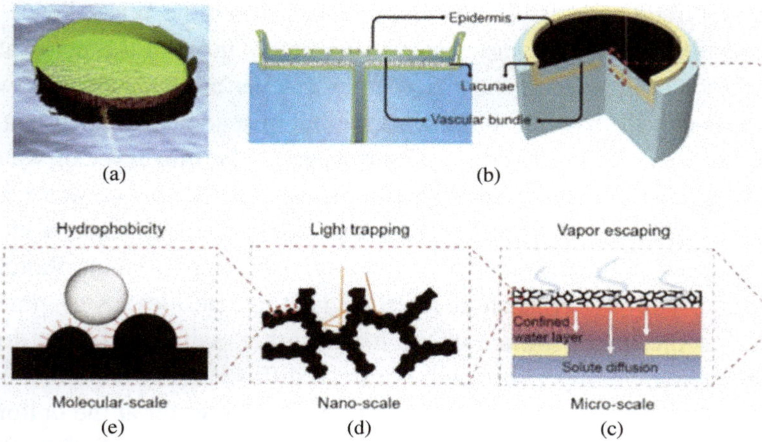

Figure 6.27. Design of a multi-stage structure (WHS) device inspired by a water lily leaf. (a) and (b) are schematic diagrams of a water lily leaf and a WHS device, respectively. They share common features: the hydrophobic surface layer absorbs sunlight and provides stomata for vapor escape; the gas chamber inside the leaf can provide more buoyancy for the leaf to float; and the rhizome can provide a localized water channel to reduce conduction heat loss. (c) Shows a micron-scale schematic of a thin layer of water sandwiched between the absorber and the base and (d) shows the effect of nanostructures on the absorber on light absorption, while (e) shows the effect of modification of the absorber surface at the molecular chain scale to make the sample surface hydrophobic. Reproduced under the terms of the CC BY 4.0 from [23]. Copyright 2019, the authors.

The WHS device designed by the research team consists of a top solar absorber and a base, which have similar properties to water lily leaves (Figure 6.27(b)). As shown in Figure 6.27(c), the top absorber, like the surface of a water lily leaf, has a rich structure of micron pores that act like artificial "stomata" to provide an effective escape path for vapor. The absorber surface has many nanoscale protrusions (Figure 6.27(d)), which can effectively trap sunlight and also provide the structural basis for the hydrophobic properties of the surface (Figure 6.27(e)).

The researchers imitated the internal chamber structure of leaf blades and installed a perforated polystyrene base beneath a copper foam, allowing the entire device to float on the water's surface. The polystyrene base possesses a very low thermal conductivity (<0.04 W m^{-1} K^{-1}), which inhibits the diffusion of heat into the water column. Finally, similar to the rhizomes of lotus leaves, a quasi-one-dimensional channel (local porosity of the polystyrene base) is used to supply water. When the WHS device processes saline water and wastewater, water enters through the pores of the base. Since both sides of the absorber's top layer are hydrophobic, water does not enter the interior of the absorber but instead gets trapped in a thin space between the base and the absorber part. The thin layer of water is crucial for stable and efficient treatment of concentrated saline water. Solar energy is absorbed by the absorber and converted into heat, which is locally absorbed in the thin layer of water, achieving efficient photothermal–vapor conversion, such as interface heating. Besides, as evaporation occurs on the lower surface of the top absorber, salt and solutes do not precipitate on the upper surface but instead diffuse downward and are discharged through the pores of the base into the bulk water at the bottom.

The research team performed performance tests of the photo-vapor conversion for this WHS device. For pure water, 10 wt% brine, and 30 wt% wastewater, the WHS device achieved 79.8%, 78.5%, and 77.2% photothermal conversion efficiency, and 1.31 kg $m^{-2}h^{-1}$, 1.28 kg $m^{-2}h^{-1}$, and 1.27 kg m^{-2} h^{-1} steam yield, respectively. During the treatment of 10 wt% concentrated brine, the researchers conducted comparative experiments with a conventional absorber. With one sun, water treatment using the WHS device, whose surface was continuously clean (on Figure 6.28(a))

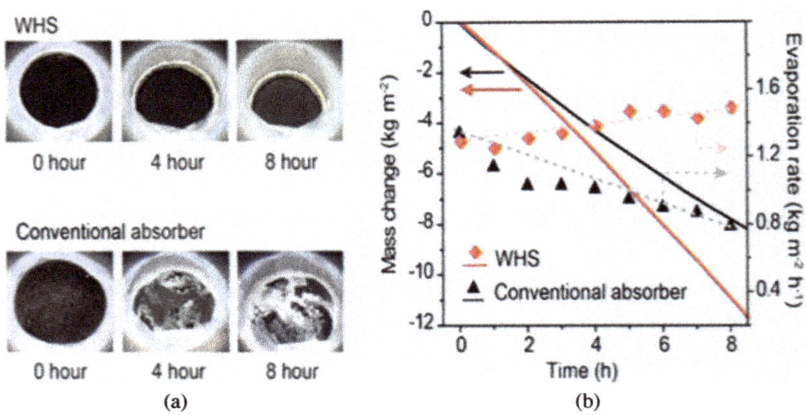

Figure 6.28. (a) Surface optical photographs and (b) water evaporation performance curves of WHS devices and conventional absorbers in treated brine (with 10 wt% starting concentration). Reproduced under the terms of the CC BY 4.0 from [23]. Copyright 2019, the authors.

and its evaporation remained at a consistently high level (Figure 6.28(b)). For the conventional absorber, the salt gradually precipitates on the surface (Figure 6.28(a) bottom), which hinders the absorption of sunlight, so its evaporation also gradually decreases (Figure 6.28(b)). Under identical conditions, the WHS device demonstrates an average evaporation rate of 1.39 kg m^{-2} h^{-1} over an 8-h duration, representing an approximate 1.4-fold increase compared to the average evaporation rate of 0.97 kg m^{-2} h^{-1} observed in conventional absorbers.

When processing saline water or wastewater, the WHS device achieves complete solid–liquid separation. As shown in Figures 6.29(a) and 6.29(b), when the WHS device floats on the surface of saline water (or wastewater), it gradually moves downward as the water evaporates, until all the water has evaporated, leaving behind the salt or solutes. During this process, the WHS device maintains a consistent and stable water evaporation efficiency, with no contamination on its surface (the deposition of salt or solutes). Surface photos of the WHS device during this process are displayed in Figure 6.29(a) and 6.29(b). After complete evaporation of the water, the remaining salt or solutes can be easily removed, as shown in the lower right inset of Figure 6.29(a) and 6.29(b).

Pro. Zhu's group has done a series of work around efficient interfacial photothermal conversion: firstly, the blackest iso-excited absorber was successfully prepared; on this basis, the first solar water desalination device based on the enhancement effect of iso-excited elements was realized [24].

Currently, traditional desalination devices suffer from issues such as large size, high energy consumption, high carbon emissions, and low desalination efficiency. The utilization of solar-driven water distillation technology is a low-carbon and environment-friendly approach for seawater desalination. However, over the years, this technology has been constrained by low photothermal conversion efficiency (approximately 30–45%), making it difficult to be applied on a large scale. Zhu's group [24] has achieved the first high efficiency solar water desalination in the international arena by utilizing the plasma exciton enhancement effect, with energy transfer efficiency equal to approximately 90% and salinity reduction of four orders of magnitude before and after desalination (Figure 6.30). The study found that the three-dimensional aluminum

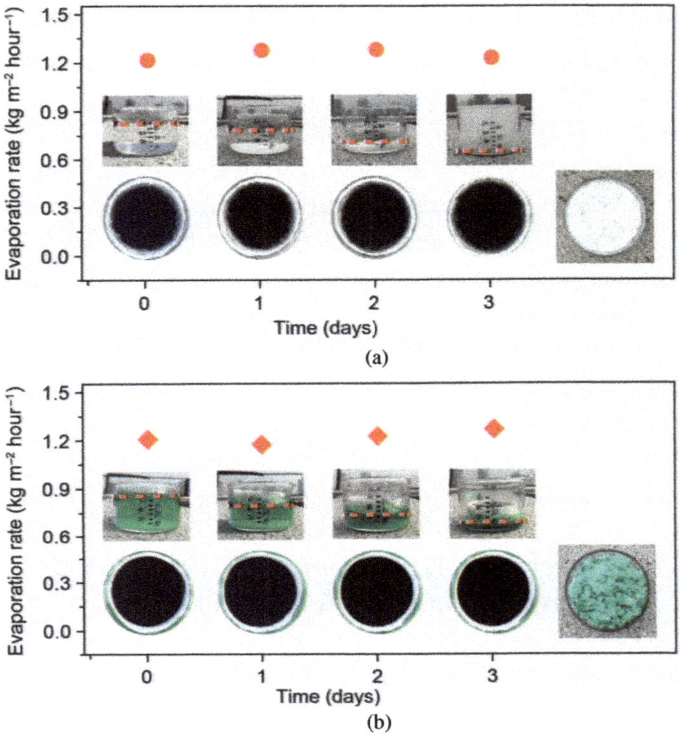

Figure 6.29. Stable and efficient photothermal steam conversion for water and solute separation. (a) and (b) demonstrate the daily evaporation efficiency of solar desalination and wastewater treatment, respectively. The first and second rows of insets display the side and top views of the samples during each operational day, respectively. The right most column showcases the salt and metal solute samples collected after achieving complete solid–liquid separation. Reproduced under the terms of the CC BY 4.0 from [23]. Copyright 2019, the authors.

particles equilibrium excitonic blackbody material is an excellent system to achieve high-efficiency solar desalination, and the material system has ultra-high light absorption efficiency in the broad solar spectrum, with an average absorption efficiency greater than 96% in the broad solar spectrum range between 400 nm and 2500 nm, ensuring a significant increase in the photothermal conversion efficiency during desalination; the local equilibrium excitonic optical resonance effect of aluminum nanoparticles

Figure 6.30. Schematic design of equipartition exciton enhanced solar desalination. Reprinted with permission from Ref. [24]. Copyright 2016, Springer Nature.

makes floating on the water surface closely aligned metal nanoparticle film local temperature rise acceleration and electromagnetic field enhancement effect, which is very conducive to the rapid warming of seawater at the membrane–liquid interface, generating freshwater vapor, and the porous structure of the film provides an effective vapor escape channel. Figure 6.31 shows the design schematic of aluminum particle blackbody material for iso-excited element enhanced solar water desalination. This aluminum particulate iso-excited element blackbody film material was prepared using low-cost metallic aluminum as the only raw material and a simple self-assembly preparation method that can be produced at scale, as shown in Figure 6.31.

Measurements show that the water quality after desalination is better than WHO standard drinkable water, and the desalination performance shows good stability and durability, which will have important significance for the large-scale practicalization of high-efficiency solar desalination technology.

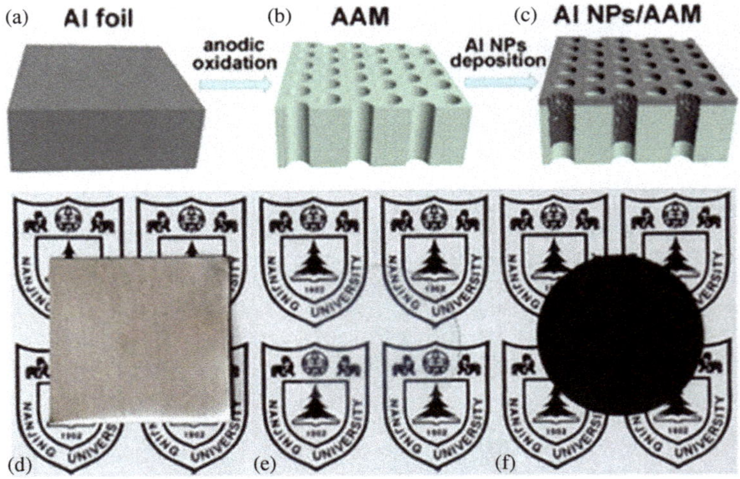

Figure 6.31. Schematic diagram of the self-assembly process of three-dimensional aluminum particles with equipartite excitonic blackbody: (a) aluminum foil, (b) porous alumina, (c) aluminum particles self-assembled in alumina template; (d), (e), (f) are the corresponding optical photos of (a), (b), (c). Reprinted with permission from Ref. [24]. Copyright 2016, Springer Nature.

Subsequently, for the first time, a bio-organism, the mushroom before and after carbonization, was used as a photothermal conversion device (as shown in Figure 6.32), making full use of its natural excellent macro- and microstructure: the black umbrella-shaped cap, the porous flesh, and the small cross-sectional area of the fibrous stalk to achieve efficient light absorption, effective water supply and vapor escape, and at the same time greatly reduce heat dissipation. The bottom of the mushroom stalk is in contact with the water body, confining the water transfer and heat conduction to this quasi one-dimensional channel (channel cross-sectional area is less than one-tenth of the light absorption cross-sectional area), ensuring effective water supply while greatly inhibiting the conduction dissipation of heat transfer to the water body; the umbrella-shaped curved cap produces a larger evaporation area and lowers the working temperature, thus effectively controlling the heat convection and heat radiation losses (as shown in Figure 6.33), thus achieving ~62% and ~78% light-to-vapor conversion efficiencies for mushrooms and carbonized mushrooms, respectively, without external focusing conditions [25].

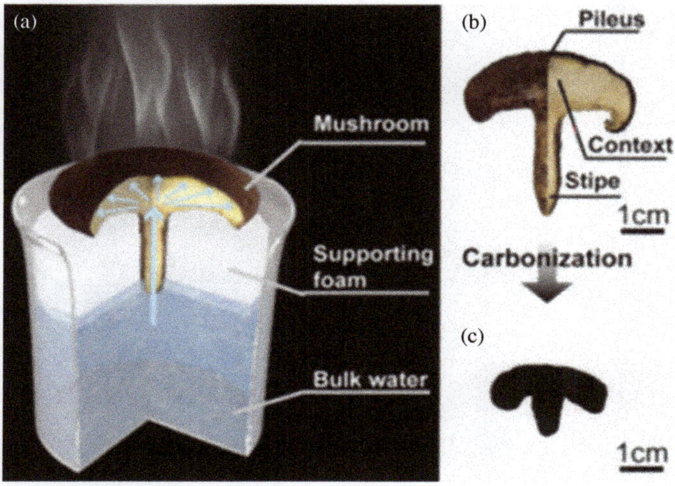

Figure 6.32. (a) Schematic diagram of the photothermal conversion device based on mushroom, (b, c) macroscopic structure of mushroom before and after carbonization. Reprinted with permission from Ref. [25]. Copyright 2017, Wiley.

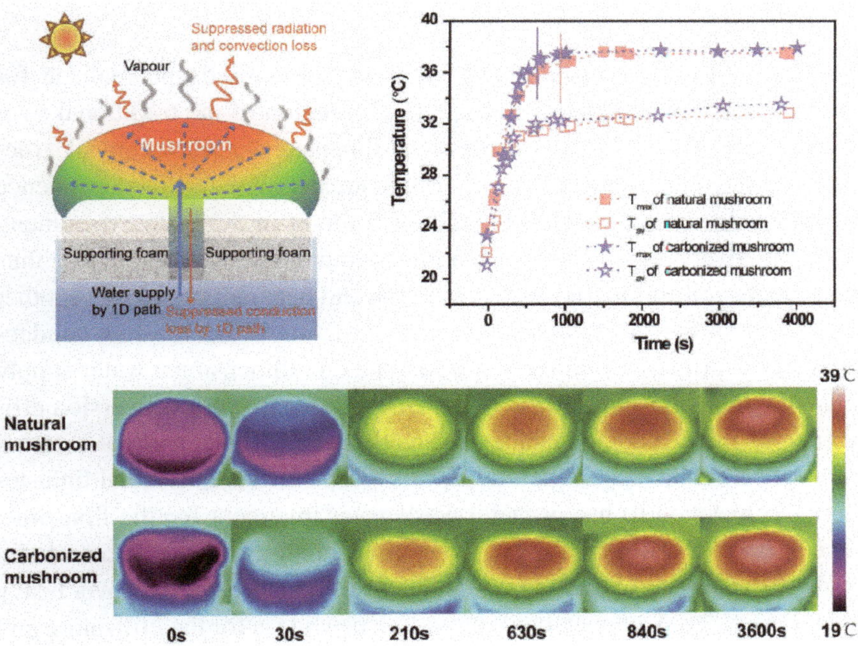

Figure 6.33. Thermal behavior analysis of mushroom structures under the sun. Reprinted with permission from Ref. [25]. Copyright 2017, Wiley.

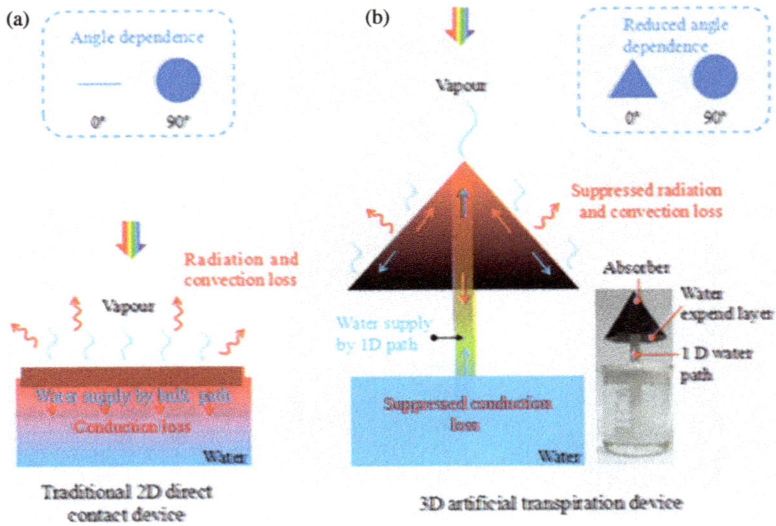

Figure 6.34. (a) Schematic diagram of a conventional photo-thermal (vapor) conversion device, (b) schematic diagram of a three-dimensional (3D) artificial transpiration device. Reprinted with permission from Ref. [26]. Copyright 2017, the author(s).

The "artificial transpiration" structure is proposed and realized for the first time, which breaks through the limitation of the traditional two-dimensional planar device and the three-dimensional hollow conical structure (as shown in Figure 6.34), which can greatly alleviate the dependence of the device on the incidence angle of sunlight in the real environment, and effectively reduce the evaporation temperature of the device, thus effectively controlling the heat convection and heat radiation loss. Another advantage of the "artificial evaporation" structure is that the heat conduction loss is effectively controlled through a one-dimensional water supply channel, thus achieving for the first time a light-to-vapor conversion efficiency of more than 85% under normal light conditions without external assistance. At the application level, this work extends interfacial light-to-vapor conversion to heavy metal wastewater treatment for the first time. The experimental results show that this technology not only can obtain water that meets drinking standards, but also can effectively recover heavy metals (e.g., gold, copper, etc.), providing new ideas for the utilization and development of efficient solar light-thermal (steam) [26].

So far, research on solar interfacial evaporation systems has been broadly divided into (1) preparation of photothermal conversion materials, (2) design and optimization of structures/configurations, and (3) construction of condensation systems. However, most solar-driven interfacial evaporation systems can only work during the daytime due to the alternation of day and night, and this intermittent work is not conducive to the practical application of evaporators. Based on this, Li's group at Northwest Normal University has prepared an all-weather solar-driven interfacial evaporator by coupling phase change energy storage and photo-thermal conversion [27]. A carbonized ultra-long poly-pyrrole nanotube aerogel loaded with octadecane is placed inside a spherical container as the phase change energy storage component, while a nylon thread impregnated with poly-pyrrole is wrapped around the surface of the spherical container as the photothermal conversion component. The prepared aerogel Cppy has rich porosity for efficient loading of octadecane (OCC). The evaporator performs photothermal evaporation during daytime while storing energy in phase change and releasing energy at night, thus achieving continuous evaporation around the clock with evaporation rate and photo-vapor efficiency as high as 2.62 kg m^{-2} h^{-1} and 92.7%, respectively, for 4 h and 10 min after turning off simulated sunlight after 6 h of continuous light exposure. In addition, the Cppy-O evaporator can collect ~8.42 kg m^{-2} during daytime and ~1.22 kg m^{-2} of clean water during nighttime in outdoor evaporation experiments. This study provides a new approach for the design and preparation of novel all-weather solar-driven interfacial evaporators and lays the foundation for breaking the diurnal limitation in the solar interfacial evaporation process.

As shown in Figure 6.35, this work used the sol–gel method to obtain ultra-long polypyrrole nanotube hydrogels by mixing methyl orange, pyrrole, and ferric chloride and then pouring them into spherical molds for aging, followed by freeze-drying and carbonization to obtain ultra-long polypyrrole nanotube aerogels. Subsequently, octadecane was loaded into the prepared aerogel by melt impregnation method to obtain Cppy-O composite PCM. Finally, the Cppy-O evaporator was obtained by placing it in a spherical vessel and wrapping the surface of the spherical vessel with polypyrrole-impregnated nylon wire. The evaporator absorbs heat during daytime and stores it in the Cppy-O composite PCM for nighttime

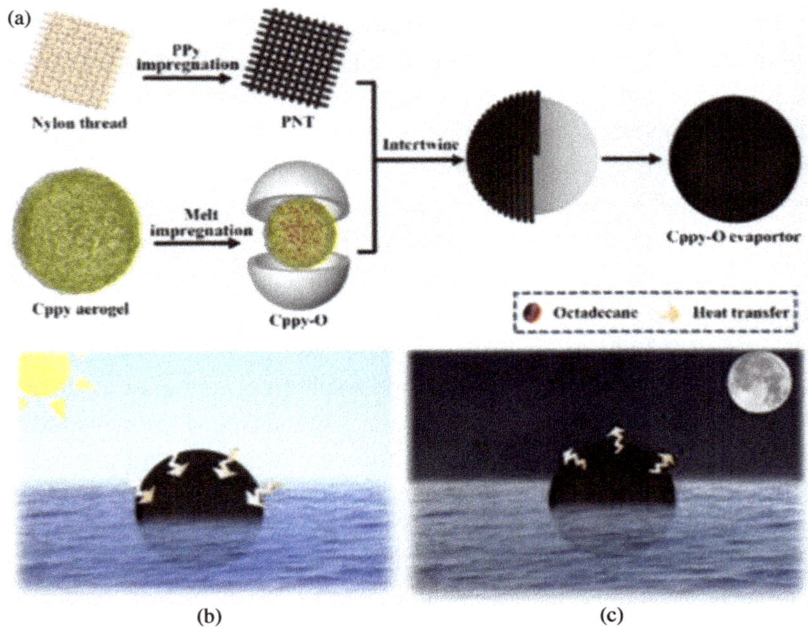

Figure 6.35. (a) Schematic diagram of the preparation of Cppy-O evaporator. (b) Daytime mode for all-weather evaporation. (c) Schematic diagram of nighttime mode. Reprinted with permission from Ref. [27]. Copyright 2022, Royal Society of Chemistry.

evaporation to achieve continuous evaporation around the clock. Subsequently, octadecane was loaded into the prepared aerogel using a melt impregnation method, resulting in the Cppy-O composite PCM. Finally, the Cppy-O evaporator was placed inside a spherical container, with the surface of the container wrapped in nylon thread impregnated with polypyrrole, to obtain the Cppy-O evaporator. The evaporator absorbs heat during the day and stores it in the Cppy-O composite PCM for nighttime evaporation, enabling continuous evaporation throughout the day and night.

This work couples phase change energy storage and photothermal conversion and designs and prepares a novel all-weather solar interface evaporator using octadecane as the phase change energy storage material and polypyrrole-impregnated nylon thread as the photothermal conversion layer. Attributed to its high light absorption, fast water transport, and excellent energy storage performance, the evaporator has excellent evaporation rate and light-vapor efficiency, with water collection up to 9.64 L m^{-2} d^{-1} under natural light (Figure 6.36). This study provides a new strategy for constructing all-weather solar-driven interfacial evaporators, laying the foundation for breaking the diurnal cycle and achieving efficient solar steam generation.

Numerous solar-driven interfacial evaporators with excellent desalination performance have been developed. However, most of these devices still have limitations in terms of intermittent solar energy utilization and salt accumulation, which remain major obstacles for their practical applications. To improve this problem, the group of Xiaodong Wang and Huan Liu at Beijing University of Chemical Technology microencapsulated *n*-docosane PCM in TiO_2/Fe_3O_4 shell and deposited polydopamine on the shell, as well as prepared a kind of multi-layer MXene/polydopamine/magnetic phase change microencapsulated composites by encapsulating MXene nanosheets through hydrogen bonding interactions, and combined it with natural wood to obtain novel interfacial evaporation system [28].

In this paper, a novel solar-driven interfacial evaporator was developed, utilizing natural wood as a water transport vehicle and incorporating multilayered MXene/poly(dopamine)/magnetic phase change microcapsules as a solar absorber for efficient desalination [28]. Multilayered MXene/polydopamine/magnetic phase change microcapsules were successfully prepared by microencapsulating n-docosane in TiO_2/Fe_3O_4 composite shell material, followed by deposition of polydopamine on the shell, and encapsulation of MXene nanosheets by hydrogen bonding interactions (Figure 6.37). The multilayer structure makes the prepared microcapsule composites exhibit excellent optical absorption capabilities, with an average absorption rate exceeding 90% across the entire solar spectrum. In addition, the multilayer structure with high hydrophilicity, good heat storage performance, and the enthalpy of phase transformation exceeds 150 J g^{-1}. The evaporator based on multilayered MXene/polydopamine/magnetic phase change microcapsules exhibited superior solar photothermal performance with an evaporation rate of 2.09 kg m^{-2} h^{-1} under one solar light intensity irradiation (Figure 6.38). The presence of narrow gaps between the microchannels of natural wood and the microcapsule composites facilitated efficient water transport and effectively prevented salt accumulation in the evaporation system.

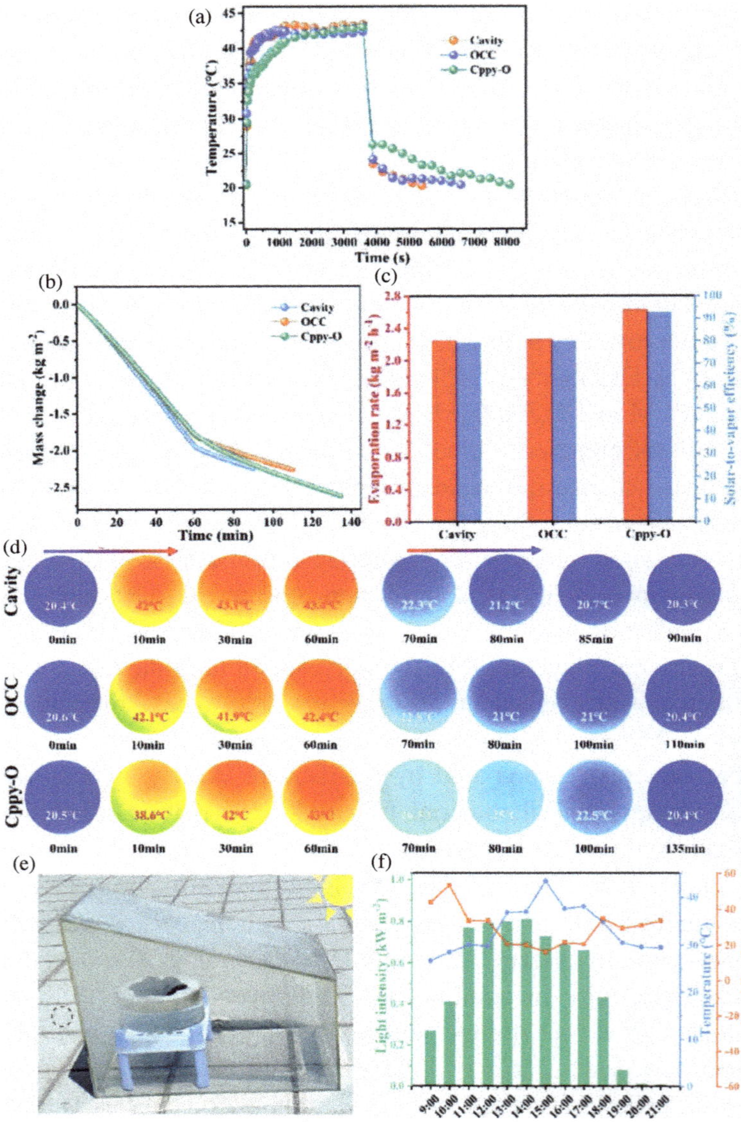

Figure 6.36. (a–d) Surface temperature of the evaporator, water mass variation curve over time, evaporation rate, light-vapor efficiency, and infrared thermogram during a 1-h period under 1 kW m^{-2} light intensity. (e) Optical image of the desalination unit installed on the laboratory building's roof. (f) Changes in sunlight intensity, ambient temperature, and ambient humidity from 9:00 to 21:00 hrs. (g) Water production of Cppy-O evaporator for seven consecutive days. (h) Ion concentration before and after the treatment of simulated seawater in the Cppy-O evaporator. Reprinted with permission from Ref. [27]. Copyright 2022, Royal Society of Chemistry.

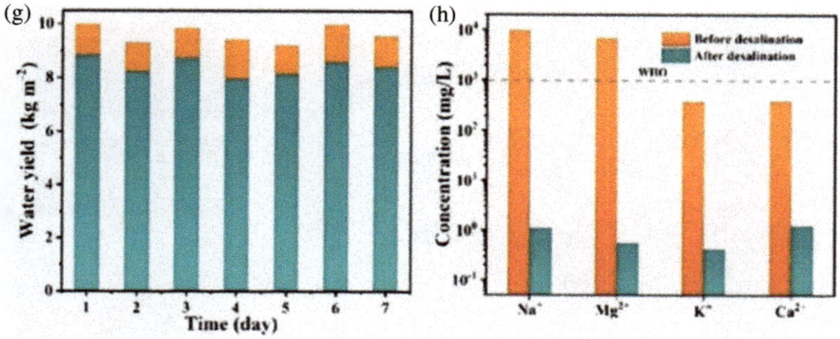

Figure 6.36. (*Continued*)

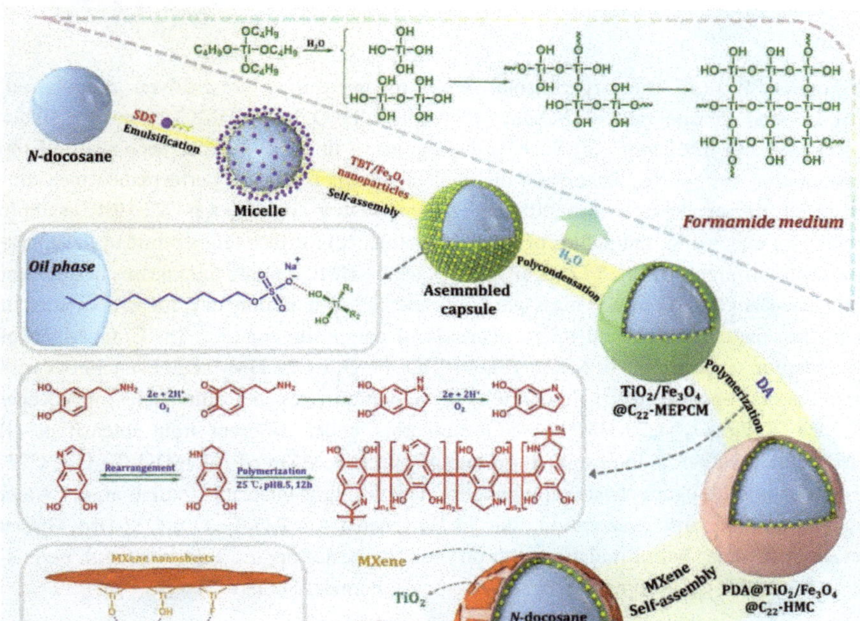

Figure 6.37. The preparation method and reaction mechanism of MXene/PDA@TiO_2/Fe_3O_4@C22-HMC The reaction mechanism is shown. Reprinted with permission from Ref. [28]. Copyright 2022, American Chemical Society.

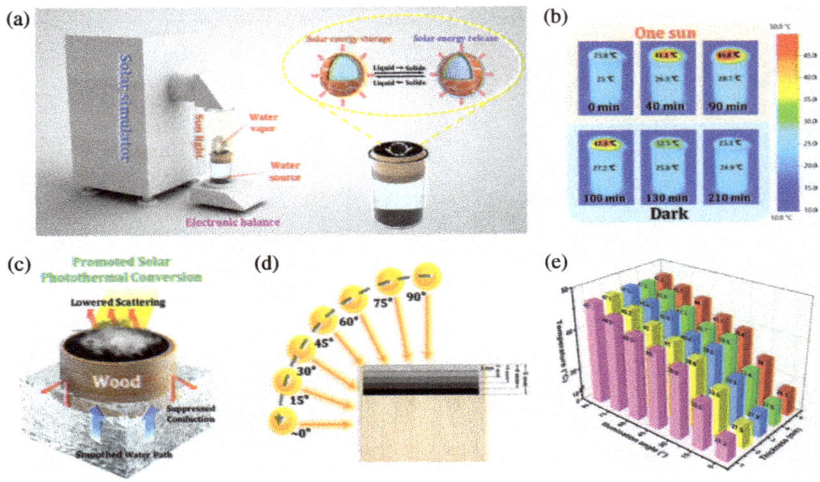

Figure 6.38. (a) The experimental setup for measuring solar-driven evaporation. (b) Infrared thermogram of MXene/ PDA@TiO_2/Fe_3O_4@C22-HMC evaporator under solar intensity irradiation. (c) Schematic diagram of the desalination mechanism of the evaporation system. (d) Schematic diagram of the photothermal performance measurement of the evaporator with different MXene/PDA@TiO_2/Fe_3O_4@C22-HMC sample thickness under different angles of light irradiation. (e) Surface temperatures of evaporators with different MXene/PDA@TiO_2/Fe_3O_4@C22-HMC sample thicknesses at different incidence angles. (f) Water mass variation and (g) evaporation rate for evaporators of aqueous, natural wood, and microencapsulated composite samples. (h) Comparison of evaporation performance of MXene/PDA@TiO_2/Fe_3O_4@C22-HM-based evaporators with evaporators reported in the literature. (i) Evaporation rates and efficiencies of MXene/ PDA@TiO_2/Fe_3O_4@C22-HMC-based evaporators under different light intensities. (j) Water mass variation and evaporation rate of different MXene/PDA@TiO_2/Fe_3O_4@C22-HMC sample thickness evaporators under one solar light intensity irradiation. (k) Water mass variation and evaporation rate of MXene/PDA@TiO_2/Fe_3O_4@C22-HMC-based evaporator under light irradiation with different incidence angles. Reprinted with permission from Ref. [28]. Copyright 2022, American Chemical Society.

The experimental findings demonstrated that, under simulated light illumination at three different intensities and evaporation in a 10 wt.% NaCl solution for a duration of 8 h, no salt crystals were observed on the surface of the evaporator. Furthermore, the addition of Fe_3O_4 nanoparticles in the shells of the microcapsules introduced magnetic responsiveness to the multilayered phase change microcapsules. This magnetic property enhanced

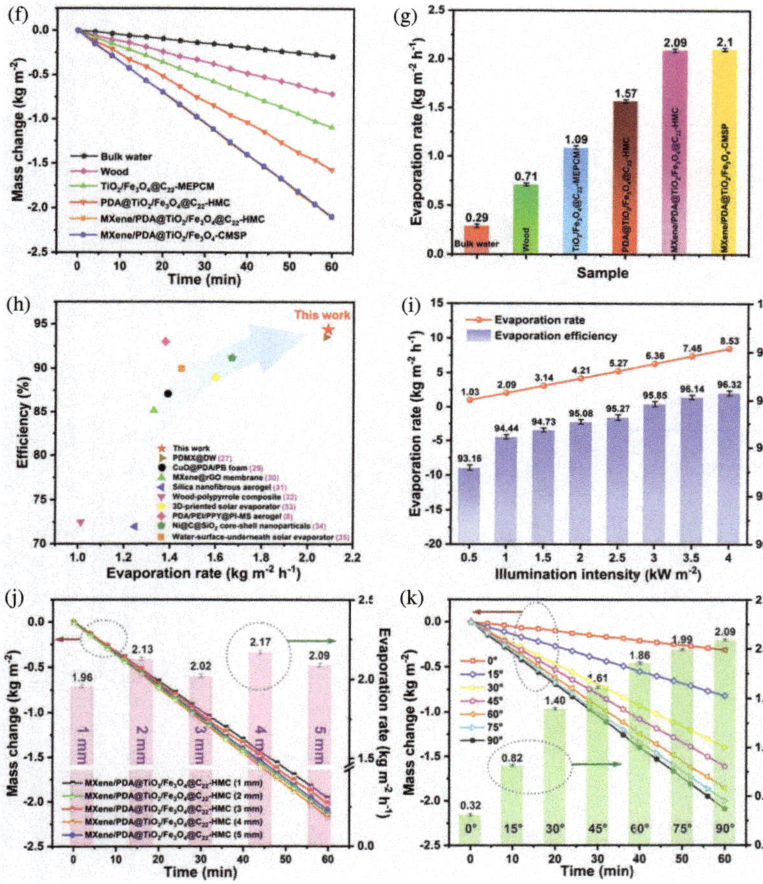

Figure 6.38. (*Continued*)

their reusability and facilitated their separation from salt crystals, resulting in improved efficiency and effectiveness of the developed evaporator in sustaining multiple cycles of salt precipitation and washing processes. As a result, the evaporator can be used for long-term interfacial evaporation in desalination applications while maintaining optimal performance.

More importantly, the inclusion of n-docosane in the evaporation system resulted in a noteworthy increase in total water production of the developed evaporator by 1.17 kg m^{-2} under semi-cloudy weather conditions, when compared to the evaporator without n-docosane (Figure 6.39).

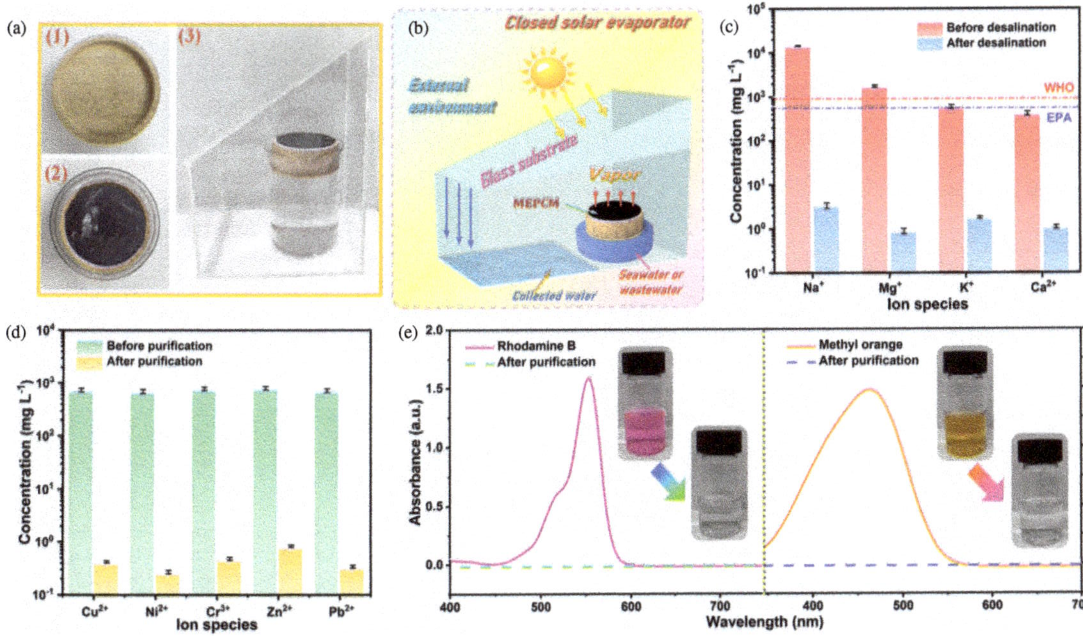

Figure 6.39. (a) digital photographs of different samples, including (1) bare wood, (2) wood filled with MXene/PDA@TiO_2/Fe_3O_4@C22-HMC, and (3) a custom-designed device for solar-driven clear water collection. (b) The schematic diagram illustrates the process of collecting condensed water. (c) Concentrations of metal ions in real seawater are compared before and after desalination. (d) Concentrations of heavy-metal ions in simulated industrial wastewater are shown before and after purification. (e) UV–Vis spectra and digital photographs depict simulated organic wastewater before and after purification. (f) Variations of solar intensity and ambient temperatures presented over time on a sunny day. (g) The evaporation performance of an MXene/PDA@TiO_2/Fe_3O_4@C22-HMC-based evaporator compared to an MXene/PDA@TiO_2/Fe_3O_4-CMSP-based evaporator on a sunny day. (h) Variations of ambient solar intensity and temperatures shown over time on a semi-cloudy day. (i) The evaporation performance of an MXene/PDA@TiO_2/Fe_3O_4@C22-HMC-based evaporator compared to an MXene/PDA@TiO_2/Fe_3O_4-CMSP-based evaporator on a semi-cloudy day. Reprinted with permission from Ref. [28]. Copyright 2022, American Chemical Society.

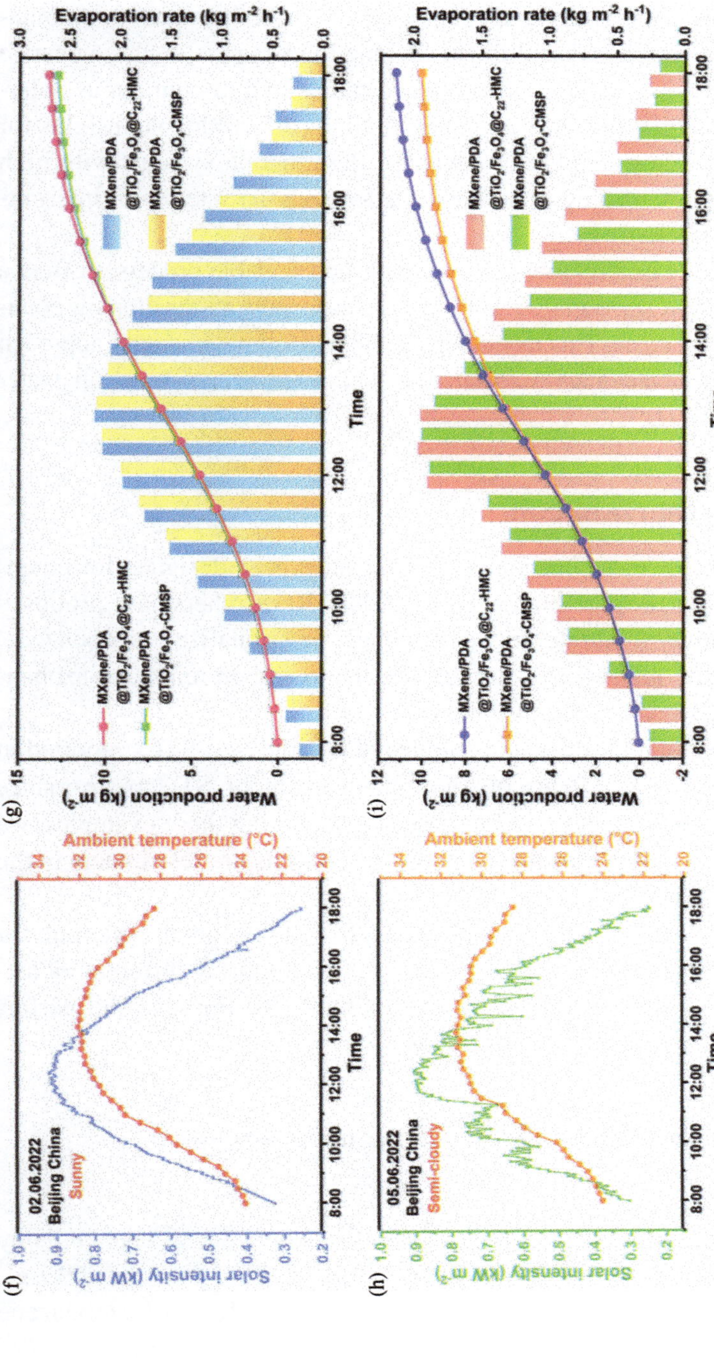

Figure 6.39. (*Continued*)

This enhancement highlights the significance of the innovative combination of natural wood and multilayer MXene/PDA/magnetic phase change microcapsule composites in achieving sustainable and continuous solar-driven desalination. Additionally, these findings offer valuable insights for the design and development of advanced desalination systems with high evaporation rates, aiming to address challenges associated with intermittent solar insolation.

Equipment based on low-cost materials and scalable processes offers a valuable solution for portable water needs, especially in developing countries and remote areas. This technology provides a complementary approach to address the challenges associated with limited access to clean water sources.

6.5 Solar thermal cell

To date, numerous azobenzene derivatives have been designed for potential applications in solar thermal cells (STCs). Researchers have proposed several feasible methods for developing functionalized azobenzene, including pure azobenzene derivatives, nano-carbon-template azobenzene, and polymer-template azobenzene.

In this section, we present examples of notable Azo-STCs reported in the literature, with a focus on the enhancement of STC performance through molecular design engineering. By examining these examples, we aim to provide a comprehensive review that can inspire researchers in the field to design efficient Azo-STCs.

The development of efficient Azo-STCs holds great potential for advancing solar thermal energy conversion. By exploring different molecular design strategies, researchers can continue to improve the performance and efficiency of these systems.

6.5.1 *Solar thermal cells based on azobenzene derivatives*

In 1978, Adamson *et al.* were the first to determine the storage enthalpy of azobenzene, which was found to be $\Delta H_{storage} = 49$ kJ mol^{-1} or 269 J g^{-1} or 75 Wh kg^{-1} [29]. They also reported that the half-life of *cis*-azobenzene was 4.2 days. Subsequently, Olmsted and colleagues conducted research on the potential storage of solar energy using substituted azobenzenes [30]. However, their research findings led them to conclude that azobenzene is not an ideal choice for photochemical solar energy storage due to its fast thermal reduction rate and limited solubility in polar solvents. They found that *trans*-azobenzene releases heat at higher temperatures unless the azobenzene solution is stored in a cool and dark environment. To address this issue, preventive measures can be taken to enable long-term storage and repeated use of azobenzene-STC [31]. These preventive measures include storing the azobenzene solution in a cool, dark room to minimize thermal backflow and ensure its stability.

In order to overcome the challenges faced by molecular azobenzene in solar thermal storage, researchers have developed a solvent-free STC called a liquid azobenzene derivative (1), as reported by Kimizuka and colleagues [32]. To enable fluidization of azobenzene at room temperature, a branched 2-ethylhexyl group was introduced at the para-position of the azobenzene molecule (Figure 6.40(a)). Despite its relatively small molecular weight, the 2-ethylhexyl group allows for the achievement of condensed phase photoisomerization while maintaining a relatively high energy density. Under UV irradiation, this compound undergoes *trans* to *cis* photoisomerization, resulting in a color change from bright orange to dark red (Figure 6.40(b)). The typical photoisomerization behavior of azobenzene photo-switches is illustrated in Figure 6.40(c), where the absorption peaks at around 344 nm and 443 nm correspond to π–π^* and n–π^* transitions, respectively. The figure shows the differential scanning calorimetry (DSC) curves of *cis*-Azo with varying degrees of photoisomerization (Figure 6.40(d)). The exothermic peaks observed at around 120°C allowed for the calculation of an energy density of 35 kJ mol^{-1} (113 J g^{-1} or 32 Wh kg^{-1}) for a *cis*-Azo content of 68%. Furthermore, it was found that the enthalpy of *cis*-to-*trans* isomerization showed a linear dependence on the molar content of *cis*-Azo. Extrapolating the data, the ΔH value corresponding to 100% *cis*-Azo was determined to be 52 kJ mol^{-1} (168 J g^{-1} or 47 Wh kg^{-1}) (Figure 6.40(e)). This study showcased the potential of a low molecular weight liquid azobenzene derivative as a new candidate for molecular solar thermal storage materials. However, challenges such as limited exothermicity, short half-life, and the difficulty of

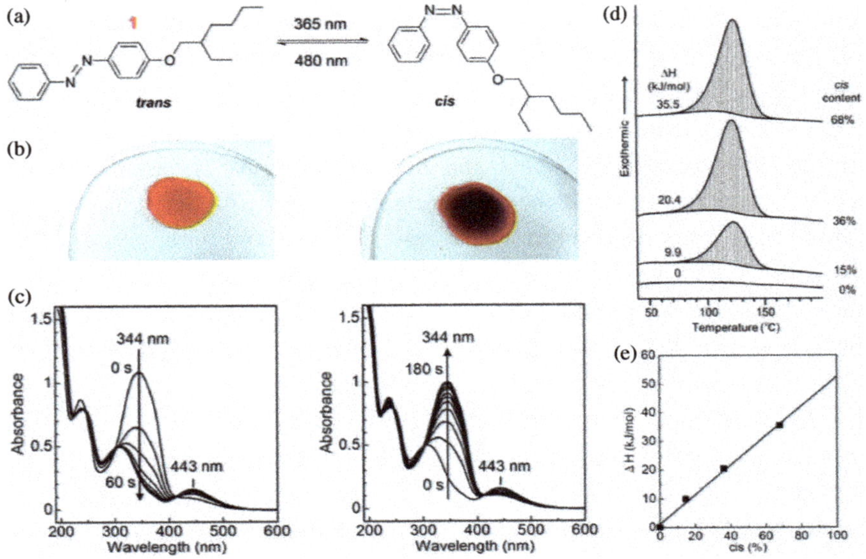

Figure 6.40. (a) The molecular structure of liquid azobenzene derivative 1 and illustrates the photoisomerization processes it undergoes. (b) Pictures of azobenzene derivative 1 in both the *trans* and *cis* states, with the structural changes upon photoisomerization. (c) The UV–Vis absorption spectra of liquid azobenzene derivative 1 under different conditions, including UV (365 nm) and visible-light (480 nm) irradiation. (d) The differential scanning calorimetry (DSC) curves of *ci,s*-Azo 1 at various molar ratios, with a heating rate of 50 K min^{-1}. (e) The dependence of the *cis*-to-*trans* isomerization enthalpy (ΔH) on the molar content of the *cis*-isomer, providing insights into the energetics of the photoisomerization process. Reprinted with permission from Ref. [32]. Copyright 2014, Royal Society of Chemistry.

obtaining pure *cis*-isomers still need to be addressed. Further research is required to tackle and overcome these challenges.

To address the limitations mentioned above, Grossman's research group reevaluated the design of pure azobenzene derivatives for solid-state applications. They introduced bulky aromatic groups such as biphenyl, phenyl, and tert-butyl phenyl onto azobenzene derivatives to achieve tunable solid-state switching thermochromics (STCs) (Figure 6.41(a–c)) [33]. The addition of these bulky phenyl groups increased the energy difference between the *trans* and *cis* forms by raising the energy state of *cis*-Azo through repulsion and steric hindrance. Besides enhancing

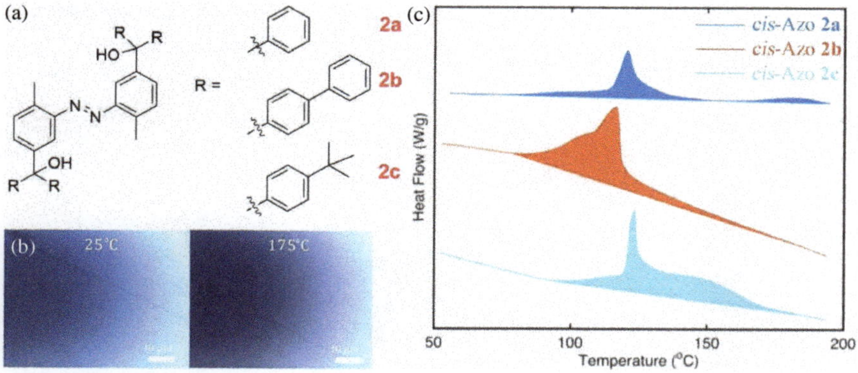

Figure 6.41. (a) Molecular structure of azobenzene derivatives 2a–c functionalized with bulky phenyl groups. (b) Optical images showing the films of azobenzene derivative 2c on silicon at 25°C (left) and 175°C (right). (c) DSC curves of *cis*-azobenzene derivatives 2a–c, arranged from top to bottom. The area under the flat baseline represents the integrated energy release. Reprinted with permission from Ref. [33]. Copyright 2017, American Chemical Society.

the energy density, incorporating such groups facilitated the formation of a solid-state molecular film. The exceptional thermal stability of the azobenzene molecular thin film is depicted in Figure 6.41(b). The bulky phenyl groups effectively prevented π–π stacking of the azobenzene derivative under thermal annealing conditions. The DSC curves of the three *cis*-form azobenzene derivatives are shown in Fig. 6.41(c). The heat release energy density per molecule for 2(a–c) is 77 kJ mol^{-1} (134 J g^{-1} or 37 Wh kg^{-1}), 88 kJ mol^{-1} (100 J g^{-1} or 28 Wh kg^{-1}) and 87 kJ mol^{-1} (108 J g^{-1} or 30 Wh kg^{-1}), respectively. It is important to note that the DSC curves exhibit non-smooth peaks due to variations in activation energies of different grafted functional groups caused by differences in the local environment. This study demonstrated that structural engineering of pure azobenzene molecules is an effective approach to achieve high energy density and thermally stable molecular Azo-STCs.

Azobenzene molecules (3) were combined with tridecanoic acid, an organic PCM, to create photo-controlled thermal energy storage materials (Figure 6.42(a)) [34]. This work was conducted by Grossman and his co-workers. The inclusion of azobenzene molecules enables the manipulation of the thermal properties of PCMs. Figure 6.42(b) illustrates a cycle

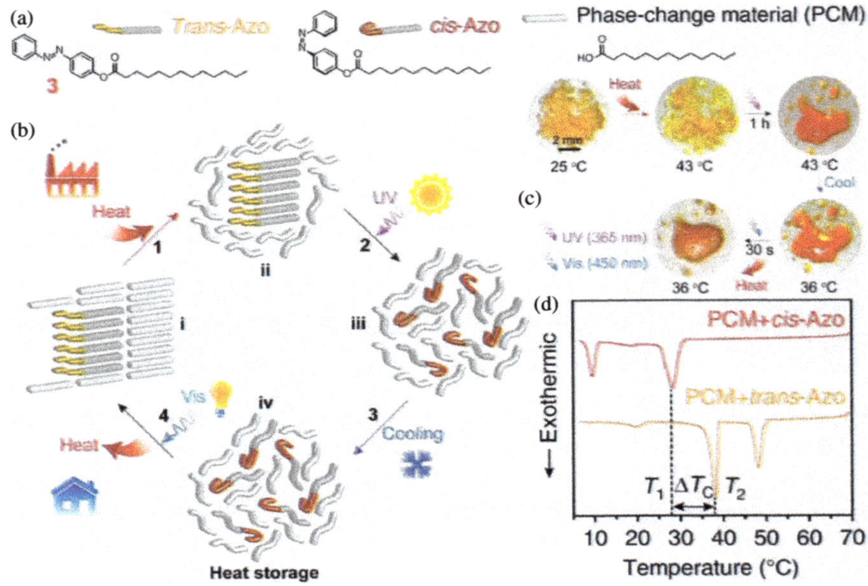

Figure 6.42. (a) The chemical structures of the azobenzene dopant and the phase change material (tridecanoic acid). (b) A diagram illustrates the light-operated cycle for thermal energy storage and release, which includes the absorption of heat by the composite (1), UV irradiation causing photoisomerization of the azobenzene dopants (2), cooling of the liquid composite (3), and visible light triggering the reverse isomerization and heat release (4). The straight and curved rods represent different levels of translational and rotational freedom. Phase i represents solid solid-PCM + *trans*-Azo (25°C), phase ii represents liquid-PCM + *trans*-Azo at 43°C, phase iii represents liquid PCM + *cis*-Azo at 43°C, and phase iv represents liquid PCM + *cis*-Azo at 36°C. (c) Photos corresponding to different stages of the azobenzene-doped PCM composites during solid-state heat absorption, UV charging, cooling, visible-light-induced discharging, and heat release processes. (d) DSC curves of charged and uncharged azobenzene-doped PCM composites obtained while cooling from 70°C at a rate of 5°C min^{-1}. T_1, T_2, and ΔT_c represent different crystallization points and the temperature difference, respectively. Reproduced under the terms of the CC BY 4.0 from [34]. Copyright 2017, the authors.

of light-controlled thermal energy storage and release. Initially, the crystalline composite (solid-PCM + *trans*-Azo, Figure 6.42(b), (i) absorbs external thermal energy. As the temperature reaches the melting point (T_m, 43°C), the composite transforms into a mixture of liquid PCM and crystalline aggregates of the azobenzene dopant (Figure 6.42b, (ii). Under

UV irradiation, the *trans*-Azo dopants convert to their *cis* conformation, dispersing evenly in the liquid PCM (Figure 6.42b, (iii). By cooling the composite below the original crystallization temperature (T_c, 38°C), it remains in a liquid state while retaining the stored latent thermal energy (Figure 6.42b, (iv). The photographs in Figure 6.42(c) depict the different stages of the azobenzene-doped PCM composites, corresponding to the schematic cycle shown in Figures 6.42(b). These results demonstrate that the azobenzene-doped PCM composites can maintain stored latent heat even at lower temperatures. Additionally, the T_c values of the uncharged (PCM + *trans*-Azo) and charged (PCM + *cis*-Azo) composites were measured using DSC (Figure 6.42(d)). For the uncharged composite, the first exothermic peak occurs at 48°C, attributable to the crystallization of *trans*-Azo, followed by PCM crystallization at 38°C. In contrast, the charged composite crystallizes at 28°C, followed by *cis*-Azo crystallization at 9°C. It is worth noting, as shown in Figure 6.42(d), that the Tc and Tm of pristine *trans*-Azo (60°C and 73°C, respectively) significantly decrease for this composite due to the solvation effect of the liquid- state PCM. In summary, by controlling the doping level and cooling rate, the phase-transition temperatures of azobenzene-doped PCM composites can be adjusted. This study demonstrated that the introduction of azobenzene dopants into conventional organic PCMs not only alters intermolecular dynamics and provides stability to the phase storing thermal energy but also enables optical control of energy release. These materials can retain thermal energy of approximately 200 J g^{-1} for over 10 h at temperatures below the original crystallization point. The azobenzene dopants can be effectively trapped in the liquid phase of PCMs by reducing their crystallization temperature through manipulation of the doping level and cooling rate. This approach presents exciting possibilities for the development of portable thermal energy storage systems.

6.5.2 *Solar thermal cells based on nanocarbon-templated azobenzene*

Graphene, an atomic-scale two-dimensional sheet made up of carbon atoms arranged in a hexagonal lattice with sp^2 hybridization, possesses exceptional optical, mechanical, and electronic characteristics. These distinctive properties make graphene extremely appealing for applications in optoelectronic devices [35]. One of the key advantages of graphene is that it can be directly fabricated on substrates without the need for complex purification processes or separation steps, as it produces no impurities during growth. Graphene oxide is a single-layer carbon membrane consisting of different functional groups such as epoxide, carboxyl, and hydroxyl groups. These functional groups provide avenues for customizing the properties of graphene by modifying or altering the attached components. For instance, the introduction of photo-responsive azobenzene molecules onto graphene sheets enables the attainment of optically adjustable electronic properties. Notably, the photoisomerization process of azobenzene can be facilitated by efficient charge transfer between the azobenzene molecule and the graphene surface [36]. This opens up exciting possibilities for utilizing the unique properties of graphene in conjunction with photoactive molecules for optoelectronic applications.

Our research group has dedicated extensive efforts to optimizing the molecular structure of graphene-based azobenzene hybrids, utilizing graphene as a template [37]. Initially, we designed and constructed various models of these hybrids by covalently attaching photo-responsive azobenzene molecules with different functional groups (ortho-, meta-, and para-) at different positions on the graphene framework [38]. By employing density functional theory calculations, we were able to predict the impact of factors such as intermolecular distance between adjacent azobenzene molecules on graphene, electronic interactions, steric hindrance, and intermolecular or intramolecular hydrogen bonding on the change in enthalpy storage ($\Delta H_{storage}$). Consequently, we identified graphene nanosheets as an ideal platform for supporting the nanocarbon-templated azobenzene-Azo-STCs.

Utilizing the results of our calculations, we proceeded to design and successfully synthesize a range of azobenzene-based materials templated on graphene (Figure 6.43) [38–43]. Our primary objective was to enhance the energy density and half-life of these materials through various effective strategies. These included increasing the number of molecular interactions, improving the grafting density, optimizing steric configurations, and strategically attaching suitable substituents at different positions of the azobenzene molecule. The resulting graphene-templated azobenzene hybrids, optimized using these approaches, offer

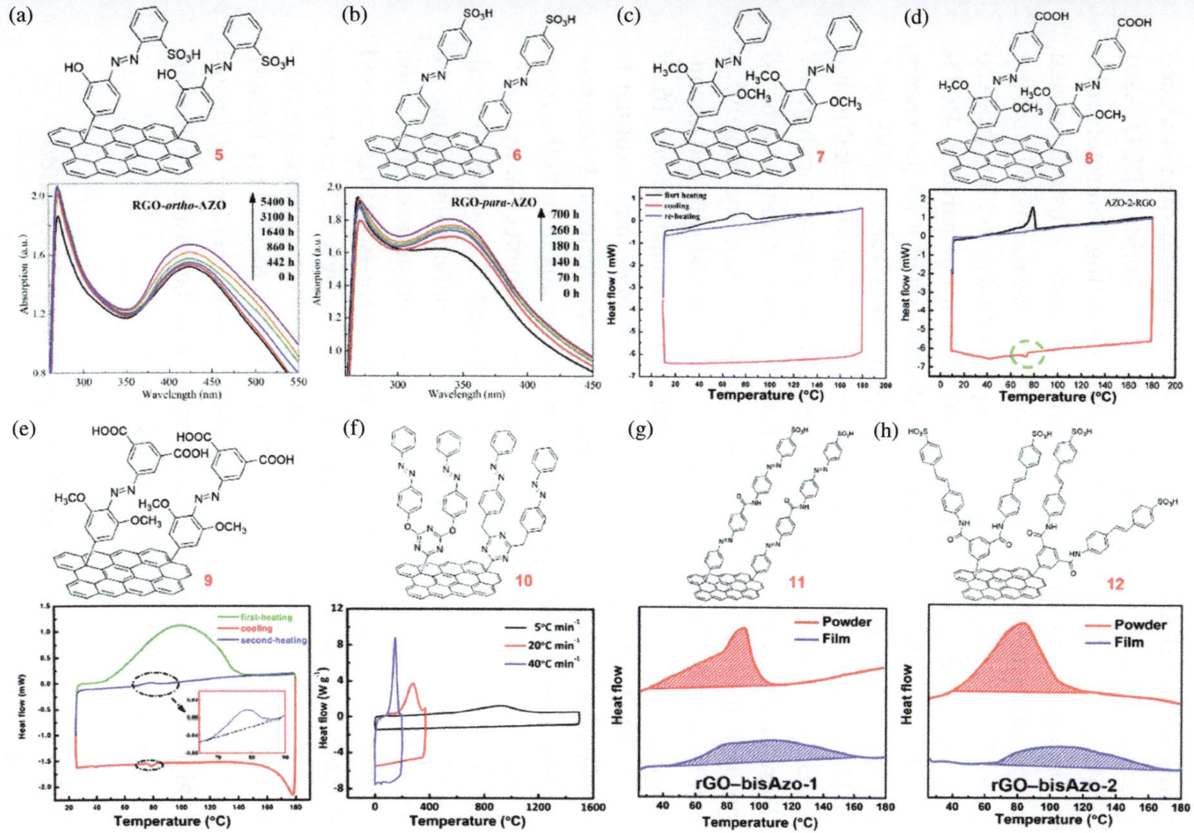

Figure 6.43. (a, b) The chemical structures and time-evolved absorption spectra of RGO-ortho-Azo and RGO-para-Azo hybrids. (c–f) The chemical structures and DSC curves of RGO-ortho-Azo, RGO-para-Azo, and RGO-bis-Azohybrids. (g, h) The chemical structures and DSC curves of the first heating stage of RGO-bis-Azo-1 and RGO-bis-Azo-2. Reprinted with permission from Ref. [37]. Copyright 2018, Royal Society of Chemistry.

tremendous potential for achieving enhanced performance in optoelectronic applications.

Additionally, we detailed the synthesis of graphene oxide-reduced reduced graphene oxide-templated azobenzene hybrids (RGO-Azo) through covalent functionalization. By incorporating intermolecular hydrogen bonding, controlled by the ortho or para substitution and close-packing arrangement of azobenzene on RGO, we achieved improved enthalpy storage ($\Delta H_{storage}$) and half-life ($t_{1/2}$). Notably, RGO-ortho-Azo exhibited a significantly longer half-life of 4540 h compared to RGO-para-Azo (5116 h), while RGO-para-Azo demonstrated a higher energy density (ED) of 75 Wh kg^{-1} compared to RGO-ortho-Azo (42 Wh kg^{-1}, Figures 6.43(a) and 6.43(b)). To gain further insights into the relationship between ED, graft density, and intermolecular hydrogen bonding, we designed three RGO-Azo hybrid materials with methoxyl and/or carboxyl groups positioned ortho-, meta-, or para- to the azobenzene moiety, as depicted in Figures 6.43(c–e) (6.8) [40,41]. Thermal stability and chemical structure characterization revealed that approximately every 16.19 carbon atoms of RGO supported one azobenzene molecule, resulting in densely packed and highly grafted azobenzene-graphene hybrids that promoted the formation of intermolecular hydrogen bonds in both the *trans* and *cis*-isomers of azobenzene. The existence of intermolecular hydrogen bonds was confirmed by the DSC curve obtained during the second-heating cycle (Figures 6.43(d) and 6.43(e)). These findings underscore the significance of optimizing the molecular structure and steric configuration of azobenzene on the graphene template to enhance isomerization and intra/intermolecular interactions. More recently, our research group presented three templated assemblies of closely packed bis-azobenzene (bis-Azo) covalently grafted onto RGO (Figures 6.43(f–h) (9–11)) [42,43]. Theoretical analyses suggest that the intramolecular coupling of bis/tri-Azo not only reduces the energy of the *trans*-isomer but also stabilizes the *cis*-isomer due to considerable steric hindrance.

6.5.3 *Solar thermal fuels based on polymer-templated azobenzene*

The polymer-templated azobenzenes provide the benefit of producing uniform and flat films with adjustable thickness, making them highly promising for applications requiring large-area exothermic. These uniform azopolymer films efficiently capture photons, enabling heat release through the process of *cis*-to-*trans* isomerization. This paves the way for the integration of these materials into a wide range of heating devices, offering exciting possibilities for practical implementation.

In recent decades, scientists have devoted substantial research endeavors to investigate the intriguing characteristics of polymer-templated azobenzenes. These properties encompass photoinduced phase transition, photoinduced birefringence and dichroism, optical nonlinearities, as well as light-controlled reversible changes [44]. As a result, a diverse array of photo-active polymer-templated azobenzenes has been synthesized, thereby facilitating continuous exploration and progress in this rapidly evolving domain. Therefore, the use of polymer-templated azobenzenes as smart thermal control materials has garnered substantial research interest.

Recently, a novel polymer design was introduced by Grossman *et al.*, wherein an alkyl chain served as the main chain and azobenzene acted as the pendant group (Figure 6.44(a)) [45]. These materials exhibited relatively long half-lives at room temperature, with the monomer exhibiting a half-life of 92 ± 1 h and the polymer exhibiting a half-life of 55 ± 1 h. Figure 6.44(b) illustrates the DSC curves of the monomer (42 ± 2 Wh kg^{-1} or 151 J g^{-1}) and polymer (29 ± 2 Wh kg^{-1} or 104 J g^{-1}) species following UV irradiation in a toluene solution (charging process).

The disparity in electronic density between the monomer and polymer can be attributed to several factors, including variations in light absorption efficiency, photoisomerization quantum yield, the influence of the photostationary state, and thermal reversion barriers [45]. In the solid state, steric hindrance can impede photo-switching due to the presence of rotational (steric-sensitive) and inversion mechanisms for isomerization, thereby restricting the performance of the STFs [46]. A photograph of a large-area freestanding polymer film obtained through a simple solution spin-coating process is depicted in Figures 6.44(c) and 6.44(d). During this process, a liquid cross-linking polymer was applied to mitigate the dissolution of the underlying film. To assess heat release, an experimental setup was devised, incorporating a heating stage to induce *cis*-to-*trans* thermal isomerization (Figure 6.44(e)). Simultaneously, an infrared camera was utilized to measure the spatial temperature profile. Figure 6.44(f) illustrates a notable temperature spike, with an average temperature

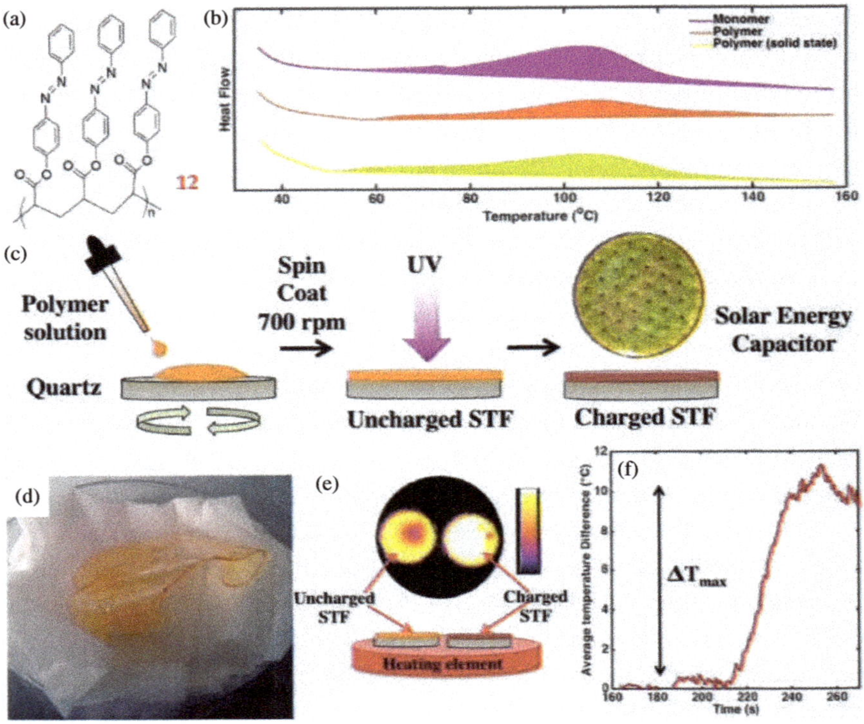

Figure 6.44. (a) The chemical structure of Azo-polymers. (b) DSC curves depicting the charged monomers and polymers. (c) Schematic illustrating the spin-coating process of a polymer solution in toluene onto a transparent quartz substrate. The charging process conducted under UV irradiation using a lamp with a wavelength of 365 nm. A photograph showcasing a large-area freestanding polymer film prepared through the cross-linking approach. Heat release in STF polymer films: The top-view displays an infrared heat map of charged and uncharged STFs, with the color bar indicating the relative magnitude of heat. The bottom-view illustrates the experimental setup. (f) The average temperature differences between the two STFs. A temperature difference of at least 10°C observed between the two STFs due to heat release. Reprinted with permission from Ref. [46]. Copyright 2011, Royal Society of Chemistry.

variation of 10°C observed between the charged and uncharged STFs. Additionally, the rapid temperature difference achieved within seconds is crucial for certain applications that necessitate swift heat dissipation. However, the incorporation of excessive groups and polymers may lead to limited heat release and propagation.

Although significant advancements have been made in enhancing and managing the characteristics of solid-state STCs (surface-relief diffraction gratings) through the utilization of polymer templates and molecular engineering, limited efforts have been dedicated to regulating the precise packing arrangement to augment the electro-optic efficiency of Azo-STCs. To improve the storage properties of solid-state STCs and address the decreased stored electrical energy density observed in polymer systems, it is essential to delve into a more comprehensive comprehension of the close-packing behavior exhibited by azobenzene molecules.

Recently, Morikawa's research group has introduced a novel perspective on the collective *cis*-to-*trans* isomerization of Azo-STF materials, suggesting that it can be observed as a macroscopic crystal-to-liquid phase transition. This phase transition provides additional energy to the azobenzene molecules through an exothermic process [47]. In this study, Morikawa *et al.* developed ionic crystals (Ics) of azobenzene derivatives, comprising spacers with varying methylene lengths (*m*) and alkyl tails (*n*), along with an oligo (ethylene oxide)-based ammonium group and a counter ion (X) (Figure 6.45(a)). Upon UV irradiation, the polymer-templated azobenzene Ics undergo a phase transition from IC to an ionic liquid (IL). Conversely, visible-light irradiation causes the *cis*-azobenzene Ils to crystallize back into *trans*-azobenzene Ics. This transformation is accompanied by a total exothermic enthalpy of 97.1 kJ mol^{-1} (128 J g^{-1} or 36 W h kg^{-1}), which is more than twice the energy stored in pure *cis*-Azo. Figure 6.45(b) presents X-ray diffraction (XRD) profiles and polarizing optical microscopy (POM) images of the 1(6,4)-Br azobenzene derivative subjected to different treatments. The POM image of the as-prepared film displays a birefringence phenomenon, consistent with the crystalline lamellar structure revealed in the XRD pattern. Upon UV irradiation, the yellow crystalline sample melts into a red liquid phase, accompanied by a conformational change of azobenzene from *trans* to *cis*.

It is important to note that the area exposed to irradiation exhibited a dark appearance when observed under a polarizing optical microscope with crossed polarizers. These observations confirmed that the photoisomerization of *trans*-1(6,4)-Br azobenzene led to its conversion into an isotropic liquid phase (IL). Subsequently, upon visible-light irradiation, the *cis*-azobenzene IL underwent reversible crystallization, as indicated by the characteristic peak of *trans*-azobenzene observed in the XRD profile. This

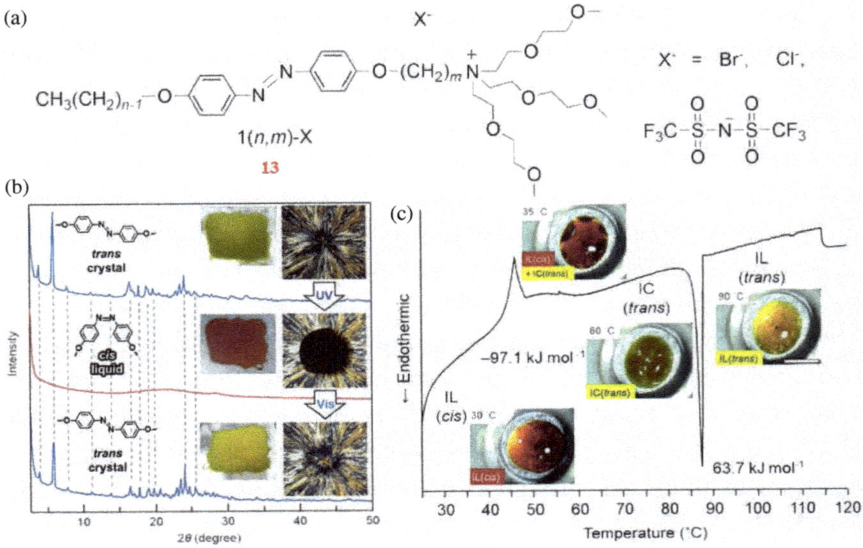

Figure 6.45. (a) The chemical structure of cationic azobenzene derivatives 1(n,m)-X. (b) X-ray diffraction (XRD) profiles of 1(6,4)-Br prepared on a silicon wafer: the top profile represents the as-prepared film, the middle profile corresponds to the UV-light-irradiated film, and the bottom profile represents the visible-light-irradiated film. Images of each sample are shown as insets. (c) Differential scanning calorimetry (DSC) curves of *cis*-1(6,4)-Br. Photographs of the corresponding states are shown as insets. Reprinted with permission from Ref. [47]. Copyright 2015, Wiley.

provides compelling evidence for the reversible photocrystallization and photoliquefaction of the 1(6,4)-Br azobenzene derivative. Figure 6.45(c) presents the DSC curves of the *cis*-azobenzene IL. In contrast to previous studies, two distinct exothermic peaks were observed within the temperature range of 34–65°C. The broader peak was attributed to the thermal *cis*-to-*trans* isomerization of azobenzene, while the sharper exothermic peak around 46°C corresponded to the enthalpy associated with the phase transition from the *cis*-azobenzene IL to the *trans*-azobenzene IC. As the heating process continued, a distinct endothermic peak emerged at 87°C, indicating the melting of *trans*-azobenzene ICs. As a result, the photoinduced *trans*-IC to *cis*-IL phase-transition in Azo-STC technology presents a promising advancement in current energy storage capacity.

The chain structure of polymers presents challenges in achieving perfect crystallization compared to small molecules, which readily form crystals. As a result, macromolecules often exist in an amorphous state or have a combination of amorphous and crystalline regions. The amorphous portion of the polymer provides ample free volume for photoisomerization to occur. The glass transition temperature (T_g) of a polymer determines its state at room temperature. The T_g is influenced by the polymer's structure and offers the potential for photogenic solid-liquid transitions in the polymer.

Si Wu's group at University of Science and Technology of China synthesized azobenzene macromolecules 14–16 (Figure 6.46a) by reversible addition-break chain transfer polymerization (RAFT) [48]. These poly(meth)acrylate azobenzene macromolecules are capable of photoisomerization and undergo reversible solid-liquid transformation under UV and visible light (Figure 6.46(b)).

To determine that the solid–liquid transition of the macromolecule is caused by the change in T_g due to the photoisomerization of the Azo phenyl group, a series of experiments were designed to investigate. First, the polymer 14 was processed according to Figure 6.46(c) to obtain four samples: *trans*-azobenzene polymer 14, *trans*-azobenzene polymer 14 containing 2.5 wt% CH_2Cl_2, *trans*-azobenzene polymer 14 dissolved in CH_2Cl_2 and then dried, and *cis*-azobenzene polymer 14. The T_g of each sample was tested by differential scanning calorimetry (DSC). The T_g of *trans*-azobenzene polymer 14 was 48°C and that of *cis*-azobenzene polymer 14 was −10°C (Figure 6.46(d)). In addition, the surface temperature of the polymer 14 after liquefaction by UV light was measured by an infrared thermometer to be 35°C, which is lower than the T_g of the *trans* polymer 14. This indicates that the photothermal effect is not sufficient to liquefy it. Therefore, we demonstrate that the light-induced solid–liquid transition is not a photothermal process. The *trans*-azophenyl groups are rod-like molecules with large axial diameter ratio and can be used as liquid crystal mesocrystals. The *trans*-azobenzene groups are arranged in liquid crystal phase in the membranes prepared by polymer 14. The photochemical phase transition from anisotropic to isotropic state was induced by UV light. To verify whether the liquid crystal phase transition is an important factor, we quenched the polymer 14 at isotropic temperature and obtained the amorphous *trans*-azobenzene polymer 14. The amorphous also underwent a solid–liquid transition under UV illumination. This indicates that the photogenic solid-liquid transition was also

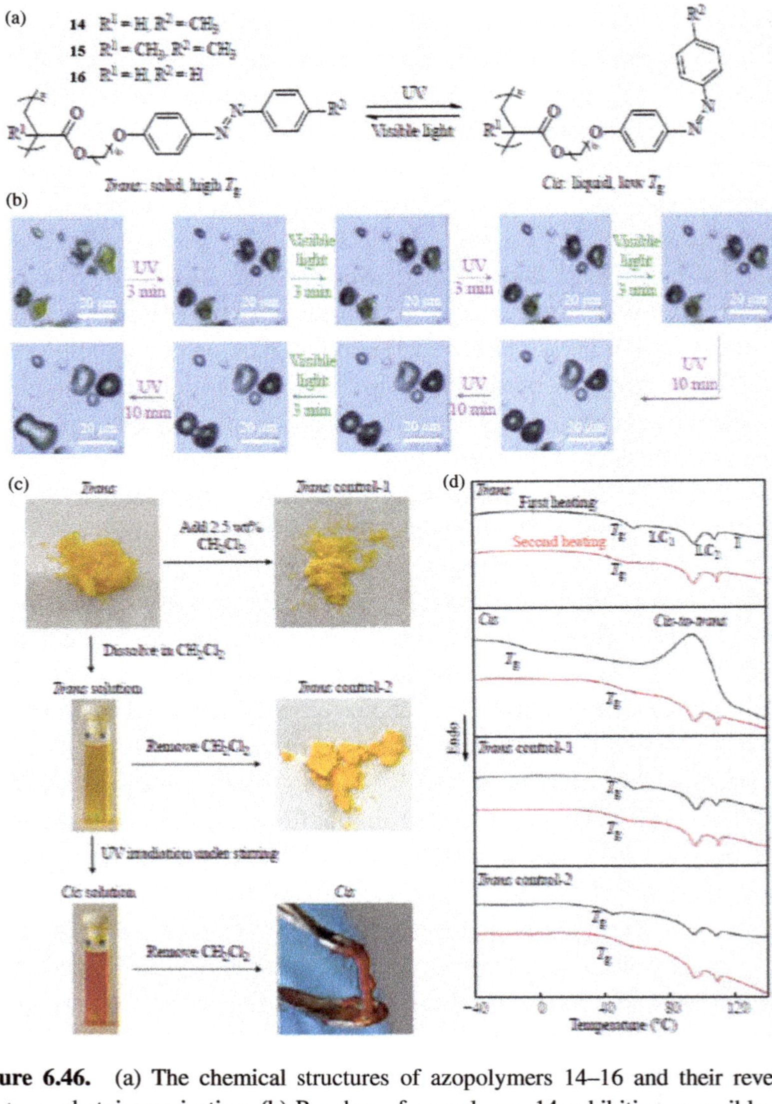

Figure 6.46. (a) The chemical structures of azopolymers 14–16 and their reversible *cis–trans* photoisomerization. (b) Powders of azopolymer 14 exhibiting reversible solid-to-liquid transitions when irradiated with alternating UV light (365 nm, 67 mW cm^{-2}) and visible light (530 nm, 5 mW cm^{-2}). (c) Photographs illustrating the sample preparation processes of azopolymer 14. (d) Differential scanning calorimetry (DSC) curves of four samples: *trans* 14, *cis* 14, and two control samples of *trans* 14. The *cis* 14 sample contained over 90% *cis*-isomers. The arrow indicates the direction of the endothermic processes. Reprinted with permission from Ref. [48]. Copyright 2016, Springer Nature.

not induced by the liquid crystal phase but by the photo T_g-triggered transition. The photoisomerization of azobenzene macromolecules induces two kinds of transitions: (1) one is the photochemical phase transition observed by Ikeda *et al.* in earlier experiments, and (2) the other is the transition from the high to the low T_g state proposed by us.

In conventional Azo-STCs, UV light is typically used for the *trans–cis* isomerization process. However, since sunlight only contains a small percentage of UV light (around 5%), the absorption efficiency of Azo-STCs is limited. Therefore, it is important to develop Azo-STCs that can effectively utilize the full solar spectrum to enhance their performance. Previous studies have shown that introducing substituents to azobenzene can modify the absorption bands [49–51]. Expanding on these findings, Wu *et al.* [52] proposed an Azo-based solar thermal fuel (Azo-STF) device that can efficiently store both UV and visible light under full solar irradiation (Figures 6.47(a) and 6.47(b)). The structure of the device, as shown in Figure 6.47(c), comprises PmAzo (17-a), Pazo (17-b), Coumarin 314, and a UV-pass filter. This hybrid device consists of four layers, including two active layers of Pazo and PmAzo, which enable solar energy storage through the photoisomerization of the Azo groups present in the polymer side chains. The other two filtering layers are utilized to effectively modify the incident spectrum.

For the experiment, spin-coated films of Pazo (approximately 35 nm) and PmAzo (approximately 20 nm) were used to create the solid layers of the STC device. The UV–Vis spectra shown in Figure 6.47(d) illustrate the expected increase in the n–π^* transition and decrease in the π–π^* transition when the amorphous polymer PmAzo undergoes UV irradiation, indicating the *trans*-to-*cis* isomerization process. Interestingly, the *trans*-to-*cis* photoisomerization of Pazo caused an increase in the intensity of the π–π^* transition. This phenomenon is attributed to the unique photochemical phase transition observed in liquid crystalline azopolymers like Pazo. In liquid crystalline azopolymers, the *trans*-Azo groups tend to align perpendicular to the substrate. Upon UV irradiation, the azopolymer transitions from a "frozen" liquid crystalline state to an amorphous state. This transition has two distinct effects on the UV–Vis spectra. Firstly, the intensity of the π–π^* transition band decreases due to partial photoisomerization of *trans*-isomers

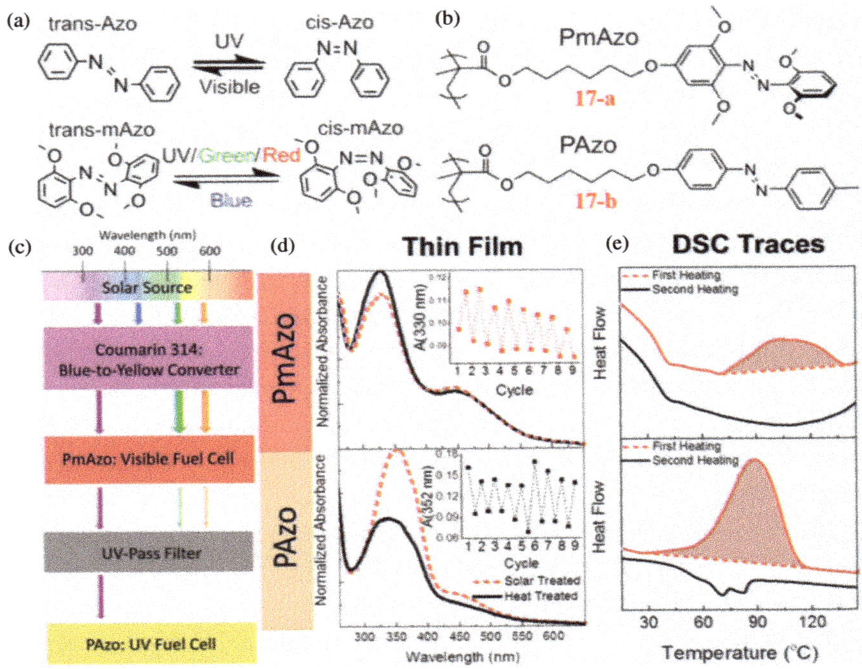

Figure 6.47. (a) The Azo and mAzo chromophores undergo reversible photoisomerization. (b) The chemical structures of UV-responsive polymer Pazo and visible-light-responsive polymer PmAzo. (c) A schematic illustration of a four-layered solar thermal cell. (d) UV–Vis spectra of Pazo and PmAzo films after 1 h of solar irradiation followed by 2.5 h of heating in a vacuum oven at 90°C. The insets show the absorbance changes of the π–π* band for Pazo and PmAzo with different cycles. (e) Differential scanning calorimetry (DSC) curves of PAzo (365 nm, 2 mW cm^{-2}, 20 min) and PmAzo (530 nm, 4 mW cm^{-2}, 40 min). Reprinted with permission from Ref. [52]. Copyright 2017, Wiley.

into *cis*-isomers. Secondly, the π–π* transition band increases because the azobenzene chromophores deviate from their perpendicular orientation, resulting in a more random orientation. The DSC measurements presented in Figure 6.47(e) determined that the enthalpy difference between *cis* and *trans* isomers was calculated to be 12.2 kJ mol^{-1} (14.4 J g^{-1} or 4.4 W h kg^{-1}) for PmAzo and 54.8 kJ mol^{-1} (125.2 J g^{-1} or 34.8 W h kg^{-1}) for Pazo. The experimental results demonstrated a conversion rate of 60–70% from *trans*-Azo to *cis*-Azo when the azopolymer was irradiated with monochromatic light. This high conversion rate was achieved by selectively blocking wavelengths of light that could induce back-isomerization from *cis* to *trans*.

In conclusion, the use of polymer-templated Azo-STCs offers numerous benefits in improving solar energy storage capacity, light absorption efficiency, and photoisomerization degree. They enable the absorption of photons across a wide range of the solar spectrum. However, in comparison to carbon-templated Azo-STCs, polymer-templated ones generally exhibit lower EDs and shorter half-lives. Nonetheless, polymer-templated Azo-STCs hold significant potential for the advancement of innovative energy storage and heat-release applications. One notable advantage is their easy integration into existing heating systems. Moreover, Azo-based polymers can undergo a controlled liquid-to-crystal transition, providing additional stored energy and the ability to be triggered for heat release. These characteristics make polymer-templated Azo-STCs highly promising for various solid-state applications and offer versatility in efficient energy utilization.

In summary, the utilization of azobenzene-based STFs has shown great promise as an effective and efficient approach for solar-thermal conversion and storage solutions. Various forms of azobenzene derivatives, including pure compounds, nanocarbon-templated structures, and polymer-templated materials, have garnered significant interest in recent years. However, the rapid advancement of this research field is hindered by challenges such as low ED, low quantum yield, short half-life, limited absorption spectrum overlaps with solar radiation and irreversible degradation. Researchers have dedicated significant effort to address the limitations of azobenzene-based STF (Azo-STC) materials. They strive to counterbalance these shortcomings through several approaches. Firstly, functional chemical groups are grafted onto the materials to improve their ED and thermal stability. Additionally, researchers are investigating the incorporation of PCMs and carbon nanomaterials to facilitate the practical implementation of Azo-STCs. Furthermore, meticulous adjustments are made to the microstructures of the monomers and polymer backbones to maximize absorption efficiency. It is evident that molecular design engineering and microstructure optimization are fundamental in achieving Azo-STCs with high-energy storage capacities. To enable the practical utilization of

Azo-STCs, researchers have identified a set of criteria known as the “three high and three low” requirements, which must be met:

(i) The storage energy density ($\Delta H_{storage}$) should be maximized.
(ii) The photoisomerization degree from *trans*-Azo to *cis*-Azo should be elevated to enhance heat release.
(iii) The energy barrier (ΔE_a) from *cis*-Azo to *trans*-Azo should be sufficiently high for long-term storage capability.
(iv) The absorption spectrum should be optimized to match the solar emission spectrum, ensuring maximum efficiency in light absorption with minimal illumination.
(v) The photoactive materials should demonstrate excellent stability, exhibiting low degradation and strong cyclability even under extreme conditions including high temperatures, freezing cold, strong acidity, high salinity, etc.
(vi) The cost of Azo-STC materials should be competitive.

It is important to note that the field of Azo-STCs (Azo-azobenzene-based solid-to-liquid transitions) is still in its early stages of development. To prepare Azo-STC materials, new approaches such as controlled self-assembly, electrospinning, electro-polymerization, 3D printing, injection molding, and calendar molding should be explored. Continued research on solid-state Azo-STCs has the potential to enable their integration into existing devices for various applications, including deicing, seawater desalination, salt production, and thermal-driven generators. Furthermore, Azo-STCs hold promise in both military and consumer-oriented sectors, with potential applications in deep space probes, thermal power systems, temperature control systems in space stations, warm clothing, solar blankets, thermal-powered cars, and solar water heaters.

Moreover, future investigations may uncover new opportunities. For instance, combining Azo-STCs with directional heat-conductive materials can enable controlled heat energy transfer. Integration with thermoelectric materials has the potential to enhance the thermoelectric figure of merit (ZT). Overall, there is ample room for advancements in the fundamental understanding and practical applications of Azo-STCs.

References

[1] Wang Z, Roffey A, Losantos R, Lennartson A, Jevric M, Petersen AU, Quant M, Dreos A, Wen X, Sampedro D, Boriesson K, Moth-Poulsen K. (2019). Macroscopic heat release in a molecular solar thermal energy storagesystem. *Energy Environ Sci*, **12**(1), pp. 187–193.

[2] Dreos A, Wang Z, Udmark J, Ström A, Erhart P, Börjesson K, Nielsen MB, Moth-Poulsen K. (2018). Liquid norbornadiene photoswitches for solar energy storage. *Adv Energy Mater*, **8**(18), pp. 1703401.

[3] Kashyap V, Sakunkaewkasem S, Jafari P, Nazari M, Eslami B, Nazifi S, Irajizad P, Marquez MD, Lee TR, Ghasemi H. (2019). Full spectrum solar thermal energy harvesting and storage by a molecular and phase-change hybrid material. *Joule*, **3**(12), pp. 3100–3111.

[4] Han GGD, Li H, Grossman JC. (2017). Optically-controlled long-term storage and release of thermal energy in phase-change materials. *Nat Commun*, **8**(1), pp. 1446.

[5] Shi Y, Gerkman MA, Qiu Q, Zhang S, Han GG. D. (2021). Sunlight-activated phase change materials for controlled heat storage and triggered release. *J Mater Chem A*, **9**(15), pp. 9798–9808.

[6] Liu H, Tang J, Dong L, Wang H, Xu T, Gao W, Zhai F, Feng Y, Feng W. (2020). Optically triggered synchronous heat release of phase-change enthalpy and photo-thermal energy in phase-change materials at low temperatures. *Adv Funct Mater*, **31**(6), pp. 2008496

[7] Hu J, Huang S, Yu M, Yu H. (2019). Flexible solar thermal fuel devices: composites of fabric and a photoliquefiable azobenzene derivative. *Adv Energy Mater*, **9**(37), pp. 1901363.

[8] Cai F, Song T, Yang B, Lv X, Zhang L, Yu H. (2021). Enhancement of solar thermal fuel by microphase separation and nanoconfinement of a block copolymer. *Chem Mater*, **33**(24), pp. 9750–9759.

[9] Saydjari AK, Weis P, Wu S. (2017). Spanning the solar spectrum: azopolymer solar thermal fuels for simultaneous UV and visible light storage. *Adv Energy Mater*, **7**(3), pp. 1601622

[10] Xu X, Zhang P, Wu B, Xing Y, Shi K, Fang W, Yu H, Wang G. (2020). Photochromic dendrimers for photoswitched solid-to-liquid transitions and solar thermal fuels. *ACS Appl Mater Interfaces*, **12**(44), pp. 50135–50142.

[11] Wang H, Feng Y, Gao J, Fang W, Ge J, Yang X, Zhai F, Yu Y, Feng W. (2022). Metallic-ion controlled dynamic bonds to co-harvest isomerization

energy and bond enthalpy for high-energy output of flexible self-heated textile. *Adv Sci*, **9**(20), pp. e2201657.

[12] Feng Y, Feng W, Noda H, Sekino T, Fujii A, Ozaki M, Yoshino K. (2007). Synthesis of photoresponsive azobenzene chromophore-modified multi-walled carbon nanotubes. *Carbon*, **45**(12), pp. 2445–2448.

[13] Kolpak AM, Grossman JC (2011). Azobenzene-functionalized carbon nanotubes as high-energy density solar thermal fuels. *Nano Lett*, **11**(8), pp. 3156–3162.

[14] Kucharski TJ, Ferralis N, Kolpak AM, Zheng JO, Nocera DG, Grossman JC. (2014). Templated assembly of photoswitches significantly increases the energy-storage capacity of solar thermal fuels. *Nat Chem*, **6**(5), pp. 441–447.

[15] Ganguly G, Sultana M, Paul A. (2018). Designing efficient solar-thermal fuels with [n.n](9,10)anthracene cyclophanes: a theoretical perspective. *J Phys Chem Lett*, **9**(2), pp. 328–334.

[16] Gerkman MA, Gibson RSL, Calbo J, Shi Y, Fuchter MJ, Han GGD. (2020). Arylazopyrazoles for long-term thermal energy storage and optically triggered heat release below 0°C. *J Am Chem Soc*, **142**(19), pp. 8688–8695.

[17] Zhang ZY, He Y, Wang Z, Xu J, Xie M, Tao P, Ji D, Moth-Poulsen K, Li T. (2020). Photochemical phase transitions enable coharvesting of photon energy and ambient heat for energetic molecular solar thermal batteries that upgrade thermal energy. *J Am Chem Soc*, **142**(28), pp. 12256–12264.

[18] Griffiths K, Halcovitch NR, Griffin JM. (2020). Long-term solar energy storage under ambient conditions in a MOF-based solid-solid phase-change material. *Chem Mater*, **32**(23), pp. 9925–9936.

[19] Wang H, Feng Y, Yu H, Dong L, Zhai F, Tang JW, Ge J, Feng W. (2021). Utilisation of photo- thermal energy and bond enthalpy based on optically triggered formation and dissociation of coordination bonds. *Nano Energy*, **89**, pp. 106401.

[20] Hu J, Huang S, Yu M, Yu H. (2019). Flexible solar thermal fuel devices: composites of fabric and a photoliquefiable azobenzene derivative. *Adv Energy Mater*, **9**(37), pp. 1901363.

[21] Cai F, Yang B, Lv X, Yu H. (2022). Mechanically mutable polymer enabled by light. *Sci Adv*, **8**(34), pp. eabo1626.

[22] Fei L, Yin Y, Yang M, Zhang S, Wang C. (2021). Wearable solar energy management based on visible solar thermal energy storage for full solar spectrum utilization. *Energy Storage Mater*, **42**, pp. 636–644.

[23] Xu N, Li J, Wang Y, Fang C, Li X, Wang Y, Zhou L, Zhu B, Wu Z, Zhu S, Zhu J. (2019). A water lily–inspired hierarchical design for stable and efficient solar evaporation of high-salinity brine. *Sci Adv*, **5**(7), pp. eaaw7013.

[24] Zhou L, Tan Y, Wang J, Xu W, Yuan Y, Cai W, Zhu S, Zhu J. (2016). 3D self-assembly of aluminium nanoparticles for plasmon-enhanced solar desalination. *Nat Photon*, **10**(6), pp. 393–398.

[25] Xu N, Hu X, Xu W, Li X, Zhou L, Zhu S, Zhu J. (2017). Mushrooms as efficient solar steam-generation devices. *Adv Mater*, **29**(28), pp. 1606762.

[26] Li X, Li NR, Ni G, Xu N, Hu X, Zhu B, Lv G, Li J, Zhu S, Zhu J. (2018). Three-dimensional artificial transpiration for efficient solar waste-water treatment. *Natl Sci Rev*, **5**(1), pp.70–77.

[27] Geng L, Li L, Zhang H, Zhong M, Mu P, Li J. (2022). Interfacial solar evaporator synergistic phase change energy storage for all-day steam generation. *J Mater Chem A*, **10**(29), pp. 15485–15496.

[28] Zheng Z, Li W, Liu H, Wang X. (2022). Sustainable interfacial evaporation system based on hierarchical MXene/Polydopamine/Magnetic phase-change microcapsule composites for solar-driven seawater desalination. *ACS Appl Mater Interfaces*, **14**(45), pp. 50966–50981.

[29] Adamson AW, Vogler A, Kunkely H, Wachter R. (1978). Photocalorimetry. Enthalpies of photolysis of trans-azobenzene, ferrioxalate and cobaltioxalate ions, chromium hexacarbonyl, and dirhenium decarbonyl. *J Am Chem Soc*, **100**(4), pp. 1298–1300.

[30] Olmsted III J, Lawrence J, Yee GG. (1983). Photochemical storage potential of azobenzenes. *Sol Energy*, **30**(3), pp. 271–274.

[31] Taoda H, Hayakawa K, Kawase K, Yamakita H. (1987). Photochemical conversion and storage of solar energy by azobenzene. *J Chem Eng Jpn*, **20**(3), pp. 265–270.

[32] Masutani K, Morikawa M, Kimizuka N. (2014). A liquid azobenzene derivative as a solvent-free solar thermal fuel. *Chem Commun*, **50**(99), pp. 15803–15806.

[33] Cho EN, Zhitomirsky D, Han GGD, Liu Y, Grossman J. (2017). Molecularly engineered azobenzene derivatives for high energy density solid-state solar thermal fuels. *ACS Appl Mater Interfaces*, **9**(10), pp. 8679–8687.

[34] Han GGD, Li H, Grossman JC. (2017). Optically-controlled long-term storage and release of thermal energy in phase-change materials. *Nat Commun*, **8**(1), pp. 1–10.

[35] Du X, Skachko I, Duerr F, Luican A, Andrei EY. (2009). Fractional quantum Hall effect and insulating phase of Dirac electrons in graphene. *Nature*, **462**(7270), pp. 192–195.

[36] Zhao J, Liu C, Ma J. (2017). A light-driven modulation of electric conductance through the adsorption of azobenzene onto silicon-doped-and pyridine-like N 3-vacancy graphene. *Nanoscale*, **9**(48), pp. 19017–19025.

[37] Dong L, Feng Y, Wang L, Feng W. (2018). Azobenzene-based solar thermal fuels: design, properties, and applications. *Chem Soc Rev*, **47**(19), pp. 7339–7368.

[38] Li M, Feng YY, Liu EZ, Qin C, Feng W. (2016). Azobenzene/graphene hybrid for high-density solar thermal storage by optimizing molecular structure. *Sci China Technol Sci*, **59**(9), pp. 1383–1390.

[39] Feng Y, Liu H, Luo W, Liu E, Zhao N, Yoshino K, Feng W. (2013). Covalent functionalization of graphene by azobenzene with molecular hydrogen bonds for long-term solar thermal storage. *Sci Rep*, **3**(1), pp. 1–8.

[40] Luo W, Feng Y, Cao C, Li M, Liu E, Li S, Qin C, Hu W, Feng W. (2015). A high energy density azobenzene/graphene hybrid: a nano-templated platform for solar thermal storage. *J Mater Chem A*, **3**(22), pp. 11787–11795.

[41] Luo W, Feng Y, Qin C, Li M, Li S, Cao C, Long P, Liu E, Hu W, Yoshino K, Feng W. (2015). High-energy, stable and recycled molecular solar thermal storage materials using Azo/graphene hybrids by optimizing hydrogen bonds. *Nanoscale*, **7**(39), pp. 16214–16221.

[42] Feng W, Li S, Li M, Qin C, Feng Y. (2016). An energy-dense and thermal-stable bis-azobenzene/hybrid templated assembly for solar thermal fuel. *J Mater Chem A*, **4**(21), pp. 8020–8028.

[43] Zhao X, Feng Y, Qin C, Yang W, Si Q, Feng W. (2016). Controlling heat release from a close-packed bisazobenzene-reduced-graphene-oxide assembly film for high-energy solid-state photothermal fuels, *ChemSusChem*, **10**(7), pp. 1395–1404.

[44] Wang H, Bisoyi HK, Wang L, Urbas AM, Bunning TJ, Li Q. (2018). Photochemically and thermally driven full-color reflection in a self–organized helical superstructure enabled by a halogen-bonded chiral molecular switch. *Angew Chem Int Ed Engl*, **57**(6), pp. 1627–1631.

[45] Zhitomirsky D, Cho E, Grossman JC. (2016). Solid-state solar thermal fuels for heat release applications. *Adv Energy Mater*, **6**(6), pp. 1502006.

[46] Bandara HMD, Burdette SC. (2012). Photoisomerization in different classes of azobenzene. *Chem Soc Rev*, **41**(5), pp. 1809–1825.

[47] Ishiba K, Morikawa M, Chikara C, Yamada T, Iwase K, Kawakita M, Kimizuka N. (2015). Photoliquefiable ionic crystals: a phase crossover approach for photon energy storage materials with functional multiplicity. *Angew Chem Int Ed Engl*, **54**(5), pp. 1532–1536.

[48] Zhou H, Xue C, Weis P, Suzuki Y, Huang S, Koynov K, Auernhammer GK, Berger R, Butt HJ, Wu S. (2017). Photoswitching of glass transition temperatures of azobenzene-containing polymers induces reversible solid-to-liquid transitions. *Nat Chem*, **9**(2), pp. 145–151.

[49] Siewertsen R, Neumann H, Buchheim-Stehn B, Herges R, Nather C, Renth F, Temps F. (2009). Highly efficient reversible Z-E photoisomerization of a bridged azobenzene with visible light through resolved S1 ($n\pi^*$) absorption bands. *J Am Chem Soc*, **131**(43), pp. 15594–15595.

[50] Beharry AA, Sadovski O, Woolley GA. (2011). Azobenzene photoswitching without ultraviolet light. *J Am Chem Soc*, **133**(49), pp. 19684–19687.

[51] Weis P, Wang D, Wu S. (2016). Visible-light-responsive azopolymers with inhibited π–π stacking enable fully reversible photopatterning. *Macromolecules*, **49**(17), pp. 6368–6373.

[52] Saydjari AK, Weis P, Wu S. (2017). Spanning the solar spectrum: azopolymer solar thermal fuels for simultaneous UV and visible light storage. *Adv Energy Mater*, **7**(3), pp. 1601622.

Index

www.ingramcontent.com/pod-product-compliance
Ingram Content Group UK Ltd.
Pitfield, Milton Keynes, MK11 3LW, UK
UKHW022106170926
8624IPUK00024B/154

9 789811 285844